Advanced Studies on Steel Structures

Advanced Studies on Steel Structures

Guest Editors

Fangxin Hu
Shujun Hu
Hai-Ting Li

MDPI

Basel • Beijing • Wuhan • Barcelona • Belgrade • Novi Sad • Cluj • Manchester

Guest Editors
Fangxin Hu
School of Civil Engineering and Transportation
South China University of Technology
Guangzhou
China

Shujun Hu
Department of Civil Engineering
Nanchang University
Nanchang
China

Hai-Ting Li
School of Naval Architecture Ocean & Civil Engineering
Shanghai Jiao Tong University
Shanghai
China

Editorial Office
MDPI AG
Grosspeteranlage 5
4052 Basel, Switzerland

This is a reprint of the Special Issue, published open access by the journal *Buildings* (ISSN 2075-5309), freely accessible at: https://www.mdpi.com/journal/buildings/special_issues/ST270M500K.

For citation purposes, cite each article independently as indicated on the article page online and as indicated below:

Lastname, A.A.; Lastname, B.B. Article Title. *Journal Name* **Year**, *Volume Number*, Page Range.

ISBN 978-3-7258-4305-3 (Hbk)
ISBN 978-3-7258-4306-0 (PDF)
https://doi.org/10.3390/books978-3-7258-4306-0

Contents

buildings

MDPI

Article

Cyclic Behavior of the Column-Tree Moment Connection with Weakened Plates: A Numerical Approach

Pablo Matthews and Eduardo Núñez *

Department of Civil Engineering, Universidad Católica de la Santísima Concepción, Concepción 4070128, Chile; pablo.matthews@ucsc.cl

* Correspondence: enunez@ucsc.cl

Abstract: The use of column-tree connections is common in controlled shop environments due to their cost-effectiveness in achieving ductile welds. Field bolts are also easy to install and inspect. However, there is currently no prequalification available for these connections, their performance is not fully understood, and the cost of aftermath repairs is still a major concern for owners. In this research, analytical and numerical studies were performed to assess the cyclic behavior considering the effects of the bolted splice location, bolt slippage, and splice plate thickness. Fourteen numerical models using the finite element method in ANSYS software were analyzed to evaluate the nonlinear behavior of moment connection configurations in terms of the strength, stiffness, ductility, energy dissipation, and overall cyclic response. The results showed that appropriately proportioned bolted splice connections can meet the requirements for prequalified moment connections. The models complied with the criteria established in AISC 358 and achieved flexural resistance that was higher than 80% of the beam plastic moment at 4% of the interstory drift ratio. The weakened plates concentrated the inelastic action, which allowed us to prevent the brittle behavior and damage to the column, welding, and other components of the moment connection. Complex geometries or specially fabricated parts were not required, providing a cost-effective way to control seismic-related damage. Also, required repairs are based on the replacement of standard parts, reducing operational detentions in facilities. Finally, the moment connection studied is classified as partially restrained (PR) according to the requirements established in AISC 360.

Keywords: steel moment connection; plates; cyclic behavior; bolted connections; finite elements

1. Introduction

Steel moment frames are well-known for their ability to withstand intense seismic events by absorbing energy. However, the ductility of these frames mainly depends on the performance of their connections, which can be either fully or partially restrained [1]. These connections are where the maximum moment demand occurs.

In 1994, the Northridge Earthquake caused unexpected and severe damage to these moment frame connections. The typical connection configuration at the time was a wide flange rolled beam site welded directly to a wide flange column, and this event prompted a closer look at ductility-related issues such as the welding details and material toughness [1]. The findings of the investigation revealed brittle fractures around the welds, leading to improvements in welding technologies and the standardization of prequalified connections [2].

In 1995, the Kobe Earthquake revealed that steel moment frames had similar failures to those seen in previous earthquakes [3]. In this case, the use of welded or cold-formed box-columns or cross-shaped columns was observed. Additionally, instead of site-welded beams, the column typically had shop-welded beam stubs, and the main beam was field-bolted. This design is referred to as "column-tree" [4]. Although the shop-welded connection has better conditions for achieving strong welds and related inspections, there were still instances of brittle fractures at the welds. According to Whittaker [5], "Japanese framing systems tend to be more redundant than similar systems in the US".

Research efforts were then dedicated to enhancing the technology of welding consumables, as well as improving the material toughness and joint preparation geometry [6]. The significance of proper ductile welds at the joint is underscored by these two seismic events. One strategy to achieve this is to reinforce the welding area and provide a greater cross-section, thereby reducing tensile stresses at the welds. Chen et al. [7] examined the impact of widened beam flange specimens, while Kang et al. [8] introduced external stiffeners at the connection with HSS columns. This design philosophy enhances the overall ductility, but brittle weld fractures may still occur during the final load cycles. Schneider and Teeraparbwong [9] tested connections with bolted flange cover plates of varying bolt geometries and quantities, resulting in a high rotational capacity and typically forming the girder hinge near the cover plates.

An alternative method is to relocate the area where the inelastic action occurs from the welded joints and instead place it in the steel beam [10]. One approach to achieving this is by designing a reduced beam section (RBS) [2], which moves the plastic hinge away from the column face. Researchers like Zhang et al. [11], Morshedi et al. [12], and Vetr et al. [13] explored different designs such as beams with openings at the web, double-RBS beams, and beams with drilled holes of various diameters at the beam flanges. By deliberately weakening a specific section of the beam, a ductile mechanism can be predicted and controlled for high cycle loads.

There are various ways to configure structures using different concepts. One example is the end-plate connection, which is explained in AISC 358 [2]. This type of connection involves plates that are prewelded in the shop, sometimes with stiffeners. In a study by Nuñez et al. [14], the seismic performance of end-plate moment connections on HSS columns was analyzed. The test results showed a maximum reaction of 1.9 Mp at 4% rad drift. Tao et al. [15] also conducted tests on end-plate connections that were applied to concrete-filled HSS columns, where the bolts passed through the entire column section.

While all of these moment connection improvements effectively resolve previous brittle performance challenges (like weld cracking, beam buckling, or beam fracture), the problem of connection ductile damage still lingers as an economic setback for building owners and industrial facilities that must detain operations if complex repair activity is required. This challenge invites us to revisit the column-tree connection as a possible solution to find a cost-effective way to limit ductile damage and repair costs.

In a comprehensive examination conducted by Astaneh-Asl [16], the focus was on the connections between column-trees, considering a variety of factors, including the number and size of bolts, the thickness of the connection plate, and the distance from the column. The objectives were to avoid any potential for brittle failure mechanisms and to enhance the structure's seismic performance by creating a balanced connection design that could act as a fuse to minimize seismic forces.

In a study conducted by Baharamast et al. [17], various iterations of bolted column-tree connections were explored, including reinforcements and reduced sections. Similarly, Gharebaghi and Hosseini [18] investigated the impacts of bolt pretension and clearance. However, both studies utilized a fully rigid splice that did not incorporate connection plates for energy dissipation. Alternatively, Palizzolo and Vazzano [19] and Chen et al. [20] proposed specially fabricated connection devices that concentrated energy dissipation through plastic strain or friction. These devices had the added benefit of being replaceable after a severe earthquake. In New Zealand, Yeow [21] examined the advantages and drawbacks of utilizing friction connections to decrease the expenses and repair time after earthquakes.

Oh et al. [22] conducted a study on the performance of three column-tree specimens, each with different plate thicknesses and bolt quantities, under the AISC 341 [23] qualification loading protocol. The connections were designed to have a strength of 95% of the beam plastic moment Mp for several intended failure mechanisms: bolt slip (CT), bolt bearing (CT-B), and plate yielding (CT-R). The test results indicated that the typical failure mode was beam flange local buckling, with ultimate strengths ranging from 1.3 Mp to 1.5 Mp and a higher level of energy dissipation when compared to fully rigid connections. Specimen CT-R, which had fewer bolts, experienced early bolt slip that prevented significant yielding of the splice plates. This suggests that energy dissipation occurred through friction and plastic strain, which are complementary. Furthermore, Oh et al. [24] compared the responses of composite and noncomposite column-tree connections and found that the presence of a concrete slab reduced the system's ductility.

In a study conducted by Lin et al. [25], it was discovered that the design strength can be represented by Md = FCR × My, where FCR is a parameter that ranges from 0.85 to 0.95 to account for a decreased required strength at a connection situated away from the column. In this equation, My = S_x × Fy is used in place of Mp = Z_x × Fy. The connection plates are designed to function as replaceable damage-control fuses, following the RBS philosophy. The study's results successfully showed that the plastic strain is concentrated at the splice plates, which decreases damage to the beam and column. Karami et al. [26] expanded on Oh's [22] research by using numerical nonlinear models to add three instances where the reduced strength connection type CT-R was located at various distances from the column. The findings revealed that the connection's ultimate strength is not affected by its position. However, as the distance increases, the connection becomes less effective at dissipating energy and reduces the weld. The main objective of this report was to determine if a connection could be prequalified for SMRFs. The results indicate that it is achievable, but it depends on the location and stiffness of the bolted splice. A connection located near the column and/or with a weakened splice will be classified as semi-rigid and cannot be used as a prequalified connection in the SMRF.

The engineering community does not have a complete understanding of the seismic performance of bolted column-tree connections. The response can be affected by variables such as the location, bolt configuration, and plate thickness. Following the Kobe Earthquake in 1995, doubts were raised about the design of fully rigid bolted connections, and a more ductile response was favored. Previous research is very clear regarding the ductility of a weakened bolted splice: the connection plates are a very good source of energy dissipation through plastic strain and plate friction [27]. However, their use in SMRFs worldwide is limited due to the lack of a straightforward design procedure that can predict the connection's cyclic behavior and ensure compliance with seismic requirements.

Another interesting aspect of designing weakened splices is the concentration of plastic strain after a severe earthquake, which protects the beam and column from damage. The behavior of bolted steel plates under fire conditions has also been studied by Akagwu et al. [28]; Yeow et al. [21] demonstrated that replacing damaged plates and bolts is a cost-effective way to repair the connection and reduce time and labor. Saravanan et al. [29] reviewed this approach of replaceable fuses and highlighted the need for further research in this area.

The characterization of the cyclic behavior of column-tree moment connections has not been completely developed, and the use of cover plates as fuse components in the connection has not been studied. The goal of this research is to assess the cyclic behavior of the bolted splice connection, considering the cover plates as fusible components. Therefore, beams and columns are protected. Additionally, the failure mechanisms used to define the nonlinear response of the column-tree moment connection are obtained. In this sense, the influences of the beam stub length, number of bolts, and plate thickness on the cyclic response, with a particular focus on the dissipation in the beam plates as the primary failure mechanism, are developed. Posteriorly, a design procedure considering the failure mechanisms of the connection is proposed. Robust numerical models using the finite element method (FEM) are used to evaluate the cyclic response in terms of the strength, ductility, and energy dissipation capacity. Finally, this conceptualization of the column-tree moment connection allows us to achieve a cost-effective replacement with lower detention times for the facilities.

2. Analytical Study

According to research conducted by Astaneh-Asl [16], prioritizing ductile failure modes over brittle ones is crucial. This philosophy must be integrated into the design process with the key parameters highlighted in Figure 1. The single splice cover plates carry both tension and compression forces, while the flange bolts are pretensioned to provide initial strength against the relative slip of the flange plates.

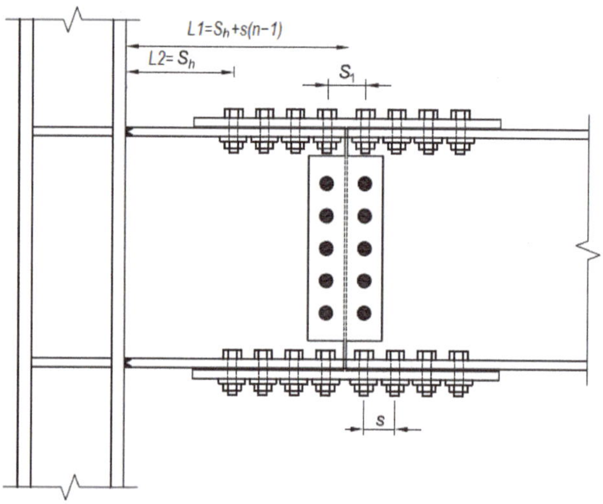

Figure 1. Details of the column-tree moment connection.

To achieve the goal of this study, is necessary to decrease the tensile stress in the welds between the beam stub and the column face using splice plates as a damage-control fuse. Therefore, any prequalified joint preparation for complete penetration welds is suitable, requiring no further discussion. However, for a typical moment frame that experiences lateral forces, there is a lineal moment distribution along the beam, with a maximum value at the column joint and a value near zero at the center of the span for seismic forces (as shown in Figure 2). Thus, the location of the bolted splice directly affects the required strength for the splice. Astaneh-Asl [16] recommends a distance for the splice in the range of 1/10 to 1/8 of the span.

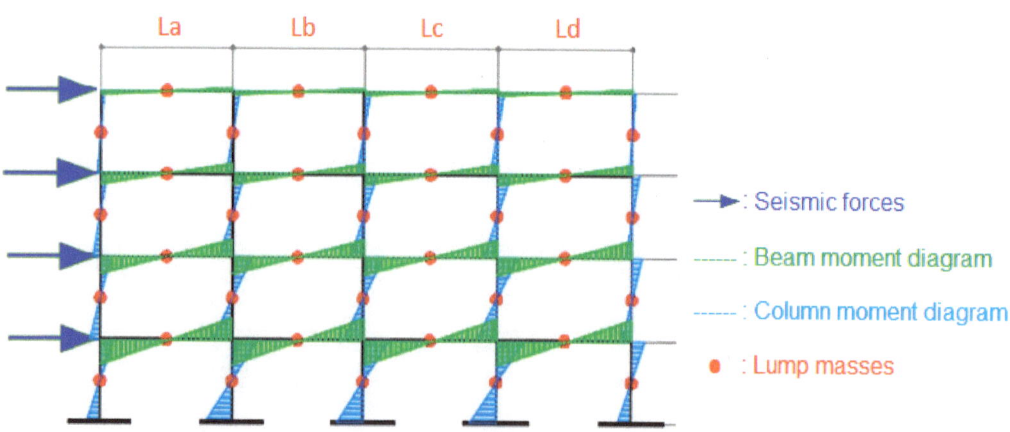

Figure 2. Distribution of seismic forces in steel moment frames.

The moment diagram factor "f" for typical frames ranges from 0.75 to 0.80. However, for atypical cases, this value may range from 0.60 to 0.85 of this value. A simple formulation can be used to consider this effect.

$$f = \left(1 - 2 * \frac{L_1}{L}\right) \tag{1}$$

where "L_1" is the beam stub length and "L" is the beam clear span.

The required splice strength "Mps" is calculated based on the moment diagram factor "f" and the following material properties:

$$\begin{array}{ll} \text{beam yielding strength} & \text{Fyb} \\ \text{splice plate yielding strength} & \text{Fys} \\ \text{beam elastic module} & \text{Sxb.} \\ \text{splice plastic module} & \text{Zxs} \\ \text{beam elastic moment strength} & \text{Myb} = \text{Fyb} \times \text{Sxb} \end{array} \tag{2}$$

$$\text{splice plastic moment strength} \quad \text{Mps} = \text{Fys} \times \text{Zxs}, \tag{3}$$

$$\text{splice design condition} \quad \text{Mps} > \text{f} \times \text{Myb} \tag{4}$$

$$\text{splice design condition} \quad \text{Zxs} > \text{f} \times \text{Sxb} \times \text{Fyb/Fys} \tag{5}$$

Moreover, the maximum size of the bolts must be determined, and then the minimum bolt quantity must be obtained. The bolt size and corresponding standard hole size affect the failure mode of the drilled beam flange and splice plates. To ensure a ductile response, the maximum bolt size depends on the beam flange geometry and material properties. We followed the AISC 360 [30] recommendations for expected yielding and expected tensile fracture at the beam flange gross area "Agf" and beam flange net area "An". Sufficient spacing between bolts must be provided to avoid block shear tear-out.

$$\text{The yielding strength is obtained as} \quad \text{Rny} = \text{Ry} \times \text{Agf} \times \text{Fyb}. \tag{6}$$

$$\text{The fracture strength is estimated as} \quad \text{Rnu} = \text{Ru} \times \text{Anf} \times \text{Fub}. \tag{7}$$

For two (2) lines of bolts across the beam flange, the net section is the

$$\text{beam flange net area} \quad \text{Ans} = \text{tf} \times (\text{B} - 2 \times \text{dh}) \tag{8}$$

where tf is the beam flange thickness, B is the beam flange width, "dh" is the bolt hole size (diameter), and the ductile mode is achieved if Rnu > Rny; therefore

$$\text{dh} < \text{B}/2 \times (1 - (\text{Ry} \times \text{Fyb})/(\text{Rt} \times \text{Fub})) - 2\ \text{mm}. \tag{9}$$

The minimum quantity of bolts is calculated based on the bearing strength; however, the slip strength must be checked, since its value is roughly 55–85% of the bolt shear strength (depending on the surface friction coefficient). To avoid the occurrence of slippage too early with low cycle loads, a great number of bolts is desirable, but avoiding slippage at all must not be pursued, since the connection would lose its energy dissipation capacity.

$$\text{required tension at the shear plane} \quad \text{Tr} = \text{Mps/d} \tag{10}$$

$$\text{required flange bolts} \quad \text{N} > \text{Tr}/0.75 \times \text{Fn} \times \text{Ab} \tag{11}$$

$$\text{slip strength} \quad \text{Rs} = 1.13 \times \mu \times \text{Tb} \tag{12}$$

where d is the beam height, Fn is the bolt nominal shear stress, Ab is the gross area of the bolt, Tb is the bolt pretension, and μ is the mean slip coefficient. A proper and representative slip factor has to be used. Then, the overall slip strength for the group of bolts should be 50–85% of the splice plate yielding strength to provide energy dissipation during medium-range load cycles [16].

The plate thickness must fulfill the same requirements as the beam flange regarding the net area and minimum bolt pitch in order to avoid brittle modes. Special attention should be paid to the distance S1 and the plate compression buckling strength (with K = 0.5 for a continuous plate)

$$\text{radius of gyration} \qquad r = e \times \sqrt{12} \qquad \text{for a rectangular plate} \tag{13}$$

$$\text{plate slenderness} \qquad \lambda = K \times S1/r \tag{14}$$

$$\text{plate slenderness} \qquad \lambda = \sqrt{12} \times S1/e = 3.5 \times S1/e \tag{15}$$

As Lin et al. [25] suggested a maximum slenderness of 35, it has been established a "rule of thumb" for the minimum plate thickness as e > S1/10.

Finally, the gap between the beam stub and the link beam must provide enough clearance to avoid the clash of both members, affecting the plastic response of the connection and damaging the parts. As shown in Figure 3, the minimum gap is $\varnothing_U = 2 \times \delta_0/h$. Since the target rotation is at least 0.04 rad for ductile frames, the minimum gap is

$$\delta_0 = \varnothing_U * h/2 > 0.02 \times h = h/50 \tag{16}$$

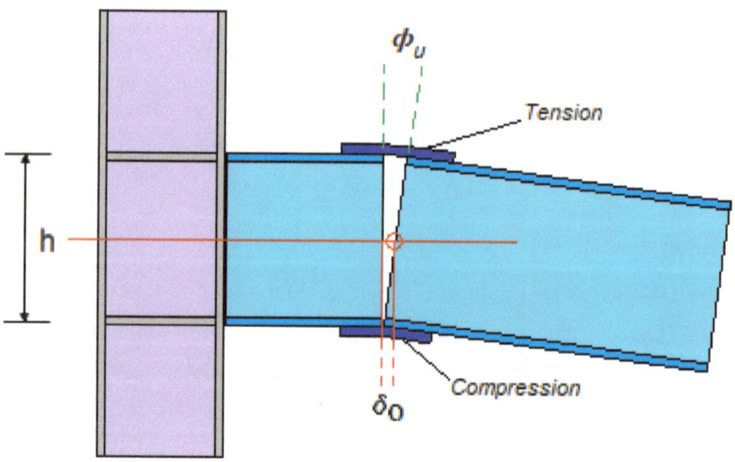

Figure 3. Relationship between the connection rotation, beam height, and required separation.

3. Numerical Study

The numerical study was based on the modeling of beam–column joint configurations in their test specimen condition (see Figure 4), which will be reported in a future article. The modeling and analysis were performed using finite elements by means of ANSYS software v2022 [31]. An explicit model of the elements was used. The elements and components of the connection were modeled explicitly. The materials used were introduced as input variables from the constitutive law of the material, while the load was considered from the application of cyclic displacements, as established by AISC 341 [23].

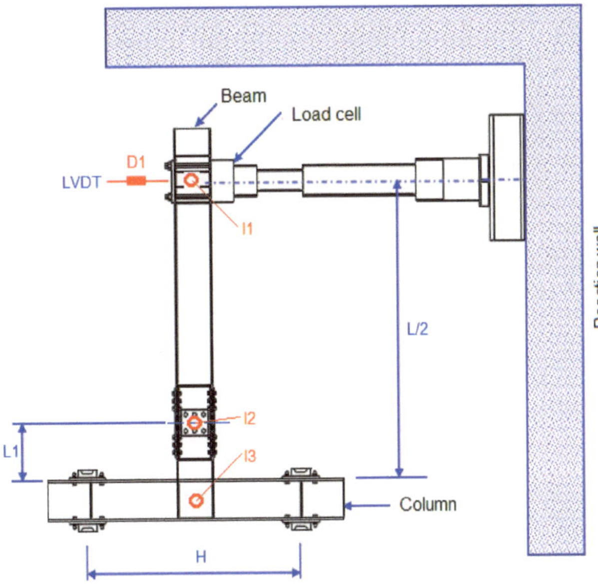

Figure 4. Representation of the test assembly.

3.1. Assembly Description

The assembly consists of welded shapes made from ASTM A36 plates. The column section is H300 × 10 × 16 × 300 mm (length = 3000 mm), and the beam section is H250 × 6 × 10 × 150 mm (clear length = 2350 mm). According to the proposed design procedure, the diameter of the selected bolts is M20 (5/8″), and the bolt pretension applied is 85 kN each, as recommended by AISC 360 [30]. The typical dimensions of the connections, as shown in Figure 1, are S = 60 mm and S1 = 100 mm for the flange bolt pattern. The size and length of the structural members were preliminary selected for a building with steel moment frames located in Santiago, Chile, with three-bay and three-story levels, designed according to seismic loads established in Chilean Code.

The two different locations for the connection (L1 = 375 mm and L1 = 500 mm) reflect the change in the required design load, the two plate thicknesses (e = 8 mm and e = 10 mm) represent both rigid and weakened approaches, and the three different bolt lines affect the energy dissipation (n = 3, 4, and 5 lines). A model M0 consists of a welded beam-to-column connection in order to obtain base results for the maximum moment and energy dissipation; two additional models with L1 = 250 mm and n = 3 are used to analyze the effect of connections as close as possible to the column. Table 1 shows a summary of all analyzed models.

Discretization of the elements was performed, using finer meshing in areas where stress concentration is expected. The boundary conditions simulated the effect of displacement constraints at the column ends. The iteration method used for the nonlinearities was the incremental Newton–Raphson method [31]. In addition, the Augmented Lagrangian method was used to achieve numerical convergence in the contact zone, in accordance with the investigations performed by [32]. Welds were not included in the model because elastic behavior is expected from the design by capacity in the welds. The diameter of the holes used is standard, according to the requirements given in AISC 360 [30].

Table 1. Summary of the models studied.

Model	Distance L1 (mm)	Plate Thickness "e" (mm)	Bolt Lines "n" in Flange Splice
M1	375	8	3
M2	375	8	4
M3	375	8	5
M4	375	10	3
M5	375	10	4
M6	375	10	5
M7	500	8	3
M8	500	8	4
M9	500	8	5
M10	500	10	3
M11	500	10	4
M12	500	10	5
MA	250	8	3
MB	250	10	3
M0	-	-	-

3.2. Material Constitutive Law

A nonlinear, bilinear, kinematic constitutive law and von Mises yield criterion were used. The materials were considered from coupon tests performed according to studies conducted by Nuñez et al. [33] and Herrera et al. [34]. The values are shown in Table 2.

Table 2. Constitutive law of materials.

Material	Fy (MPa)	εy (mm/mm)	Fu (MPa)	εu (mm/mm)
ASTM A36	294.3	0.00146608	568.5	0.243246
ASTM A-325	786	0.00393	923.897	0.16

3.3. Loading and Boundary Conditions

The numerical model is representative of real experimental conditions. In this sense, a pinned restriction was applied to the base connection of the column. The end of the beam was deemed to be a simple support with displacement allowed in the vertical direction. Bolt pretension (70% of bolt tension strength) was applied according to the value established in AISC 360 [30]. Also, according to AISC 341 [23], the load was applied to the tip of the link beam through horizontal displacements (Table 3), and lateral supports to the beam were assigned to satisfy the stability criteria. The boundary conditions are shown in Figure 5.

Table 3. Load protocol used in the numerical model [23].

No.	Number of Cycles	Drift Angle (θ) [rad]
1	6	0.00375
2	6	0.00500
3	6	0.00750
4	4	0.0100
5	2	0.0150

Table 3. *Cont.*

No.	Number of Cycles	Drift Angle (θ) [rad]
6	2	0.0200
7	2	0.0300
8	2	0.0400
9	1	0.0500

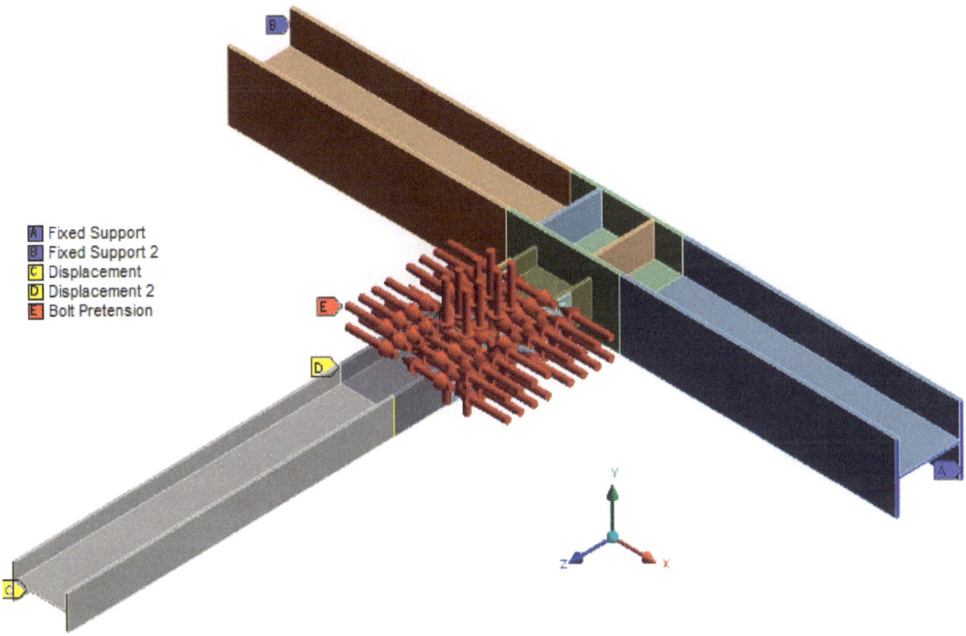

Figure 5. Numerical model with boundary conditions.

3.4. Elements, Mesh, and Contacts

The elements and components of the connections in the numerical models were modeled explicitly using the SOLID 186 element. This element is defined by 20 nodes and three degrees of freedom per node in the three principal directions (X, Y, Z). Additionally, this element has a higher capacity to reproduce plasticity, hyperelasticity, creep, stress stiffening, large deflection, and large strain phenomenon. In order to discretize the interest zones, different mesh sizes were considered, ensuring a rational computational cost. Consequently, the solution implemented was to apply a fine mesh in zones with great stresses and coarser mesh in the other regions, as shown in Figure 6.

The contacts used for welded parts are of the "bonded" type, which is equivalent to a fully restrained condition; a "frictional" contact was used to simulate the possible displacement between the components in tight contact, and "frictionless" contacts were used between the bolt shank and the inner face of the holes for the bolts. The friction coefficient varies depending on the surface as-built condition and hole clearance [30]; also, the bolt tightening method could affect the actual applied pretension [18]. However, these parameters are intended to prevent slip at the service load level [35,36], and the static or monotonic response is different than that under cyclic loads. The FEA models only allows the definition of a single mean slip coefficient, in this case, $\mu = 0.2$, to account for a clean mill scale. Table 4 show the type of contacts used in the FEA models.

Figure 6. Contacts and mesh in numerical models.

Table 4. Type of contacts by element in the numerical model.

No.	Region Interaction	Type of Contact
1	Splice plates–beam	Frictional μ = 0.2
2	Bolt shank–plates	Frictionless
3	Bolt shank–beam	Frictionless
4	Bolt shank–bolt head	Bonded
5	Bolt head–splice plates	Frictional μ = 0.2
6	Bolt shank–nut	Bonded
7	Nut–splice plates	Frictional μ = 0.2
8	Beam stub–column	Bonded
9	Column–stiffeners	Bonded

4. Results

4.1. Previous Calibration of the Numerical Model

In order to ensure a representative numerical model, previous calibration of the numerical model was performed based on the experimental research conducted by Oh et al. [22]. Figure 7 shows the mesh, and Figure 8 shows the equivalent stress at the end of the 5% rad cycle. An H600 × 200 × 11 × 7 beam with a length of L = 3500 mm, H400 × 400 × 13 × 21 column, and L1 = 900 mm was used as an assembly in a beam–column joint. The high-strength bolts were of the ASTM A490 type with a pretension of 179 kN and a friction coefficient of 0.2.

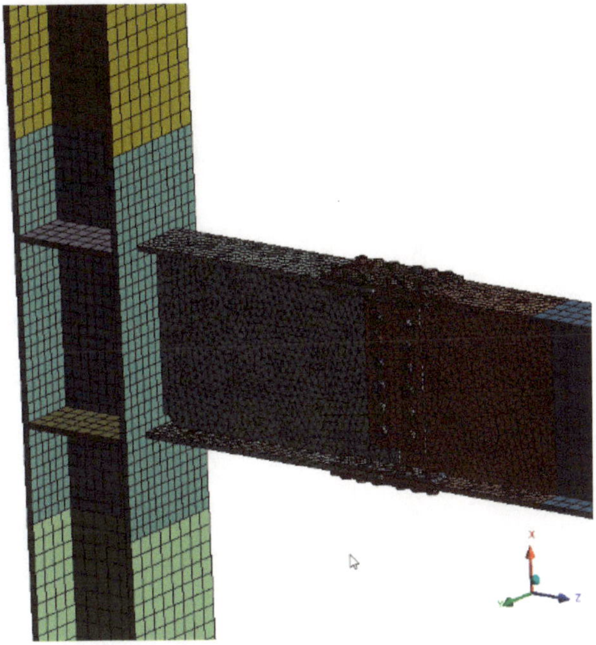

Figure 7. Mesh of the calibration numerical model.

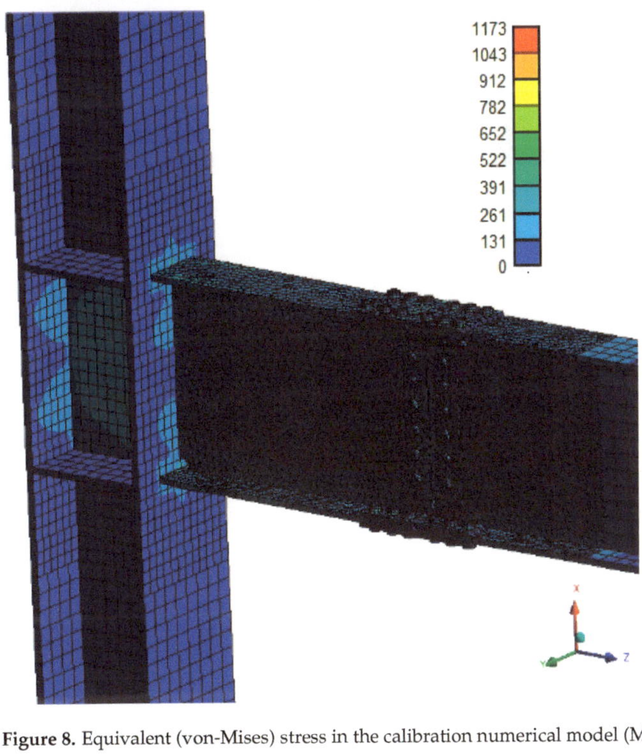

Figure 8. Equivalent (von-Mises) stress in the calibration numerical model (MPa).

The calibration hysteresis curve was compared with three specimens tested by Oh et al. [22], obtaining an acceptable adjustment between the numerical model and experimental specimens. Figure 9 shows the results of the numerical model for specimen CT-B performed by Karami et al. [26], the test by Oh et al., and the present study as a comparison to validate the proposed numerical models.

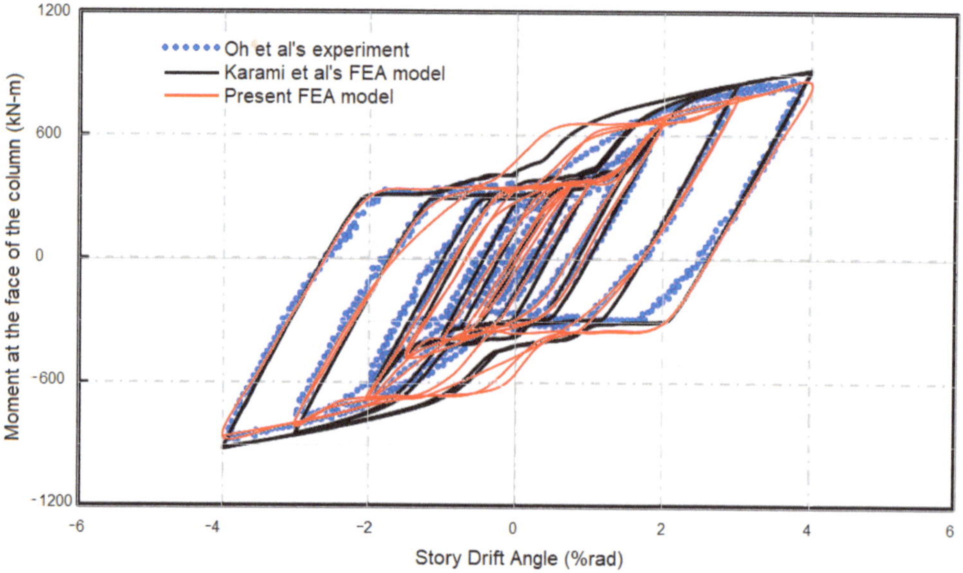

Figure 9. Calibration of the numerical model; adapted from Karami et al. [26].

4.2. Results of Numerical Models

In the following text, the hysteretic behavior is reported through a normalized moment–rotation curve. The normalized tangent and secant stiffness are analyzed with the purpose of analyzing the loss of stiffness in the hysteric behavior of the connection. On the other hand, the performance by type of configuration can be analyzed from the dissipated energy, which is an area closed by a loop.

4.2.1. Hysteretic Curve

The summary of normalized moment–rotation for each model is shown in Figures 10–13; all figures include a curve for a beam with a welded connection (base model M0), for comparative purposes. These curves show stable behavior without any significant losses in strength and stiffness. In all cases, the number of bolts does not affect the ultimate strength of the connection. The plate thickness has little effect on the connection strength (but is not proportional to the thickness increase). However, the effect of the number of bolts plateaus at marked different load levels. A summary of the maximum moment at the face of the column is shown in Table 5.

Table 5. Summary of maximum normalized moment M/Mp at the column face.

		at 0.04 Rad		at 0.05 Rad	
Position	**Bolts**	**PL8**	**PL10**	**PL8**	**PL10**
L1 = 375 mm	n = 3	0.98	1.07	1.07	1.15
	n = 4	1.00	1.09	1.08	1.16
	n = 5	1.01	1.11	1.09	1.19

Table 5. *Cont.*

		at 0.04 Rad		at 0.05 Rad	
Position	**Bolts**	**PL8**	**PL10**	**PL8**	**PL10**
L1 = 500 mm	n = 3	1.07	1.11	1.13	1.17
	n = 4	1.07	1.12	1.14	1.18
	n = 5	1.07	1.19	1.18	1.22
Welded	None	1.21		1.22	

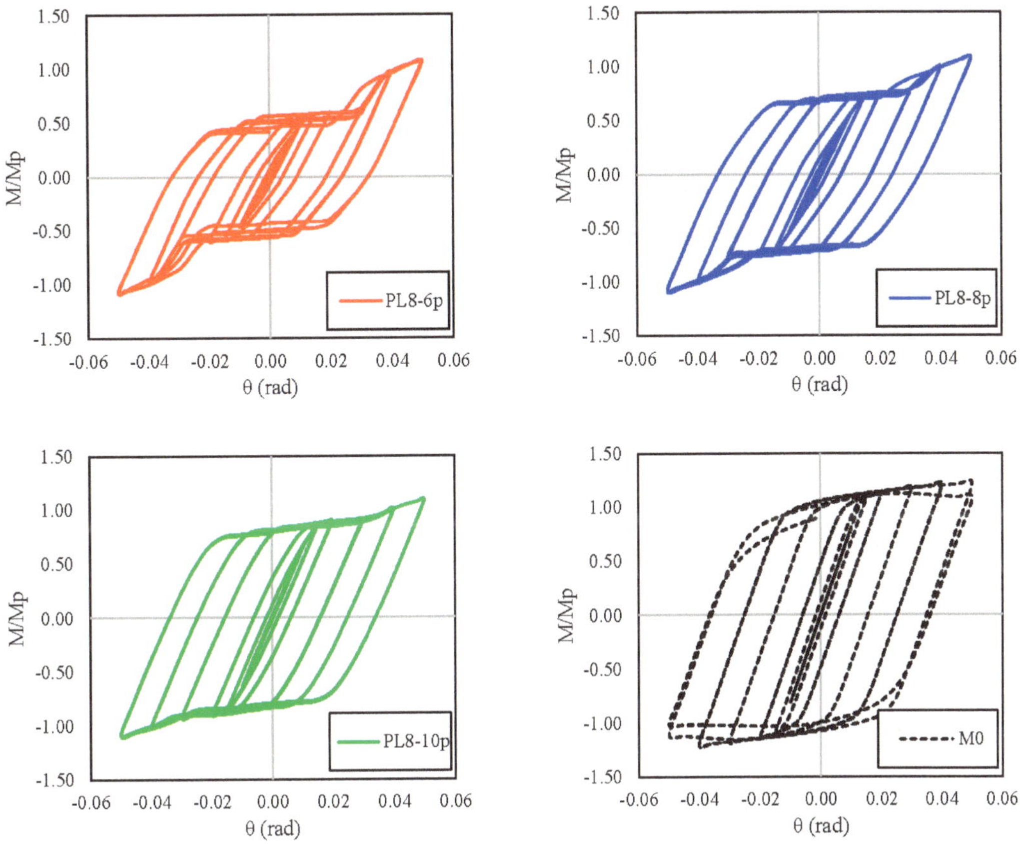

Figure 10. Summary of hysteretic curves for models M1 to M3 (L1 = 375 mm and 8 mm plates).

In all cases, the maximum moment at the column face is higher than 1.0 Mp, except for in model M1, which was designed as the weakest connection. For model M12, designed as the strongest case, the hysteretic curve shows that it behaves similarly to the base model M0 with a connection strength of 1.22 Mp.

The connection's ultimate strength depends directly on the selected plate thickness, shown in Figure 14 at the end of the 5% cycle. After the 1% cycle, the relative slip between connection components determines a horizontal segment for every curve, and its magnitude depends on the bolt quantity. Therefore, models with the same bolt quantity (i.e., M1 and M4 or M7 and 10, for example) have similar curves, differing only after the 3% cycle as they have different plate thicknesses. For these "weakened" connections, these two effects can clearly work independently from each other.

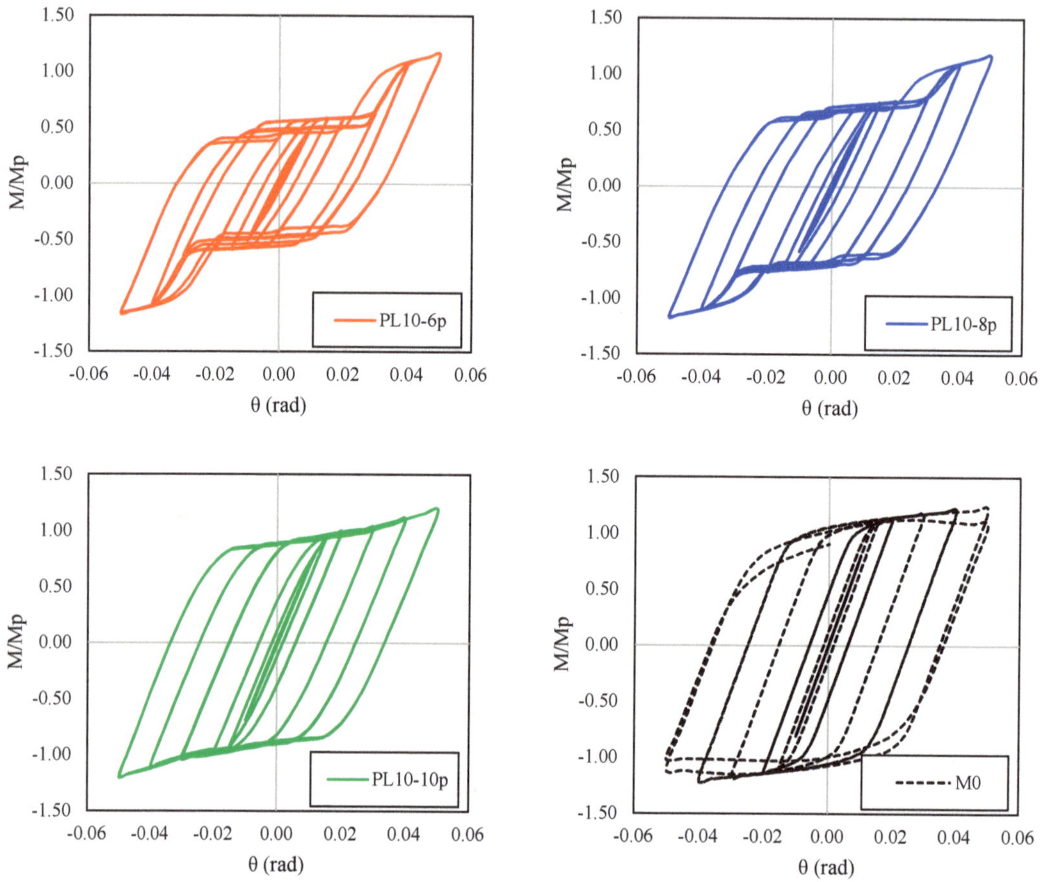

Figure 11. Summary of hysteretic curves for models M4 to M6 (L1 = 375 mm and 10 mm plates).

However, for models M6 and M12 (with 10 mm plates and 10 bolts), a synergy effect occurs between both mechanisms, and the horizontal part of the curve for 10 mm plates is located higher than for 8 mm plates. Also, the curve for model M12 (located further from the column) almost matches the curve for the base model M0.

4.2.2. Plastic Strain

Figure 15 shows the plastic deformation of the base model (M0) at the end of the 5% drift cycle. The beam flange local buckling began at the end of the 4% drift as result of a high stress concentration. However, the local buckling was not reported in the models with bolts and cover plates. The plastic strain was concentrated at the splice plates, and in some cases, there was evidence of bearing around the bolt holes at the beam flange. Table 6 shows a qualitative description of the location of the plastic strain concentration.

Table 6. Location of the plastic strain concentration at 5% drift.

Model	Concentration of Plastic Strain	Damage in Stub?
M1	Plate (εpmax = 0.08)	Not reported
M2	Plate (εpmax = 0.08)	Low

Table 6. *Cont.*

Model	Concentration of Plastic Strain	Damage in Stub?
M3	Plate (εpmax = 0.12)	Low
M4	Plate/Stub (εpmax = 0.09)	Medium
M5	Stub (εpmax = 0.10)	High
M6	Stub (εpmax = 0.10)	High
M7	Plate (εpmax = 0.08)	Not reported
M8	Plate/Stub (εpmax = 0.08)	Medium
M9	Plate (εpmax = 0.09)	Medium
M10	Stub (εpmax = 0.09)	Medium
M11	Stub (εpmax = 0.10)	Stub
M12	Stub (εpmax = 0.06)	Stub

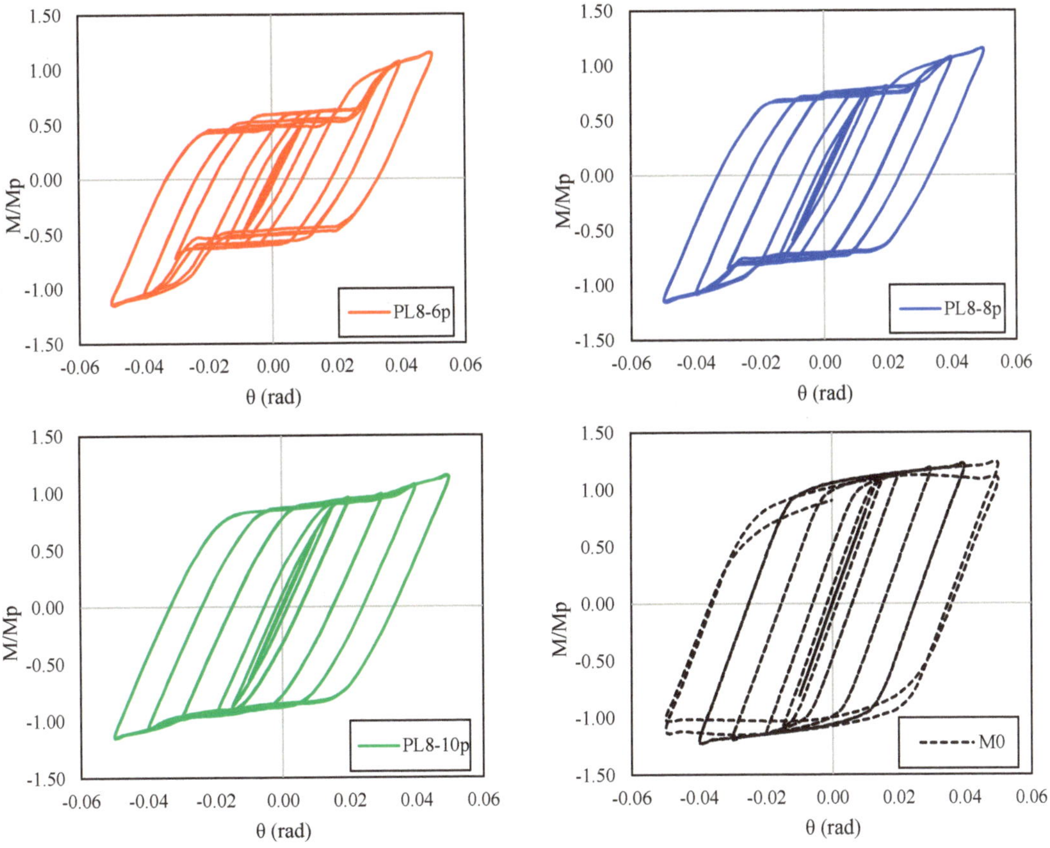

Figure 12. Summary of hysteretic curves for models M7 to M9 (L1 = 500 mm and 8 mm plates).

Figure 16 shows the plastic strain at the end of the 5% cycle for the models with L1 = 375 mm. Models M1, M2, and M3 with 8 mm plates present moderate damage for the cases with fewer bolts, while model M3 shows no damage at all, meaning that most of the energy is dissipated through friction. Figure 17 shows that models M4, M5, and M6 with

10 mm plates are stiffer, and some damage occurs at the beam stub flanges instead of at the splice plates. Plastic strain in the stub was reached for models M4 and M6, because the cover plates are more rigid and closer to the column. This effect was not reported in models M1, M2, and M3.

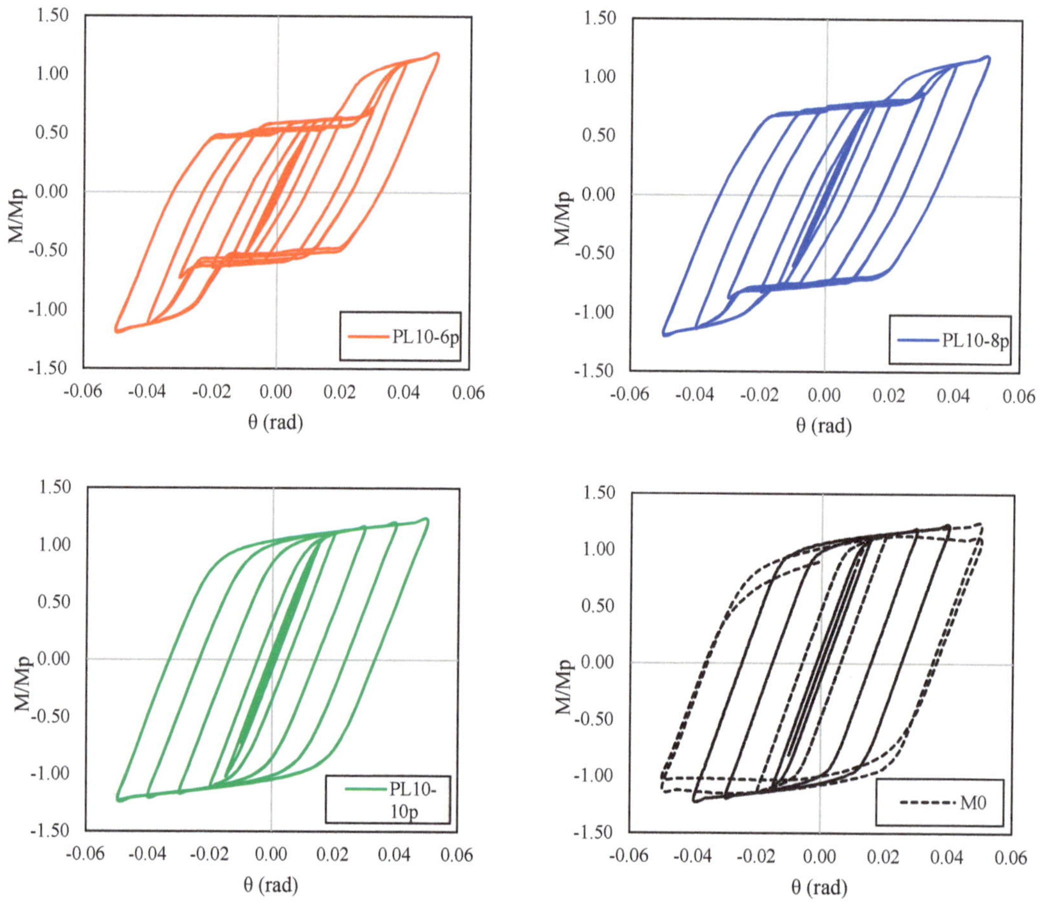

Figure 13. Summary of hysteretic curves for models M10 to M12 (L1 = 500 mm and 10 mm plates).

Figure 18 shows the plastic strain at the end of 5% cycle for the models with L1 = 500 mm. Models M7, M8, and M9 with 8 mm plates present moderate damage for the cases with fewer bolts, while model M9 shows less damage, meaning that most of the energy dissipated through friction. Figure 19 shows that models M10, M11, and M12 with 10 mm plates are stiffer, and little damage occurs at the beam stub flanges in the weld's vicinity. However, model M12 presents a high level of damage in the stub. These results are similar to those obtained by Oh et al. [22] for specimens designed with higher strength (specimen CT).

It is observed that when the extension of the connection plate (with more bolts) results in a smaller gap or distance to the column face, damage concentration occurs at the beam stub (models M6 and M12).

4.2.3. Stiffness Curves

According to the AISC specification [30], the nonlinear behavior of the connection manifests, even at low moment–rotation levels, or exists only for a limited moment–rotation range. The secant stiffness, $Ks = Ms/\theta s$, at service loads is taken as an appropriate parameter for the connection, where Ms is the moment at the service loads and θs is the rotation at the service loads. Implicit in the moment–rotation curve is the definition of the connection as being a region of the column and beam, along with the connecting elements. Then, the ratio between the connection rotational stiffness Ks and the beam rotational stiffness (IE/L) is used to classify the connection as fully rigid (FR), partially rigid (PR), or flexible (F). Figure 14 shows an overview of the tangent and secant stiffness from the backbone curves of the assemblies. The connection relative rotation due to bolt slip does not allow for the direct determination of the tangent stiffness, although a tangent stiffness tendency is shown after the 4% cycle when the bolt clearance limits further slippage.

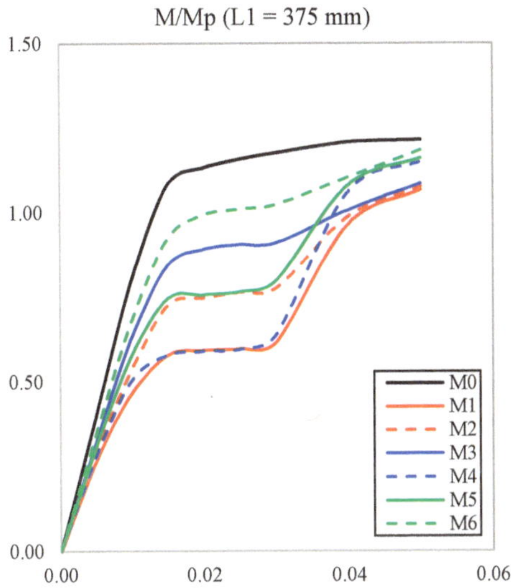

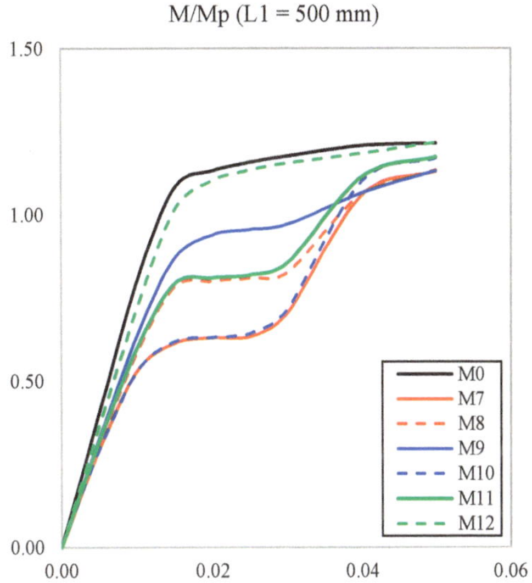

Figure 14. Summary of backbone curves, M/Mp, at the column face.

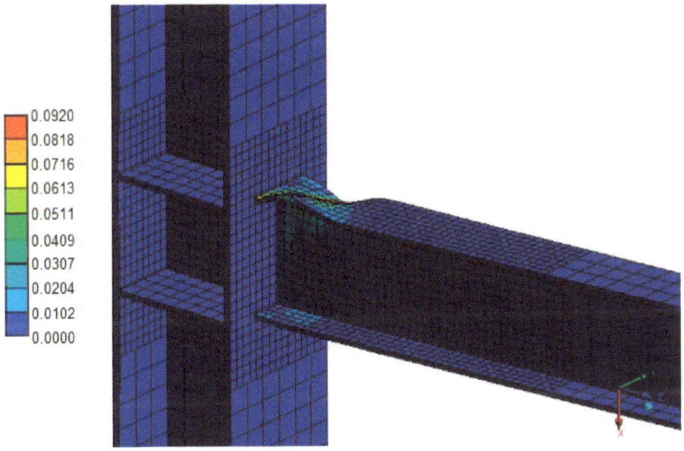

Figure 15. Plastic strain for the base model M0 (mm/mm).

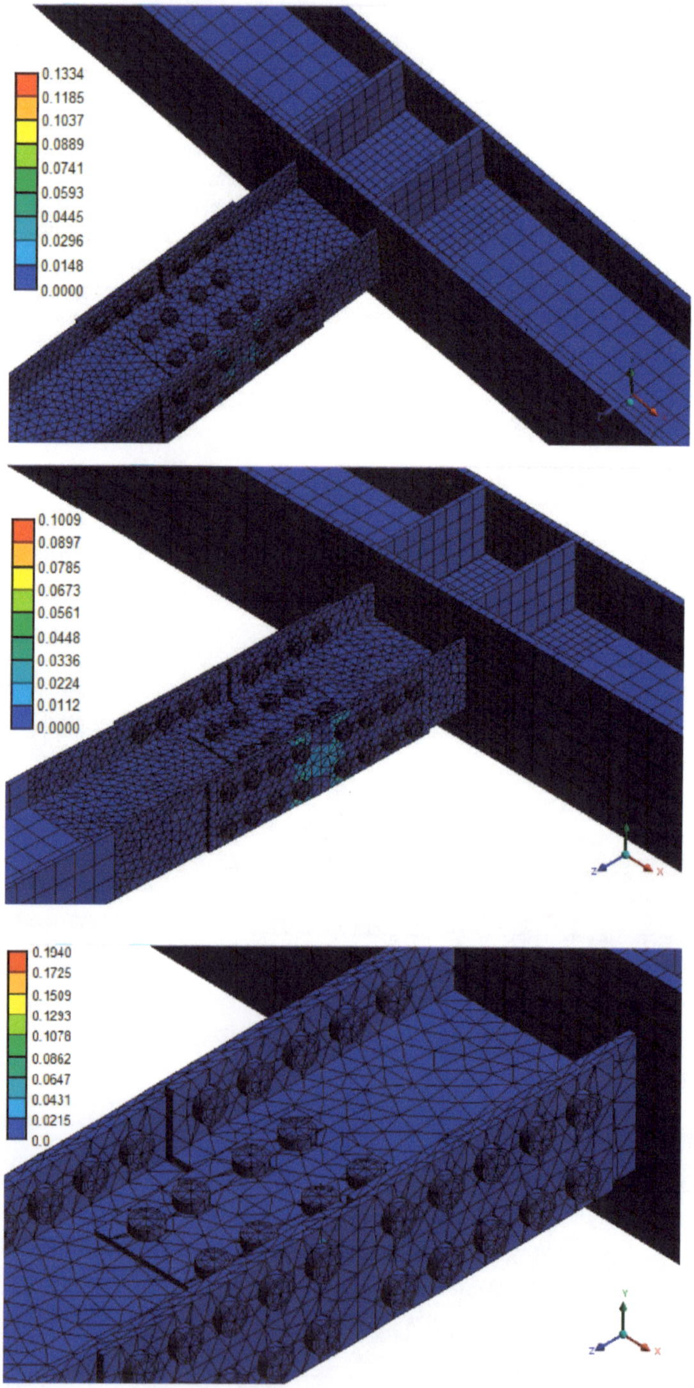

Figure 16. Plastic strain for models M1, M2, and M3 (from top to bottom) (mm/mm).

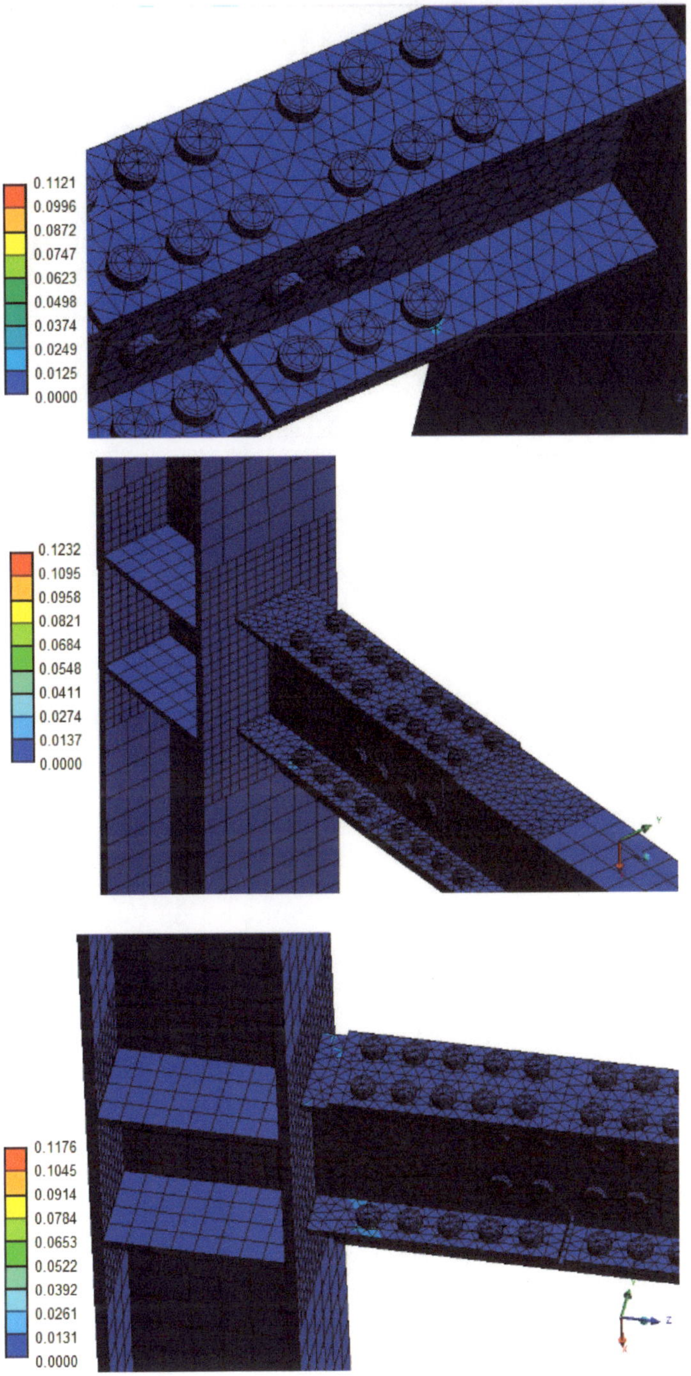

Figure 17. Plastic strain for models M4, M5, and M6 (from top to bottom).

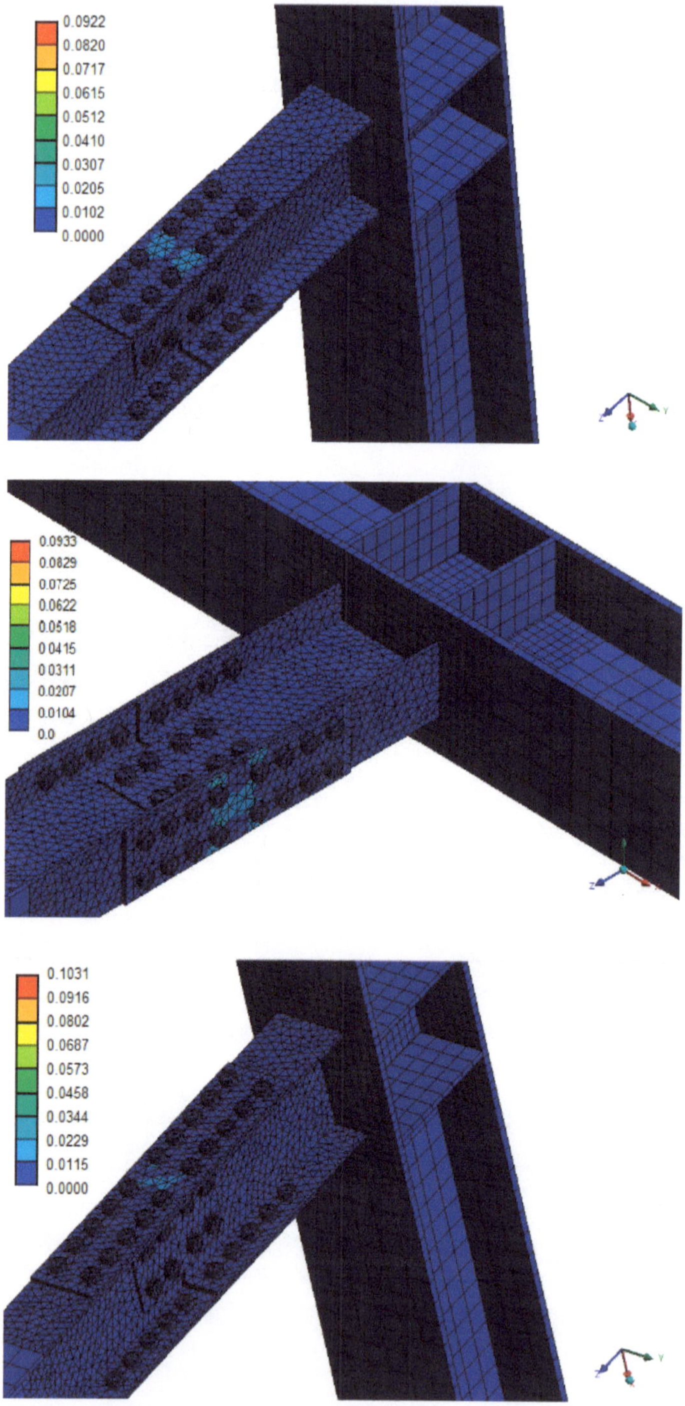

Figure 18. Plastic strain for models M7, M8, and M9 (from top to bottom) (mm/mm).

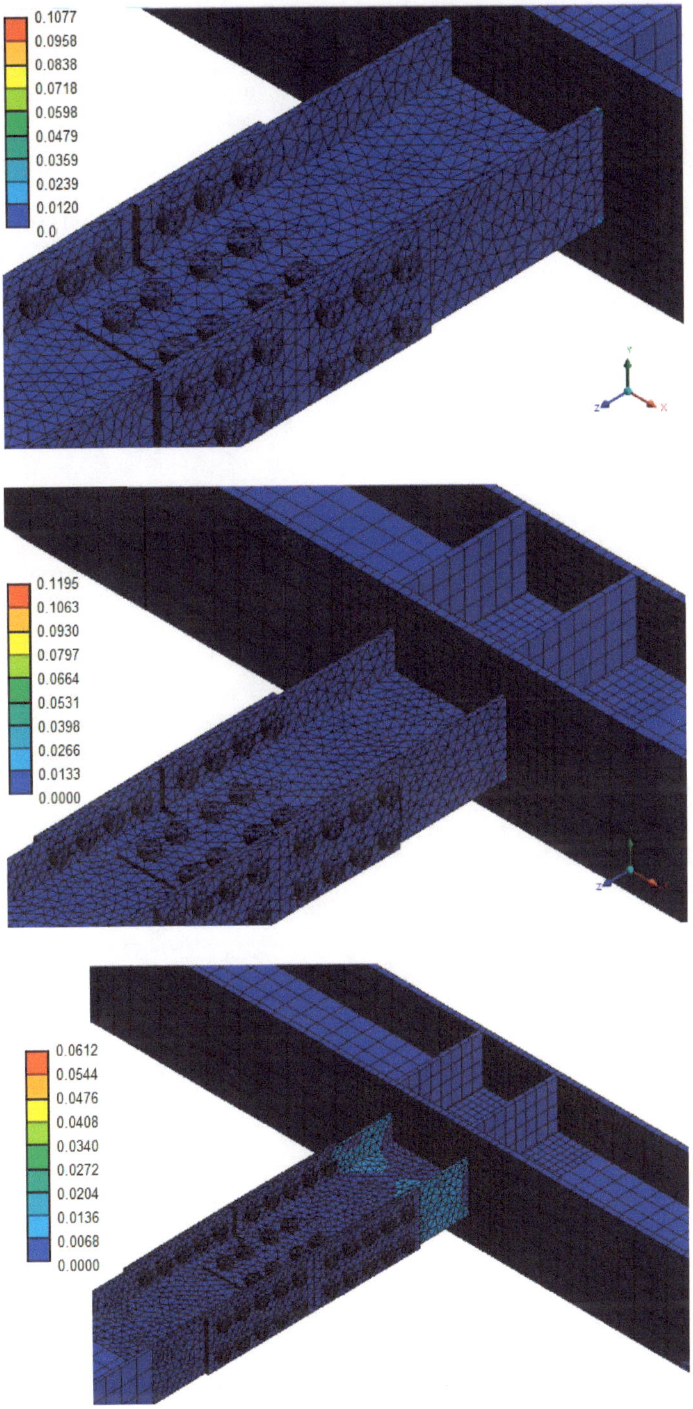

Figure 19. Plastic strain for models M10, M11, and M12 (from top to bottom) (mm/mm).

Astaneh-Asl [16] suggests that, for column-tree systems where the length of the beam stub is less than 0.15 of the span length L, the rotational stiffness of the bolted splice Ksp should be used instead. Figure 20 shows the backbone curves for the bolted splice for moment and rotation at the splice location, where the secant stiffness Ksp can be calculated. However, the "service load" level is not clearly defined for the seismic demand; this study assumes that Ms = Sx × Fy, which is the moment that causes the initial yielding of the beam flange; in this way, the service load aims to avoid damage at the beam stub.

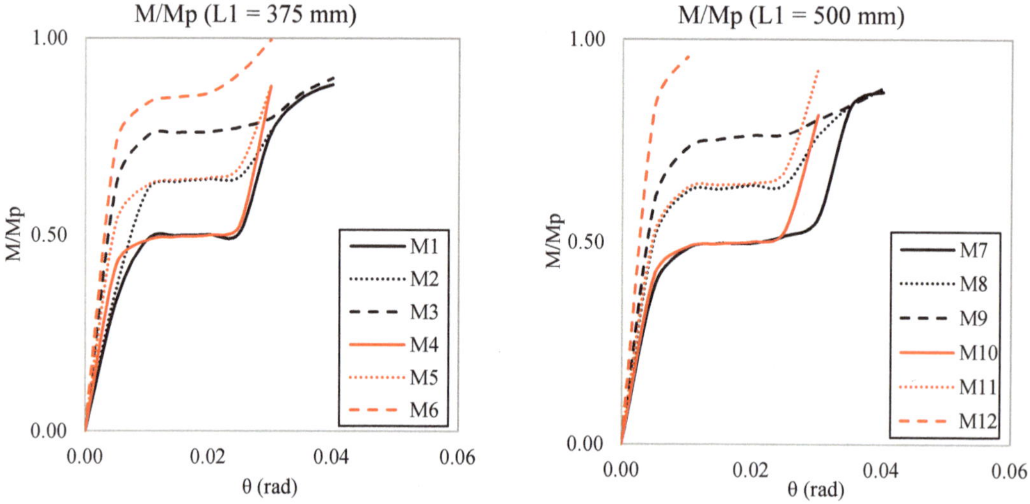

Figure 20. Summary of backbone curves for bolted connections.

Table 7 shows the parameters used for the calculation of the rotational stiffness of the models included in the study.

Table 7. Parameters of the H250 × 6 × 150 × 10 beam.

Beam Mp	109.4	kN-m
Beam My	98.2	kN-m
Beam Ixx	437.5	cm4
Beam length	4.7	m
Steel E	200,000	Mpa
EI/L	2159	kN-m/rad

Table 8 shows the rotational stiffness parameters for the analyzed models. Those with six bolts per flange are the most flexible with m-s < 2. They are classified as simple connections due to the early relative slippage of the components. The rest of the models are classified as PR and have a tendency to increase their stiffness when using thicker plates (i.e., 12 mm) or more bolts (i.e., 12 per flange).

Table 8. Stiffness parameters for the H250 × 6 × 150 × 10 beam.

Model	Mreq	Mplate	Mslip	Ksp-e	Ksp-s	m-e	m-s	Type
M1	92	77.1	31.9	11,239	2859	5.21	1.32	Simple
M2	92	77.1	42.5	13,800	6748	6.39	3.13	PR

Table 8. *Cont.*

Model	Mreq	Mplate	Mslip	Ksp-e	Ksp-s	m-e	m-s	Type
M3	92	77.1	53.1	16,222	15,414	7.51	7.14	PR
M4	92	97.1	31.9	12,856	2808	5.95	1.30	Simple
M5	92	97.1	42.5	16,143	6824	7.48	3.16	PR
M6	92	97.1	53.1	22,631	23,028	10.48	10.67	PR
M7	86	77.1	31.9	13,004	2918	6.02	1.35	Simple
M8	86	77.1	42.5	15,915	11,910	7.37	5.52	PR
M9	86	77.1	53.1	17,638	13,417	8.17	6.21	PR
M10	86	97.1	31.9	13,851	3247	6.42	1.50	Simple
M11	86	97.1	42.5	17,363	13,833	8.04	6.41	PR
M12	86	97.1	53.1	26,858	29,514	12.44	13.67	PR

Mreq: required moment strength at the splice location (kN-m); Mplate: moment strength of the splice plates—yielding (kN-m); Mslip: moment strength of the bolt group—slip (kN-m); Ksp-e: bolted splice rotational stiffness at the elastic load level (kN-m/rad); Ksp-s: bolted splice rotational stiffness at the service load level (kN-m/rad); m-s: stiffness parameter for the connection classification (m = Ksp-s/(EI/L).

4.2.4. Energy Dissipation

In Figure 21, the results show that an increment in the quantity of bolts has a notorious positive effect on the cumulative energy dissipation over time. However, an increment in the plate thickness does not have a relevant effect. In Figure 22, the results show that the location of the connection is not relevant for energy dissipation, with a slight increment for L1 = 500 mm.

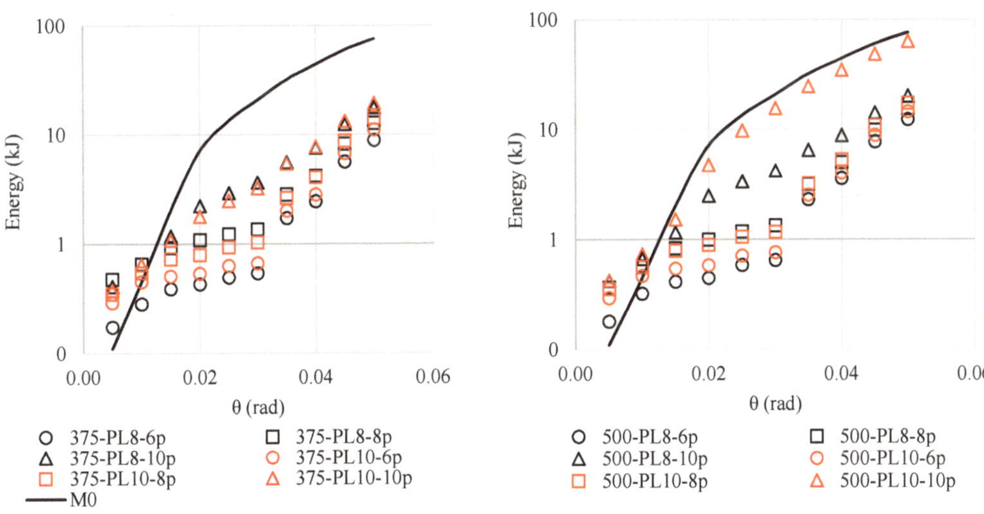

Figure 21. Summary of the energy dissipation (kJ) for all models analyzed for models M1@M6 (**left**) and models M7@M12 (**right**). Black line represents the model M0 (welded beam).

4.2.5. Bolted Splice Ductility

The bolted splice ductility m is the ratio between the inelastic rotation (θu–θy) and the elastic limit θy. For the present study, the connection strength was set at a rotation equal to θu = 4% drift. The results are summarized in Table 9. As the connection stiffens (more bolts, thicker plates), its ductility reduces. An increase in the beam stub length causes a slight increase in the connection ductility.

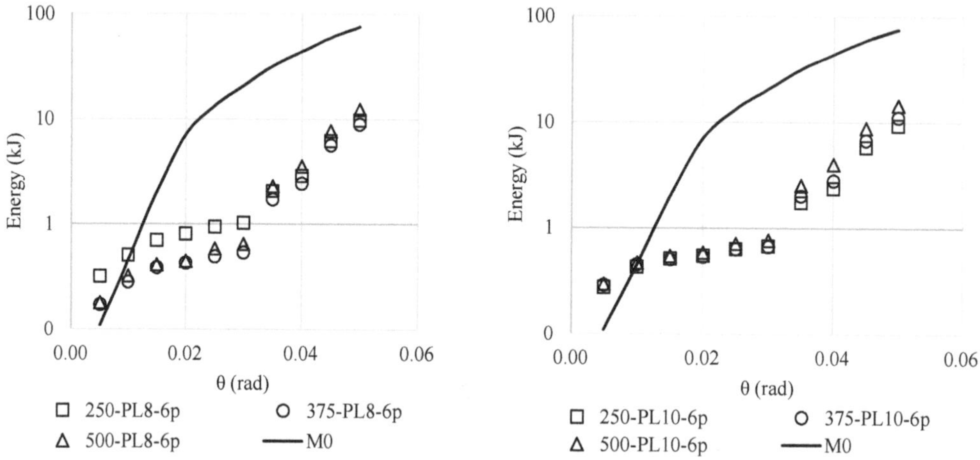

Figure 22. Summary of the energy dissipation for different locations of connection (kJ). Models M0, M1, and M7 (**left**) and models M0, M4, and M10 (**right**). Black line represents the model M0 (welded beam).

Table 9. Summary of the connection ductility.

Model	θy	θu	μ
M1	0.37%	3.37%	8.1
M2	0.46%	3.35%	6.3
M3	0.71%	3.41%	3.8
M4	0.33%	3.12%	8.5
M5	0.41%	3.01%	6.3
M6	0.56%	2.75%	3.9
M7	0.34%	3.44%	9.1
M8	0.41%	3.36%	7.2
M9	0.66%	3.35%	4.1
M10	0.48%	3.07%	5.4
M11	0.74%	2.94%	3.0
M12	0.41%	0.81%	1.0

Ductility is also related to damage, in this case, in the form of plastic deformation. Model M0 is shown in Figure 15, with evident damage at the flange buckling. Table 10 shows that, for a connection closer to the column face (L1 = 375 mm), the quantity of bolts is the main factor that affects the connection ductility. For the connections located far from the column (L1 = 500 mm), the ductility is affected by both the bolt quantity and the plate thickness.

Table 10. Ductility for different bolt quantities and plate thicknesses.

L1 = 375 mm	e = 8 mm	e = 10 mm	L1 = 500 mm	e = 8 mm	e = 10 mm
6 bolts	8.1	8.5	6 bolts	9.1	5.4
8 bolts	6.3	6.3	8 bolts	7.2	3.0
10 bolts	3.8	3.9	10 bolts	4.1	1.0

5. Discussion

A comparison between the different parameters of the connection behavior is developed, as follows.

5.1. Location Effect

As the connection is located away from the column face, the moment required at the bolted splice decreases, and its relative stiffness increases. The overall effect is an increase of 1% to 6% in the tension at the welds and the beam stub flanges, as shown in Table 11. There is no significant increase in the energy dissipation.

Table 11. Comparison of the moments at the column face for different splice locations with a 5% drift cycle.

L = 375 mm	M/Mp	L = 500 mm	M/Mp	M500/M375
M1	1.07	M7	1.13	106%
M2	1.08	M8	1.13	105%
M3	1.09	M9	1.13	104%
M4	1.15	M10	1.17	102%
M5	1.16	M11	1.17	101%
M6	1.18	M12	1.22	103%

For practical purposes, this variable can be useful to set a desired connection stiffness.

5.2. Plate Thickness Effect

The main effect of the plate thickness is the connection stiffness. Figure 23 shows that the plate thickness does not have a significant effect on the ultimate connection strength; however, a thinner plate does lower the reaction at the column face, decreasing the plastic strain (damage) at the beam stub flanges.

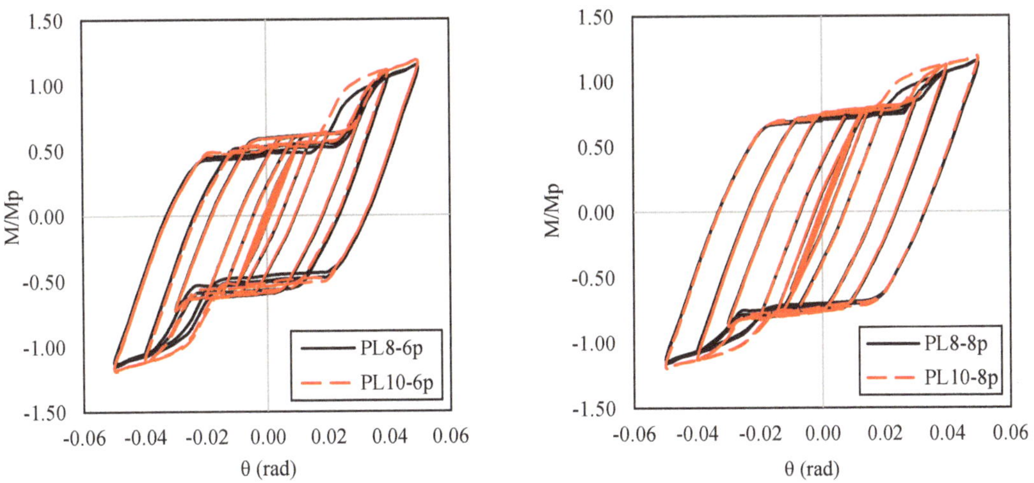

Figure 23. Effect of the splice plate thickness on the connection strength for six bolts (**left**) and eight bolts (**right**), for L1 = 500 mm.

The plate thickness plays a secondary role to the bolt quantity and is harder to predict. A thinner plate is more prone to deformation and deflection, increasing the tension at the flange bolts closer to the connection center. The result is a loss in the friction resistance.

5.3. Bolt Quantity Effect

The main effect of the bolt quantity is the energy dissipation (see Figure 21). Given a minimum quantity to provide a bearing shear capacity, the plates slip at low cycle loads preventing the connection from taking advantage of the friction work. A higher number of bolts means more friction and a higher potential for energy dissipation. However, if the slip strength of the bolt group reaches or surpasses the plastic moment of the splice plates, slippage takes place only at the last load cycles.

Therefore, a specific feature effect of the bolt quantity is the staggered shape of the hysteretic curve (see Figure 14), which means that a lower number of bolts reduces the stiffness of the connection after it reaches its slip strength, protecting the beam stub from developing early plastic strain. More studies on this topic may lead to a procedure being found to determine the optimum number of bolts to maximize the energy dissipation of the connection while minimizing the damage to the beam and the column.

6. Conclusions

In column-tree moment connections, the bolted splice is usually designed to be stronger than the connected beams and does not affect the seismic performance of the frame. However, if the moment connection is designed as semi-rigid in a selected location away from the column, the connection can develop the ability to effectively reduce the seismic forces and provide a good source of energy dissipation. The goal of this research was to assess the cyclic performance of column-tree moment connections using weakened plates. The newly proposed approach was evaluated in analytical and numerical studies using the finite element model with ANSYS. A total of fourteen nonlinear models were developed, simulating a full-scale test of joints with a built-up H-column to I-beam subjected to the AISC seismic provisions loading protocol. A previous calibration from the experimental test was performed. The main conclusions obtained are described as follows:

(1) The 80% of plastic moment was reached at an interstory drift ratio of 0.04 for all cases analyzed. However, a pinching of the hysteretic curve was obtained for models with a combined failure mechanism (failure in plates and stub) and failure in plates.
(2) The selected location for the bolted splice does not affect the overall performance of the connection. However, models with a length stub of 250 mm (MA, MB, and M6) showed that, when located too close to the column, a collateral effect of over-stiffness induced a stress concentration in the vicinity of the beam-to-column weld.
(3) The weakened connections effectively reduced the plastic strain at the beam-to-column welds and at the beam stub itself. Hence, the connection plates performed well as damage-control fuses that can be replaced easily after a severe seismic event. Also, the thickness plate has a marginal effect on the ultimate strength of the connection, and the bolt quantity does not play a major role in the connection strength.
(4) According to the classification of moment connections established in AISC 360, the column-tree moment connection using weakened plates proposed in this research is classified as partially restrained (PR). Therefore, the seismic design of steel buildings with moment frames using this type of moment connection should consider the elastic stiffness of the connection to verify the drift limits.
(5) Finally, damage after a severe earthquake will mostly develop as plastic deformation of the connection plates, which can be replaced in a cost-effective manner. This change in philosophy regarding the use of replaceable components as fuse components over the use of a beam as a fuse member was demonstrated in the experimental tests carried out by the authors.

Author Contributions: Conceptualization, E.N.; methodology, E.N.; software, P.M.; validation, E.N. and P.M.; formal analysis, P.M.; investigation, P.M.; resources, P.M. and E.N.; data curation, E.N.; writing—original draft preparation, P.M.; writing—review and editing, P.M.; visualization, E.N.; supervision, E.N.; project administration, E.N.; funding acquisition, P.M. and E.N. All authors have read and agreed to the published version of the manuscript.

Funding: The research reported herein was possible thanks to the funding of Proyectos Ingeniería 2030 (ING222010004) and InES—Ciencia—Abierta (INCA210005).

Data Availability Statement: Data are contained within the article.

Conflicts of Interest: The authors declare no conflict of interest.

References

1. FEMA. *State of the Art Report on Connection Performance*; SAC Joint Venture; FEMA: Washington, DC, USA, 2000.
2. AISC. *AISC 358—Prequalified Connections for Special and Intermediate Steel Moment Frames for Seismic Applications*; American Institute of Steel Construction: Chicago, IL, USA, 2005.
3. Horikawa, K.; Sakino, Y. Damages to Steel Structures Caused by the 1995 Kobe Earthquake. *Struct. Eng. Int.* **1996**, *6*, 181–182. [CrossRef]
4. Nakashima, M.; Roeder, C.W.; Maruoka, Y. Steel Moment Frames for Earthquakes in United States and Japan. *J. Struct. Eng.* **2000**, *126*, 861–868. [CrossRef]
5. Whittaker, A.; Moehle, J.; Higashino, M. Evolution of seismic building design practice in Japan. *Struct. Des. Tall Build.* **1998**, *7*, 93–111. [CrossRef]
6. Nakashima, M.; Inoue, K.; Tada, M. Classification of damage to steel buildings observed in the 1995 Hyogoken-Nanbu earthquake. *Eng. Struct.* **1998**, *20*, 271–281. [CrossRef]
7. Chen, C.-C.; Lin, C.-C. Seismic performance of steel beam-to-column moment connections with tapered beam flanges. *Eng. Struct.* **2013**, *48*, 588–601. [CrossRef]
8. Kang, C.-H.; Kim, Y.-J.; Shin, K.-J.; Oh, Y.-S. Experimental Investigation of Composite Moment Connections with External Stiffeners. *Adv. Struct. Eng.* **2013**, *16*, 1683–1700. [CrossRef]
9. Schneider, S.P.; Teeraparbwong, I. Inelastic Behavior of Bolted Flange Plate Connections. *J. Struct. Eng.* **2002**, *128*, 492–500. [CrossRef]
10. Malley, J.O.; Carter, C.J.; Saunders, C.M. Seismic Design Guidelines and Provisions for Steel-Framed Buildings. *Earthq. Spectra* **2000**, *16*, 179–203. [CrossRef]
11. Zhang, X.; Zheng, S.; Zhao, X. Seismic performance of steel beam-to-column moment connections with different structural forms. *J. Constr. Steel Res.* **2019**, *158*, 130–142. [CrossRef]
12. Morshedi, M.A.; Dolatshahi, K.M.; Maleki, S. Double reduced beam section connection. *J. Constr. Steel Res.* **2017**, *138*, 283–297. [CrossRef]
13. Vetr, M.; Miri, M.; Haddad, A. Seismic Behavior of a New Reduced Beam Section Connection by Drilled Holes Arrangement (RBS_DHA) on the Beam Flanges through Experimental Studies. In Proceedings of the 15th World Conference on Earthquake Engineering, Lisbon, Portugal, 24–28 September 2012.
14. Nuñez, E.; Torres, R.; Herrera, R. Numerical Study on Cyclic Response of End-Plate Biaxial Moment Connection in Box Columns. *Metals* **2017**, *25*, 523. [CrossRef]
15. Tao, Z.; Li, W.; Shi, B.-L.; Han, L.-H. Behaviour of bolted end-plate connections to concrete-filled steel columns. *J. Constr. Steel Res.* **2017**, *134*, 194–208. [CrossRef]
16. Astaneh-Asl, A. *Seismic Design of Steel Column-Tree Moment-Resisting Frames*; Departement of Civil and Environmental Engineering—University of California: Berkeley, CA, USA, 1997.
17. Baharmast, A.; Razzaghi, J.; Kamouneh, S. Modification of pre-Kobe column-tree moment resisting connection. *Asian J. Civ. Eng.* **2018**, *19*, 223–237. [CrossRef]
18. Gharebaghi, R.; Hosseini, A. Local behavior of column-tree connections with considering bolt-hole clearance and bolt pretension effects. *Int. J. Steel Struct.* **2017**, *17*, 379–388. [CrossRef]
19. Palizzolo, L.; Vazzano, S. Iterative Optimal Design of Special Moment Resisting Devices for Steel Frames. *IOP Conf. Ser. Mater. Sci. Eng.* **2021**, *1203*, 032079. [CrossRef]
20. Chen, Y.; Tong, L.; Chen, Y.; Zhou, X. Experimental study on cyclic behavior of cast steel connectors for beam-to-column joints. *Adv. Struct. Eng.* **2016**, *19*, 1677–1695. [CrossRef]
21. Yeow, T.Z.; Orumiyehei, A.; Sullivan, T.J.; MacRae, G.A.; Clifton, G.C.; Elwood, K.J. Seismic performance of steel friction connections considering direct-repair costs. *Bull. Earthq. Eng.* **2018**, *16*, 5963–5993. [CrossRef]
22. Oh, K.; Li, R.; Chen, L.; Hong, S.-B.; Lee, K. Cyclic testing of steel column-tree moment connections with weakened beam splices. *Int. J. Steel Struct.* **2014**, *14*, 471–478. [CrossRef]
23. AISC. *AISC 341—Seismic Provisions for Structural Steel Buildings*; American Institute of Steel Construction: Chicago, IL, USA, 2016.
24. Oh, K.; So, J.; Ha, H.; Lee, K. Seismic performance evaluation of Korean column-tree steel moment connections. *Int. J. Steel Struct.* **2016**, *16*, 1287–1298. [CrossRef]
25. Lin, X.; Chen, Y.; Yan, J.-B.; Hu, Y. Seismic Behavior of Welded Beam-to-Column Joints of High-Strength Steel-Moment Frame with Replaceable Damage-Control Fuses. *J. Struct. Eng.* **2020**, *146*, 04020143. [CrossRef]
26. Karami, B.; Hashemi, B.H.; Ramhormozian, S.; Zarabimanesh, Y. Numerical study on prequalification and cyclic performance of column-tree connections. *Structures* **2022**, *41*, 331–348. [CrossRef]

27. Kim, T.; Yu, E. Seismic analysis of steel moment frames with column-tree connections. *J. Constr. Steel Res.* **2020**, *168*, 105871. [CrossRef]
28. Akagwu, P.; Ali, F.; Nadjai, A. Behaviour of bolted steel splice connections under fire. *J. Constr. Steel Res.* **2020**, *170*, 106103. [CrossRef]
29. Saravanan, M.; Goswami, R.; Palani, G.S. Replaceable Fuses in Earthquake Resistant Steel Structures: A Review. *Int. J. Steel Struct.* **2018**, *18*, 868–879. [CrossRef]
30. AISC. *AISC 360—Specification for Structural Steel Buildings*; American Institute of Steel Construction: Chicago, IL, USA, 2016.
31. ANSYS Inc. *ANSYS 2022*; ANSYS Inc.: Canonsburg, PA, USA, 2022.
32. Vilela, P.M.L.; Carvalho, H.; Grilo, L.F.; Montenegro, P.A.; Calçada, R.B. Unitary model for the analysis of bolted connections using the finite element method. *Eng. Fail. Anal.* **2019**, *104*, 308–320. [CrossRef]
33. Nuñez, E.; Parraguez, G.; Herrera, R. Bidirectional Response of Weak-Axis End-plate Moment Connections: Numerical Approach. *Metals* **2020**, *10*, 964. [CrossRef]
34. Herrera, R.; Salas, C.; Beltran, J.F.; Nuñez, E. Experimental performance of double built-up T moment connections under cyclic loading. *J. Constr. Steel Res.* **2017**, *138*, 742–749. [CrossRef]
35. Grondin, G.; Jin, M.; Josi, G. *Slip Critical Connections—A Reliability Analysis for Design at the Ultimate Limit State*; University of Alberta: Edmonton, AB, Canada, 2007.
36. Maison, B.F.; Speicher, M.S.; Maison, M.B.F.; Eeri, M. Loading Protocols for ASCE 41 Backbone Curves. *Earthq. Spectra* **2016**, *32*, 2513–2532. [CrossRef]

buildings

MDPI

Article

Mechanical Properties and Influence Factors of Ordinary Shear Links

Shujun Hu [1], Shangwen Liu [1], Sizhi Zeng [2] and Tiefeng Shao [1,*]

1 School of Infrastructure Engineering, Nanchang University, Nanchang 330031, China; hushujun@ncu.edu.cn (S.H.); 402500220020@email.ncu.edu.cn (S.L.)
2 Zhongmei Engineering Group Co., Ltd., Nanchang 330001, China; zengsizhi1986@163.com
* Correspondence: shaotf8293624@ncu.edu.cn; Tel.: +86-15970614128

Abstract: The current specification requires the same limiting values of inelastic rotation and the overstrength factor for shear links with a length ratio less than 1.6. However, recent studies have shown that the mechanical properties of ordinary shear links with a length ratio ranging from 1.0 to 1.6 are obviously different from those of very short shear links with a length ratio less than 1.0. Additionally, shear links made of different steel materials have differences in mechanical properties. Based on Q345 steel, three ordinary shear links with a length ratio of 1.36 were designed to intensively explore the influence of stiffener configurations and spacing on mechanical properties. Under cyclic loading tests, the failure modes, hysteresis curves, skeleton curves, secant stiffness curves and energy dissipation capacities of shear link specimens were recorded. The results show that the overstrength factor and inelastic rotation of specimens SL-1 and SL-2, which had different stiffener configurations, reached 1.59 and 0.10, while those of specimen SL-3, which had wider stiffener spacing, reached 1.48 and 0.07, which showed that varying the stiffener configuration has no obvious effect, while relaxing stiffener spacing can result in severe buckling of the web. Additionally, its bearing capacity, inelastic rotation, secant stiffness and energy dissipation capacity reduced. Hence, the stiffener spacing should satisfy the requirements of the specification and not be too wide. Based on ABAQUS software, finite element models of ordinary shear links proved to be accurately consistent with test specimens in terms of mechanical properties. On this basis, 114 numerical models of ordinary shear links with different length ratios, stiffener spacings, flange-to-web area ratios, flange strengths, web depth-to-thickness ratios and stiffener thicknesses were designed to study the influence on the overstrength factor.

Keywords: eccentrically braced frame; ordinary shear links; length ratio; overstrength factor; hysteresis curve

1. Introduction

Eccentrically braced steel frame structures have been widely applied at home and abroad, and shear links are a typical energy-dissipating component of an eccentrically braced frame [1,2]. Eccentrically braced steel frame structures dissipate seismic energy through inelastic deformation of shear links and effectively ensure that columns, supports, non-energy-dissipating beams and other components are in the elastic stage [3–7]. A significant characteristic of the eccentrically braced steel frame structure is that damage occurs in predesigned shear links, protecting the connecting nodes from damage. Only replacing the shear link can restore the structure to its original function, which brings great convenience to post-earthquake repairs [8–11]. The overstrength factor and inelastic rotation are used to evaluate the lateral stiffness and energy dissipation capacity of shear links. According to specification [12], shear links are divided into shear type ($\rho \leq 1.6$), bending type ($1.6 < \rho < 2.6$) and bending–shear type ($\rho \geq 2.6$) according to the length ratio ρ (as shown in Equation (1)). Studies have shown that shear links have the best ductility and energy-dissipating capacity and are thus used preferentially in eccentrically braced steel frame structures [12]. The current specification requires the same limiting values of inelastic rotation and the overstrength factor for shear links with a length ratio less than 1.6, while the limiting values are 0.08 and 1.50, respectively [12].

Existing research has shown that shear links made of different steel materials have different mechanical properties. Yin [13] studied shear links based on LYP100 steel with a length ratio of 0.4 and pointed out that the overstrength factor could reach 3.29. Zhuang [14] studied shear links based on LYP160 steel with a length ratio of 0.97, and the overstrength factor could reach 2.0. Ghadami [15] studied shear links using LYP225 steel with a length ratio of 1.4, showing that the average overstrength factor was 1.55. Elgammal [16] created numerical models of shear links based on S275 steel with a length ratio of 0.97 and pointed out that the overstrength factor could reach 1.78. Yao [17] studied shear links using Q235 steel with a length ratio of 1.3, showing that the average overstrength factor was 1.85. Hu [18] conducted a study on numerical models of shear links based on Q235 steel with length ratios less than 1.0, and pointed out that the average inelastic rotation and overstrength factor were 0.13 and 1.90, respectively. Liu [19] studied shear links based on Q345 steel with a length ratio of 1.53, and the overstrength factor could reach 1.47. Men [20] studied a shear link whose flange was made of Q345 and whose web was made of Q235 with a length ratio of 1.40 and pointed out that the overstrength factor was 1.97. Mansour [21] studied shear links using CSA G40.21-350W-grade steel with a length ratio of 1.40 and pointed out that the maximum value of the overstrength factor was 1.68, and the average value was 1.50. Okazaki [22] studied shear links using A992 grade steel with a length ratio of 1.49, and the overstrength factor and inelastic rotation were 1.37 and 0.08, respectively.

Existing research [23] has shown that when the length ratio of shear link is higher than 1.0, its mechanical properties are significantly different from those of very short shear links with length ratios less than 1.0. Ozkilic [24] designed shear links with length ratios of 1.0 and 1.43 made of S275, whose overstrength factor could reach 1.76 and 1.58, respectively. Hu [18] recorded the overstrength factor of shear links based on Q235 steel with a length ratio less than 1.0 as 1.90, while Liu [19] studied shear links based on Q345 steel with a length ratio of 1.53 and pointed out that the overstrength factor could reach 1.47. To distinguish it from very short shear links, a shear link with a length ratio higher than 1.0 is defined as an ordinary shear link. In addition, existing research has shown that stiffener spacing [25,26], flange strength [27,28], flange-to-web area ratio [29], etc., have an obvious impacts on the overstrength factor of shear links.

To intensively explore the mechanical properties and influence factors of ordinary shear links made of Q345 steel with length ratios higher than 1.0, this research conducted a series of cyclic loading tests and abundant numerical modeling. Firstly, three ordinary shear link specimens with a length ratio of 1.36 were designed to study the influence of the stiffener configuration and stiffener spacing on the mechanical properties. Secondly, ABAQUS 2021 software was used to simulate the hysteresis curves and failure modes of the experimental specimens. Thirdly, parametric numerical research was performed to study the influencing factors of ordinary shear links on the mechanical properties.

2. Experimental Program

2.1. Basic Definitions of Shear Links

- Length ratio is denoted by ρ as shown in Equation (1), where e is the length of the shear link, V_p is the plastic shear capacity of the shear link, and M_p is the plastic moment capacity of the shear link. See the equations to define these parameters below:

$$\rho = eV_p / M_p \tag{1}$$

$$V_p = 0.58 h_w t_w f_y \tag{2}$$

$$M_p = W_p f_y \quad (3)$$

- Overstrength factor: V_u/V_p, where V_u is the ultimate shear capacity of the shear link, and V_p is the plastic shear capacity of the shear link;
- Inelastic rotation: Δ_u/e, where Δ_u is the maximum lateral displacement of the shear link.

When the length ratio of a shear link is higher than 1.0, it is defined as an ordinary shear link. The specification stipulates that the overstrength factor and inelastic rotation of shear links shall not be less than 1.50 and 0.08 [12].

2.2. Test Material

In individual ordinary shear link specimens, all plates are made of Q345 steel. The composition of Q345 steel is shown in Table 1. The thicknesses of the web, flange and stiffener are 6 mm, 8 mm and 10 mm, respectively. Based on tensile property tests on steel materials of different thicknesses, the specific material property results are shown in Table 2.

Table 1. Composition of Q345 steel.

Element	C	Mn	Si	P	S	Nb	V	Ti	Ni	Cr	Mo
Content	≤0.20	≤1.70	≤0.50	≤0.35	≤0.35	≤0.07	≤0.15	≤0.20	≤0.012	≤0.30	≤0.10

Table 2. Mechanical properties of steel plates used in the specimens.

Part	Thickness t (mm)	Yield Strength f_y (MPa)	Ultimate Tensile Strength fu (MPa)	Elastic Modulus E_s (GPa)	Elongation δ (%)
Web	6	356	535	203	21.7
Flange	8	361	547	206	22.2
Stiffener	10	368	552	208	23.3

2.3. Test Specimens

Based on Q345 steel, three specimens with different stiffener configurations and stiffener spacings, SL-1, SL-2 and SL-3, were designed to study the influence of stiffener configuration and stiffener spacing on the mechanical properties of ordinary shear links. The dimensions of the three specimens are shown in Figure 1, and the stiffener parameters of the specimens are shown in Table 3.

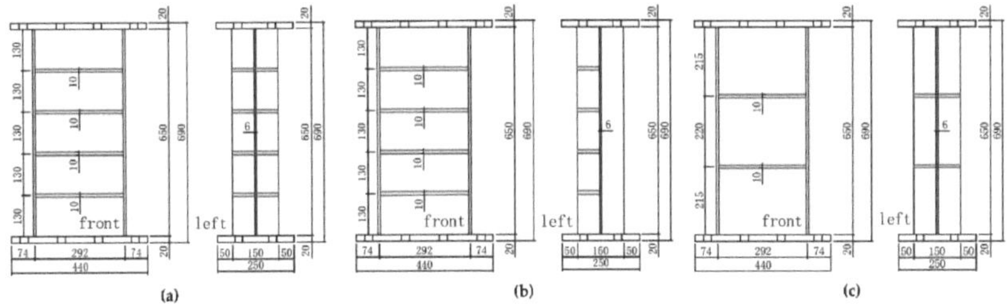

Figure 1. Dimensions of the test specimens: (**a**) specimen SL-1; (**b**) specimen SL-2; (**c**) specimen SL-3.

Table 3. Design parameters of the test specimens.

Specimen	Length Ratio ρ	Section (mm)	Stiffener Spacing (mm)	Stiffener Configuration
SL-1	1.36	300 × 150 × 6 × 8	4@130	Both sides
SL-2	1.36	300 × 150 × 6 × 8	4@130	One side
SL-3	1.36	300 × 150 × 6 × 8	2@220	Both sides

2.4. Test Protocol

The cyclic loading protocol was displacement-controlled. In Step 1, we used the general moment method [30] to determine the yield load P_y and the yield displacement Δ_y of individual specimens. In Step 2, according to the requirements of the specification, the displacement step-by-step loading method was used to perform the cyclic displacement-controlled loading protocol on the specimens. The loading displacements are the multiples of the yield displacement Δ_y which are carried out in the order of $\Delta_y/4$, $\Delta_y/2$, $3\Delta_y/4$, Δ_y, $2\Delta_y$, $3\Delta_y$, $4\Delta_y$, etc. The loading displacement at each loading level cycles once.

2.5. Test Setup and Instrumentation

This test was conducted in the Structural Laboratory of Nanchang University, Jiangxi Province, China. The test setup and instrumentation are shown in Figure 2. The servo actuator was installed between the reaction wall and the loading beam to provide horizontal cyclic displacement. The upper end plate of the specimen was connected to the loading beam with high-strength friction-type bolts, while the lower end plate was connected to the ground beam with high-strength friction-type bolts. The high-strength bolts were 10.9 M22-grade. The yield strength and ultimate strength of the high-strength bolts were 940 MPa and 1040 MPa, respectively. To eliminate the outward movement, the loading beam and the ground beam were connected with two parallel pin-connected columns, and the loading beam was fixed by two reaction frames.

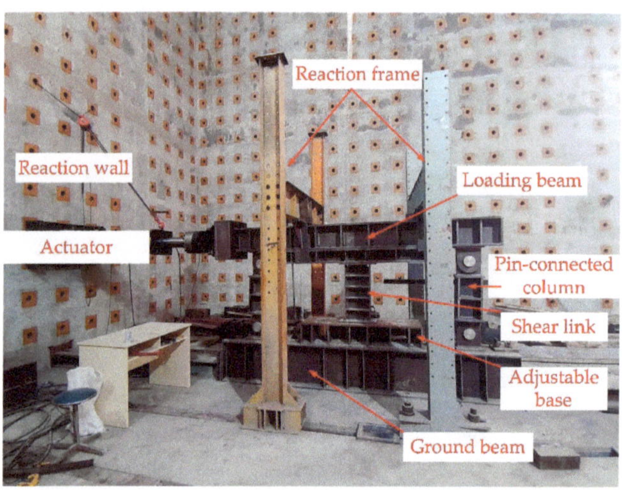

Figure 2. Experiment setup.

2.6. Measurement Distribution

This test measurement included the load and displacement in the shear direction and the strain in the dangerous area. The load and displacement in the shear direction were displayed on the monitor of the servo actuator. The measurement schemes of the displacement and strain in the dangerous area are shown in Figure 3.

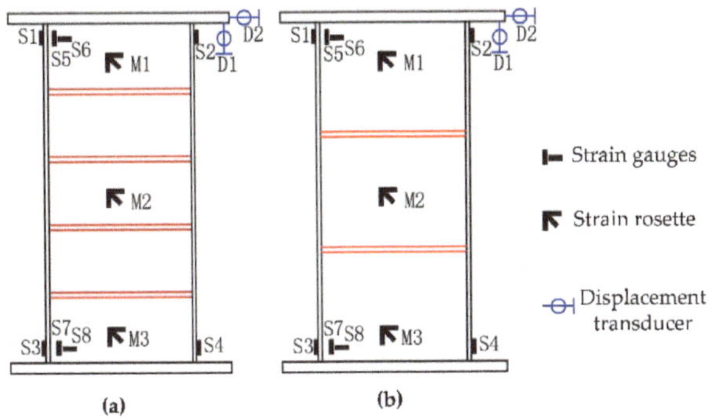

Figure 3. Measurement distribution: (**a**) SL-1 and SL-2; (**b**) SL-3.

(1) Displacement measurement. To ensure the displacement value showed by the actuator was close to the actual displacement of the specimen, the vertical and horizontal displacement transducers, D1 and D2, were distributed on the upper end plate of the specimen.

(2) Strain measurement. Based on the stress characteristics and symmetry, the strain gauges S1, S2, S3, and S4 were distributed in the middle of the upper and lower ends of the flange, while the transverse strain gauges S6 and S8 and the longitudinal strain gauges S5 and S7 were set at two corners of the web, respectively. The strain rosettes M1, M2 and M3 were distributed at the end and in the middle areas of the web.

3. Test Results

3.1. Failure Modes

3.1.1. Specimen SL-1

The failure process of SL-1 as determined using the cyclic loading protocol is shown in Figure 4. It mainly went through five stages: shear yielding, flange buckling, micro-cracking at the weld connection between the flange plate and the end plate, micro-crack extension and weld tearing.

As shown in Figure 4a, in the early stage of loading, the horizontal displacement of the specimen was small, and when the yield displacement Δ_y was reached, the shear strain values at S1, S2, S3 and S4 all reached the expected yield (Figure 4a). When the loading displacement was 22.67 mm, the end of the flange panel buckled (Figure 4b). When the displacement was −30.87 mm, a micro-crack appeared at the weld connection between the flange plate and the end plate (Figure 4c). When the loading displacement reached 44.18 mm, the buckling of the flange intensified, and the micro-crack at the weld extended (Figure 4d). When loaded to 63.31 mm, the maximum load value of the specimen reached 560.02 kN. One side of the weld between the flange plate and the end plate was completely torn, and the flange buckled violently (Figure 4e). The specimen was considered to have lost its bearing capacity. The test was finally stopped (Figure 4f).

3.1.2. Specimen SL-2

The stiffeners of specimen SL-2 were only distributed on the signal side, and stiffener spacing and other parameters were consistent with specimen SL-1. The aim was to explore the influence of stiffener configurations on the mechanical properties of ordinary shear link. Under cyclic loading, the failure mode of specimen SL-2 was basically the same as that of SL-1, which mainly went through five stages: shear yield, flange buckling, micro-cracking at the weld, micro-crack extension and crack tearing.

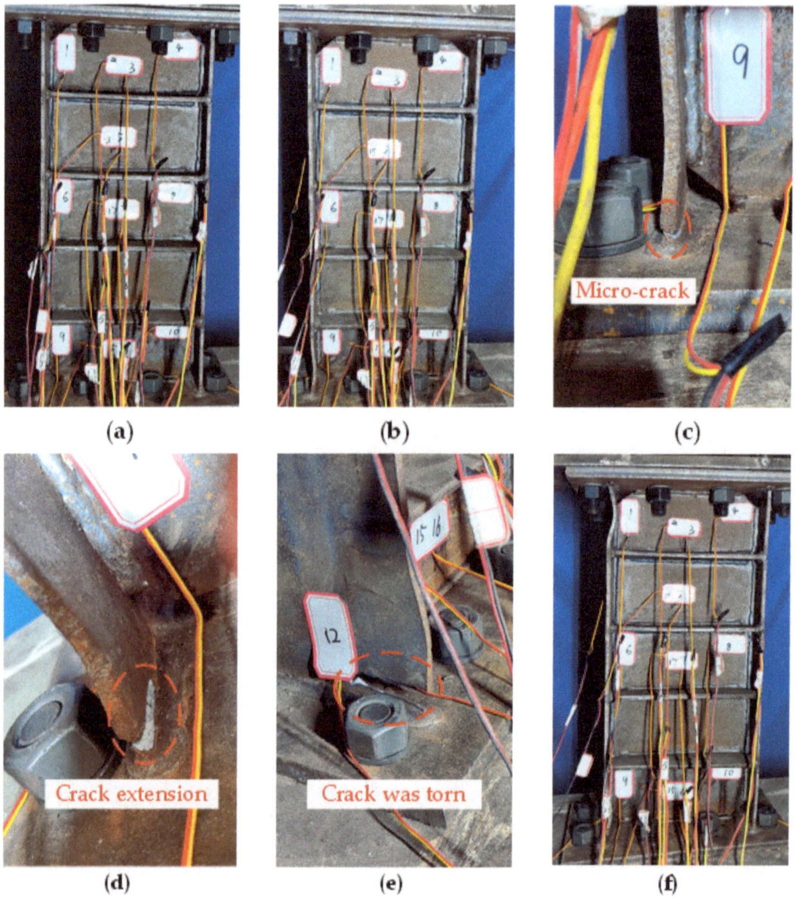

Figure 4. Failure process of SL-1: (**a**) shear yield; (**b**) flange buckling; (**c**) micro-crack appearing at the weld connection between the flange plate and the end plate; (**d**) micro-crack extension; (**e**) a crack was torn; (**f**) failure mode of SL-1.

As shown in Figure 5a, in the early stage of loading, the horizontal displacement of the specimen was small. When the yield displacement Δ_y was reached, the shear strain values at S1, S2, S3 and S4 all reached the expected yield (Figure 5a). When the loading displacement was −22.21 mm, the end plate of the flange buckled (Figure 5b). When the displacement was 30.57 mm, a micro-crack appeared at the weld connection between the flange plate and the end plate (Figure 5c). When the loading continued until the displacement reached 43.98 mm, the micro-crack extended (Figure 5d). When loaded to 63.12 mm, the maximum load value of the specimen was 556.20 kN, and the crack extended into a 30° diagonal crack (Figure 5e). One side of the weld was completely torn. The flange plate buckled violently and lost its bearing capacity, and the specimen was damaged. The test was finally stopped (Figure 5f).

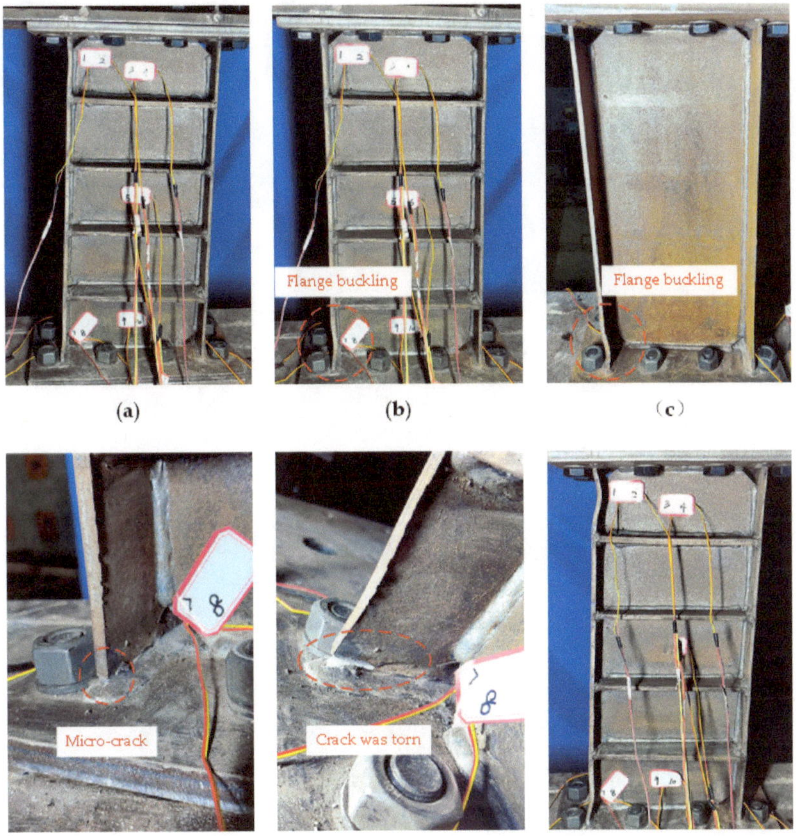

(**a**) (**b**) (**c**)

(**d**) (**e**) (**f**)

Figure 5. Failure process of SL-2: (**a**) shear yield; (**b**) flange buckling; (**c**) rear view of flange buckling; (**d**) micro-crack appearing at the weld connection between the flange plate and the end plate; (**e**) a crack was torn; (**f**) failure mode of SL-2.

3.1.3. Specimen SL-3

The stiffeners of specimen SL-3 were distributed on both sides, but the stiffener spacing exceeded the limit of the specification. The aim was to study the impact of relaxing the stiffener spacing on the mechanical properties of ordinary shear links. The early loading stage was basically the same as that of SL-1, but the final damage pattern was different. It mainly went through five stages: shear yield, flange buckling, micro-cracking at the weld connection between the flange plate and the end plate, web buckling and specimen damage.

As shown in Figure 6a, in the early stage of loading, the horizontal displacement of the specimen was small. When the yield displacement Δ_y was reached, the shear strain values at S1, S2, S3 and S4 all reached the expected yield (Figure 6a). When the loading displacement was 20.71 mm, the end of the flange plate buckled (Figure 6b). When the displacement was 30.75 mm, a micro-crack appeared at the weld connection between the flange plate and the end plate (Figure 6c). When the loading continued until the displacement reached 42.18 mm, the crack extended, and the web plate underwent slight buckling (Figure 6d). When the displacement reached 44.57 mm, the maximum load value was 518.73 kN, and the bearing capacity of specimen SL-3 suddenly decreased, while the displacement continued to increase. With continuous loading, the shear force value instantly dropped to 460.31 kN, the entire shear link experienced a large lateral shift, the web plate severely buckled, and the specimen lost its bearing capacity, so the specimen was damaged. The test was finally stopped (Figure 6e).

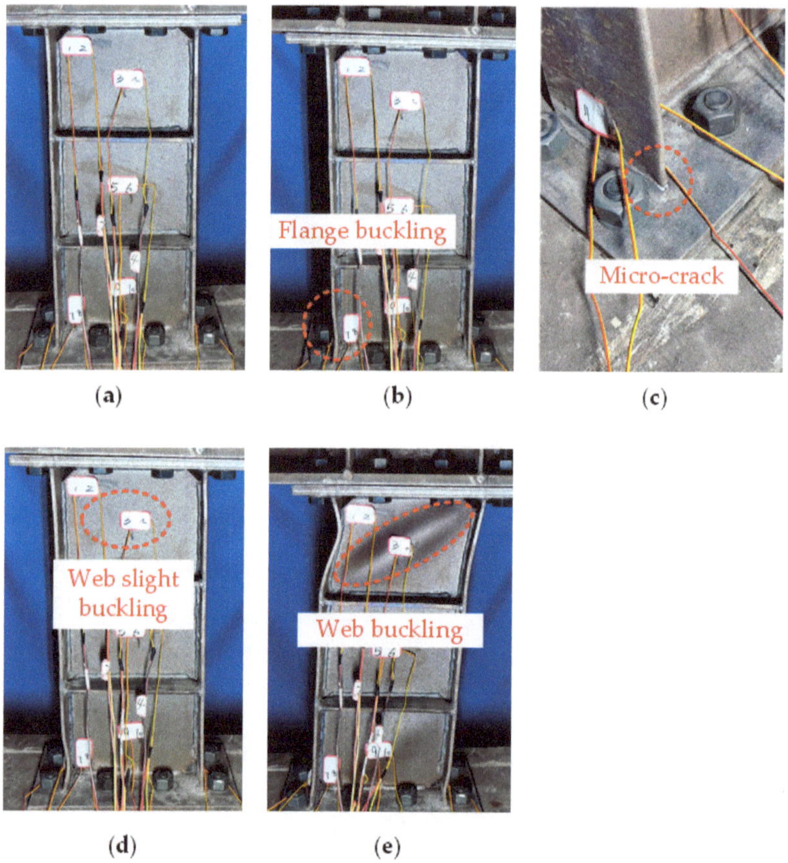

Figure 6. Failure process of SL-3: (**a**) shear yield; (**b**) flange buckling; (**c**) micro-crack appearing at the weld connection between the flange plate and the end plate; (**d**) web buckling; (**e**) failure mode of SL-3.

To sum up, the failure processes of the three specimens were roughly the same. The failure modes of SL-1 and SL-2 were basically the same, indicating that the stiffener configuration has no significant impact on the mechanical properties of ordinary shear links. However, the crack in SL-2 extended into a 30° diagonal crack, which had a negative impact on the eccentrically braced steel frame. In seismic design, it is recommended that stiffeners be distributed on both sides of the web. Compared with SL-1, SL-3 had a decrease in the bearing capacity and an overall area plastic phenomenon, in which both the web and flange turned plastic, indicating that relaxing stiffener spacing will decrease the bearing capacity and ductility of ordinary shear links. Therefore, whether it is a very short shear link or an ordinary shear link, the stiffener spacing should satisfy the specification requirements and not be too wide.

3.2. Hysteresis Curves

Figure 7 shows the hysteresis curves of individual specimens under the cyclic loading protocol. The hysteresis curves are wide, and the load increased with the increase in displacement. There is an obvious smooth section before the load increased to the maximum value. According to the hysteresis curves, when the shear force reached 70% of the maximum load, the force remained unchanged, and the displacement continued to increase. When the inelastic rotation was 0.03 rad, the shear force slowly increased to the maximum load value as the displacement increased. This phenomenon is caused by the full-section plasticity of the flange end and tearing of the weld at the connection between the flange plate and the end plate. As shown in Figure 7a, when the loading displacement was 26.95 mm, the shear force in the hysteresis curve was delayed in reaching the maximum value. In the experiment (Figure 4), a micro-crack in the weld appeared when the displacement was 30.87 mm, which is consistent with the hysteresis curve. Therefore, the plasticity of the entire flange end will cause the hysteresis curve to be delayed in reaching the maximum value.

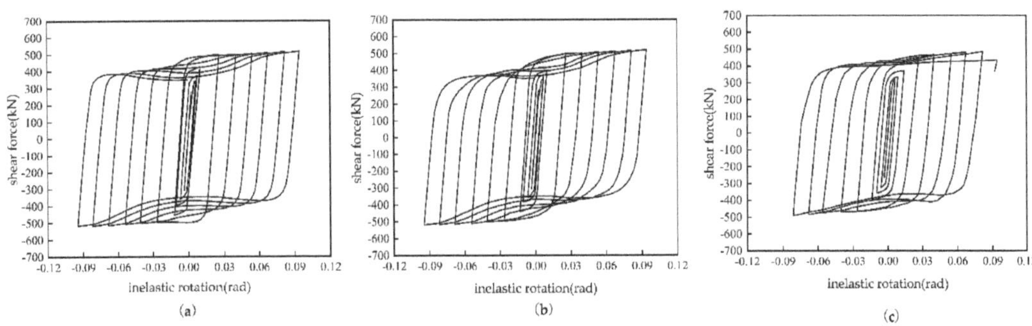

Figure 7. Hysteresis curves of the specimens: (**a**) specimen SL-1; (**b**) specimen SL-2; (**c**) specimen SL-3.

The hysteresis curves of SL-1 and SL-2 are always in the strengthening stage, with the ultimate shear forces of specimens SL-1 and SL-2 being 560.02 kN and 556.20 kN, respectively, and the ultimate inelastic rotations were 0.10 rad for both. Therefore, distributing stiffeners on one side or both sides has little impact on the mechanical properties of ordinary shear links, and both specimens have good ductility and energy-dissipating capacity. For specimen SL-3, which had a large stiffener spacing, the ultimate shear force was only 517.31 kN when the inelastic rotation reached 0.07 rad. During the subsequent loading process, the shear force decreased with the increase in displacement, and when the inelastic rotation was 0.08 rad, the shear force was only 436.28 kN, so relaxing the stiffener spacing will cause both the bearing capacity and ductility of ordinary shear links to decrease violently. The corresponding ultimate bearing capacity and inelastic rotation of specimen SL-3 were 517.31 kN and 0.07 rad, respectively. The mechanical properties of individual specimens are shown in Table 4.

Table 4. Mechanical properties of the specimens.

Specimen	Ultimate Shear Force V_u (kN)	Yield Shear Force V_p (kN)	Overstrength Factor		Inelastic Rotation	
			Test	Specification	Test	Specification
SL-1	560.02	350.85	1.60	1.50	0.10	0.08
SL-2	556.20	350.85	1.59	1.50	0.10	0.08
SL-3	517.73	350.85	1.48	1.50	0.07	0.08

To sum up, the overstrength factor and inelastic rotation of SL-1 and SL-2 reached 1.59 and 0.10, while those of SL-3 only reached 1.48 and 0.07, which are lower than the existing design standard values of 1.50 and 0.08. It is necessary to meet the required stiffener spacing. Considering the 30° diagonal crack in the weld of SL-2, distributing stiffeners on one side is not recommended.

3.3. Skeleton Curves

Figure 8 shows the skeleton curves of each specimen. It can be seen from the figure that the initial stage of the three specimens was the elastic stage, after which all specimens entered the elastic–plastic and plastic stages. The shear force of specimens SL-1 and SL-2 always increased gradually with the increase in displacement, and there was no obvious difference between these two specimens in any loading step; however, the shear force of specimen SL-3 reached a maximum when the inelastic rotation was 0.07 rad, followed by a sharp decrease in the load-bearing capacity due to the buckling of the web and flange. Compared with SL-1, SL-3 had a decreased load-bearing capacity after reaching the maximum shear force. Therefore, the distribution of stiffeners has no significant impact on the mechanical properties of ordinary shear links, but relaxing the stiffener spacing will significantly affect their bearing capacity and ductility.

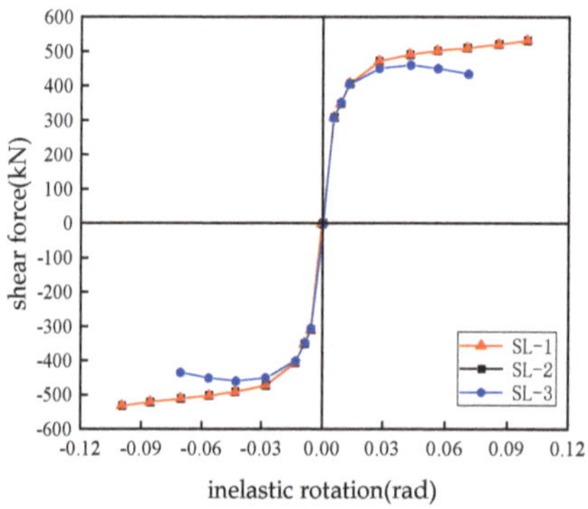

Figure 8. Skeleton curves of all test specimens.

3.4. Secant Stiffness Curves

The secant stiffness refers to the stiffness of a structure when it reaches the maximum load point from the origin when subjected to external forces. The decrease in the secant stiffness will cause the structure to deform greatly during the stress process, which may exceed the allowable deformation range and lead to damage to the function of the structure, so it is vital to investigate the secant stiffness of ordinary shear links.

Figure 9 shows the secant stiffness curves of individual specimens. During the entire loading process, the secant stiffness values of individual specimens showed gradually decreasing trends. At the beginning of loading, the secant stiffness values of individual specimens were basically the same. As the displacement increased, the secant stiffness values of specimens SL-1 and SL-2 were basically equal, but the secant stiffness value of specimen SL-3 decreased, which was mainly caused by relaxing the stiffener spacing of the ordinary shear link, leading to greater buckling of the web and flange.

3.5. Energy-Dissipating Capacity

The energy dissipation value is an important indicator for evaluating the energy dissipation capacity. When the energy dissipation capacity of a structure is insufficient, it cannot effectively absorb and dissipate the energy of external loads, causing the remaining energy to be concentrated in a certain part of the structure. This will cause the structure to undergo greater stress and deformation under earthquakes or other loads, increasing the severity of the structure's damage, so it is vital to investigate the energy dissipation capacity of ordinary shear links.

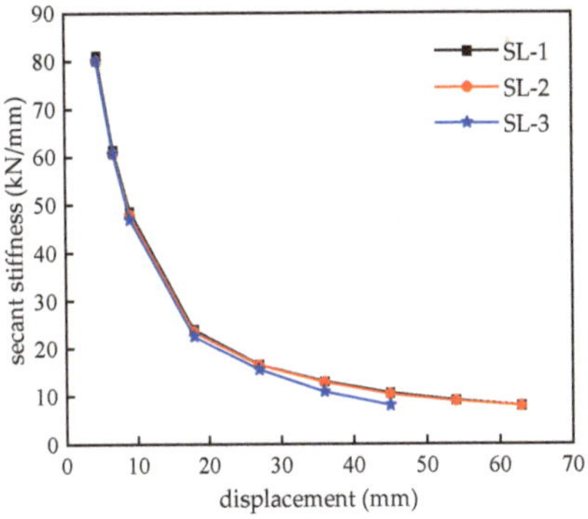

Figure 9. Secant stiffness curves of all test specimens.

Figure 10 shows the energy dissipation values of each specimen. As the displacement increased, the energy dissipation values of three specimens all increased steadily, indicating that the three specimens all show good energy dissipation capacities. The energy dissipation values of SL-1 and SL-2 at the maximum displacement of 63 mm were 99.84 kJ and 97.92 kJ, respectively, with only a 3% decrease. However, the energy dissipation value of SL-3 at the maximum displacement 45 mm was only 66.62 kJ, while the energy dissipation value of SL-1 under the same displacement was 75.36 kJ, with a decrease of 13%, which shows that distributing stiffeners on one side will reduce the energy dissipation capacity by 3%, while relaxing the stiffeners will reduce the energy dissipation capacity by 13%.

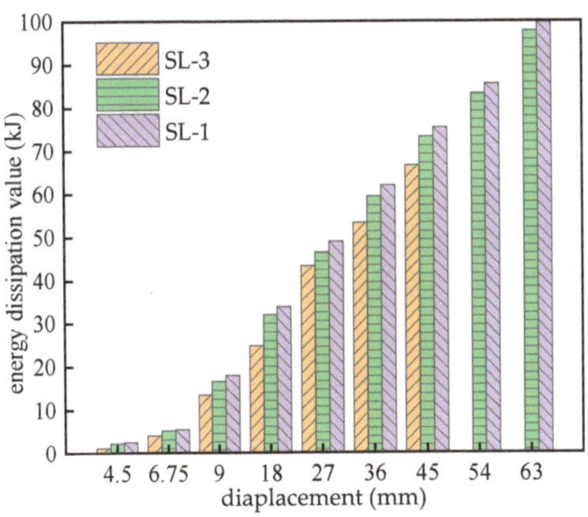

Figure 10. Energy dissipation curves of all test specimens.

4. Numerical Modeling of Ordinary Shear Links

4.1. Basic Procedure of Numerical Modeling

(1) Material properties. The material properties of the steel plates of the numerical shear link models are shown in Table 2. The material constitutive curves of the numerical models are shown in Figure 11.

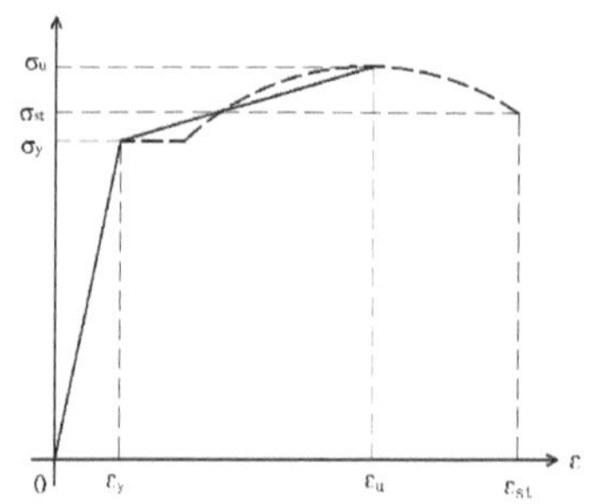

Figure 11. Material constitutive curves of the numerical models.

(2) Boundary conditions. The specimen was in the loading environment described in Section 2.5. In the horizontal direction, the loading beam and ground beam were connected with two parallel pin-connected columns, and the loading beam was fixed by two reaction frames to eliminate outward movement, and the in-plane displacement was provided by the actuator. In the vertical direction, the shear link was connected to the loading beam and ground beam through friction high-strength bolts. Therefore, in finite element models, the displacement of the lower end plate of the shear link is restricted in all directions, and the horizontal in-plane displacement of the upper end plate of the shear link is allowed.

(3) Loading protocol. In finite element models, where horizontal displacement is induced on the upper end plate of the shear link, the moment and shear force at the end plate of the shear link can be simulated in the same way as the actual shear force in the eccentrically frame structure. The specific loading method is the same as that used in the experiment, as described in Section 2.4.

4.2. Comparison of Finite Element and Test Results

The SL-1, SL-2 and SL-3 models were established using ABAQUS software. The parameters and boundary conditions were consistent with those used in the test.

4.2.1. Ultimate Shear Force and Overstrength Factor

The ultimate shear force and overstrength factor obtained by finite element models are shown in Table 5. The errors between the test and finite element analysis results were less than 5%. Therefore, the finite element model is reliable to simulate actual specimens.

Table 5. Comparison of the mechanical properties.

Specimen		Yield Shear Force F_y (kN)	Yield Displacement Δ_y (mm)	Ultimate Shear Force F_u (kN)	Overstrength Factor	Errors (%)
SL-1	Test	340.97	8.93	560.02	1.59	0.71
	FEM	340.97	9.10	564.31	1.60	
SL-2	Test	340.97	8.87	556.20	1.58	0.69
	FEM	340.97	9.01	560.09	1.60	
SL-3	Test	340.97	8.72	518.73	1.49	1.29
	FEM	340.97	8.93	525.40	1.51	

4.2.2. Hysteresis Curves

The hysteresis curves are shown in Figure 12. The hysteresis curves of the test specimens are in good agreement with the hysteresis curves of the finite element models, and both had good energy dissipation capabilities. Additionally, the hysteresis curves of finite element models indicate that shear force led to a delay in reaching the maximum load, which is consistent with the test results.

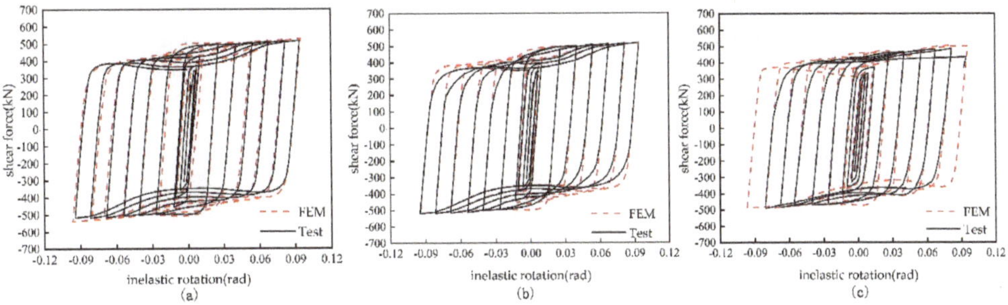

Figure 12. Hysteresis curves of the finite element models: (**a**) specimen SL-1; (**b**) specimen SL-2; (**c**) specimen SL-3.

4.2.3. Failure Process

During the test, the flange of specimen SL-1 buckled; then, the weld between the flange and the end plate was torn, and finally, the specimen lost its bearing capacity. As shown in Figure 13, the flange first buckled after the web experienced shear yielding. As the displacement continued to increase, the stress between the flange and the end plate of specimen SL-1 reached the maximum, indicating that the weld tearing phenomenon occurs between the flange and the end plate, which is consistent with the failure mode during the test. The finite element results of specimen SL-2 are basically the same as those of SL-1. The finite element results of specimen SL-3 showed the most severe flange buckling, and the ultimate displacement was only $5\Delta_y$, which is consistent with the test, proving the finite element model is correct.

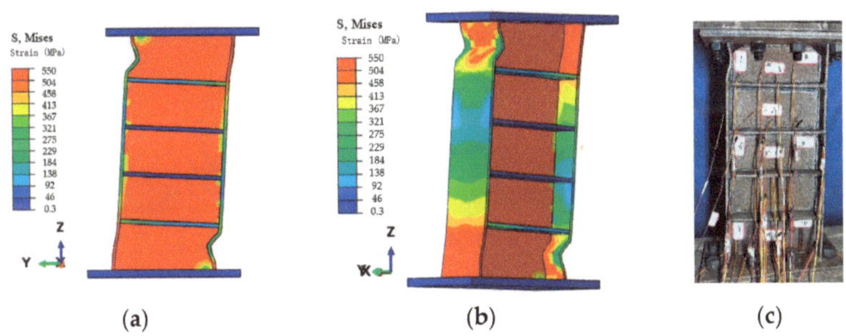

Figure 13. Comparison of the failure processes of specimens SL-1 and SL-2 with finite element models: (**a**) FEM: Δ = 27 mm; (**b**) FEM: Δ = 63 mm; (**c**) test: Δ = 63 mm.

From Figures 13 and 14, it can be observed that the full-section stress at the connection between the flange plate and the end plate reached the maximum, which resulted in the tearing of the weld between the flange plate and the end plate of the ordinary shear link. For very short shear links, the connection between the flange and the end plate of the did not reach full-section maximum stress initially [13]. Therefore, the control section of an ordinary shear link with a length ratio close to 1.6 was the weld connection between the flange plate and the end plate, and the failure characteristic was the tearing of the weld between the flange and the end plate.

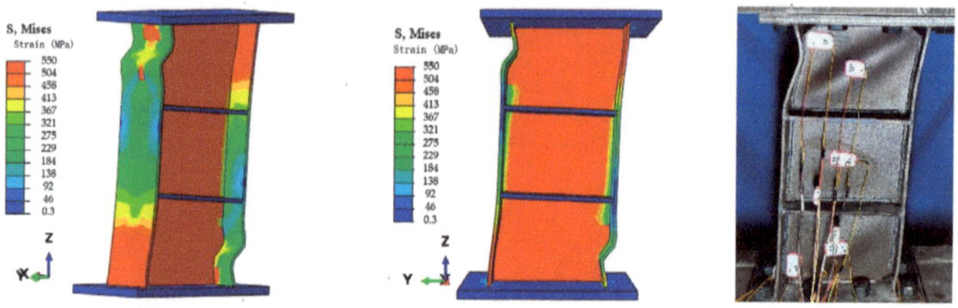

Figure 14. Comparison of the failure process of specimen SL-3 with the finite element models: (**a**) FEM: Δ = 27 mm; (**b**) FEM: Δ = 45 mm; (**c**) test: Δ = 45 mm.

4.3. Parametric Study on the Factors Influencing the Overstrength of Ordinary Shear Links

In order to intensively explore the influencing factors and changing rules of the overstrength factors of ordinary shear links made of Q345 steel, multiple sets of ordinary shear link models were designed to study the influence of six parameters on the overstrength factor. The research models were all based on two basic sections: the narrow flange section (H300 × 150 × 6 × 8) and middle flange section (H362 × 250 × 14 × 20). The finite element models adopted the cyclic loading protocol in the seismic specification for steel structures, and the boundary conditions were the same as those of the finite element models in Section 4.1. The properties of Q235 and Q345 steel are shown in Table 2. All stiffeners were distributed on both sides, and the stiffener spacing of all shear link sections met the specification requirements. The stiffener thickness of the basic section H300 × 150 × 6 × 8 was 10 mm, and the stiffener thickness of the basic section H362 × 250 × 14 × 20 was 14 mm.

4.3.1. Length Ratio

An existing study showed that the overstrength factor of shear links was significantly affected by the length ratio. As shown in Table 6, overstrength factors of different length ratios were obtained by changing the length of the ordinary shear links while keeping the cross-section unchanged. The web, flange and stiffeners were all made of Q345 steel, and the stiffener thicknesses were 10 mm and 14 mm, respectively. There were four stiffeners equidistantly distributed on both sides.

Table 6. Section properties for investigating effect of length ratio.

Section (mm)	Length *e* (mm)	Length Ratio
H300 × 150 × 6 × 8	650, 670, 690, 710, 730, 750	1.36, 1.40, 1.45, 1.49, 1.53, 1.57
H362 × 250 × 14 × 20	1080, 1110, 1140, 1170, 1200, 1230	1.36, 1.40, 1.44, 1.48, 1.51, 1.55

It can be seen from Figure 15 that the overstrength factor has an obvious relationship with the length ratio. The overstrength factor of the models decreased with the increase in the length ratio; that is, the ultimate shear force decreased with the increase in the length ratio.

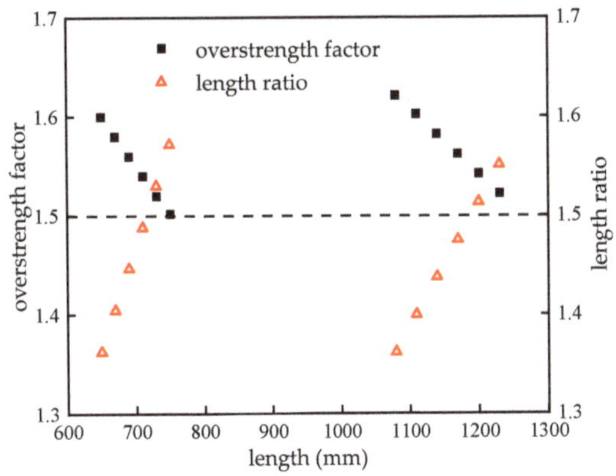

Figure 15. Relationship between overstrength factor and length ratio.

4.3.2. Stiffener Spacing

Existing research showed that the stiffener spacing has a significant impact on the mechanical properties of shear links. In order to intensively explore the impact of stiffener spacing on the overstrength factor of ordinary shear links with length ratios higher than 1.0, 24 finite element models were designed. The length-to-stiffener spacing ratio is denoted by ns:

$$ns = e/a \tag{4}$$

where e is the length of shear links, and a is the stiffener spacing. The flange, web and stiffeners were all made of Q345, and the stiffener thicknesses were 10 mm and 14 mm, respectively. Other dimensions of the finite element models are shown in Table 7.

Table 7. Section properties for investigating the effect of stiffener spacing.

Section (mm)	Length e (mm)	ns
H300 × 150 × 6 × 8	650, 670, 690, 710, 730, 750	5 6
H362 × 250 × 14 × 20	1080, 1110, 1140, 1170, 1200, 1230	5 6

In Figure 16, the solid square represents ns = 6, and the hollow triangle represents ns = 5. Whether it is Figure 16a or Figure 16b, as ns changes from 6 to 5, the overstrength factor significantly decreases. Hence, relaxing the stiffener spacing will significantly reduce the overstrength factor of ordinary shear links.

4.3.3. Flange-to-Web Area Ratio

On the basis of sections H300 × 150 × 6 × 8 and H362 × 250 × 14 × 20, except for changing the flange extension width and link length, other conditions remained unchanged. For sections H300 × b_f × 6 × 8 and H362 × b_f × 14 × 20, 24 finite element models with different flange extension widths and link lengths were designed to study the influence of the flange-to-web area ratio on the overstrength factor of ordinary shear links. Their parameters are shown in Table 8. The flange width-to-thickness ratio of all models meet the 10ε value proposed by Hu [16], and the stiffener spacing meets the specification requirements. The web, flange and stiffeners were all made of Q345 steel, and the stiffener thicknesses were 10 mm and 14 mm, respectively.

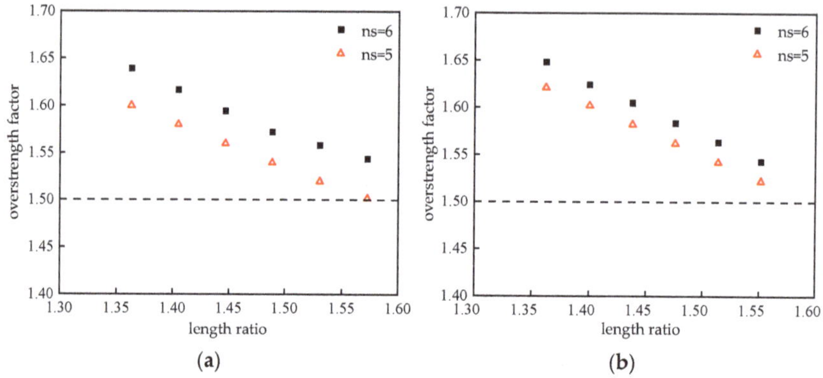

Figure 16. Relationships between overstrength factor and stiffener spacing: (**a**) Section H300 × 150 × 6 × 8; (**b**) Section H362 × 250 × 14 × 20.

Table 8. Section properties for investigating the effect of flange-to-web area ratio.

Section (mm)	b_f (mm)	$A_{f/}A_w$	Length e (mm)	ns
H300 × b_f × 6 × 8	150, 170, 190, 210	1.41, 1.60, 1.78, 1.97	710, 730, 750	5
H362 × b_f × 14 × 20	230, 250, 270, 290	1.66, 1.71, 1.75, 1.77	1170, 1200, 1230	5

As shown in Figure 17, the solid square, triangle and circle represent H300 × b_f × 6 × 8 sections with lengths of 710 mm, 730 mm and 750 mm, respectively. The hollow square, triangle and circle represent H362 × b_f × 14 × 20 sections with lengths of 1170 mm, 1200 mm and 1230 mm, respectively. The abscissa is the flange-to-web area ratio, as shown in Table 8. From the analysis in the figure, it can be seen that changing the flange-to-web area ratio has a significant impact on the overstrength factor of ordinary shear links. Two sections all show that increasing the flange-to-web area ratio will increase the overstrength factor.

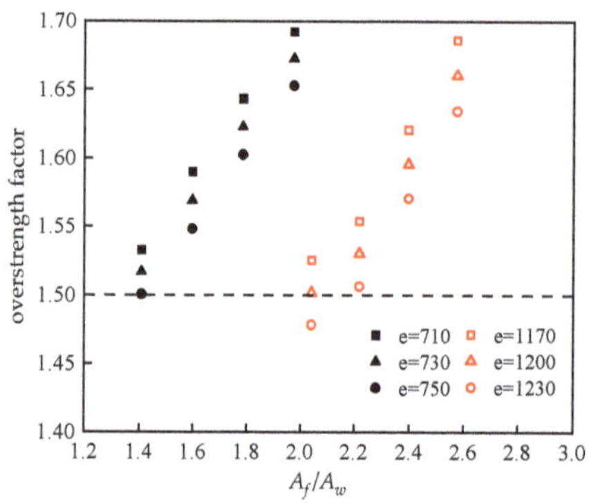

Figure 17. Relationships between overstrength factor and flange-to-web area ratio.

4.3.4. Flange Strength

On the basis of the sections H300 × 150 × 6 × 8 and H362 × 250 × 14 × 20, the flanges were made of Q235 and Q345 steel, respectively, the web and stiffeners were made of Q345 steel, and the stiffener thicknesses were 10 mm and 14 mm, respectively. The remaining parameters are shown in Table 9.

Table 9. Section properties for investigating the effect of flange strength.

Section (mm)	Length *e* (mm)	Flange Strength	ns
H300 × 150 × 6 × 8	650, 670, 690, 710, 730, 750	Q235 Q345	5
H362 × 250 × 14 × 20	1080, 1110, 1140, 1170, 1200, 1230	Q235 Q345	5

As shown in Figure 18, the solid square represents the models whose flanges are made of Q345 steel, and the hollow triangle represents the models whose flanges are made of Q235 steel. From the figure, it can be seen that the flange strength has a clear relationship with the overstrength factor of ordinary shear links. As the flange strength decreased, the overstrength factor dropped violently to below 1.50. Therefore, for ordinary shear links, increasing the flange strength will increase the overstrength factor, and it is not recommended that the flange strength be lower than the web strength.

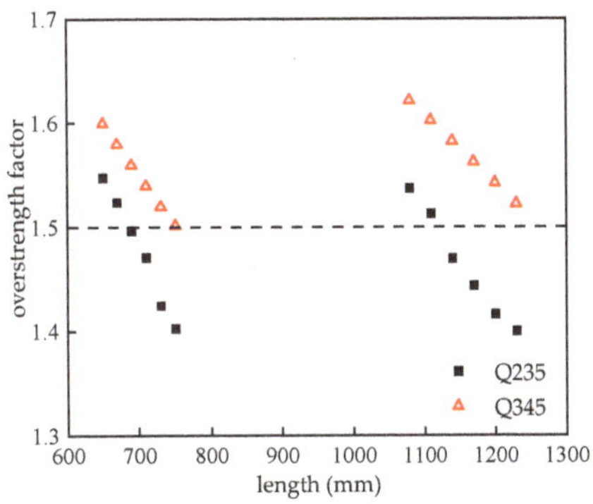

Figure 18. Relationships between overstrength factor and flange strength.

4.3.5. Web Depth-to-Thickness Ratio

On the basis of the sections H300 × 150 × 6 × 8 and H362 × 250 × 14 × 20, the web area remained unchanged, while the depth-to-thickness ratio of the web was changed, as shown in Table 10. The flange, web and stiffeners were made of Q345 steel, and the thicknesses of the stiffeners were 10 mm and 14 mm, respectively.

Table 10. Section properties for investigating the effect of web depth-to-thickness ratio.

Section (mm)	Web Depth-to-Thickness Ratio	Length *e* (mm)	ns
300 × 150 × 6 × 8	47.33		
260 × 150 × 7 × 8	34.87	650, 690, 730, 750	5
229 × 150 × 8 × 8	26.63		
362 × 250 × 14 × 20	23.00		
340 × 250 × 15 × 20	20.00	1080, 1140, 1200, 1230	5
322 × 250 × 16 × 20	17.63		

As shown in Figure 19, the solid shapes represent the $h \times 150 \times t_w \times 8$ section with lengths of 650 mm, 690 mm, 730 mm and 750 mm, respectively, while the hollow shapes represent the $h \times 250 \times t_w \times 20$ section with lengths of 1080 mm, 1140 mm, 1200 mm and 1230 mm, respectively. It can be seen from Figure 19 that reducing the web depth-to-thickness ratio can increase the overstrength factor of ordinary shear links, but the web cannot be too thin. Hence, there must be a suitable value to make the most use of materials. Corresponding research can be conducted to decide the most suitable depth-to-thickness ratio.

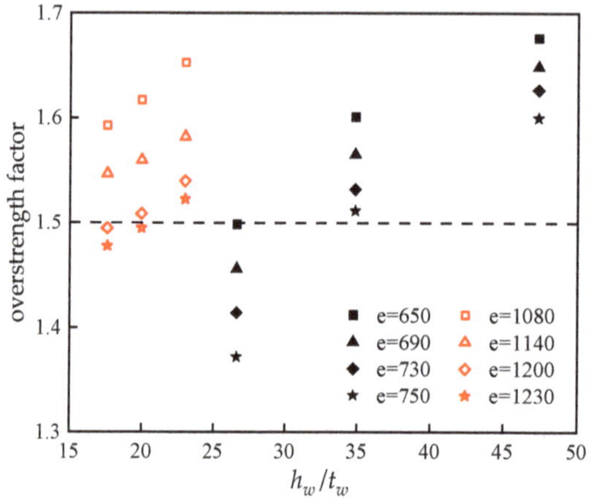

Figure 19. Relationships between overstrength factor and web depth-to-thickness ratio.

4.3.6. Stiffener Thickness

On the basis of the sections H300 × 150 × 6 × 8 and H362 × 250 × 14 × 20, the stiffener thickness was changed, and other parameters remained unchanged. As shown in Table 11, t_s is the stiffener thickness of ordinary shear links. The flange, web and stiffeners were made of Q345 steel.

Table 11. Section properties for investigating the effect of stiffener thickness.

Section (mm)	t_s (mm)	Length e (mm)	ns
H300 × 150 × 6 × 8	6, 8, 10	650, 690, 730, 750	5
H362 × 250 × 14 × 20	14, 12, 10	1080, 1140, 1200, 1230	5

As shown in Figure 20a, when the stiffener thickness was changed from 10 mm to 6 mm, the overstrength factor was only reduced by 1.08%. As can be seen in Figure 20b, the stiffener thickness was changed from 14 mm to 10 mm, but the overstrength factor was only reduced by 1.87%. The error is negligible. Therefore, changing the stiffener thickness has little influence on the overstrength factor of ordinary shear links, but the stiffener thickness should still meet the structural requirements.

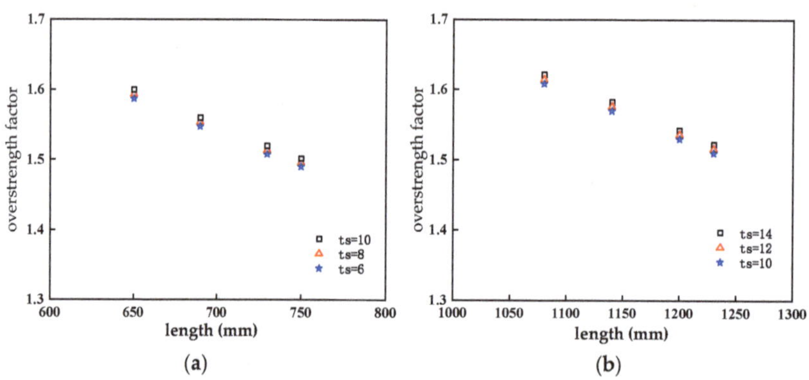

Figure 20. Relationships between overstrength factor and stiffener thickness: (**a**) section H300 × 150 × 6 × 8; (**b**) section H362 × 250 × 14 × 20.

5. Conclusions

To investigate the cyclic behavior of ordinary shear links made of Q345 steel, this research designed three ordinary shear link specimens, each having different stiffener spacings and configuration patterns, with length ratios higher than 1.0 for experimental research and finite element analysis. In addition, 114 finite element models were designed to study the factors affecting the mechanical properties of ordinary shear links. The following conclusions are drawn:

(1) The failure modes of the three specimens were basically the same. They all experienced four stages: shear yielding, flange buckling, micro-cracking in the weld between the flange plate and the end plate and weld extension, while the damage was failures in all shear yield. Specimens SL-1 and SL-2 experienced weld tearing and lost their bearing capacities. Specimen SL-3 experienced severe web buckling and lost its bearing capacity.

(2) Distributing stiffeners on one side or both sides has no significant impact on the mechanical properties of ordinary shear links, while relaxing stiffener spacing would significantly increase the buckling phenomenon of the web and decrease its ultimate shear force and ultimate displacement.

(3) The overstrength factor and inelastic rotation of Specimens SL-1 and SL-2 were almost the same, at 1.59 and 0.10, respectively, while those of Specimen SL-3 were 1.48 and 0.07, significantly lower than the specification values. Stiffener spacing should meet the specification requirements in engineering design.

(4) The results of the 114 parameter numerical models of ordinary shear links made of Q345 steel indicated that the length ratio, stiffener spacing, flange-to-web area ratio, flange strength and web depth-to-thickness ratio have significant influences on the overstrength factor of ordinary shear links.

(5) The main failure mode of causing the tearing of ordinary shear links is weld tearing between the flange plate and the end plate; the process of weld tearing is slow, and the specimen still plays a connecting role and can withstand a certain shear force after damage. In the K-type eccentrically braced steel frame, the small ultimate displacement of the ordinary shear link is beneficial to protect the floor from damage, and slow damage allows the frame to have good ductility. The application of ordinary shear links in the K-type eccentrically braced frame can be intensively studied in the future.

(6) The failure process of the specimens includes large steel plate deformation, and this deformation was observed with the naked eye. Visualization software could be used to monitor tiny deformations in future research.

Author Contributions: Conceptualization, S.H. and S.L.; methodology, S.H.; software, S.L. and T.S.; validation, S.H. and T.S.; formal analysis, S.L.; investigation, S.L.; resources, S.Z.; data curation, T.S.; writing—original draft preparation, S.L.; writing—review and editing, S.H. and T.S.; visualization, S.Z.; supervision, T.S.; project administration, S.H.; funding acquisition, S.H. and S.Z. All authors have read and agreed to the published version of the manuscript.

Funding: This research was funded by the Natural Science Foundation of China (Grant No. 51908268), Young Leader of Science and Technology Research Project for Jiangxi Province (Grant No. 20232BCJ23 065), Natural Science Foundation of Jiangxi Province (Grant No. 20224BAB204062), Young Leader of Science and Technology Research Project for Jiangxi Bureau of Geology (Grant No. 2022JXDZKJRC09) and Jiangxi Postgraduate Innovative Foundation (Grant No. YC2023-S078).

Data Availability Statement: The data used to support the findings of this study are available from the authors upon request. The data are not publicly available due to privacy.

Conflicts of Interest: Author Sizhi Zeng was employed by the company Zhongmei Engineering Group Co. Ltd. The remaining authors declare that the research was conducted in the absence of any commercial or financial relationships that could be construed as a potential conflict of interest.

References

1. Volynkin, D.; Dusicka, P.; Clifton, G.C. Intermediate web stiffener spacing evaluation for shear links. *J. Struct. Eng.* **2019**, *145*, 04018257. [CrossRef]
2. Hu, S.J.; Liao, Y.; Zeng, S.Z.; Huang, H.; Yu, Y.J. Cyclic shear test and stiffness study of a cross-section-type shear connector connection. *J. Constr. Steel Res.* **2024**, *212*, 108318. [CrossRef]
3. Tashakori, J.; Razzaghi, J.; Ansari, S. Reassessment of current design criteria of plastic hinges in shear links. *J. Constr. Steel Res.* **2019**, *158*, 350–365. [CrossRef]
4. Chen, Q.D.; Xu, L.Y.; Liu, Y.; Wu, Y.X. Application and analysis of seismic isolation design for a high-rise residential building in high seismic fortification intensity area. *J. Nanchang Univ., Eng. Technol.* **2021**, *43*, 22–29. [CrossRef]
5. Xu, R.R.; Tao, Z.; Dai, B.H.; Dai, J.S.; Zhao, C.; Yang, L. Finite element analysis of sectoral viscoelastic dampers applied to tenon joint reinforcement. *J. Nanchang Univ. Eng. Technol.* **2022**, *44*, 344–350. [CrossRef]
6. Minh, P.N.; Uyen, T.M.T.; Do, T.T.; Nguyen, V.-T.; Nguyen, V.T.T. Enhancing the fatigue strength of the weld line in advanced polymer injection molding: Gas-assisted mold temperature control for thermoplastic polyurethane (TPU) composites. *Polymers* **2023**, *15*, 2440. [CrossRef] [PubMed]
7. Hu, S.J.; Zeng, S.Z.; Zhou, Q.; Zhi, Q.; He, K. Mechanism and seismic performance of a novel precast concrete beam-to-very short shear link joint Experimental and numerical simulation. *Structures* **2023**, *56*, 104938. [CrossRef]
8. Vetr, M.G.; Ghamari, A.; Bouwkamp, J.A. Investigating the nonlinear behavior of eccentrically braced frame with vertical shear links (V-EBF). *J. Build. Eng.* **2017**, *10*, 47–59. [CrossRef]
9. Bozkurt, M.B.; Azad, S.K.; Topkaya, C. Development of detachable replaceable links for eccentrically braced frames. *Earthq. Eng. Struct. Dyn.* **2019**, *48*, 1134–1155. [CrossRef]
10. Ghadami, A.; Al-Mazaidh, R. Simple truss model to estimate the shear strength of short links in the eccentrically braced frame (EBF) steel system. *Thin Wall Struct.* **2023**, *188*, 110811. [CrossRef]
11. Hu, S.J.; Chang, L.Q.; Zhang, B.; Zeng, S.Z.; Tang, F.H.; Zhi, Q. Seismic response of resilient steel frame with self-centering SMA brace. *Int. J. Steel Struct.* **2023**, *23*, 1587–1601. [CrossRef]
12. *GB 50011-2010*; Code for Seismic Design of Buildings. Chinese Building Industry Press: Beijing, China, 2016.
13. Yin, W.H.; Sun, F.F.; Jin, H.J.; Hu, D.Z. Experimental and analytical study on plastic overstrength of shear links covering the full range of length ratio. *Eng. Struct.* **2022**, *220*, 110961. [CrossRef]
14. Zhuang, L.D.; Wang, J.J.; Nie, X.; Wu, Z.H. Experimental study on the cyclic behaviour of shear links made of BLY160 steel. *Thin Wall Struct.* **2022**, *174*, 109072. [CrossRef]
15. Ghadami, A.; Pourmoosavi, G.H.; Talatahari, S.; Azar, B.F. Overstrength factor of short low- yield-point steel shear links. *Thin Wall Struct.* **2021**, *161*, 107473. [CrossRef]
16. Elgammal, A.; Hassanein, M.F.; Seleemah, A. Enhancing the cyclic performance of shear links using longitudinal stiffeners. *J. Constr. Steel Res.* **2023**, *211*, 108200. [CrossRef]
17. Yao, Z.C.; Wang, W.; Fang, C.; Zhang, Z.Y. An experimental study on eccentrically braced beam-through steel frames with replaceable shear links. *Eng. Struct.* **2020**, *206*, 110185. [CrossRef]
18. Hu, S.J.; Xiong, J.G.; Zhou, Q.; Lin, Z.B. Analytical and numerical investigation of overstrength factors for very short shear links in EBFs. *KSCE J. Civ. Eng.* **2018**, *22*, 4473–4482. [CrossRef]
19. Liu, X.G.; Fan, J.S.; Liu, Y.F.; Yue, Q.R.; Nie, J.G. Experimental research of replaceable Q345GJ steel shear links considering cyclic buckling and plastic overstrength. *J. Constr. Steel Res.* **2017**, *134*, 160–179. [CrossRef]
20. Men, J.J.; Xiong, L.Q.; Wang, J.C.; Zhang, Q.; He, P.J. An experimental study on the seismic behavior and replaceability of the replaceable steel shear links. *Structures* **2021**, *33*, 2334–2348. [CrossRef]
21. Mansour, N.; Christopoulos, C.; Tremblay, R. Experimental validation of replaceable shear links for eccentrically braced steel frames. *J. Struct. Eng.* **2011**, *137*, 1141–1152. [CrossRef]
22. Okazaki, T.; Engelhardt, M.D. Cyclic loading behavior of EBF links constructed of ASTM A992 steel. *J. Constr. Steel Res.* **2007**, *63*, 751–765. [CrossRef]
23. Hu, S.J.; Wang, X.F.; Xiong, J.G.; Zeng, S.Z.; Dong, X.W. Analysis on influence factors and change rules of overstrength coefficient for shear links. *J. Build. Struct.* **2020**, *50*, 89–96. [CrossRef]
24. Ozkilic, Y.O.; Topkaya, C. Extended end-plate connections for replaceable shear links. *Eng. Struct.* **2021**, *240*, 112385. [CrossRef]
25. Singh, M.; Sharma, S.; Muniappan, A.; Pimenov, D.Y.; Wojciechowski, S.; Jha, K.; Dwivedi, S.P.; Li, C.; Królczyk, J.B.; Walczak, D.; et al. In situ micro-observation of surface roughness and fracture mechanism in metal microforming of thin copper sheets with newly developed compact testing apparatus. *Materials* **2022**, *15*, 1368. [CrossRef] [PubMed]
26. Özkiliç, Y.O. Interaction of flange and web slenderness, overstrength factor and proposed stiffener arrangements for long links. *J. Constr. Steel Res.* **2022**, *190*, 107150. [CrossRef]
27. Ghadami, A.; Pourmoosavi, G.H. Numerical investigation on the flange contribution in the shear strength of short LYP I-shaped links without intermediate stiffeners. *Structures* **2022**, *40*, 485–497. [CrossRef]
28. Yin, Z.Z.; Feng, D.Z.; He, Y.L. Restoring force model of end-plate bolted shear links in eccentrically braced frames. *J. Constr. Steel Res.* **2022**, *199*, 107623. [CrossRef]

29. Corte, G.D.; D'Aniello, M.; Landolfo, R. Analytical and numerical study of plastic overstrength of shear links. *J. Constr. Steel Res.* **2013**, *82*, 19–32. [CrossRef]
30. Xue, J.Y.; Liu, Y.; Zhao, H.T.; Chen, Z.P.; Sui, Y. Experimental study on seismic behavior of steel reinforced concrete special-shaped column-beam joints. *J. Build. Struct.* **2009**, *30*, 69–77. [CrossRef]

buildings MDPI

Article

Seismic Performance of Wall-Type Concrete-Filled Steel Tubular Column to H-Beam Connections with Internal Diaphragms

Hanchao Liu [1,*], Honggang Lei [1], Yuqi Huang [2], Yongchang Chen [2] and Feng Xu [3]

1 College of Civil Engineering, Taiyuan University of Technology, Wanbailin District, Taiyuan 030024, China; lhgang168@126.com
2 Xi'an University of Architecture and Technology Design and Research Institute Co., Ltd., Xi'an 710000, China; huangyuqi_1989@163.com (Y.H.); changge605@163.com (Y.C.)
3 Xi'an Longhu Real Estate Development Co., Ltd., Xi'an 710000, China; fengxu1@longfor.com
* Correspondence: sxlllhc@163.com

Abstract: This study introduces a new internal diaphragm joint to connect steel beams with wall-type concrete-filled steel tubular (WCFST) columns. This study combines experimental research with numerical simulations for analysis. Firstly, one full-scale test specimen was designed and subjected to cyclic loading. The obtained failure mode and hysteresis curves illustrate that the joint specimens exhibit ample energy dissipation capacities. Subsequently, a reliable FE model was obtained based on experimental verification, and parametric analysis was conducted. The findings indicate that axial compression ratios critically affect the load-carrying capacity and displacement at failure, with a 2% reduction in capacity for each 0.1 increment in the ratio. The thicknesses of the column web and flange in the joint area are recommended to be 0.85 to 1.2 times and 1 to 1.2 times the beam flange thickness, respectively. The length of the internal diaphragm is advised to be between 0.2 and 0.3 times the width of the cross-section. Overall, these results significantly enrich our understanding of WCFST systems and will inform future design and construction best practices.

Keywords: CFST column; internal diaphragms; seismic behavior; FE analysis

1. Introduction

In China, the drive toward industrializing building construction has significantly supported the proliferation of prefabricated steel-structured buildings [1]. This initiative is robustly supported at multiple governmental levels, including national, provincial, and municipal [2,3]. Among the technologies propelling this trend, concrete-filled steel tubular (CFST) columns are prominent due to their remarkable structural performance, cost-efficiency, and speed of construction [4,5]. These columns are extensively used in high-rise prefabricated buildings. Yet, their conventional designs often result in columns that extend beyond the wall line in residential applications, which can interfere with architectural aesthetics and functionality. Wall-typed CFST (WCFST) columns represent a design evolution aimed specifically at overcoming this challenge by aligning the column's width with that of the building wall. This innovation not only prevents protrusions, thereby preserving the interior space integrity, but also significantly upgrades the quality of assembled steel-structured houses [6–8]. This approach reflects an advanced integration of structural functionality and architectural design, yielding more efficient and aesthetically pleasing living spaces.

Beam–column joints are crucial force transmission pathways in structural connections. The internal diaphragm joint is the most common type of joint utilized in CFST columns. Pioneering research on this joint configuration has been conducted by Nishiyama and Fujimoto, among others [9–12], who have extensively studied diaphragm joints and internal diaphragm joints equipped with stiffening ribs. Their findings affirm that the initial stiffness of most joint specimens aligns well with theoretical predictions, manifested through high stiffness, robust ductility, and excellent seismic resistance capabilities. Further explorations by Dawe [13] on CFST columns connected to H-beams, featuring both tension and compression reinforcement plates, as well as reinforced end plates, reveal that such reinforcements significantly enhance joint stiffness, thereby optimizing the material properties of steel beams. In a detailed investigation, Jianguo Nie [14,15] assessed several types of joints, including external ring plate joints, internal diaphragm joints, and bolt-anchored internal joints within steel tube concrete columns. Their studies, particularly focusing on three internal diaphragm joints under low-cyclic reciprocating load tests, elucidated that internal diaphragms adeptly transmit both beam end moments and shear forces. Notably, while these joints generally demonstrate high ductility and load-bearing capacities, their mechanical performance excels, particularly during beam-end failures. Conversely, joint performance is compromised during domain or weld failures. Expanding on these research foundations, Lü Xilin [16] collaborated with Tanaka Kiyoshi from the Fujita Technological Research Institute in Japan to develop a series of cross-shaped internal diaphragm joints. Their collective experimental work, which involved analyzing strains in steel beams, internal diaphragms, and column flange areas of the joint region, provided deeper insights into the joint domain's force transmission mechanisms and the influence of various joint configurations on mechanical performance.

Traditional internal diaphragm joints, however, pose significant challenges in concrete casting within WCFST columns due to their restrictive cross-sectional widths. Addressing the unique sectional characteristics of WCFST columns, Liu and Huang [17–19] introduced an innovative side-plate connection joint, which, despite delivering commendable seismic performance, detrimentally influences the aesthetic integrity of buildings due to its external positioning on column walls. In response to these challenges, this study pioneers a novel internal diaphragm joint specifically designed for WCFST columns that adeptly resolves concrete casting difficulties. As delineated in Figure 1, this new joint configuration entails pre-welding a fixed-length internal diaphragm with the column's steel plate at the factory, notably omitting vertical diaphragms within WCFST columns. Following this initial assembly, upper and lower column segments are welded together in the factory to form a unified column structure. Onsite, steel beams are connected to the columns using traditional internal diaphragm joint techniques. This newly introduced joint replaces traditional configurations with four smaller-sized internal diaphragms, excluding vertical diaphragms from the steel tube concrete column walls. This modification not only simplifies the welding process but also enlarges the apertures for concrete pouring, thus enhancing the ease of concrete application and ensuring superior quality control over the concrete within the joint area.

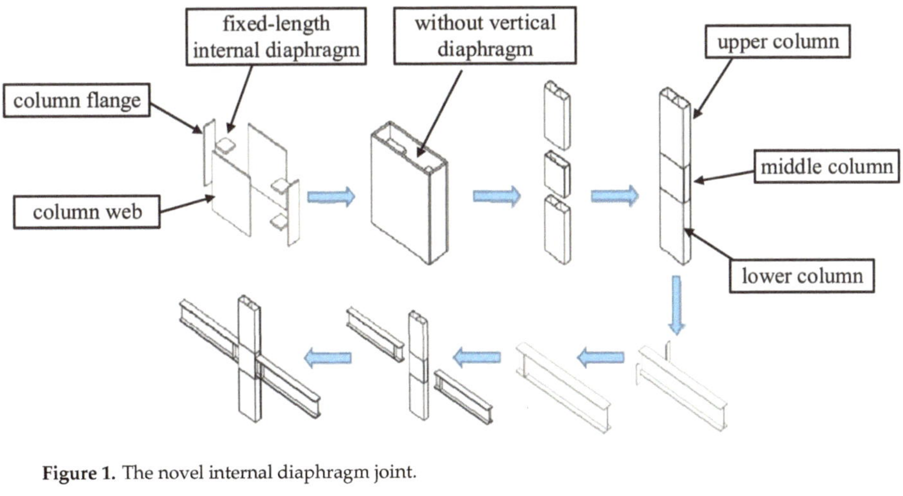

Figure 1. The novel internal diaphragm joint.

The aim of this study was to rigorously evaluate the seismic performance of WCFST column–beam joints equipped with internal diaphragms. This evaluation involved testing one full-scale specimen experimentally and analyzing 28 FE models. Comprehensive analyses were conducted to examine the failure patterns, load-bearing capacity, ductility, stiffness, and energy dissipation capabilities of these specimens. Through detailed discussions and reporting, this study provides an in-depth investigation of the structural responses and performance metrics under seismic loads. The findings from both experimental and FE analyses are instrumental in enhancing the design theory of these joints.

2. Experimental Program

2.1. Specimen Design

In this study, one specimen was designed following the "Standard for Design of Steel Structures" (GB50017-2017) [20], ensuring that their dimensions matched those used in actual engineering projects. Typical mid-column joints from these projects were selected as the primary focus of the research. Consequently, a specimen of beam-to-column joints, JD1, was meticulously fabricated for comprehensive analysis. The primary objective of this experiment was to assess whether the seismic performance of these joints satisfied the criteria established by relevant standards.

The steel column was divided into three sections and welded together at the factory. The upper and lower sections shared identical cross-sections, with dimensions D, B, t, and t_f being 550 mm, 180 mm, 8 mm, and 6 mm, respectively. The middle section, which served as the joint domain, eliminated the central diaphragm t_f and instead incorporated an internal diaphragm. The internal diaphragm measured 125 mm × 156 mm × 18 mm and was aligned with the upper and lower flanges of the steel beam. The steel beams were of H-shape, with dimensions H_b, B_b, t_{wb}, and t_{fb} specified as 400 mm, 180 mm, 8 mm, and 14 mm, respectively. The column had a total height H_c of 2600 mm, while the beam had a total length L_b of 1600 mm.

The key features of the specimens are illustrated in Figure 2. The steel materials in JD1 were Q235B. For specimen fabrication, commercial concrete with a design grade of C30 was utilized.

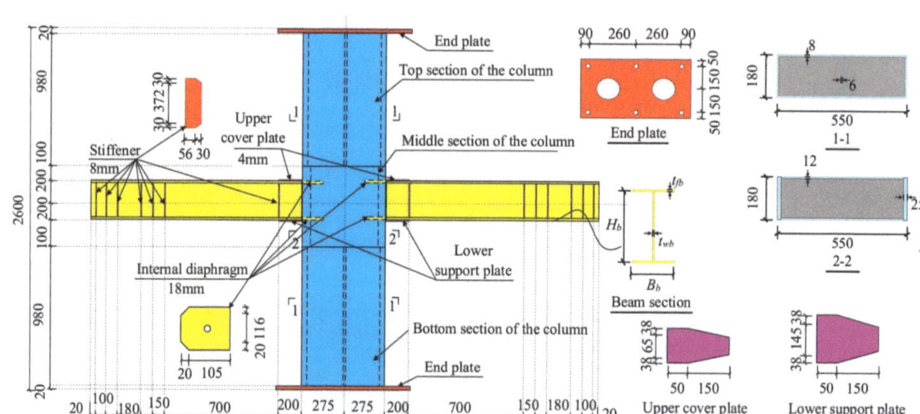

Figure 2. Drawing for manufacturing (in mm).

2.2. Material Properties

In accordance with the standard [21,22], three coupon tests were conducted for each steel plate (Figure 3). Table 1 presents the average mechanical properties obtained from the coupon tests, including the yield strength (f_y), tensile strength (f_u), elastic modulus (E), and elongation (δ). The measured yield-to-tensile strength ratios of Q235B steel consistently exceeded 1.3, and the elongation rates were all greater than 20%.

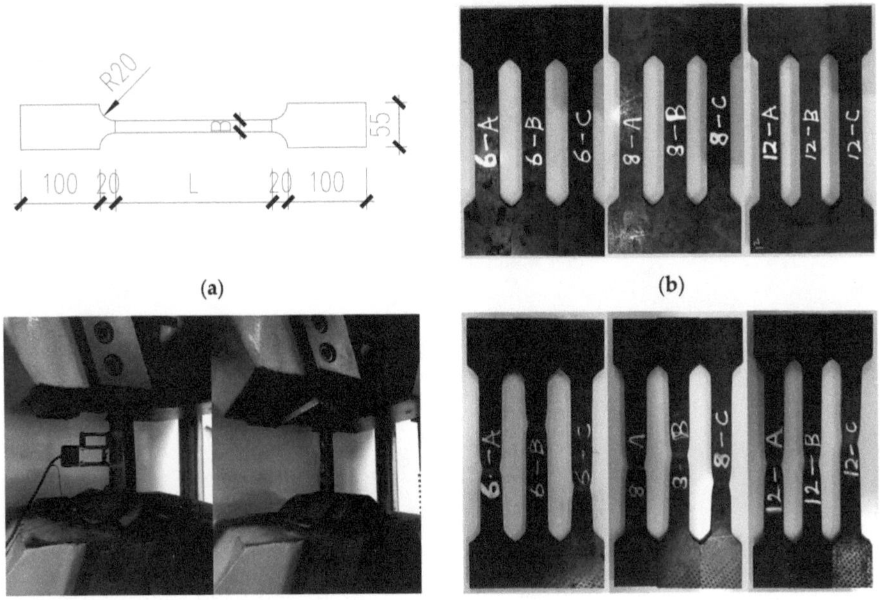

Figure 3. Test of steel material properties. (a) Standard specimen size. (b) Standard specimens before stretching. (c) Standard specimens while stretching. (d) Standard specimens after pulling off.

Table 1. Material properties of steel coupons.

Plate Thickness	Thickness/mm	f_y/MPa	f_u/MPa	δ	E/MPa	f_u/f_y
t = 6 (Q235B)	5.82	380.40	513.70	24.67%	175,763	1.35
t = 8 (Q235B)	7.68	372.29	520.37	24.97%	188,873	1.40
t = 12 (Q235B)	12.04	375.36	493.64	30.59%	217,158	1.32
t = 14 (Q235B)	14.16	383.86	533.04	30.18%	210,669	1.39
t = 18(Q235B)	17.43	260.58	411.98	35.95%	215,361	1.58
t = 25 (Q690)	24.68	388.86	548.74	33.47%	198,539	1.41

Three standard concrete cubes (150 mm × 150 mm × 150 mm) were fabricated and cured under the same conditions as the joint specimens. The concrete cubes were tested in accordance with GB50010-2010 [23]. The test results are presented in Table 2 and Figure 4, including the compressive strength test value for cubic specimens (f_{cu}), the average value from the compressive strength test for concrete specimens ($f_{cu,m}$), the standard compressive strength value for cubic specimens ($f_{cu,k}$), the compressive strength in the axial direction (f_{ck}), and the average axial compressive strength (f_{cm}). The derivation of these values follows specific formulas: Formula (1) is used to calculate the standard compressive strength value for cubic specimens ($f_{cu,k}$). Formula (2) is used to determine the compressive strength in the axial direction (f_{ck}), and Formula (3) is used to derive the average axial compressive strength (f_{cm}). In these formulas, σ represents the standard deviation of the cubic strength, and δ_c denotes the coefficient of variation for concrete.

$$f_{cu,k} = f_{cu,m} - 1.645\sigma \quad (1)$$

$$f_{ck} = 0.88 \times 0.76 f_{cu,k} \tag{2}$$

$$f_{cm} = f_{ck}/(1 - 1.645\delta_c) \tag{3}$$

Table 2. Material properties of concrete cubes.

Concrete Grade	f_{cu}/MPa	$f_{cu,m}$/MPa	$f_{cu,k}$/MPa	f_{ck}/MPa	f_{cm}/MPa
C30	32.87 31.24 32.64	32.25	28.14	18.82	21.57

Figure 4. Test of concrete material properties.

2.3. Test Setup

Quasi-static testing methods on structures are currently one of the most widely used methods for studying the force and deformation characteristics of structures or structural components. These tests apply specific loading or displacement control to subject specimens to low-cycle repeated loading, leading to a test method where specimens are subjected to forces from the beginning until failure, thus revealing the non-elastic load-deformation characteristics of structures or structural components.

This experiment utilized a quasi-static testing method, applying repeated horizontal loads under a constant axial load. Figure 5 illustrates the experimental apparatus used. For this test, with considerations from actual engineering practices, the axial compression ratio of the column was set at 0.47, while the applied vertical load amounted to 1740.58 kN. Additionally, to facilitate horizontal free sliding, a roller was positioned between the jack and the reaction beam. This arrangement ensured the maintenance of a constant vertical load even as the reciprocating horizontal load was executed. The imposed axial load corresponded to approximately 40% of the column's ultimate load capacity. To administer the lateral cyclic load at the mid-point of the loading beam, an MTS hydraulic actuator with a stroke capacity of ±250 mm was employed. The end of the beam was connected to the base beam via a linkage rod, which permitted unfettered horizontal movement of the beam end. This linkage rod was equipped with a precision sensor that effectively measured the vertical force exerted on the beam end, ensuring accurate data collection during the experiment.

The configuration of measurement points is detailed in Figure 6. Here, D1 to D5 denote linear variable differential transformers (LVDTs) strategically placed to measure the horizontal displacement along the height of the column. R1 and R2 are inclinometers tasked with assessing the rotation angle of the steel beam. Additionally, M1 represents a magnetostrictive displacement sensor designed to measure the horizontal displacement at the loading point. Furthermore, 27 strain gauges were installed in the plastic hinge zones of the column and beam. These gauges are critical for recording longitudinal, transverse, and shear strains, providing comprehensive data essential for analyzing structural behavior under varying loads. This setup ensures the precise monitoring and reporting of structural responses, which is vital for accurately evaluating the system's performance during testing.

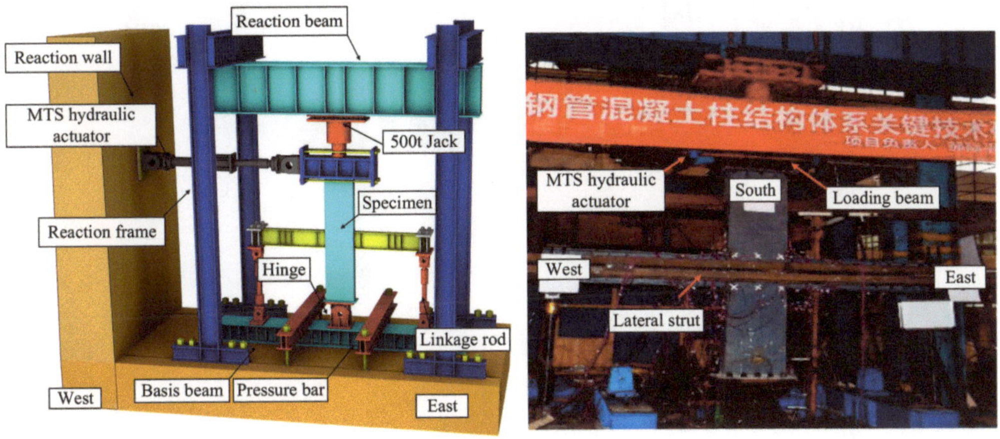

Figure 5. Experimental apparatus.

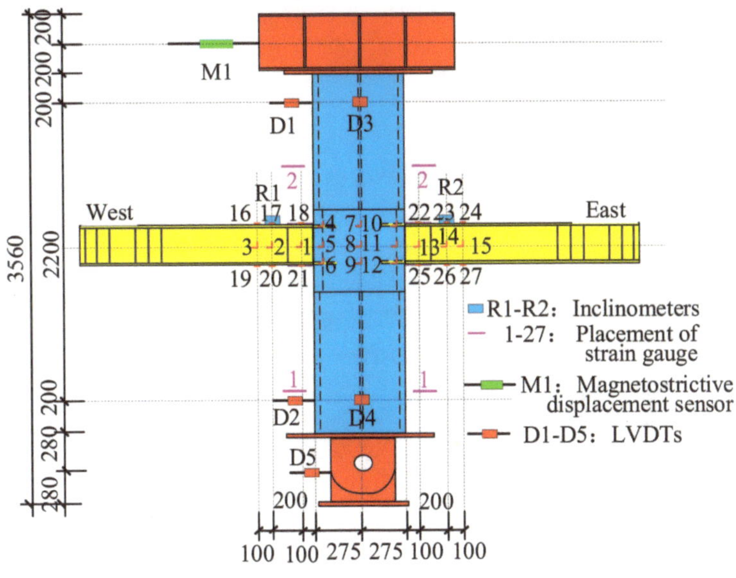

Figure 6. The layout of the measurement points.

Before formal loading, preloading was performed to eliminate gaps between different components. Subsequently, cyclic loading was carried out in accordance with the loading strategy depicted in Figure 7 (referencing [24]), where F, F_y, Δ, and Δ_y are the horizontal load, yield load, horizontal displacement, and yield displacement, respectively. The test was stopped under the following conditions: (i) if the specimen showed signs of steel plate buckling, steel plate tearing, or concrete crushing; (ii) if it could not withstand the axial load or the horizontal load; and (iii) if the horizontal load decreased below 85% of the peak horizontal force. These stopping criteria were strategically implemented to prevent excessive damage to the apparatus and ensure safety during the experimental procedure.

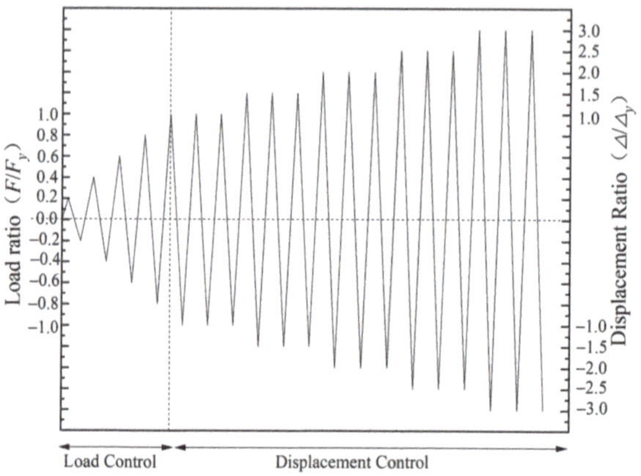

Figure 7. Loading strategy.

3. Test Results and Discussion

3.1. Failure Modes

As depicted in Figure 8, the primary failure mode for the specimens is bulging in the region where the beam flange and web are located. The dominant failure mode is the beam hinge mechanism, where the plastic hinge is situated at a distance from the column edge, specifically around 0.7 to 1.2 times the beam's height.

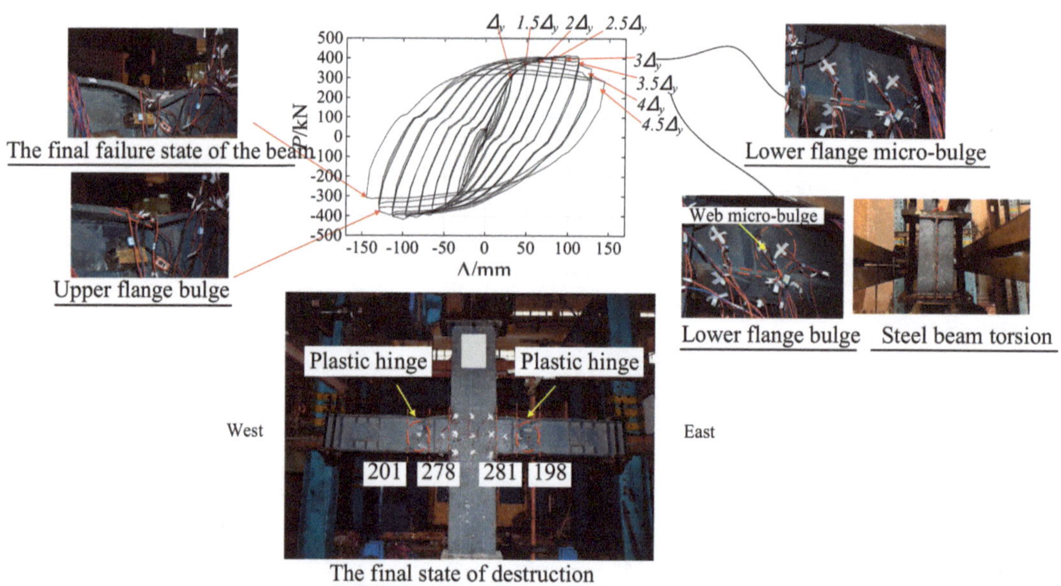

Figure 8. The destruction process of JD1.

For JD1, at a load level of $3\Delta_y$, slight twisting was observed on the outer flange of the top cover plate of the steel beam on the east side and the outer flange of the bottom support plate of the steel beam on the west side. At a load level of $3.5\Delta_y$, the load peaked. At this point, slight bulging was observed in the web plates of the steel beams on both the east and west sides. Specifically, the web plate of the east-side steel beam was concave toward the north, while that of the west-side steel beam bulged toward the south. Simultaneously, the buckling of the upper and lower flanges of the steel beams increased, and slight twisting occurred at the ends of the beams. At a load level of $4.0\Delta_y$, the buckling of the upper and lower flanges of the steel beams on both the east and west sides became more pronounced, and the bulging of the web plates of the beams was more evident. Additionally, there were six loud noises heard during the process. At a load level of $4.5\Delta_y$, the load-bearing capacity of the specimen had fallen below 80% of the peak load, indicating that the specimen had failed and could no longer sustain further loading. Consequently, the experiment was concluded.

3.2. Hysteresis Curves and Skeleton Curves

Figure 8 shows the hysteresis curves for the specimen, where P represents the lateral load and Δ signifies the lateral displacement at the loading point. The force-measuring device of the MTS system was used to obtain the values of P, and the reading of M1 was used to obtain Δ, as indicated in Figure 6. In Figure 8, the following observations can be drawn: (1) Prior to yielding, the specimen remains in the elastic phase. During this stage, the relationship between the load and deformation is nearly linear, and there is no notable change in the slope of the curve. After unloading, the joint exhibits minimal residual deformation, allowing the hysteresis loop to effectively close in a very small area. This behavior signifies no significant stiffness degradation at this stage, and the energy dissipated is minimal. (2) After yielding, the joint transitions into the plastic phase. As the cyclic displacement increases, the slope of the loading curve gradually decreases, with the rate of decrease accelerating. When comparing the slopes of the successive loading curves, it is evident that each subsequent loading curve has a reduced slope relative to its predecessor, indicating a continuous degradation in component stiffness under repeated loading. In this phase, the area of the hysteresis loop continuously enlarges, and the loop cannot fully close and instead forms a continuous hysteresis curve. With further increases in cyclic displacement, the load-bearing capacity also increases until it reaches the ultimate load-bearing capacity. Upon continued loading, a descending segment appears in the hysteresis curve, indicating a decrease in the load-bearing capacity of the specimen. (3) Throughout the loading process, each loading or unloading curve exhibits a small segment of slippage. This phenomenon occurs because the end of the steel beam in the specimen is connected to the linkage rod via lug plates. The diameter of the pin is slightly smaller than the hole in the lug plate, resulting in an installation gap that facilitates this slippage. However, the hysteresis curves of the specimen appear robust and symmetrical overall, showing no significant pinching effects. Additionally, the hysteresis loops at different loading levels largely overlap, indicating that the joint specimen exhibits excellent deformation capability, strong energy dissipation capacity, and superior seismic performance.

In Figure 9, the skeleton curves for all test specimens are presented, and the following observations can be drawn: In the initial stages of loading, the specimen remains in the elastic phase, where the relationship between load and deformation is nearly linear. As the load increases, the specimen reaches its yield point and transitions into the plastic phase, causing the backbone curve to exhibit nonlinearity. The slope of this curve decreases as the load increases, indicating a gradual degradation in the specimen's stiffness. As the loading continues to increase to the peak load, the specimen enters the failure stage, where its load-bearing capacity declines continuously. Overall, the backbone curves in both the pushing and pulling directions are very similar, highlighting consistent mechanical behavior under tensile and compressive stresses.

Table 3 provides a comprehensive summary of the detailed test results for the specimen. It indicates that the yield loads of the specimens were approximately 0.80 times the peak loads. The θ_u values of all the specimens exceed 2.0%, indicating that their inelastic deformation capacity meets the inter-story drift angle requirements (0.02 radians) of GB 50011-2010 [23] and AISC 341-10 [24].

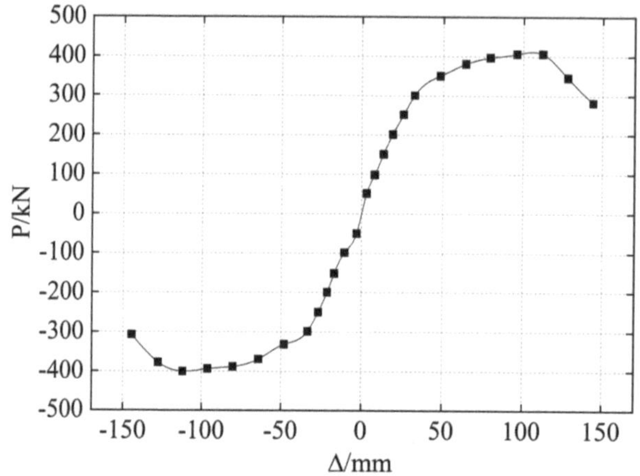

Figure 9. Specimen skeleton curve.

Table 3. Summary of the test results.

Specimen ID	Loading Direction	Yield			Peak			Failure			μ	$\zeta_{eq,p}$
		P_y/kN	Δ_y/mm	θ_y/rad	P_p/kN	Δ_p/mm	θ_p/rad	P_u/kN	Δ_u/mm	θ_u/rad		
DSPH1	Forward	330.16	40.38	0.013	407.37	105.98	0.034	345.95	128.10	0.041	3.14	0.44
	Reverse	−319.00	−42.53	0.013	400.55	111.25	0.036	−340.47	−136.91	0.044	3.23	

Note: The positive loading direction is on the east side of the specimen. The nominal yield load, peak load, and failure load are represented by P_y, P_p, and P_u, respectively. The corresponding displacements for the nominal yield load, peak load, and failure load are denoted as Δ_y, Δ_p, and Δ_u, respectively. The inter-story drift angles corresponding to Δ_y, Δ_p, and Δ_u are indicated by θ_y, θ_p, and θ_u, respectively. It is important to note that θ is calculated as the displacement Δ divided by *H*, where *H* represents the distance from the measuring point to the bottom of the column.

3.3. Ductility

The yield displacement Δ_y and ultimate displacement Δ_u listed in Table 3 were used to calculate the ductility coefficient (μ) of each specimen.

$$\mu = \frac{\Delta_u}{\Delta_y} \quad (4)$$

The μ of the specimen was about 3.1, while that reported in previous research [16–19,25] was about 3.0. This indicates that the joint possessed inherent ductility without compromising the structural functionality of the building.

3.4. Energy Dissipation Capacities

To evaluate the energy dissipation capacities of the specimens, the area enclosed by the hysteresis curve can be considered. This can be expressed using the equivalent viscous damping coefficient (ζ_{eq}) and the cumulative energy dissipation. Specifically, ζ_{eq} can be calculated by taking the ratio of the area enclosed by curve ABCD in Figure 10 to the combined area of triangles OBE and OBF.

$$\zeta_{eq} = \frac{1}{2\pi}\frac{S_{(ABC+CDA)}}{S_{(OBE+ODF)}} \quad (5)$$

The cumulative energy dissipation for all specimens is presented in Figure 11. Notably, specimen JD1 exhibited an equivalent viscous damping ratio ($\zeta_{eq,p}$) of 0.44. This performance stands out compared to previous studies in [26,27], which reported an equivalent viscous damping ratio of 0.176 for an outward-extending end-plate joint and approximately 0.3 for a double-side plate joint. Furthermore, the $\zeta_{eq,p}$ for a steel–concrete composite joint was estimated at 0.3, while an ordinary reinforced concrete joint typically exhibits a ratio of around 0.1. These comparisons illustrate that the seismic performance of JD1 is substantially better than that of the steel–concrete composite joint, indicating a superior capability in energy dissipation and overall seismic resilience.

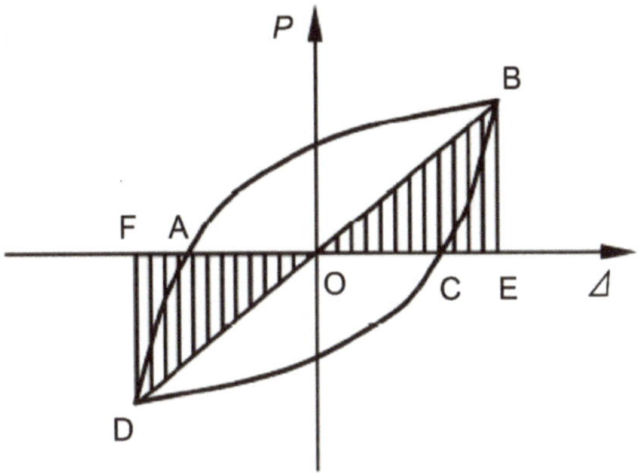

Figure 10. Calculation of the equivalent viscous damping coefficient.

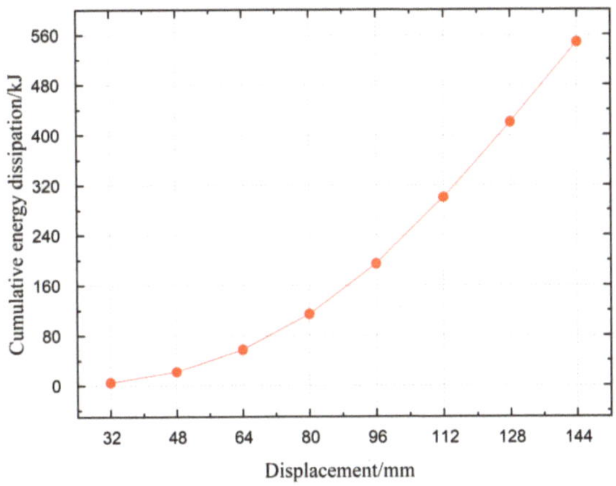

Figure 11. Cumulative energy dissipation curve.

3.5. Stiffness Degradation

Peak stiffness (K_i) is commonly used to assess stiffness degradation due to cumulative damage in structures. The calculation method for this metric is provided in Formula (6).

$$K_i = \frac{|-F_i|+|+F_i|}{|-X_i|+|+X_i|} \tag{6}$$

In the formula, K_i represents the stiffness value for the *i* cycle; $+F_i$ and $-F_i$ denote the positive and negative peak load values for the *i* cycle, respectively; and $+X_i$ and $-X_i$ are the positive and negative peak displacement values for the *i* cycle, respectively.

Figure 12 illustrates the stiffness degradation curves for the specimen. It is evident from these curves that the stiffness degradation occurs in a relatively uniform and gradual manner, with no significant abrupt reductions observed. This behavior suggests that the joint specimens possess robust energy dissipation capabilities, a desirable characteristic for enhancing seismic resilience in structural components.

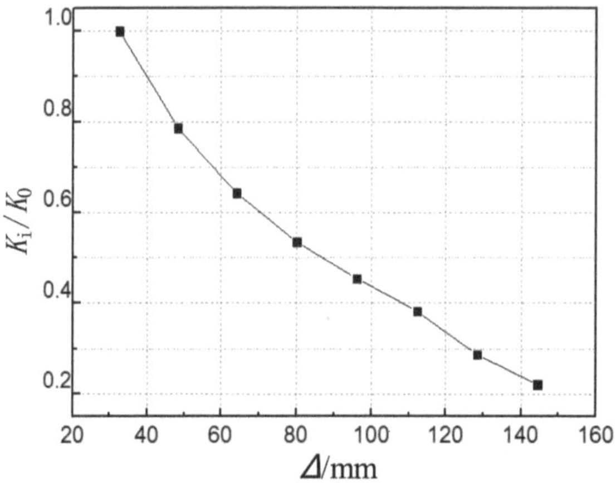

Figure 12. Stiffness degradation curve.

4. FE Analysis

4.1. FE Modeling

A sophisticated 3D finite element model of the proposed joint was developed using ABAQUS(2022), a nonlinear finite element (FE) analysis software. To effectively simulate concrete and steel, C3D8R and S4R elements were employed, respectively. These element types, S4R for shell elements and C3D8R for continuum elements, are renowned for their adaptability in most structural nonlinear analyses [28,29]. An optimal mesh density was established through a mesh convergence analysis, which balanced the need to minimize computational time while ensuring model accuracy. Consequently, small mesh sizes were selected for the FE model to enhance the resolution of the simulations. The interaction between the concrete and the inner surface of the steel tube was modeled using a surface-to-surface contact approach. A friction coefficient of 0.3 was assigned to this contact interface [30], reflecting realistic interaction conditions between these materials during stress application. This detail is critical for accurately capturing the joint's behavior under load. Both physically and geometrically nonlinear analyses were conducted, taking into account the nonlinear effects of large displacements.

4.2. Material Modeling

This study used the dual-line dynamic reinforcement model [31] to describe the elastic–plastic behavior of steel, as shown in Equation (7).

$$\sigma = \begin{cases} E\varepsilon & \varepsilon \leq \varepsilon_y \\ f_y + E_1(\varepsilon - \varepsilon_y) & \varepsilon > \varepsilon_y \end{cases} \tag{7}$$

where E and f_y are from Table 1. E_1 is the tangent modulus, which is taken as 0.01 E.

The concrete encapsulated within steel tubes is subjected to a high confining pressure owing to the constraining effect of the steel pipes, which enhances its compressive strength. Consequently, the stress–strain relationship for constrained concrete was utilized, with detailed parameters available in [32]. To accurately replicate the characteristics of the core concrete in the column, the damaged concrete plasticity model was employed, as detailed in a study by Yang et al. [33]. This model effectively simulates the nonlinear, inelastic behavior of concrete under various loading conditions, capturing both the damage and plastic deformation aspects critical to realistic structural analysis.

4.3. Applied Loads and Boundary Conditions

In the experiment, the bottom hinge of the wall-type steel pipe concrete column had a large torsional stiffness along the vertical axis, which could constrain the rotation of the specimen, and lateral support was set at the beam end. In the FE model, the bottom reference point U_y, U_z translational displacement, and the rotational displacement M_Z along the vertical axis were constrained. U_y, U_z displacement constraints were set at the beam end. The specific constraints are presented in Figure 13.

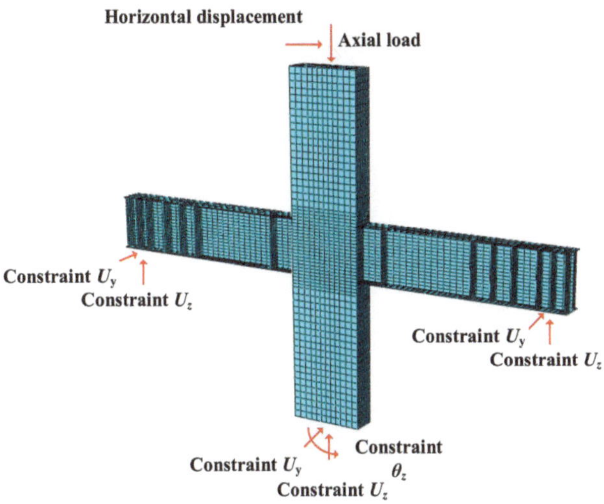

Figure 13. Model boundary conditions.

4.4. Confirmation of FE Modeling

To demonstrate the reliability of the numerical simulation, this comparison not only compares the current results but also validates them against experimental results in the WCFST column-to-beam connection detailed in the existing literature [34], namely DSP1.

Figure 14a shows the correlation between the hysteresis curves derived from both physical testing and finite element (FE) analyses of a particular specimen. Despite the finite element model demonstrating slightly improved performance in terms of load-bearing capacity, stiffness, and energy dissipation capabilities over the experimental results for the joint specimens, it is important to recognize the reasons for this difference. The FE analysis typically represents a more idealized experimental scenario, which does not account for the potential geometric initial imperfections and residual stresses that often arise during the specimen-fabrication process. Moreover, inherent limitations in laboratory testing, such as unavoidable errors stemming from friction and gaps, can further contribute to discrepancies between theoretical predictions and actual experimental outcomes. These differences underscore the importance of understanding the idealized nature of simulation results and the need for careful interpretation when comparing them to practical experiments.

Additionally, Figure 14b details the failure modes observed in specimens JD1 and DSP1, where both the experimental and FE analysis results indicated that failure was primarily due to local buckling in the beam flanges. This consistency between the two methodologies highlights critical failure mechanisms. It also reaffirms the value of FE analysis as a predictive tool, albeit with an understanding of its limitations related to idealized conditions and the necessity to integrate empirical data to validate and refine theoretical models. This synergy between experimental and theoretical approaches is crucial for advancing structural engineering practices, ensuring both reliability and efficiency in the design and assessment of structural components.

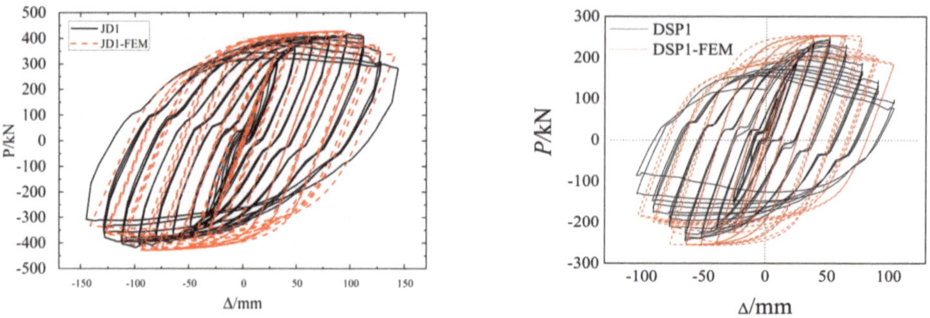

(a)

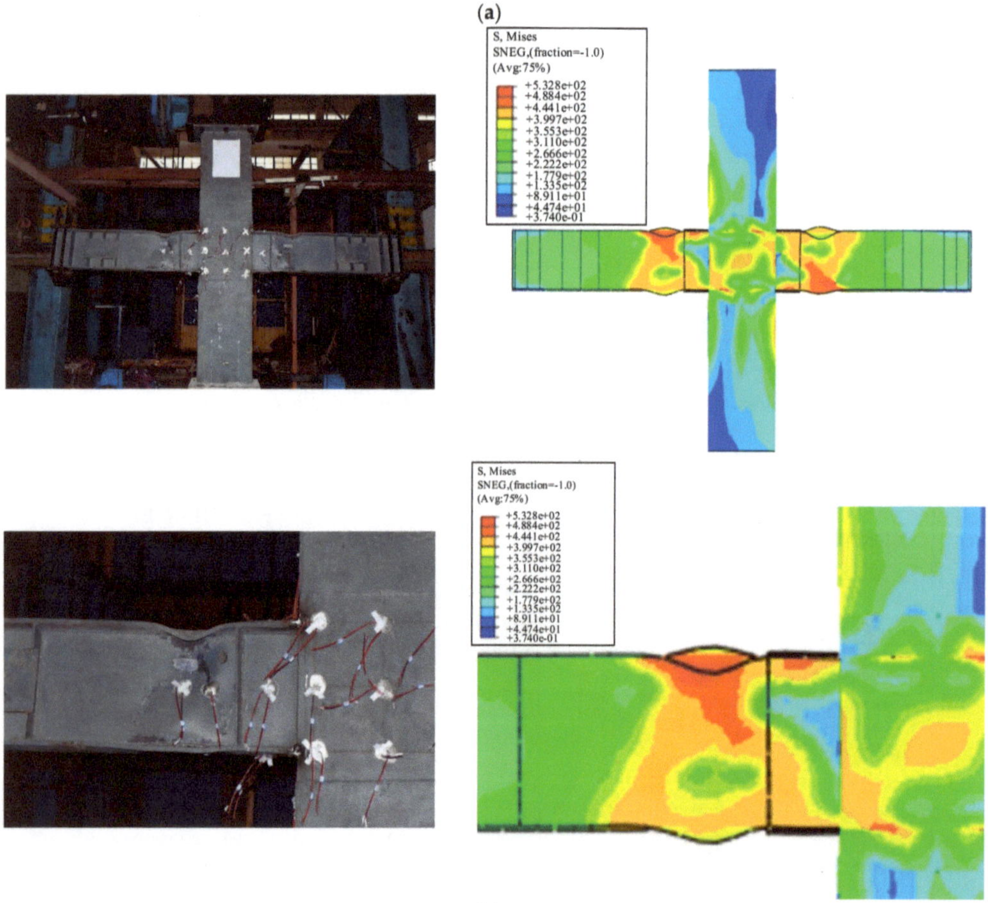

(b)

Figure 14. *Cont.*

(**c**)

Figure 14. Comparison of tests and finite element analysis results. (**a**) Comparison of hysteresis curves in the experiment and finite element analysis. (**b**) Comparison of the experiment and finite element failure modes of JD1. (**c**) Comparison of the experiment and finite element failure modes of DSP1.

5. Parametric Analysis

To evaluate the impact of variables such as the axial compression ratio, column web thickness, column flange thickness, and diaphragm length on the joint's load-bearing capacity and ductility, 28 specimens were generated using the validated finite element (FE) model. All FEA specimens had the same geometric parameters as the test specimen JD1.

5.1. *Effect of the Compression Ratio*

This section delves into the influence of varying design axial compression ratios on the mechanical performance of joint specimens. The axial ratios examined were 0.2, 0.3, 0.4, 0.5, 0.6, 0.7, and 0.8. The detailed model parameters are outlined in Table 4.

Table 4. Effect of the compression ratio.

Specimens	Compression Ratio	P_p(kN)	μ
JD1-1	0.2	270.86	6.35
JD1-2	0.3	265.91	5.14
JD1-3	0.4	261.4	4.37
JD1-4	0.5	257.02	3.86
JD1-5	0.6	252.26	3.49
JD1-6	0.7	248.05	3.18
JD1-7	0.5	243.9	2.94

Figure 15 illustrates the load–displacement curves of the joints subjected to different axial compression ratios, while Table 4 displays the associated differences in the load-bearing capacity and ductility across these ratios. A comparative analysis of these curves and the tabulated data allows for several key observations:

(1) As the axial compression ratio increases, there is a noticeable yet modest decline in the yield load, ultimate load-bearing capacity, and failure load of the specimens. The reduction in load-bearing capacity with each incremental increase of 0.1 in the axial compression ratio is less than 2%, showing a relatively uniform rate of decrease. This illustrates that while the axial compression ratios influence the structural performance, the overall impact within the observed range is not severe.

(2) Specimen ductility decreases as the axial compression ratio increases, although the rate of decrease progressively diminishes. By carefully managing the axial compression ratio in the design phase, desirable levels of ductility in the joints can be preserved. This control is critical for ensuring the structural responses remain predictable and manageable under load.

(3) There is a slight effect of the axial compression ratio on yield displacement, which contrasts with its significant impact on failure displacement. Increasing the axial compression ratio reduces the failure displacement; this trend can mainly be attributed to the increment in the second-order effects post-yielding. As the axial load rises, these second-order effects become more pronounced, influencing the deformation characteristics and susceptibility to failure. The interaction between axial loads and bending moments highlights the importance of second-order analysis in structural assessments, particularly at higher axial compression ratios.

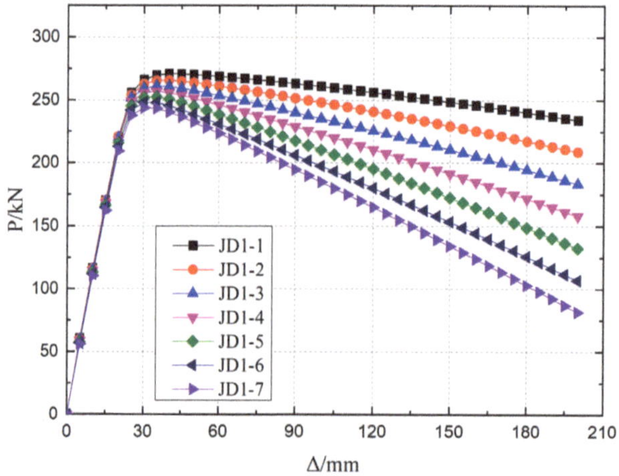

Figure 15. Load–displacement curves of joints.

These findings highlight the critical importance of incorporating the axial compression ratio in both the design and evaluation phases of structural joints. Controlling this parameter offers a way to optimize joint performance, especially in terms of balancing load-bearing capacities with the requisite ductility for safety and serviceability. Adjustments to the axial compression ratio should, therefore, be made judiciously, with these behavioral trends in mind, to tailor the joint characteristics to specific engineering demands and compliance with design standards.

5.2. Effect of the Column Web Thickness

This section methodically explores the effects of varying web thicknesses within the joint region of columns on the load-bearing capacity and ductility of joints. Details regarding model parameters are outlined in Table 5.

Table 5. Effect of the column web thickness.

Specimens	Column Web Thickness (mm)	P_p (kN)	μ
JD2-1	4	188.87	3.44
JD2-2	6	238.81	3.45
JD2-3	10	253.63	4.69
JD2-4	12	259.37	4.74
JD2-5	16	259.92	4.83
JD2-6	20	260.32	4.90
JD2-7	25	260.59	4.97

Figure 16 illustrates the load–displacement curves for joints subjected to different column web thicknesses, while Table 5 provides comparative data on the load-bearing capacity and ductility across these variations. The analysis of these curves and the tabulated data yields the following notable observations:

(1) Insufficient web thickness in the joint region leads to a marked decline in the load-bearing capacity of the joints. Specifically, specimens JD2-1 and JD2-2 display significantly lower load-bearing capacities than other samples. With incremental increases in web thickness, there is a noted, albeit modest, enhancement in both the yield load-bearing capacity and the ultimate load-bearing capacity, with improvements ranging from 0.1% to 2.3%.
(2) Initial increases in the web thickness substantially improve the ductility of the joint when starting from a relatively thin state. Nevertheless, as the web thickness approaches an optimal level, further increments yield diminishing returns in terms of enhancement in ductility, eventually leading to stabilization.

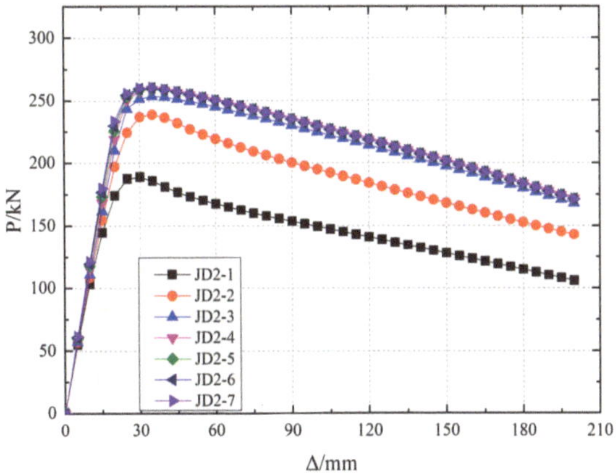

Figure 16. Load–displacement curves of joints.

Figure 17 shows the stress contour plots that reveal the damage states of JD2-1, JD2-3, JD2-5, and JD2-7 under varying thicknesses of the column web in the joint region. Analyzing the stress distribution within each joint provides additional insights:

(1) JD2-1 and JD2-2 reach the yield stress early within the webs of the column in the joint area, leading to premature failure. This condition significantly compromises the load-bearing capacity of these joints, categorizing their failure mode explicitly as joint region failure.
(2) JD2-3 features a plastic hinge that forms concurrently at the column web and beam ends. This intermediate state between joint region failure and beam end plastic hinge failure underlines the importance of ensuring adequate web thickness in the joint area during design considerations to circumvent this type of failure.
(3) JD2-4 to JD2-7 demonstrate a typical failure mode of beam–column joints, characterized by the formation of plastic hinges at the beam ends. In these instances, regions connected to the beam ends and the internal diaphragm exhibit elevated stress levels, while most of the joint region remains within the elastic phase. Comparisons between Figure 17c,d indicate that an increase in column web thickness lessens the stress concentration, reduces the extent of plastic development, and subtly enhances both the load-bearing capacity and ductility of the joints.

Based on these findings, maintaining a prudently defined range regarding the thickness of the column web in the joint area is critical during design. A web that is too thin may fail to satisfy load-bearing capacity and ductility requirements, while excessively thick webs can lead to material inefficiency and increased costs. It is, therefore, recommended that the web thickness of the column in the joint region be set between approximately 0.85 and 1.2 times the thickness of the beam flanges in design specifications. This approach strikes a balance between structural performance and material efficiency.

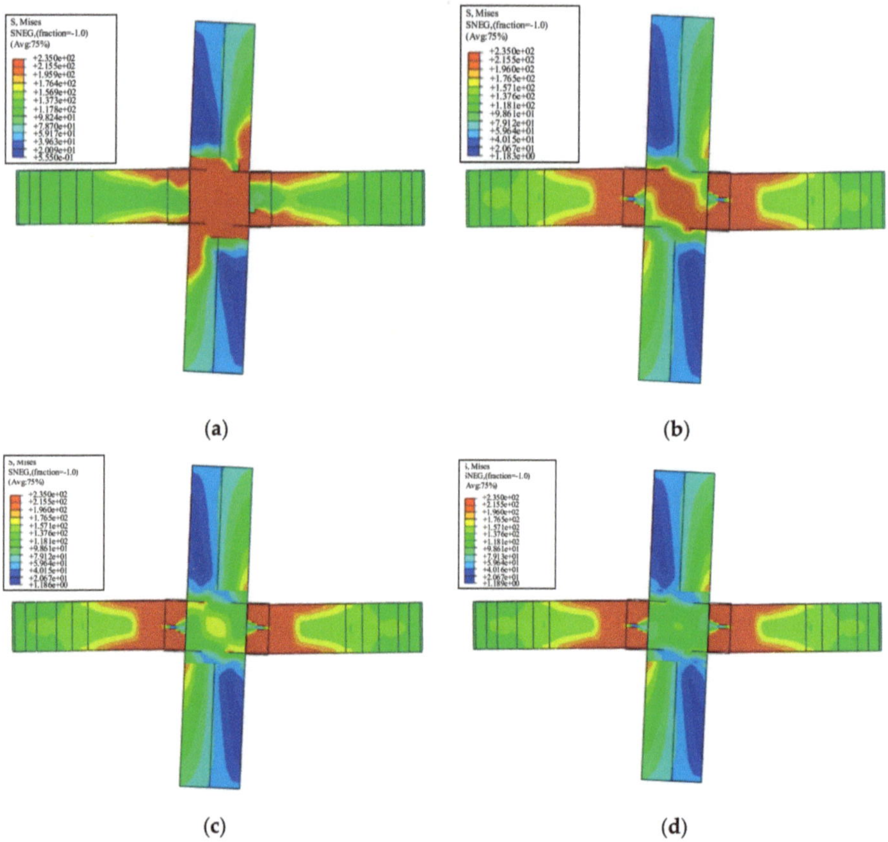

Figure 17. Stress contour plots in the damage states. (**a**) JD2-1. (**b**) JD2-3. (**c**) JD2-5. (**d**) JD2-7.

5.3. Effect of Column Flange Thickness

This section rigorously examines the effects of variations in the column flange thickness within the joint region on the load-bearing capacity and ductility of structural joints. The model parameters are comprehensively outlined in Table 6.

Table 6. Effect of the column flange thickness.

Specimens	Column Flange Thickness (mm)	P_p (kN)	μ
JD3-1	8	173.24	4.36
JD3-2	10	206.21	4.38
JD3-3	12	236.06	4.67
JD3-4	14	253.84	4.70
JD3-5	16	257.51	4.72
JD3-6	20	258.68	4.73
JD3-7	25	259.37	4.74

Figure 18 displays the load–displacement curves for the joints subjected to different column flange thicknesses, and Table 6 enumerates the corresponding variations in the load-bearing capacity and ductility. An analysis of these curves alongside the tabulated data yields several insights:

(1) There is a pronounced enhancement in the yield load, ultimate load, and failure load of the joint as the thickness of the column flange in the joint area increases. Nonetheless, the rate of increase in load-bearing capacity diminishes as the flange thickness grows, transitioning from a rate of 20% to as low as 0.27%. The considerable variability in the load-bearing capacity among the samples JD3-1 to JD3-4 suggests that an insufficient column flange thickness substantially undermines the joint's performance. Conversely, for samples JD3-5 to JD3-7, the load-bearing capacity differences are marginal, less than 0.5%, indicating that beyond a certain threshold, additional thickness does little to further bolster the load-bearing capacity.
(2) For joint areas with a minimal flange thickness, enhancing the flange thickness significantly improves ductility. However, once the flange thickness is deemed sufficient, further increases only marginally affect ductility enhancement.

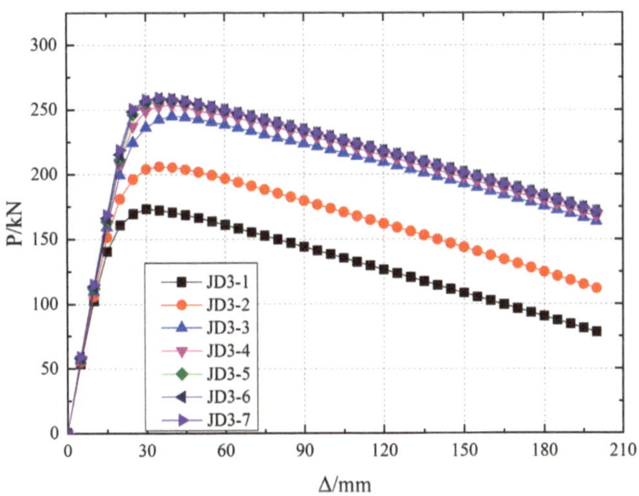

Figure 18. Load–displacement curves of joints.

Figure 19 shows the stress contour plots that depict the damage states of JD3-1, JD3-3, JD3-5, and JD3-7 under various column web thicknesses in the joint region. Analyzing the stress distributions across each joint provides further understanding.

(1) For JD3-1 and JD3-2, yielding primarily occurs in the flanges of the steel beams, with most of the web retaining its elastic properties. Significant plastic development is observed in the column flange area of the joint, accompanied by noticeable bulging of the steel plates. The ultimate failure typifies a weak joint with failure localized in this area.
(2) JD3-3 and JD3-4 exhibit considerable plastic deformation in the column flange area and slight bulging of the steel plates. Both the flanges and webs of the steel beams have transitioned to a plastic state. This condition indicates a critical state, where plastic hinges form simultaneously in the column flange area and at the beam ends.
(3) In samples JD3-5 to JD3-7, aside from minor plastic development connected to steel beams, the rest of the column flange area remains largely elastic. This scenario exemplifies a failure mode designed to introduce plastic hinges predominantly at the beam ends, adhering to the "strong column, weak beam, and stronger joint" design principle. A comparison of Figure 19e,f demonstrates that an increased flange thickness gradually restricts the extent of plastic development in the joint area, subtly enhancing both the load-bearing capacity and ductility.

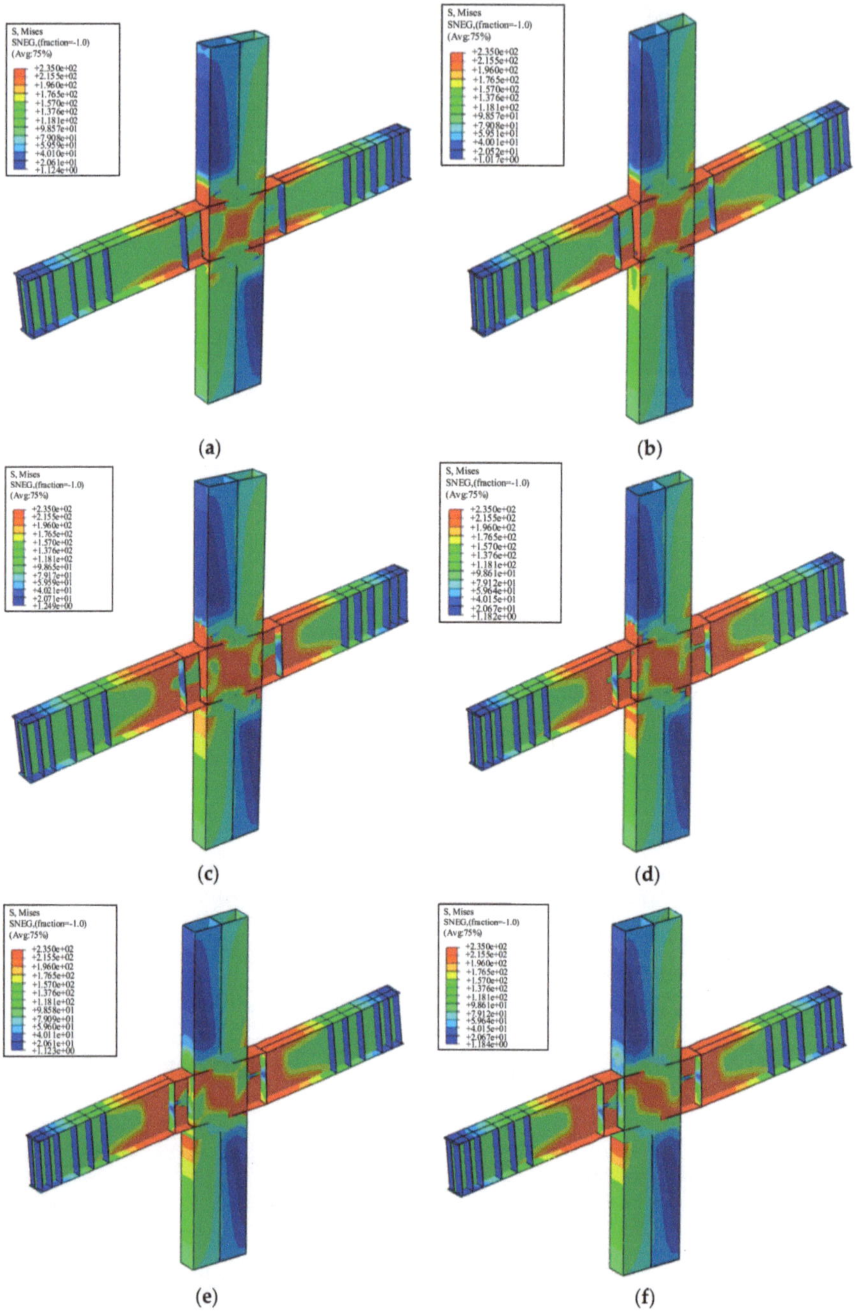

Figure 19. Stress contour plots in the damage states. (**a**) JD3-1. (**b**) JD3-2. (**c**) JD3-3. (**d**) JD3-4. (**e**) JD3-5. (**f**) JD3-7.

Given these findings, it is important to ensure that the thickness of the column flange in the joint area is neither too thin—to control excessive plastic development and uphold the design ethos of strong joints and weaker members—nor unduly thick—to prevent material overuse and waste. An optimal flange thickness of 1.0 to 1.2 times the thickness of the beam flange is advised for the column flange in the joint area, balancing structural efficacy with resource efficiency in design considerations.

5.4. Effect of the Diaphragm Length

This section extensively explores the influence of varied diaphragm lengths on the load-bearing capacity and ductility of steel structure joints. Table 7 presents the specifics of the model parameters.

Table 7. Effect of diaphragm length.

Specimens	Diaphragm Length (mm)	P_p (kN)	μ
JD4-1	60	241.41	4.58
JD4-2	50	251.65	4.63
JD4-3	100	255.94	4.67
JD4-4	120	257.74	4.69
JD4-5	140	259.08	4.71
JD4-6	160	259.16	4.72
JD4-7	180	259.25	4.73

Figure 20 depicts the load–displacement curves for joints subjected to different diaphragm lengths, and Table 7 illustrates the variations in the load-bearing capacity and ductility corresponding to these lengths. From the analysis of these curves and the data in the table, important findings emerged:

(1) An increase in the diaphragm length significantly enhances the yield load, ultimate load, and failure load of the specimens. A direct comparison between JD4-1 and JD4-2 showed considerable variability in the load-bearing capacity, with changes in the diaphragm length when it is relatively short. For instance, extending the diaphragm by 20 mm resulted in load-bearing capacity differences ranging from 1.7% to 4.3%. In contrast, for JD4-3 to JD4-7, where the diaphragm length was more optimal, further increases still improved the load-bearing capacity, though the increments were less substantial, with growth rates below 1%.

(2) At shorter diaphragm lengths, modifications to the length markedly influenced the specimen's ductility. However, once the diaphragm length reached an optimal value, increases in ductility tended to decelerate and eventually stabilize with further length increments.

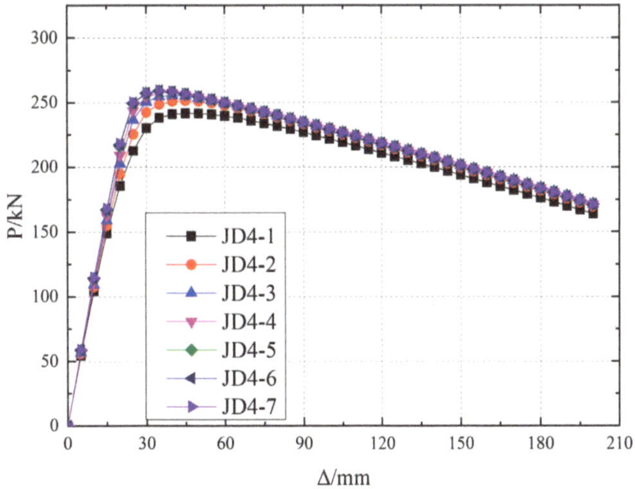

Figure 20. Load–displacement curves of joints.

Figure 21 displays stress contour plots that illustrate the damage states of JD4-1 through JD4-6 under varying diaphragm lengths. These visual comparisons of the stress distribution across each joint highlight the following insights:

(1) JD4-1 and JD4-2 showed significant plastic development in both the steel beams and joint regions. The formation of plastic hinges at both the column flanges and beam ends occurred nearly simultaneously, indicating a transitional state between joint failure and beam-end failure. This leads to these joints having a relatively lower load-bearing capacity.

(2) JD4-3 to JD4-7 demonstrated a classic joint failure mode where plastic hinges were predominantly developed at the beam ends. With longer diaphragm lengths, the plastic development within the column's web and flange in the joint region became increasingly confined, slightly enhancing the specimens' ductility and load-bearing capacity.

(3) The robustness of the joints, attributed to the columns' flange and web thickness coupled with a relatively low beam-to-column stiffness ratio, prevented premature joint area failure, even with shorter diaphragm lengths. Variations in the diaphragm length could notably influence the timing and degree of plastic phase onset in the steel components of the joint, consequently affecting the mode of failure.

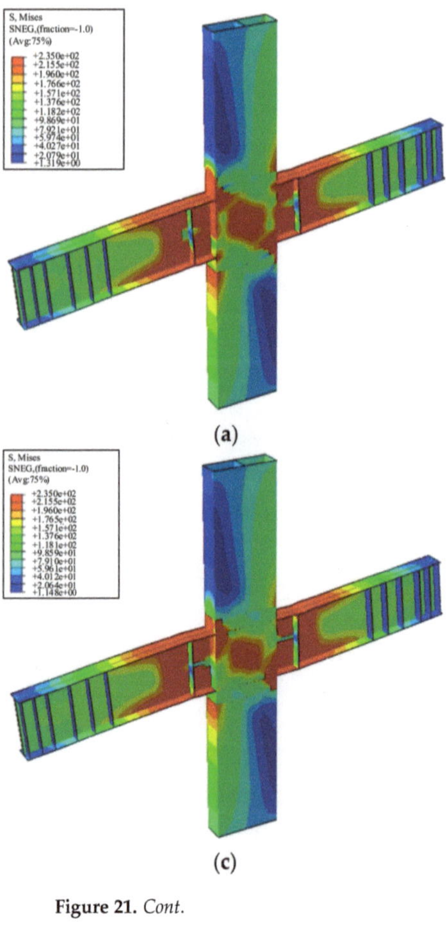

(a) (b) (c) (d)

Figure 21. *Cont.*

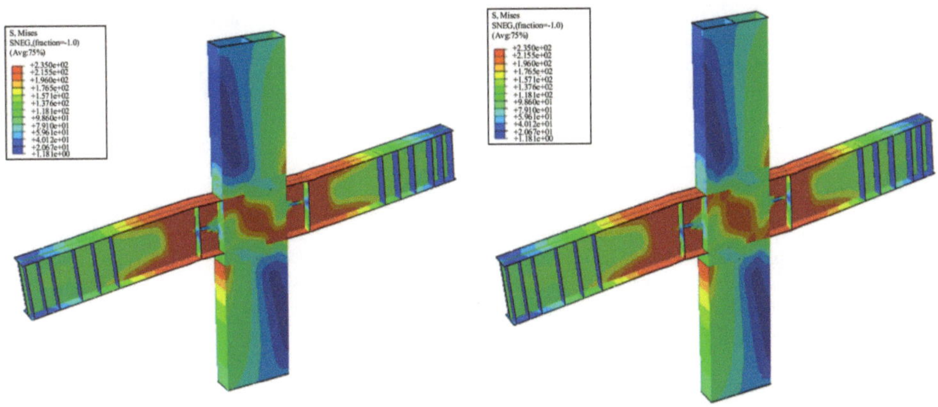

(e) (f)

Figure 21. Stress contour plots in the damage states. (**a**) JD4-1. (**b**) JD4-2. (**c**) JD4-3. (**d**) JD4-4. (**e**) JD4-5. (**f**) JD4-7.

To prevent excessive plastic development within the joint area and ensure failure via plastic hinge formation at the beam ends, diaphragm lengths should not be excessively short. However, overly large diaphragm dimensions might complicate the concrete-casting process. Diaphragm lengths should be approximately 0.2 to 0.3 times the width of the cross-section to balance effective control of plastic development with practical considerations in construction and material efficiency.

6. Conclusions

Based on experimental and numerical investigations, this study introduced and examined a joint specifically designed for beam-to-WCFST column connections. The research findings enable several key conclusions:

(1) The internal diaphragm joints designed for WCFST columns successfully met the seismic design principles of a "strong joint and weak member," as evidenced by high ductility coefficients exceeding 3.1 and an equivalent viscous damping coefficient around 0.44. The plastic hinge tended to form at a distance of about 0.7 to 1.2 times the beam height from the column edge, indicating effective energy dissipation and failure prevention at critical points.

(2) An increase in the axial compression ratio was correlated with reduced ductility in the specimens. Each 0.1 increase in the ratio resulted in a 2% decrease in the load-carrying capacity and altered displacement at failure. By carefully managing the axial compression ratio during design, enhanced ductility can be achieved, thereby optimizing joint performance. Additionally, the thickness of the column web and flange in the joint area is crucial in dictating the failure mode; an overly thin web or flange can lead to premature joint failure, whereas appropriate sizing can boost both ductility and load-bearing capacity.

(3) For optimal performance, the thickness of the column web and flange in the joint area should be about 0.85 to 1.2 times and 1 to 1.2 times the thickness of the beam flange, respectively. Moreover, the recommended length of the internal diaphragm should range between 0.2 and 0.3 times the width of the cross-section. These specifications are crucial for ensuring the structural integrity and functionality of the joints in WCFST systems.

Overall, this study's insights substantially broaden our understanding of WCFST systems and provide valuable guidance for future design and construction practices aiming to enhance safety and efficiency in seismic applications.

Author Contributions: H.L. (Hanchao Liu): methodology, investigation, software, and writing—original draft. H.L. (Honggang Lei): conceptualization and supervision. Y.H.: resources and supervision. Y.C.: methodology and writing—review and editing. F.X.: methodology. All authors have read and agreed to the published version of the manuscript.

Funding: This research received no external funding.

Data Availability Statement: Data are contained within the article.

Conflicts of Interest: Authors Yuqi Huang and Yongchang Chen were employed by the company Xi'an University of Architecture and Technology Design and Research Institute Co., Ltd. Author Feng Xu was employed by the company Xi'an Longhu Real Estate Development Co., Ltd. The remaining authors declare that the research was conducted in the absence of any commercial or financial relationships that could be construed as a potential conflict of interest.

References

1. Hao, J.; Xue, Q. *Research and Progress of Prefabricated-Steel-Structure Buildings*; China Architecture & Building Press: Beijing, China, 2022; p. 262.
2. General Office of the State Council. *Notice of the General Office of the State Council on Printing and Distributing the Action Plan for Peaking Carbon Emissions by 2030*; General Office of the State Council: Beijing, China, 2021; Volume 23.
3. Kexin, B.Z.; Consulting Co. Ltd. *China Prefabricated-Building Industry Policies [EB/OL]*; Beijing Zhiren Kexin Consulting Co., Ltd.: Beijing, China, 2022.
4. Wei, J.; Ying, H.; Yang, Y.; Zhang, W.; Yuan, H.; Zhou, J. Seismic performance of concrete-filled steel tubular composite columns with ultra high performance concrete plates. *Eng. Struct.* **2023**, *278*, 115500. [CrossRef]
5. Zhang, X.; Liu, X.; Zhang, S.; Wang, J.; Fu, L.; Yang, J.; Huang, Y. Analysis on displacement-based seismic design method of recycled aggregate concrete-filled square steel tube frame structures. *Struct. Concr.* **2023**, *24*, 3461–3475. [CrossRef]
6. Hao, J.P.; Sun, X.L.; Xue, Q.; Fan, C.L. Research and application of prefabricated-steel-structure building system. *J. Eng. Mech.* **2017**, *34*, 1–13. [CrossRef]
7. Sun, X.L.; Hao, J.P.; Xue, Q.; Fan, C.L.; Liu, H.C.; He, M.N. Experimental study on seismic behaviour of walled concrete-filled steel tubular columns. *J. Build Struct.* **2018**, *39*, 92–101. [CrossRef]
8. Hao, J.; Xue, Q.; Huang, Y.; Liu, H. Research of prefabricated building–based system theory. *Xi'An Univ. Archit. Technol. (Nat. Sci. Educ.)* **2019**, *51*, 14–20+26. [CrossRef]
9. Nishiyama, I.; Morino, S.; Sakino, K.; Nakahara, H.; Fujimoto, T.; Mukai, A.; Inai, E.; Kai, M.; Fukumoto, T.; Mori, K.; et al. *Summary of Research on Concrete Filled Structural Steel Tube Column System Carried Out under the US Japan Cooperative Research Program on Composite and Hybrid Structures*; BRI Research Paper No. 147; Building Research Institute: Tsukuba, Japan, 2002.
10. Nishiyama, I.; Fujimoto, T.; Fukumoto, T.; Yoshioka, K. Inelastic force-deformation response of jointshear panels in beam-column moment connections to concrete-filled tubes. *J. Struct. Eng.* **2004**, *130*, 244–252. [CrossRef]
11. Fujimoto, T.; Inai, E.; Tokinoya, H.; Kai, M.; Mori, K.; Mori, O.; Nishiyama, I. Behavior of beam-to-column connection of CFT column system under seismic force. In Proceedings of the 6th ASCCS International Conference on Steel-Concrete Composite Structure, Los Angeles, CA, USA, 22–24 March 2000.
12. Fujimoto, T.; Inai, E.; Kai, M.; Mori, K.; Mori, O.; Nishiyama, I. Behavior beam-to-colunm connetion of CFT colunm system. In Proceedings of the 12th WCEE World Conference on Earthquake Engineering, Auckland, New Zealand, 30 January–4 February 2000.
13. Dawe, J.L.; Grondin, G.Y. W-shape beam to RHS column connections. *Can. J. Civ. Engrg.* **1990**, *17*, 788–797. [CrossRef]
14. Nie, J.; Xu, G. Study on shear resistance properties of concrete column nodes of square steel tubes. *J. Tsinghua Univ. (Sci. Technol.)* **2009**, *49*, 798–802. [CrossRef]
15. Nie, J.; Qin, K.; Zhang, G. A study on bending bearing capacity of block nodes in square steel tubular concrete column. *J. Archit. Civ. Eng.* **2005**, *1*, 42–49+54.
16. Lü, X.; Li, X.; Xu, Y. Design method for connection of concrete column and steel beam of square steel tub. *J. Tongji Univ. (Nat. Sci.)* **2002**, *1*, 1–5.
17. Huang, Y. Mechanical behaviors of side-plate joint between walled concrete-filled steel tubular column and H-shaped steel beam. *Adv. Steel Constr.* **2020**, *16*, 346–353.
18. Liu, H.; Hao, J.; Xue, Q.; Sun, X. Seismic performance of a wall-type concrete-filled steel tubular column with a double-side-plate I-beam connection. *Thin Walled Struct.* **2021**, *159*, 107175. [CrossRef]
19. Liu, H.; Hao, J.; Xue, Q.; Huang, Y.Q. Stress response and initial stiffness of side-plate connections to WCFT columns. *Adv. Steel Constr.* **2021**, *17*, 306–317.
20. *GB50017-2017*; Standard for Design of Steel Structures. Standards Press of China: Beijing, China, 2017. Available online: https://www.chinesestandard.net/PDF/English.aspx/GB50017-2017 (12 December 2017).
21. *GB/T2975-2018*; Sampling Positions and Sample Preparation for Mechanical Property Tests of Steel and Steel Products. Standards Press of China: Beijing, China, 2018.

22. *GB/T228.1-2010;* Metallic Materials—Tensile Testing—Part 1: Method of Test at Room Temperature. Standards Press of China: Beijing, China, 2016.
23. *GB50011-2010;* Code for Seismic Design of Buildings. 2016 Version; China Architecture & Building Press: Beijing, China, 2016.
24. American Institute of Steel Construction Inc.—AISC. *Seismic Provisions for Structural Steel Buildings;* AISC: Chicago, IL, USA, 21 May 2002; Supplements 1 and 2.
25. Yang, L.; Ye, M.; Huang, Y.; Dong, J. Study on mechanical properties of displacement-amplified mild steel bar joint damper. *Iran. J. Sci. Technol. Trans. Civ. Eng.* **2023**, 1–14. [CrossRef]
26. Wang, X. Behavior Research and Aseismic Design Criterion on New-Type Connections of Concrete-Filled Square Tubular Column and Steel Beam. Master's Thesis, Xi'an University of Architecture & Technology, Xi'an, China, 2008.
27. He, Y.; Huang, P.; Leng, Q.; Xiao, A. Study on shear carrying capacity of high-strength-bolts end-plate connection for concrete-filled square steel-tube column. *Build Struct.* **2010**, *40*, 83–86.
28. *ABAQUS Standard User's Manual*, version 5.8; Hibbit, Karlsson, Sorensen: Pawtucket, RI, USA, 1998.
29. Han, L.; Yao, G.; Tao, Z. Performance of concrete-filled thin-walled steel tubes under pure torsion. *Thin Walled Struct.* **2007**, *45*, 24–36. [CrossRef]
30. Jafari, A.; Beheshti, M.; Shahmansouri, A.A.; Bengar, H.A. Cyclic response and damage status of coupled and hybrid-coupled shear walls. *Structures* **2024**, *61*, 106010. [CrossRef]
31. Shi, Y.; Wang, M.; Wang, Y. Experimental and constitutive model study of structural steel under cyclic loading. *J. Constr. Steel Res.* **2011**, *67*, 1185–1197. [CrossRef]
32. Tao, Z.; Wang, Z.B.; Yu, Q. Finite element modeling of concrete-filled steel stub columns under axial compression. *J. Constr. Steel Res.* **2013**, *89*, 121–131. [CrossRef]
33. Yang, S.L.; Zhang, L.; Zhang, J.B.; Fu, B.; Tong, G.; Jing, T. Seismic behavior of concrete-filled wide rectangular steel tubular (CFWRST) stub columns. *J. Constr. Steel Res.* **2022**, *196*, 107402. [CrossRef]
34. Liu, H.; Lei, H.; Huang, Y. Seismic performance of the WCFTC-beams joint with high-strength double-side plates. *J. Constr. Steel* **2024**, *218*, 108727. [CrossRef]

buildings

MDPI

Article

Experimental and Numerical Investigation on Stress Concentration Factors of Offshore Steel Tubular Column-to-Steel Beam (STCSB) Connections

Chao Hu [1], Qian Xia [1], Erxian Zeng [1], Jielong Zhu [2], Songsong Yu [1], Lei Zhang [2] and Fei Xu [2,*]

[1] Central Southern China Electric Power Design Institute Co., Ltd., Wuhan 430071, China
[2] School of Civil Engineering, Chongqing University, Chongqing 400030, China
* Correspondence: fei.xu@cqu.edu.cn

Abstract: Steel tubular column-to-steel beam (STCSB) connections are critical parts in offshore structures, where complex component connections and the stress concentration are of significant concern. This study conducted stress concentration tests on welded STCSB connections and subsequently developed a finite element (FE) model for the connections, with the experimental results validating the accuracy of the model. The discussion focused on the influence of parameters such as the width-to-diameter ratio of the beam to the column, the diameter-to-thickness ratio of the column, the diameter-to-thickness ratio of the column to the beam, and the height-to-thickness ratio of the beam web on the fatigue performance. The study proposed optimization methods including the addition of stiffeners and outer flange plates. The findings indicate that optimized connection configurations can effectively mitigate stress concentration in the connected areas, thereby enhancing the structural stability and fatigue life. The width-to-diameter ratio of the beam to the column and the diameter-to-thickness ratio of the column significantly affect the fatigue performance of welded STCSB connections, with an increased width-to-diameter ratio of the beam to the column or a reduced diameter-to-thickness ratio of the column leading to a substantial decrease in the maximum stress concentration factors (SCFs). The addition of stiffeners and adjustment of the outer flange plate can improve stress concentration effects in the connection area.

Keywords: connection optimization; FE analysis; parametric analysis; STCSB connection; stress concentration factor

1. Introduction

In response to the imperatives of achieving carbon neutrality and peaking, alongside the strategic deployment of offshore wind energy, this study underscores the necessity of substantial technological and economic advancements. Such evolutions are pivotal to the realization of comprehensive global decarbonization efforts [1]. The Chinese government has taken steps to undergo a transition from traditional fossil fuels to low-carbon clean energy, where offshore wind energy holds immense potential. As a result, it will be promising for China to promote the development of renewable energy and ensure the low-carbon development of the sustainable economy [1,2]. Offshore wind energy boasts distinct advantages, such as high wind speed, low wind shear, minimal turbulence intensity, stable prevailing wind direction, and no need for land occupation. The gradual development of marine energy and resources towards deep-sea areas represents a crucial direction for China to promote energy transition [3] and an essential pathway to achieving the "2060 carbon peak and carbon neutrality" targets [4]. However, the service environment of the offshore wind jacket is complex and harsh, and the offshore structure is subjected to wind, waves, currents, and other alternating loads for a long-term period. The jacket joint performance is constantly deteriorating, and then fatigue failure occurs. In engineering, the S-N curve based on the hot spot stress method is often used to evaluate the fatigue life of joints [5,6]. The S-N curve illustrates the number of cycles a joint can withstand until fatigue failure occurs at various stress levels, plotting the logarithm of stress amplitude against the logarithm of the number of cycles to failure. This method is used to predict the fatigue life of the joint by obtaining the size of the hot spot stress amplitude along the weld by means of numerical or experimental methods [7].

STCSB connections have been widely applied in offshore jacket structures, which are characterized by high stiffness better seismic performance and ductility compared to steel tubular joints [8,9]. Many researchers have conducted experiments and numerical analyses on the column to connections. Korkmaz and Tankut [10] explored the seismic performance of precast connections, assessing the behavior of specimens with both conventional and enhanced connection details to develop a "moment-resistant precast concrete beam-to-beam joint." Cao et al. [11] developed a novel seismic retrofitting technique for a precast reinforced concrete and ultra-high-performance concrete composite support frame, proposing a displacement-based design approach utilizing the stochastic capacity spectrum method (CSM). Lu et al. [12] and Kamba et al. [13] investigated the bearing capacity of the STCSB connections through experiments and numerical analyses; these studies revealed that the bearing capacity of joints is mainly affected by the ratio of beam flange width to column diameter and the column diameter to thickness ratio. When the column diameter to thickness ratio is greater than 27, the bearing capacity of the joints is less than the tensile capacity, and when the column diameter to thickness ratio is less than 22, the tensile and compressive capacities of the joints are almost the same. Using the results of parameter analysis, modified bearing capacity methods for STCSB connections were developed. Kim et al. [14] investigated the influences of the different steel plate-reinforced STCSB connections, and their investigations highlighted that the bearing capacity of the flange plate is higher than that of the cover plate. Meanwhile, Kosteski et al. [15] summarized the formulas of the bearing capacity of the STCSB connections. Further research was conducted by Kamba et al. [16], who found that the failure model of the STCSB connections is fracture failure at weld along the column wall on both sides of the beam flange. Moreover, the CIDECT design guide [17] provides a formula for calculating the welding STCSB connection considering the influence of the parameters of beam flange width to column diameter and column diameter to thickness ratio, and the corresponding failure mode is plastic failure of the column wall. Subsequently, many scholars also proposed an improved column-to-beam connection mode to improve the mechanical properties of the STCSB connections in view of the existing shortcomings of the connections [18,19]. Furthermore, many researchers have also conducted experiments on the seismic-resistant performance of STCSB connections. Dong et al. [20] proposed a novel connection of a concrete-filled steel tube column to an H-shaped beam with a slide plate, conducting six experiments under reversed cyclic loading, which indicated sufficient energy dissipation capacity and excellent ductility. Chen et al. [21] investigated the influence of cyclic hardening characteristics on the performance of a welded austenitic stainless steel H-section beam to column under combined uniaxial cyclic bending and constant compression. It was indicated that the cyclic strain hardening influenced the performance of the members under seismic loading. It can be found from the aforementioned analyses that the research on STCSB connections has mainly focused on static mechanical properties, stiffness, failure models, load-transferring mechanisms, seismic performance, and ductility. However, there are limited studies on the stress concentrations of STCSB connections. To provide a reliable design for an offshore jacket using STCSB connections, it is necessary to explore the fatigue performance of STCSB connections to support offshore structure design.

The existing literature on stress concentration and fatigue behavior of STCSB connections is limited. This research examines stress concentrations in STCSB connections via experimental and numerical methods, employing a hotspot stress approach-based numerical model to assess the influence of critical geometric parameters on the SCFs. Based on experimental investigation and parametric analysis, this study considers the impact of various geometric configurations on STCSB connections. We introduce targeted optimization strategies, such as stiffener augmentation and flange plate shape alteration, to reduce the stress concentration in the connection zone, contributing valuable insights to the field and offering enhanced design guidance for the engineering and implementation of offshore structures.

2. Materials and Methods

2.1. Specimens

Three STCSB connections with stiffening rings were fabricated using Q355 steel by Zhejiang Shengda Iron Tower Co., Ltd., a prominent manufacturer located in Zhejiang, China. The detailed dimensions of the tested STCSB connections with stiffening rings are given in Figure 1 and Table 1. The I-shape steel beam, circular steel column, and stiffening ring were welded by means of submerged arc welding. For ease of specimen installation and loading apparatus setup, the dimensions L and l were designated as 2240 mm and 1800 mm, respectively. The steel column diameter D was uniformly set at 300 mm across all connections. The steel beams were specified with dimensions of 450 × 200 × 9 × 14 mm and 450 × 120 × 9 × 14 mm ($h_w \times b_f \times t_w \times t_f$). The stiffening ring was of diamond type. All the specimens were labeled as IB450 × 200-CC300-6-SRA20-14, where IB, CC, and SR refer to the I-shape steel beam, circular steel column, and stiffening ring. The numbers and letters following IB, CC, and SR represent the dimensions of the beam, column, and stiffening ring, i.e., 450 is the height of the steel beam, 200 is the width of the steel beam, 300 is the outer diameter of the steel column, 6 is the thickness of the column, A is the type of the stiffening ring, 20 is the width of the stiffening ring, and 14 is the thickness of the stiffening ring.

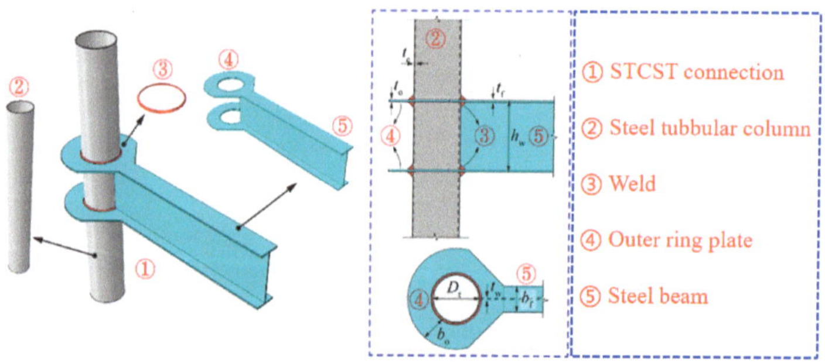

Figure 1. Schematic diagram dimensions of the STCSB connection.

Table 1. Geometric dimensions and dimensionless parameters of specimens.

Number	L	l	D_t	t_c	h_w	b_f	t_w	t_f	b_o	t_0	Type of Stiffening Ring
	mm	mm	mm	mm	mm	mm	mm	mm	mm	mm	
IB450 × 200-CC300-6-SRD20-14	2240	1800	300	6	450	200	9	14	20	14	Diamonds
IB450 × 200-CC300-5-SRD20-14	2240	1800	300	5	450	200	9	14	20	14	Diamonds
IB450 × 120-CC300-6-SRD20-14	2240	1800	300	6	450	120	9	14	20	14	Diamonds

Note: L is the length of the circular steel column, l is the length of the I-shape steel beam, D_t is the dimeter of the circular steel column, and t_c is the thickness of the circular steel column. h_w is the height of the web of the steel beam, b_f is the width of the steel beam flange, t_w is the thickness of the web of the steel beam, and t_f is the thickness of the steel beam flange. b_o is the width of the stiffening ring (the difference between the inner and outer diameters of the stiffening ring), and t_0 is the thickness of the stiffening ring.

Tensile tests were conducted utilizing a 100 kN-capable monotonic tensile testing machine at an ambient temperature of approximately 20 °C. Tensile coupons were extracted from the I-shaped steel beams and circular steel columns. The specific dimensions of these specimens, designed in accordance with ISO 6892-1 [22], are illustrated in Figure 2 and detailed in Table 2, respectively. Throughout the loading procedure, deformation at the gauge length and the corresponding applied load were captured using an extensometer and the testing machinery, respectively. In compliance with ISO 6892-1 [22] and the methodology outlined by Huang and Young [23], the test was extensometer-controlled, with the specific loading protocol delineated in Table 3. The strain rates were set at 0.000017 s^{-1} for the elastic phase, 0.00005 s^{-1} for the yield plateau, 0.0002 s^{-1} for the strain-hardening phase, and 0.00033 s^{-1} for the necking phase. The resulting engineering stress–strain curves are depicted in Figure 3, while the mechanical properties derived from the tensile coupons are enumerated in Table 4.

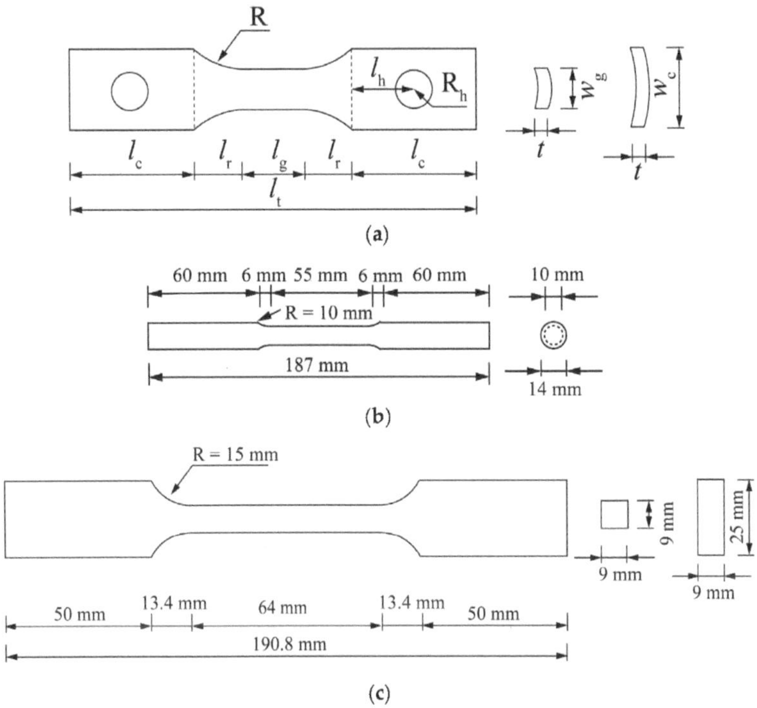

Figure 2. Dimensions of the tensile coupon specimens. (**a**) Specimen of circular steel column. (**b**) Specimen of flange of the I-shape steel beam. (**c**) Specimen of web of the I-shape steel beam.

Table 2. Detailed dimensions of the tensile coupon specimens for circular steel columns.

Member	Parameter	l_t/mm	l_r/mm	l_g/mm	l_c/mm	R/mm	t/mm	w_g/mm	w_c/mm	R_h/mm	l_h/mm	Number
Circular steel column	300 × 5	214	15.0	64	60	15	5	16	45	6	30	2
	300 × 6	214	15.0	64	60	15	6	13	40	6	30	2

Note: l_t denotes the overall length of the tensile coupon specimens. The length of the transition arc segment is represented by l_r, with l_g designating the gauge length of the specimen. The clamping segment's length is given by l_c. The radius of the transition arc is R, and the specimen's thickness is denoted by t. The width of the gauge segment is w_g, while w_c signifies the width of the clamping segment. The radius of the holes within the specimens is R_h, and l_h measures the distance from the center of the hole to the end of the clamping segment. These dimensions are critical parameters in the design and analysis of tensile test specimens for material property evaluation.

Table 3. Strain rate of the tensile coupon test.

Stage	Strain Rate/s^{-1}
Elastic stage	0.000017
Yield plateau stage	0.00005
Strain-hardening stage	0.0002
Necking stage	0.00033

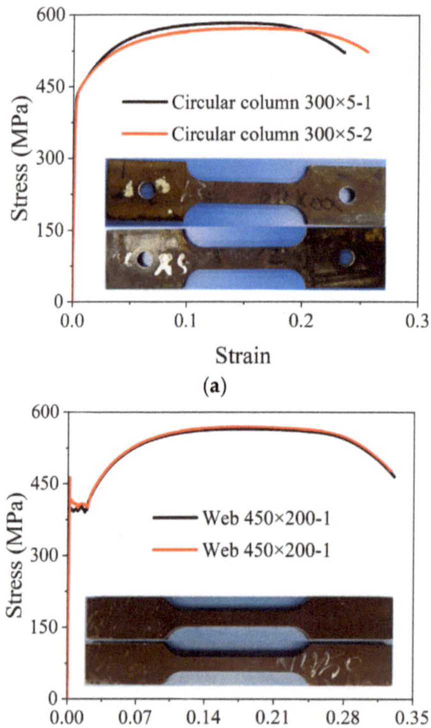

Figure 3. Typical stress–strain curve of the materials for circular steel column and I-shape steel beam. (**a**) Circular steel column. (**b**) I-shape steel beam.

Table 4. Mechanical properties of the tested Q355 steel.

Member	Specimen	Elastic Modulus/GPa	Yield Strength/MPa	Ultimate Strength/MPa
Circular steel column	300 × 5-1	220.72	435.79	582.86
	300 × 5-2	223.37	435.71	571.33
	300 × 6-1	200.76	405.37	559.31
	300 × 6-2	197.37	394.75	556.34
I-shape steel beam	Web-450 × 200-1	219.33	376.63	544.76
	Web-450 × 200-2	226.66	385.34	548.51
	Flange-450 × 200-1	207.64	357.10	552.03
	Flange-450 × 200-2	194.03	363.07	552.55

2.2. Test Setup

The schematic representation of the experimental setup for ascertaining the SCF of the STCSB connection is depicted in Figure 4. The I-shape steel beam was placed vertically, and the top end of the beam was hinged to the fatigue actuator with a loading capacity of 50 t. The other end of the actuator was fixed to the reaction wall. The circular steel column was horizontally placed, and both ends of it were hinged to the supports. Note that lateral supports were installed on both sides of the steel beam to avoid out-of-plane deformation of the steel beam during the loading process.

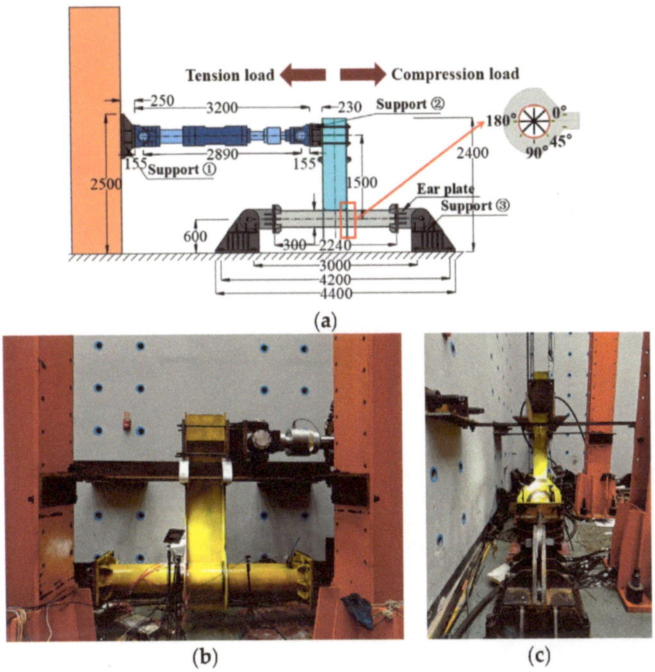

Figure 4. Schematic diagram of specimen installation and test loading mechanism. (**a**) Loading device. (**b**) Front view of the specimen installation. (**c**) Lateral view of the specimen installation.

In the experimental procedure, a gradient strain gauge was deployed to ascertain the hot spot strain in proximity to the weld toe, as illustrated in Figure 5. This gradient strain gauge was composed of five biaxial strain gauges symmetrically positioned along the longitudinal axis with a uniform center-to-center spacing of 4 mm. The positioning of the gradient strain gauge's extremity in relation to the weld toe's terminus was prescribed by the guidelines of DNV [24] and CIDECT [25]. Furthermore, four uniaxial strain gauges were systematically positioned adjacent to the circular steel column's periphery to measure the nominal column strain, thereby facilitating the computation of the actual load exerted on the specimens. The comprehensive layout of the strain gauges is delineated in Figure 6.

Figure 5. Gradient strain gauge.

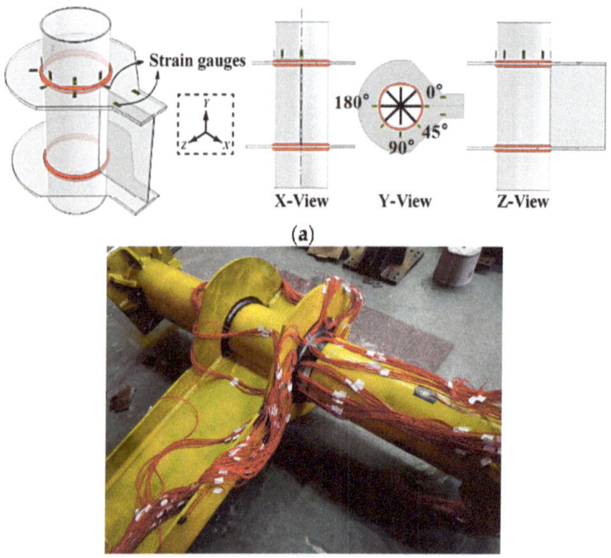

(a)

(b)

Figure 6. Arrangement of strain gauges. (**a**) Layout of the gradient strain gauges near the weld toe. (**b**) Beam–column welded joints with stiffening ring.

2.3. Measuring and Loading System

During this experimental assessment, the application of load was conducted in two phases: initial preloading and subsequent stress concentration evaluation. The preloading phase was essential to confirm that the specimen's response remained within the linear elastic regime and to mitigate any effects arising from gaps present during setup. Displacement control was implemented at a rate of 0.02 mm/s using an actuator. For the stress concentration phase, force control was utilized, maintaining an actuator loading velocity of 1 N/min throughout the duration of the test. At specific load levels—6 kN, 12 kN, 18 kN, 24 kN, 27 kN, and 30 kN—the test was paused for two minutes to acquire ample strain measurements during periods of stable loading. Strain data were consistently recorded at a frequency of 20 Hz. As depicted in Figure 4a, due to the symmetrical nature of the STCSB connection about the beam's web, strain gauges were applied to a single side only. The testing protocol encompassed two stages of loading: tensile and compressive.

3. Test Results and Discussion

When measuring the stress distribution around the welds of STCSB connections, strain gauges were placed at every 45° along the welds on both the steel column and the outer flange to measure the strains perpendicular ($\varepsilon_{\perp}$) and parallel ($\varepsilon_{//}$) to the welds. An additional strain gauge is placed in the area anticipated to have significant stress concentration prior to testing (at the 22.5° position). The arrangement of the strain gauges is detailed in Figure 6a. The hot spot strain at the weld was interpolated from the strain gauges, allowing for the calculation of the strain concentration factor (SNCF), which is the ratio of the hot spot strain to the corresponding nominal strain. The relationship between stress and strain, as per Hooke's law for a 3D problem, was simply derived to establish the relationship between the stress concentration factor (SCF) and the SNCF [26].

$$\text{SCF} = c \cdot \text{SNCF} \tag{1}$$

$$c = \frac{\left(1 + v\frac{\varepsilon_{//}}{\varepsilon_{\perp}}\right)}{(1 - v^2)} \tag{2}$$

where v is Poisson's ratio and $\varepsilon_{//}$ and $\varepsilon_{\perp}$ are the strains parallel and perpendicular to the weld, respectively.

In this study, the hot spot strain was calculated using the linear extrapolation method to determine the SNCF. Although the quadratic extrapolation method may be more efficient in data processing, it is more sensitive to data errors, potentially leading to an underestimation of the SNCF [27]. Therefore, considering its advantages in error control, the linear extrapolation method was chosen for calculating the SNCF [28].

The results of the stress concentration tests are presented in Table 5. For the specimen IB450 × 200-CC300-6-SRD20-14, strain gauge failure occurred, resulting in the acquisition of the SCF data only from the tensile side of the cylinder. As depicted in Figures 7–9, the influence of the diameter-to-thickness ratio of the column and the width-to-diameter ratio of the beam to the column on the distribution and magnitude of the SCF at STCSB connections was analyzed. In the analysis, to distinguish between the compressive- and tensile-side SCFs, the compressive side is denoted with "-SCF" in the table. Additionally, the maximum SCF was determined by the maximum absolute value. It can be observed that the SCF distribution patterns on the tensile and compressive sides of the STCSB specimens under various parameters were similar. The maximum SCF on the tensile side was consistently located at the 0° position. For all specimens, the maximum SCF on the compressive side was at the 180° position. As the column wall thickness increased, i.e., the diameter-to-thickness ratio of column decreased, the maximum SCF gradually decreased, although this trend was not significant, as shown in Figure 7. Figure 8a suggests that reducing the flange width of the beam, i.e., the width-to-diameter ratio of the beam to the column, had a significant effect on the SCF distribution and maximum SCF, and a decrease in the maximum SCF was observed. Additionally, a sharp decrease in SCF from 0° to 22.5° on the tensile side of specimen IB450 × 120-CC300-6-SRD20-14 was observed, with a similar phenomenon on the compressive side, indicating that reducing the beam flange width may have altered the stress distribution near the 0° position at the weld.

Table 5. SCFs of the STCSB connections.

Member		Tensile Side						Compressive Side					
		0°	22.5°	45°	90°	135°	180°	0°	22.5°	45°	90°	135°	180°
IB450 × 200-CC300-6-SRD20-14	Column	4.38	2.81	1.74	0.29	−1.85	−3.00	-	-	-	-	-	-
	Plate	-	-	-	-	-	-	-	-	-	-	-	-
IB450 × 200-CC300-5-SRD20-14	Column	4.76	3.77	1.70	0.35	−2.12	−3.96	−4.66	−3.69	−1.68	−0.36	2.08	3.79
	Plate	−0.47	−0.33	0.46	0.12	−0.42	−0.11	0.42	−1.24	−0.60	−0.18	0.06	0.02
IB450 × 120-CC300-6-SRD20-14	Column	3.17	0.77	0.96	0.38	−1.43	−2.62	−3.02	−0.77	−0.98	−0.41	1.36	2.54
	Plate	−0.21	0.17	0.24	0.09	−0.14	−0.11	0.03	−0.19	−0.27	−0.11	0.13	0.13

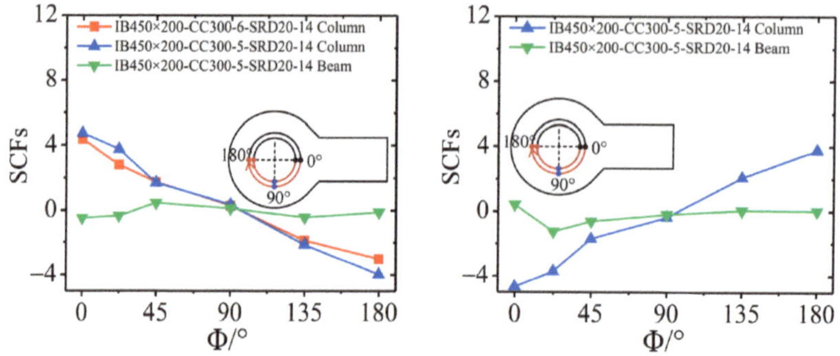

(**a**) (**b**)

Figure 7. SCF distribution of the different diameter-to-thickness ratios of the column. (**a**) SCF distribution under tension load; (**b**) SCF distribution under compression load.

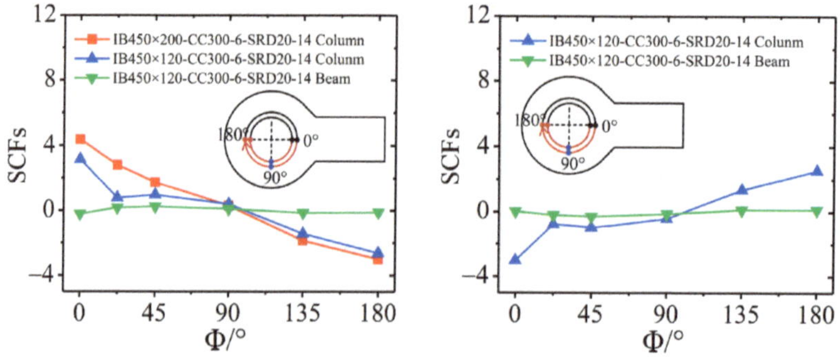

(**a**) (**b**)

Figure 8. SCF distribution of the different width-to-diameter ratios of beam to column. (**a**) SCF distribution under tension load; (**b**) SCF distribution under compression load.

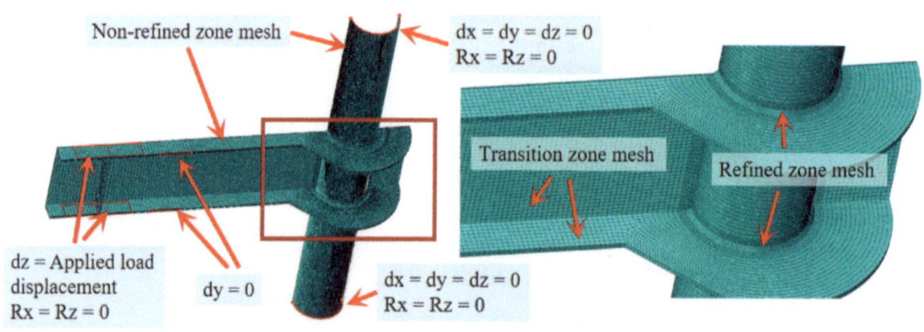

Figure 9. Finite element model of STCSB connection.

4. Finite Element Analysis

4.1. FE Models

To devise rational and feasible optimization designs, this study investigated the impact of geometric configuration on stress concentration at connections. The finite element model employs a global modeling method where components are interconnected through shared nodes. This approach ensures consistency across the model, which is essential for maintaining accuracy and integrity throughout the analysis. The material's constitutive relationship, which dictates the stress–strain behavior, is crucial for simulating the mechanical behavior of STCSB connections. The elastic modulus of Q355 steel was obtained using the material property test, and Poisson's ratio was 0.3. The model's boundary constraints and loading conditions were set in accordance with existing experimental loading and boundary constraints. Specifically, the column was pinned at the top and fixed at the bottom, with lateral constraints applied at the beam ends. The loading method was displacement-controlled. The finite element model and boundary conditions were defined as illustrated in Figure 9.

Mesh sensitivity analysis was conducted to select an appropriate mesh size that would balance computational efficiency with result accuracy. The analysis included three zones: non-refined, transition, and refined mesh areas. According to CCS specifications [29], the mesh size in the hot spot stress area, including the interpolation area, should not exceed the thickness t of the loaded component, with a recommended mesh size of t × t. The refined mesh area should extend at least 10 t in all directions from the hot spot. Preliminary analysis considered non-refined zone mesh sizes from 10 to 40 mm, a transition zone of 1.0 t, and a refined zone of 0.2 t to assess their impact on the hot spot stress concentration factor. Further analysis of mesh sizes ranging from 10 to 20 mm for points 1 to 4, as shown in Figure 10, yielded the results depicted in Figure 11a. The findings indicate that variations in the non-refined zone mesh size had minimal impact on the stress concentration; thus, a mesh size of 20 mm was selected. Subsequently, based on the 0.2 t refined zone mesh size and the 20 mm non-refined zone size, a sensitivity analysis of the transition zone mesh sizes from 0.5 t to 1.0 t was conducted, with the results shown in Figure 11b. The transition zone mesh size had a minor impact on the stress concentration, leading to the selection of 0.9 t as the transition zone mesh size. Finally, for the refined zone, the impact of mesh sizes from 0.15 t to 0.4 t on stress concentration at four selected points was analyzed, with the results illustrated in Figure 11c. It was found that mesh sizes of 0.15 t to 0.25 t in the refined zone significantly affected the stress concentration, which stabilized when the mesh size reached 0.3 t. Consequently, a refined zone mesh size of 0.3 t was chosen. The simulation employed C3D8I solid elements, which, under optimal mesh quality, yielded results comparable to a quadratic element (e.g., C3D20R), with the hot-spot stress and strain approximated to a quadratic element.

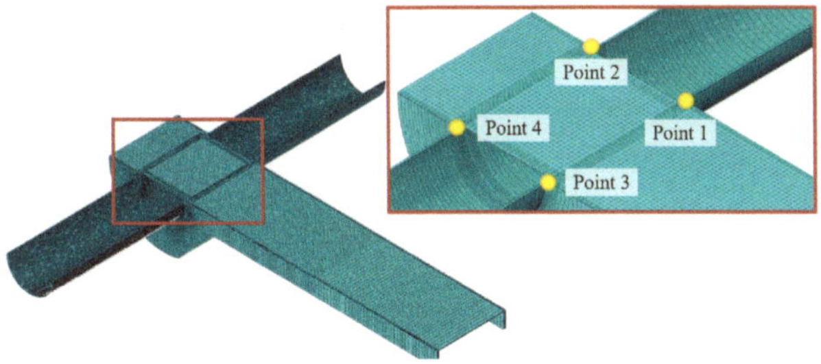

Figure 10. Schematic diagram of points selected for grid sensitivity analysis.

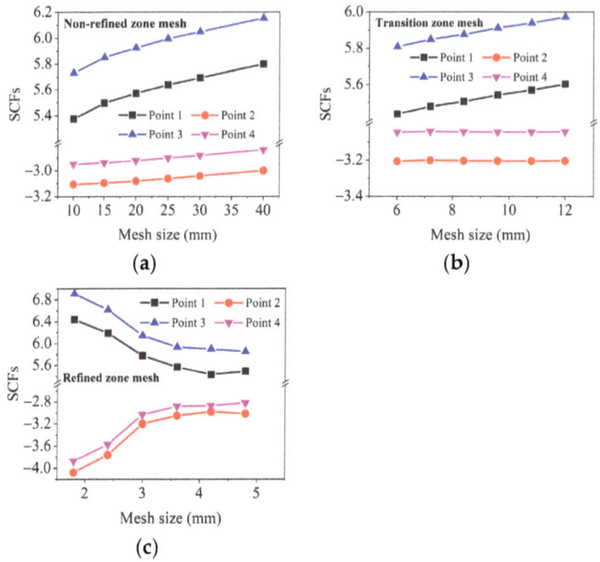

Figure 11. Grid sensitivity analysis results in different zones. (**a**) Non-refined zone mesh. (**b**) Transition zone mesh. (**c**) Refined zone mesh.

4.2. Validations

Experimental data from the current research were utilized to validate the findings. Table 6 presents the computational results for the STCSB connections, where SCF_{FE} denotes the SCF obtained from finite element analysis and SCF_{Test} represents the data derived from experiments. It should be noted that the SCFs recorded on the flange plate were relatively low. To ensure the accuracy of the validation results, considering factors such as gauge placement errors and equipment precision, the comparative analysis was conducted solely on the SCF values from the column. This focused approach facilitated a more precise evaluation while mitigating potential sources of discrepancy in the analysis.

Table 6. Comparison of results between SCF_{FE} and SCF_{Test}.

Member		SCF_{FE}						SCF_{Test}					
Tensile Side		**0°**	**22.5°**	**45°**	**90°**	**135°**	**180°**	**0°**	**22.5°**	**45°**	**90°**	**135°**	**180°**
IB450 × 200-CC300-6-SRD20-14	Column	5.13	3.26	1.87	0.20	−2.03	−3.04	4.38	2.81	1.74	0.29	−1.85	−3.00
IB450 × 200-CC300-5-SRD20-14	Column	5.07	3.93	2.44	0.26	−2.47	−3.73	4.76	3.77	1.70	0.35	−2.12	−3.96
IB450 × 120-CC300-6-SRD20-14	Column	3.51	2.24	1.07	0.35	−1.40	−2.54	3.17	0.77	0.96	0.38	−1.43	−2.62
Compressive side													
IB450 × 200-CC300-6-SRD20-14	Column	−5.10	−3.23	−1.89	−0.61	2.03	3.06	-	-	-	-	-	-
IB450 × 200-CC300-5-SRD20-14	Column	−5.02	−3.89	−2.45	−0.28	2.46	3.77	−4.66	−3.69	−1.68	−0.36	2.08	3.79
IB450 × 120-CC300-6-SRD20-14	Column	−3.49	−2.23	−1.08	−0.36	1.39	2.06	−3.02	−0.77	−0.98	−0.41	1.36	2.54

Figure 12 presents a comparative analysis between the experimental results and finite element (FE) simulations, demonstrating a high level of correlation between the empirical data and computational predictions. However, notable discrepancies were observed in specific data points, such as the specimen IB450 × 120-CC300-6-SRD20-14, where, at the 22.5° position, the FE-predicted SCF_{FE} were considerably higher than the experimentally determined ones (SCF_{Test}) for both the tension and compression sides. These deviations may be attributed to the differences between the calculated weld dimensions in the FE model and the actual specimens; variations in welding quality and bead appearance due to manual welding proficiency; and potential inaccuracies in the placement of strain gauges, leading to measurement errors.

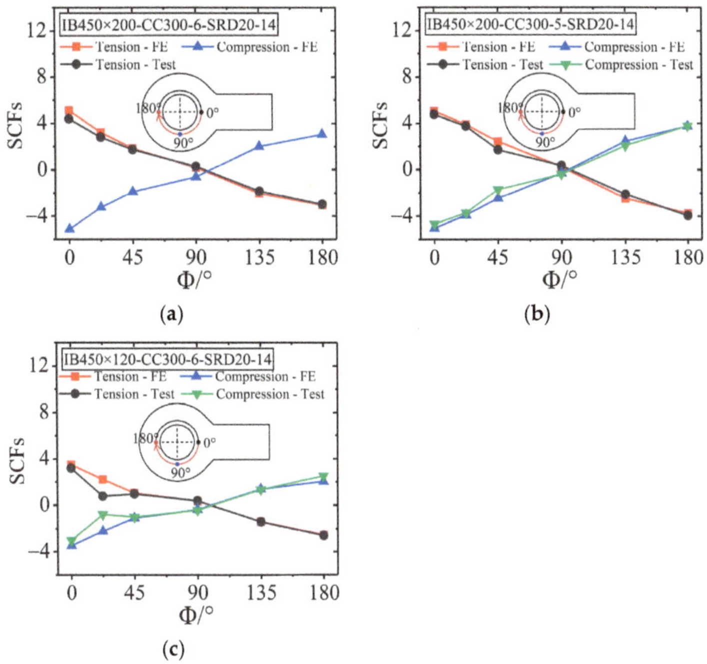

Figure 12. Comparison of results between SCF_{FE} and SCF_{Test}. (**a**) IB450 × 200-CC300-6-SRD20-14. (**b**) IB450 × 200-CC300-5-SRD20-14. (**c**) IB450 × 120-CC300-6-SRD20-14.

For the purpose of substantiating the precision of numerical simulations and affirming the validity of the stress concentration factors obtained through the hot spot stress method, Table 7 delineates the ratios of finite element analysis results to experimental data (SCF_{FE}/SCF_{Test}) alongside their average values. Moreover, to account for data variability, the table includes the coefficient of variation (CoV) for the stress concentration factor ratios, which were ascertained by employing the hot spot stress approach. The average SCF_{FE}/SCF_{Test} ratio for the STCSB connections ranged from 0.99 to 1.09, indicating a general overestimation by the FE analysis. Upon meticulous review of each SCF_{FE}/SCF_{Test} ratio, it was observed that, aside from the significant discrepancy at the 22.5° position for the specimen IB450 × 120-CC300-6-SRD20-14, there was a substantial error at the 45° position across all specimens. This could be due to the minimal stress concentration effects at 45° and the challenges associated with accurately measuring the SCF at this position, influenced by factors such as strain gauge placement and the precision of measuring instruments.

In summary, while the mean ratios fell within an acceptable numerical range and the CoV values ranged from 0.07 to 0.16, indicating a low degree of variability and suggesting greater stability and predictability in the data, certain data points were excluded from the computation of the statistical mean and CoV. Specifically, the data points at the 45° position, which exhibited significantly lower SCF values, and the outlier at the 22.5° position for specimen IB450 × 120-CC300-6-SRD20-14 were not considered. This exclusion was a prudent approach to ensure the reliability and integrity of the statistical analysis, taking into account the inherent complexities of experimental and numerical stress concentration assessments.

Table 7. Comparison of errors between SCF_{FE} and SCF_{Test} results.

Member		SCF_{FE}/SCF_{Test}						Average	CoV
Tensile Side		0°	22.5°	45°	90°	135°	180°		
IB450 × 200-CC300-6-SRD20-14	Column	1.17	1.16	1.07	0.69	1.10	1.01	1.03	0.12
IB450 × 200-CC300-5-SRD20-14	Column	1.07	1.04	1.44	0.74	1.17	0.94	1.07	0.16
IB450 × 120-CC300-6-SRD20-14	Column	1.11	2.91	1.11	0.92	0.98	0.97	1.02	0.07
Compressive side									
IB450 × 200-CC300-6-SRD20-14	Column								
IB450 × 200-CC300-5-SRD20-14	Column	1.08	1.05	1.46	0.78	1.18	0.99	1.09	0.15
IB450 × 120-CC300-6-SRD20-14	Column	1.16	2.90	1.10	0.88	1.02	0.81	0.99	0.12

5. Parametric Study

5.1. General

This study examines the effects of geometric configuration on the fatigue performance of structural connections by analyzing the influence of critical geometric parameters on SCF distribution. The abrupt geometric transition at the interface between the gusset outer flange plate and the column, which is prone to stress concentration, was targeted for analysis. SCFs were calculated along a path encircling the column–gusset plate junction. The parameters under investigation included the width-to-diameter ratio of the beam to the column, the diameter-to-thickness ratio of the column, the diameter-to-thickness ratio of the column to the beam, and the height-to-thickness ratio of the beam web, as outlined in Table 8. Given the relatively low SCFs observed for the beam, the analysis was concentrated on the SCF distribution of the steel tubular column.

Table 8. Detailed dimensions of the STCSB connections.

Member	Geometrical Parameter	Column		Plate		Beam				Parameter Analysis Range
		D	t_c	h_s	t_s	b_f	t_f	h_w	t_w	
1 2 3 4	Width-to-diameter ratio of beam to column (b_f/D)	300	6	200	12	130 160 190 220	12	440	6	0.43 0.53 0.63 0.73
5 6 7 8	Diameter-to-thickness ratio of column (D/t_c)	300	4 6 8 10	200	12	160	12	440	6	30.0 37.5 50.0 72.0
9 10 11 12	Diameter-to-thickness ratio of column to beam (D/t_f)	300	6	200	12	160	4 8 12 15	440	6	75.0 37.5 25.0 20.0
13 14 15	Height-to-thickness ratio of beam web (h_w/t_w)	300	6	200	12	160	12	280 360 440	6	46.67 60.00 73.33

5.2. Influence of the Width-to-Diameter Ratio of Beam to Column

The influence of the width-to-diameter ratio of the beam to the column on the stress concentration factor of the connections is delineated in Figure 13. The ratio was examined over a range from 0.43 to 0.73, with other parameters being held constant. Models 1 to 4, as per Table 8, were analyzed. Figure 13a illustrates the SCF distribution along the weld seam, and Figure 13b details the SCFs at specific points. Symmetry in SCF distribution was observed around the 90° position relative to the weld circumference. A significant rise in the SCF was noted beyond a width-to-diameter ratio of 0.63, with consistent trends across various measurement points.

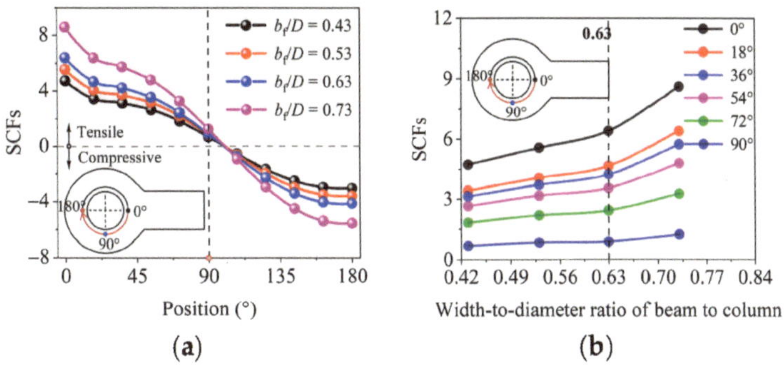

Figure 13. Effect of the width-to-diameter ratio of beam to column on SCFs. (**a**) SCF distribution along the weld. (**b**) SCF at different points.

5.3. Influence of the Diameter-to-Thickness Ratio of the Column

The influence of the column's diameter-to-thickness ratio on the stress concentration factors of the connections is delineated in Figure 14. The ratio was examined from 30 to 72, with other parameters being unchanged. Models 4 to 8, as detailed in Table 8, were analyzed. Figure 14a shows the SCF distribution along the weld seam, and Figure 14b presents the SCFs at fixed points. The connection's SCF increased significantly with the diameter-to-thickness ratio, with an accelerated growth rate at the 0° position beyond a ratio of 37.5. On the tensile side, a sharp decline in SCF values was observed from 0 to 22.5°, with a more gradual decrease between 22.5 and 45°, with the pattern becoming increasingly evident as the diameter-to-thickness ratio of the column increased.

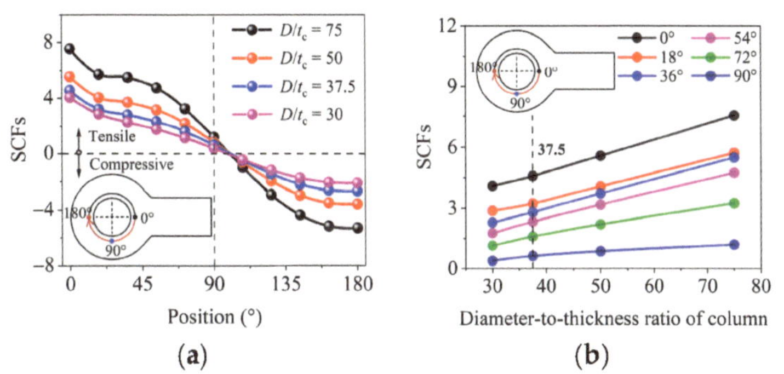

Figure 14. Effect of the diameter-to-thickness ratio of the column on SCFs. (**a**) SCF distribution along the weld. (**b**) SCF at different points.

5.4. Influence of Column Diameter-to-Beam Thickness Ratio

The influence of the column-to-beam diameter-to-thickness ratio on the stress concentration factor of the STCSB connections is detailed in Figure 15. The diameter-to-thickness ratio was examined across a range of 20 to 75, with other parameters being unchanged. Models 9 through 12, as depicted in Table 8, were analyzed. Figure 15a displays the SCF distribution along the weld seam, while Figure 15b shows the SCFs at discrete points. The connection's SCF diminished with an increasing column-to-beam diameter-to-thickness ratio. Beyond a ratio of 25, the SCF exhibited a stabilizing trend after an initial reduction. Furthermore, thinner steel beams resulted in more pronounced geometric discontinuity at the gusset plate connection (with a constant flange plate thickness), causing increased stress concentration.

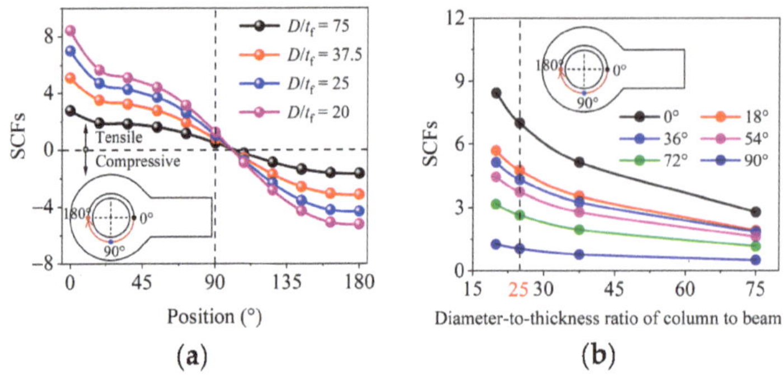

Figure 15. Effect of the diameter-to-thickness ratio of column to beam on SCFs. (**a**) SCF distribution along the weld. (**b**) SCF at different points.

5.5. *Influence of the Height-to-Thickness Ratio of the Beam Web*

The influence of the height-to-thickness ratio of the beam web on the connection's stress concentration factor is illustrated in Figure 16. The height-to-thickness ratio of the beam web varied from 46.67 to 73.33, with other variables held constant. Models 13 to 15, as listed in Table 8, were evaluated. Figure 16a presents the SCF distribution along the weld seam, while Figure 16b shows the SCFs at fixed points. The connection's SCF increased with the height-to-thickness ratio of the beam web, yet the increment rate attenuated with an increasing ratio.

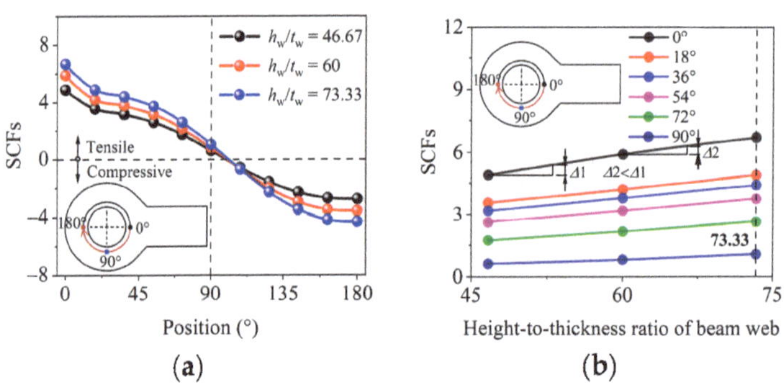

Figure 16. Effect of the height-to-thickness ratio of beam web on SCFs. (**a**) SCF distribution along the weld. (**b**) SCF at different points.

6. Optimization of STCSB Connections

Experimental and finite element analysis results indicate that the SCF distribution around the weld seam of these STCSB connections exhibited significant variation and was symmetrically distributed at 90° intervals. The maximum SCF was observed at either the 0° or 180° position, with notable disparities in the distribution near these peak stress locations. The shape of the outer flange may have also influenced the SCF distribution and magnitude. Stress concentration in welded connections is commonly attributed to geometric discontinuities at the weld locations. To alleviate stress concentration at the connections, the optimization of the outer flange configuration was attempted, and to ensure a smoother transition, the strategic placement of stiffeners was employed to reduce the SCFs. This section explores the enhancement of the overall SCF distribution at the connections through the modification of stiffener shapes as an optimization technique.

As depicted in Figure 17, significant stress concentrations were evident at the junction of the column with the outer flange and the contact area between the beam and the outer flange. Consequently, stress concentration could be mitigated by reinforcing the outer flange area with additional stiffeners and by optimizing the connection between the outer flange and the column to improve stress transfer pathways. Figure 18 illustrates two approaches. Method 1 involved the addition of stiffeners to the outer sides of the upper and lower outer flanges of the column, while Method 2 entailed the installation of a circular stiffener at the central position of the STCSB connection. The optimization of the outer flange is shown in Figure 18a,b. Analysis was conducted on three different parameters of diamond, arc, and circular shapes for the outer flange in order to determine the optimal form and parameters. The parameters for the stiffeners and outer flanges are detailed in Tables 9 and 10, respectively.

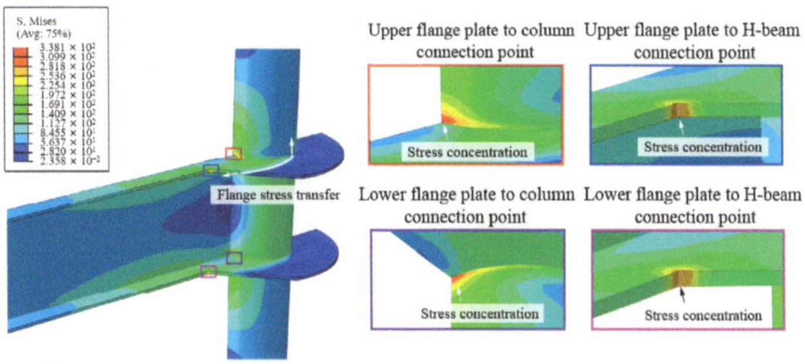

Figure 17. Stress nephogram of STCSB connection elastic stage.

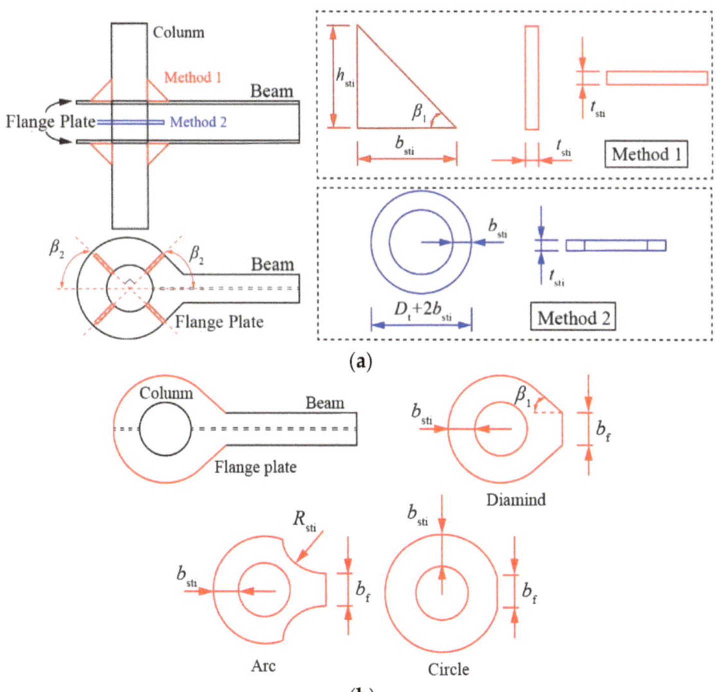

Figure 18. Schematic diagram of stiffener. (**a**) Addition of stiffeners. (**b**) Outer flange Optimization.

Table 9. Stiffener dimensions.

Member	Stiffener	b_{sti}/mm	h_{sti}/mm	t_{sti}/mm	β_1/°	β_2/°
S1	Method 1	190.73	190.73	6	45	-
S2		190.73	190.73	12	45	-
S3		190.73	330.35	6	60	-
S4		190.73	330.35	12	60	-
S5		88.52	330.35	6	75	-
S6		147.51	147.51	6	45	45
S7	Method 2	100	-	6	-	-
S8		100	-	12	-	-

Table 10. Details of the outer flange.

Member	Outer Flange Configurations	b_{sti}/mm	b_f/mm	R_{sti}/mm	β_1/°
S9	Diamond	200	200	-	45
S10	Arc	200	200	1167	
S11		200	200	540	-
S12		200	200	365	-
S13		200	200	285	-
S14	Circle	200	200	-	--

The maximum SCFs for specimens S1–S14 are illustrated in Figure 19. Specimens S1–S6 incorporated stiffeners on the outer sides of the upper and lower outer ring plates of the column; S7–S8 featured circular stiffeners at the connection's central position; S9 utilized the diamond-shaped ring plate, which is the standard STCSB connection; S10–S12 utilized arc-shaped ring plates; and S13–S14 employed circular ring plates. The computational results indicate that the addition of stiffeners at various locations and forms led to a fluctuating distribution of the maximum SCFs. Notably, specimen S4 exhibited the lowest maximum SCF, at 4.97, whereas specimen S2 recorded the highest SCF, reaching up to 7.05. This can be attributed to the stiffeners facilitating stress transfer at the connection between the ring plate and the column, significantly reducing the stress concentration effect at this location and causing a shift in the position of maximum stress concentration at the connection. Furthermore, the diamond and circular outer ring plates demonstrated better improvements in the maximum stress concentration effect. Collectively, these findings suggest that the addition of stiffeners can effectively mitigate stress concentration effects in the connection area and enhance the flexural performance of the STCSB connection.

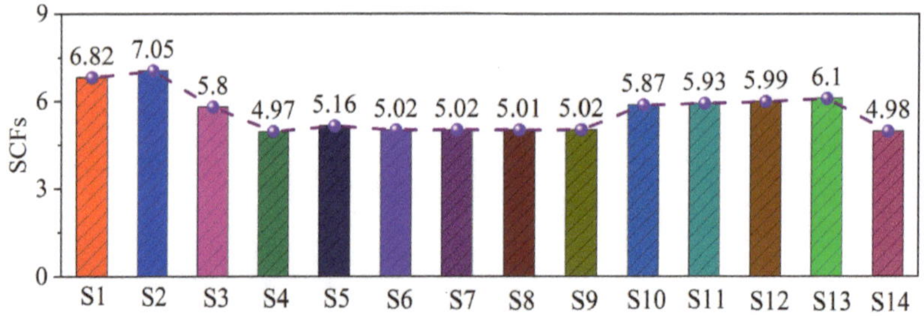

Figure 19. Maximum SCFs of specimens with different outer ring plates and stiffeners.

7. Conclusions

This study presents a comprehensive investigation into the fatigue performance optimization of STCSB connections, which are pivotal in the construction of offshore structures. Through experimental testing and FE analysis, the research meticulously examined the impact of various geometrical parameters on SCFs. The study introduced optimization methods such as the addition of stiffeners and outer flange plates, demonstrating their efficacy in mitigating stress concentration and enhancing fatigue life.

(1) The stress concentration tests on welded STCSB connections revealed significant variations in the SCF distribution. The maximum SCF was consistently identified at the 0° or 180° position, highlighting the areas of peak stress.

(2) The developed FE models, incorporating detailed geometrical and material properties, were validated against the experimental results. The congruence between FE predictions and experimental data confirmed the accuracy and reliability of the numerical approach.

(3) A comprehensive parametric study was conducted to assess the impact of various geometrical parameters on the fatigue performance of STCSB connections. Key parameters such as the width-to-diameter ratio of the beam to column, the diameter-to-thickness ratio of the column, and the diameter-to-thickness ratio of the column to the beam were found to significantly influence the SCF distribution and magnitude. The SCF increased with both the width-to-diameter ratio of the beam to the column and the diameter-to-thickness ratio of the column. Conversely, it decreased as the diameter-to-thickness ratio of the column to the beam increased.

(4) The installation of a circular stiffener at the center of STCSB connections or the addition of stiffeners on the outer sides of the column's upper and lower outer flanges effectively alleviated stress concentration. This optimization enhanced the connection's bending resistance, highlighting the significance of flange plate design in the optimization of fatigue performance.

Currently, there is a scarcity of research on the stress concentration and fatigue performance of STCSB connections. This study analyzed the influence of various parameters on the stress concentration and proposed corresponding optimization schemes, thereby providing a reference for this field of research. While this study provides substantial contributions to the field, there are limitations. The research primarily focused on fatigue life assessment under uniaxial load, which may not encapsulate the full complexity of real loading scenarios. Future research could extend this study by incorporating multi-axial loading conditions. Moreover, long-term performance studies and further experimental validations could provide additional layers of reliability to the proposed optimization methods.

Author Contributions: Conceptualization, C.H.; Methodology, C.H. and S.Y.; Validation, J.Z.; Investigation, J.Z.; Data curation, Q.X. and E.Z.; Writing—original draft, L.Z.; Writing—review & editing, F.X.; Supervision, F.X. All authors have read and agreed to the published version of the manuscript.

Funding: This work was financially sponsored by the Natural Science Foundation of Chongqing (CSTB2023NSCQ-MSX0576), China Power Engineering Consulting Group Corporation Limited through the research program "Research on Key Technologies of Offshore Wind Power Converter Platform Structure" (Project No: DG2-T01-2022) and the Fundamental Research Funds for the Central Universities (2022CDJQY-009).

Data Availability Statement: The original contributions presented in the study are included in the article; further inquiries can be directed to the corresponding author.

Conflicts of Interest: Authors Chao Hu, Qian Xia, Erxian Zeng, and Songsong Yu were employed by the company Central Southern China Electric Power Design Institute Co., Ltd. The remaining authors declare that the research was conducted in the absence of any commercial or financial relationships that could be construed as potential conflicts of interest. The authors declare that this study received funding from China Power Engineering Consulting Group Corporation Limited. The funder was not involved in the study design; the collection, analysis, or interpretation of data; the writing of this article; or the decision to submit it for publication.

References

1. Allan, B.; Lewis, J.I.; Oatley, T. Green industrial policy and the global transformation of climate politics. *Glob. Environ. Politics* **2021**, *21*, 1–19. [CrossRef]
2. Zou, C.N.; Xiong, B.; Xue, H.Q.; Zheng, D.W.; Ge, Z.X.; Wang, Y.; Jiang, L.Y.; Pan, S.Q.; Wu, S.T. The role of new energy in carbon neutral. *Pet. Explor. Dev.* **2021**, *48*, 480–491. [CrossRef]
3. Paraschiv, S.; Paraschiv, L.S.; Serban, A.; Cristea, A.G. Assessment of onshore wind energy potential under temperate continental climate conditions. *Energy Rep.* **2022**, *8*, 251–258. [CrossRef]

4. Oh, K.Y.; Nam, W.; Ryu, M.S.; Kim, J.Y.; Epureanu, B.I. A review of foundations of offshore wind energy convertors: Current status and future perspectives. *Renew. Sustain. Energy Rev.* **2018**, *88*, 16–36. [CrossRef]
5. Ahmadi, H.; Lotfollahi-Yaghin, M.A. Effect of SCFs on SN based fatigue reliability of multi-planar tubular DKT-joints of offshore jacket-type structures. *Ships Offshore Struct.* **2013**, *8*, 55–72. [CrossRef]
6. Ahmadi, H.; Lotfollahi-Yaghin, M.A.; Aminfar, M.H. Effect of stress concentration factors on the structural integrity assessment of multi-planar offshore tubular DKT-joints based on the fracture mechanics fatigue reliability approach. *Ocean Eng.* **2011**, *38*, 1883–1893. [CrossRef]
7. Ahmadi, H.; Lotfollahi-Yaghin, M.A.; Aminfar, M.H. The development of fatigue design formulas for the outer brace SCFs in offshore three-planar tubular KT-joints. *Thin Walled Struct.* **2012**, *58*, 67–78. [CrossRef]
8. Wang, W.; Chen, Y.Y.; Li, W.Q.; Chen, X. *Behaviour of Steel Structures in Seismic Areas*; Trans Tech Publications: London, UK, 2009.
9. Li, J.F.; Mou, B.; Wang, Z. Cyclic behavior of column-to-column connections in novel prefabricated H-shaped steel beam to CFST column joint. *J. Constr. Steel Res.* **2023**, *200*, 107657. [CrossRef]
10. Korkmaz, H.H.; Tankut, T. Performance of a precast concrete beam-to-beam connection subject to reversed cyclic loading. *Eng. Struct.* **2005**, *27*, 1392–1407. [CrossRef]
11. Cao, X.-Y.; Feng, D.-C.; Wang, C.-L.; Shen, D.; Wu, G. A stochastic CSM-based displacement-oriented design strategy for the novel precast SRC-UHPC composite braced-frame in the externally attached seismic retrofitting. *Compos. Struct.* **2023**, *321*, 117308. [CrossRef]
12. Lu, L.H.; Wardenier, J. The static strength and behaviour of uniplanar and multiplanar connections between I-beams and RHS columns. In *Connections in Steel Structures III*; Pergamon: Elmsford, NY, USA, 1996; pp. 127–135. [CrossRef]
13. Kamba, T.; Taclendo, C. CHS column connections without stiffener. In *Tubular Structures VIII*; A. A. Balkema: Rotterdam, The Netherlands, 1998; pp. 567–576.
14. Kim, T.; Whittaker, A.S.; Gilani, A.S.J.; Bertero, V.V.; Takhirov, S.M. Cover-plate and flange-plate steel moment-resisting connections. *J. Struct. Eng.* **2002**, *128*, 474–482. [CrossRef]
15. Kosteski, N.; Packer, J.A.; Puthli, R.S. A finite element method based yield load determination procedure for hollow structural section connections. *J. Constr. Steel Res.* **2003**, *59*, 453–471. [CrossRef]
16. Kamba, T.; Taclendo, C.; Namba, H. Tubular Column to H-Beam Connections without Diaphragms. In *Tubular Structures X*; Routledge: London, UK, 2002.
17. Packer, J.; Wardenier, J.; Zhao, X.-L.; van der Vegte, G.; Kurobane, Y. *Design Guide for Rectangular Hollow Section (RHS) Joints under Predominantly Static Loading*; CIDECT: Geneva, Switzerland, 2009.
18. Goswami, R.; Murty, C.V.R. Externally reinforced welded I-beam-to-box-column seismic connection. *J. Eng. Mech.* **2010**, *136*, 23–30. [CrossRef]
19. Couchaux, M.; Vyhlas, V.; Kanyilmaz, A.; Hjiaj, M. Passing-through I-beam-to-CHS column joints made by laser cutting technology: Experimental tests and design model. *J. Constr. Steel Res.* **2021**, *176*, 106298. [CrossRef]
20. Dong, X.T.; Wang, L.; Xu, Z.F.; Guan, S.S.; Liu, L.Y.; Chen, H.T. Seismic behaviour of concrete-filled steel tubular column to H-shape steel beam connection with side plate. *Structures* **2023**, *50*, 1608–1624. [CrossRef]
21. Chen, Y.; Zhou, F.; Zheng, S.; Li, J.Q.; Shang, C.Z.; Lin, X.X. Influence of cyclic hardening characteristic on seismic performance of welded austenitic stainless steel H-section beam-column. *Eng. Struct.* **2023**, *288*, 116210. [CrossRef]
22. *ISO-6892-1*; Metallic Materials–Tensile Testing—Part 1: Method of Test at Room Temperature. ISO: Geneva, Switzerland, 2019.
23. Huang, Y.; Young, B. The art of coupon tests. *J. Constr. Steel Res.* **2014**, *96*, 159–175. [CrossRef]
24. *DNV-RP-C203*; Fatigue Design of Offshore Steel Structures. Det Norske Veritas: Bærum, Norway, 2011.
25. Zhao, X.-L.; Herion, S.; Packer, J.A.; Puthli, R.; Sedlacek, G.; Wardenier, J. *Design Guide for Circular and Rectangular Hollow Section Joints under Fatigue Loading*; CIDECT Publication; TUV-Verlag: Köln, Germany, 2000.
26. Shao, Y.B.; Lie, S.-T. Experimental and numerical studies of the stress concentration factor (SCF) of tubular K-joints. *Eng. Mech.* **2006**, *23*, 79–85.
27. van Wingerde, A.M.; Packer, J.A.; Wardenier, J. Criteria for the fatigue assessment of hollow structural section connections. *J. Constr. Steel Res.* **1995**, *35*, 71–115. [CrossRef]
28. Tong, L.W.; Xu, G.W.; Zhao, X.-L.; Zhou, H.M.; Xu, F. Experimental and theoretical studies on reducing hot spot stress on CHS gap K-joints with CFRP strengthening. *Eng. Struct.* **2019**, *201*, 109827. [CrossRef]
29. CCS. *Guidelines for the Fatigue Strength Assessment of Offshore Engineering Structures*; China Classification Society (CCS): Beijing, China, 2022.

buildings

MDPI

Article

Cyclic Behavior of Concrete-Filled Tube Columns with Bidirectional Moment Connections Considering the Local Slenderness Effect

Ramón Mata and Eduardo Nuñez *

Department of Civil Engineering, Universidad Católica de la Santísima Concepción, Concepción 4061735, Chile; ramon.mata@uss.cl
* Correspondence: enunez@ucsc.cl

Abstract: In this research, the cyclic behavior of concrete-filled thin tube (CFTT) columns with bidirectional moment connections was numerically studied within the context of thin-walled structures. Novel considerations in the design of CFTT columns with slenderness sections are proposed through a parametric study. A total of 70 high-fidelity finite element (FE) models are developed using ANSYS software v2022 calibrated from experimental research using similar 3D joint configurations. Furthermore, a comparison of different width-to-thickness ratios in columns was considered. The results showed that the models with a high slenderness ratio reached a stable cyclic behavior until 0.03 rad of drift, and a flexural strength of 0.8 Mp was reached for 4% of the drift ratio according to the Seismic Provisions. However, this effect slightly decreased the strength and the dissipated energy of the moment connection in comparison to columns with a high ductility ratio. Moreover, an evaluation of concrete damages shows concrete cracked for cyclic loads higher than 3% of drift. Finally, the joint configurations studied can achieve a good performance, avoiding brittle failure mechanisms and ensuring the plastic hinges in the beams.

Keywords: steel moment connection; bidirectional effect; cyclic behavior; CFTT columns; finite element method

1. Introduction

The findings from the damages incurred in past earthquakes highlight the need for implementing earthquake-resistant structures, particularly in the form of steel and composite moment frames. The investigations conducted by the authors of [1,2] involved a survey of observed damages in steel buildings resulting from the Tohoku earthquake and tsunami in 2011. This survey specifically took place in the city of Sendai, where the recorded maximum acceleration exceeded 1 g (PGA > 1 g) in the Miyagi and Fukushima prefectures. The results revealed that steel structures not only withstood the seismic event but also the tsunami without experiencing collapses. However, significant damages were identified in elements such as bracings, connections, and H-type columns with local buckling, all attributable to design deficiencies not in compliance with regulatory provisions. In general, in Japan, most steel buildings use moment-resisting frames (MRFs) with hollow structural section (HSS) columns [3], which experienced fewer damages during these seismic events. Considering this evidence, the advantages of using tubular profiles for earthquake-resistant structures can be demonstrated, as shown in [4].

On the other hand, the 1995 Kobe earthquake revealed that the primary damages in steel structures with tubular profiles were concentrated mainly by local buckling phenomena of HSS profiles and their connections with beam elements [5]. Subsequently, as a mitigation alternative for the effects of local buckling in tubular columns, the concrete-filled tube (CFT) concept was proposed, as demonstrated in the research by the authors of [6]. Additionally, the experimental study conducted by the authors of [7] proposed a configuration for CFT columns to concentrate damages in a zone between plates. The results indicated that failure mechanisms were associated with local buckling within the rigidized zone, as expected. However, the study did not evaluate the use of slender box sections that could influence section rupture and the bidirectional effect on the column. Similarly, another study [8] evaluated the hysteresis performance of thin-walled circular tubular profiles filled with concrete employing the finite element method (FEM). The findings showed that these columns exhibited high strength and ductility, demonstrating that the FEM results aligned with experimental findings, revealing local buckling and metal fracture phenomena. Nevertheless, this investigation was limited to the use of hollow structural section columns with a specific width-to-thickness ratio. An experimental study developed by the authors of [9] explored the performance of CFT columns subjected to cyclic loads. Square HSS and circular CFT sections with two different thicknesses were tested under constant axial load and cyclic lateral load. The results indicated that the failure pattern of specimens was controlled by the width-to-thickness ratio of the elements, regardless of the type of concrete filling. Similarly, the experimental research developed by the authors in [10] demonstrates that the use of concrete-filled columns in fiber-reinforced plastic tubular profiles, even if not made of steel, significantly improved the behavior in terms of strength and stiffness. However, the evaluation of only two width-to-thickness ratios did not establish a direct relationship between these variables.

The research conducted by the authors of [11] identified parameters for three hysteretic models simulating the response of CFT columns to lateral loads using the finite element method. The implemented hysteretic models were Bouc–Wen, Ramberg–Osgood, and Al–Bermani. This parametric study involved 64 CFT columns with different width-to-thickness ratios, shapes, and strengths. The results indicated that the proposed concentrated plasticity models could provide good predictions of the force–deformation response of these columns. However, the study did not analyze the variation in the cyclic response in the CFT columns with moment connections based on their width-to-thickness ratio and was limited to the use of HSS profiles. Other variations in the behavior of CFT columns have been explored, such as the one proposed by the authors of [12], using ultra-high-strength steel for the development of CFT columns, based on an experimental study using HSS columns. The results showed that the local buckling phenomenon persisted in the columns, and the high strength of the steel prevented the early occurrence of this phenomenon, allowing for higher drifts. Nevertheless, this research was limited to the use of HSS profiles and did not consider the influence of the width-to-thickness ratio on the response variation. Another experimental study performed by the authors of [13] proposed the addition of binding bars between the walls of the profile to confine the interior concrete. The results showed increased lateral stiffness and energy dissipation of specimens with the addition of binding bars. Additionally, it was observed that the deformation capacity decreased as axial load levels increased. However, this research did not consider bidirectional effects on the column or study box-type sections.

The current Seismic Provisions AISC 341-22 [14] provides guidelines for width-to-thickness ratio limits for elements of high and moderate ductility. However, a notable limitation exists in the code specifications for the bidirectional connection effects in CFT columns. On this basis, the authors of [15] conducted a numerical study suggesting an assessment of limits based on plastic rotation normalized for axial loads below 30% of the compression capacity of the section. Although the study showed the potential limitations of Codes, it primarily focused on H-type steel profiles, thus restricting the generalizability of its findings to other CFT column configurations. A similar numerical and experimental study performed by the authors of [16] analyzed stainless steel box-section columns, revealing a reduced ductility and inelastic deformation capacity for higher width-to-thickness ratios. However, the study did not consider the influence of concrete filling in box-section columns and applied only a constant axial load. Likewise, the authors of [17] evaluated the lateral cyclic response in CFT columns filled with ultra-high-strength concrete, showing a decrease in ductility with increased axial load or width-to-thickness ratio. However, the study was focused on only three width-to-thickness ratios, and its proximity to the limits set by the authors of [14] may limit the generalizability of its conclusions. Moreover, the study did not explore a broader range of width-to-thickness ratios, leaving a gap in understanding the full spectrum of CFT column behavior. In this sense, these investigations have demonstrated that the cyclic behavior of CFT columns is strongly influenced by factors such as width-to-thickness ratio, axial load, and material characteristics.

Despite the limitations in characterizing the cyclic behavior of CFT columns composed of box sections, many applications using seismic-resistant systems have been developed. One of the main applications is the use of these columns in moment-resisting frame systems, which require moment connections that, according to the current code [14], must be tested and prequalified according to [18]. However, this code [18] only contemplates two proposed moment connections that meet this prequalification requirement [19,20], both of which are patented. Therefore, multiple connection proposals for composite elements have been developed to facilitate their implementation and use. Among the initial proposals, the authors of [21] assessed a CFT column connection with W-type beams welded to the column face, with transverse stiffeners. The results of the study showed that although the resistance exceeded the plastic moment, failure mechanisms such as local buckling and welding fractures were obtained. Similarly, the authors of [22] explored a different connection typology, finding reduced ductility as the width-to-thickness ratio increased. However, neither study considered the bidirectional effects and the use of box-section profiles for columns.

An experimental study conducted by the authors of [23] investigated the hysteretic performance of a proposed connection with concrete-filled tubular (CFT) columns formed by hollow structural sections (HSSs), utilizing a bolted connection to directly connect W beams to the column face. The results revealed failure mechanisms associated with bolt shank fracture and yielding of the CFT column face, mechanisms not expected in moment frame systems. Similarly, the authors of [24] experimentally studied a connection by bolting the beam directly to the column face using through bolts. In this case, failure mechanisms were associated with the local buckling of the beam flanges and web, along with fractures in the weld connecting the end plate. Nevertheless, these studies did not evaluate the effect of the width-to-thickness ratio on connection behavior, and they did not consider the use of box sections for CFT formation.

Other proposals have focused on the use of internal diaphragms in the column for the development of moment connections, as demonstrated in the experimental research in [25]. This connection involves welding the beam flanges directly to the column face with continuity plates in the inner area of the column. The findings indicated a failure mechanism related to the fracture of the beam flange and part of the web at a drift level of 4%. Subsequently, the authors of [26] extended the experiments from the aforementioned study to CFT columns with a large steel thickness, incorporating an exterior diaphragm. This connection was configured using bolts in the beam web and welding the flange directly to the stiffeners. The results demonstrated that the use of the exterior diaphragm contributes to a more uniform stress distribution, improving cyclic performance. Nevertheless, the connection was assessed for use in intermediate moment frames, revealing a significant degradation of connection stiffness, limiting its fully restrained behavior. Additionally, these studies did not evaluate CFT columns formed by box sections for different width-to-thickness ratios. A similar experimental study [27] analyzed CFT column connections with an exterior diaphragm but considered a reduced section (RBS) in the beam and utilized high-ductility HSS columns. The results showed that the initially developed damage occurred in the RBS zone of the beam due to yielding and local buckling. However, the hysteretic behavior of this welded connection exhibited a loss of strength with increasing load cycles. Furthermore, the study did not address box sections or the influence of the width-to-thickness ratio on the connection response, including the bidirectional effect. Likewise, the authors of [28] numerically studied the ConXL® [19] connection, emphasizing ductility differences with and without concrete infill. In addition, the limitations of the study include the consideration of only one beam subjected to cyclic loading, neglecting the bidirectional effect.

Another proposal, i.e., analyzing CFT columns composed of circular profiles with beams directly welded to the column face, was analyzed numerically by the authors of [29]. The results showed that as the beam flange thickness increased, ductility decreased, and the damage was concentrated in the column. However, this proposal was limited to welded connections with circular profiles without analyzing the cyclic behavior and the bidirectional effect on the column. In this context, the authors of [30] conducted a review of experimental studies on moment connections in CFT columns with both internal and external diaphragms, highlighting failure mechanisms in welded connections and suggesting the benefits of external diaphragms. Likewise, an extensive review and database of beam-column moment connections for the seismic design of special composite moment frames (C-SMFs) were developed by the authors of [31]. The review recommended Double-split T (DST) connections for C-SMFs with rectangular CFT (Concrete-Filled Tube) columns based on the analyzed behavior. External diaphragm connections were recommended for C-SMFs with rectangular and circular CFT columns. Furthermore, bolted connections were recommended to reduce the damage mechanism common in welded connections. Nevertheless, these reviews did not emphasize experiments considering the bidirectional effect of multiple connected beams, and most of the evaluated connections were welded.

On the other hand, a numerical and experimental study developed by the authors of [32] analyzed a welded connection with an exterior diaphragm and a CFT column using HSS profiles, with perforations in the beam flanges. These perforations aimed to simulate a reduction in the beam section based on the concept of the RBS connection. The results showed that the use of perforations in the beams reduced stress concentrations in the exterior diaphragm or column connection zone, increasing connection ductility. However, for rotations higher than 0.04 rad, fractures occurred in the joint area between the beam flange and the stiffener, revealing the disadvantage of welded connection failure mechanisms. Moreover, this investigation did not consider the bidirectional effect on the column and the use of box-section profiles for CFT fabrication. Subsequently, a numerical study based on the proposal in [32] was conducted by the authors of [33] to evaluate different patterns of perforations in the beam and their influence on the connection's performance. The results indicated that the number and arrangement of bolts strongly influenced the cyclic performance, reducing ductility and stress concentration at the connection node.

Furthermore, the authors of [34] studied a welded connection composed of an interior diaphragm passing through the beam-column joint and connected to the beams. The width-to-thickness ratios of the two columns were studied to assess their effect on connection behavior. The results showed that, for all cases, the damage was primarily concentrated in the beams, with no damage to the columns or other components. Additionally, it was observed that the concrete infill of the profile increased the strength of the connection by up to 20% and stiffness by 4.9 times. However, the study did not consider the bidirectional effect of multiple beams connected in orthogonal planes on the column or the use of box-section profiles. Another study [35] compared the reduced beam section (RBS) connection to the proposed perforated beam connection (DFC) using numerical analysis. The study varied parameters such as the number of holes in the plates and beam size, demonstrating that the DFC had higher strength and ductility than the RBS in CFT columns. However, due to the welded nature of the proposed connection, stress concentrations and plastic deformations were observed in the column face. Similarly, the authors of [36] presented a proposal for a connection between CFT columns and open-web beams by bolting the beam directly to the column face. An experimental study with six specimens was conducted, considering the use of tube bolts and EHB bolts to connect the end plate to the column. The results showed that the tube bolt-enhanced connection exhibited improved resistance, ductility, and energy dissipation compared to the EHB-bolted connection. However, a loss of strength was evident after loading cycles with rotations exceeding 0.02 radians, along with damage to the column area and inner concrete. Nevertheless, this research did not account for the bidirectional effect of beams connected orthogonally to the column and did not consider the influence of varying the slenderness ratio of the column on connection performance. On the other hand, the authors of [37] conducted an experimental evaluation of a similar connection proposal, considering the use of through bolts in the CFT column node and comparing V-cut and RBS sections in the beam. The results indicated that using the V-cut section in the beam improved the connection's cyclic performance in terms of resistance, ductility, and reduction of torsional deformation compared to the RBS section. Moreover, plastic deformations in the column node area were also shown, along with a loss of up to 30% of elastic stiffness. Thus, due to limited energy dissipation capacity, this connection did not meet the prequalification criteria established in [14]. Additionally, this study did not consider the bidirectional effect on the connection or the use of box sections to manufacture the CFT column.

An experimental assessment of the seismic behavior of bolted pass-through beam connections in CFT columns was conducted in [38] for implementation in prefabricated modular moment frames. Six specimens, divided into three test groups, were subjected to cyclic loading and constant axial loading, with one group using RBS sections in the beam. The results showed significant damage concentration in the panel and column node zone for all specimens. Furthermore, this research did not address the use of box sections for manufacturing CFT columns and the bidirectional effect resulting from beams connected in both planes of the column. Recently, in a numerical and experimental study [39], the researchers proposed the use of a welded connection with an interior diaphragm composed of plates and an inner steel tube within the CFT column. The experimental results showed a failure mechanism characterized by fractures in the web of the beam at a rotation of 0.04 radians. Moreover, both studies demonstrated that the use of internal diaphragms with steel tubes increased the bending capacity and energy dissipation of the connection by 20% and 22%, respectively. However, the last proposal did not meet the prequalification requirements for moment connections [14], and the bidirectional effect and axial load acting concurrently in the column were not considered in these studies. In addition, the authors of [40] conducted an experimental study evaluating the experimental behavior of a new tube-in-tube (TIT) welded connection using an internal diaphragm in the column composed of tubes and plates. The results showed that the proposed connection achieved the specified bending strength and rotation capacity according to [14], demonstrating that the use of this diaphragm improved the behavior of the connection. However, plastic deformations in the panel area were evident in this study, and the effects of bidirectional actions on the column were not considered, which could lead to greater damage in the column area.

The abovementioned investigations demonstrate that the proposed solutions for CFT columns have primarily relied on welded and bolted connections directly to the column face. The use of internal or external diaphragms in the connection has been implemented to improve performance. However, despite this implementation, mechanisms of brittle failure are still evident due to the use of welded connections between diaphragms and beams. Moreover, few studies have addressed the variation in the width-to-thickness or slenderness ratio of the column, and none have considered the bidirectional effect on the column. Other studies on tubular profiles considering the bidirectional effect on the column have been conducted numerically and experimentally. Initially, the authors of [41] experimentally demonstrated that, for HSS column connections with welded external diaphragms, the bidirectional effect reduces the connection's capacity by 20% of its uniaxial capacity. Additionally, a higher concentration of stresses and plastic deformations in the column node zone were evident. Similarly, the authors of [42,43] analyzed connections similar to a proposal by the authors of [44] for tubular columns with external diaphragms and bolted end plates. These studies investigated different beam configurations, axial load, and L/d ratio, finding that the bidirectional effect and the level of axial load reduce the connection's strength, stiffness, and dissipation capacity. However, these studies also showed that this proposal for the connection concentrates plastic deformations entirely in the beam without brittle failure mechanisms or damage to the column. Subsequently, the authors of [45] continued research in this area, evaluating the influence of the width-to-thickness ratio of the box-section column on the cyclic performance of the connection through a numerical study calibrated from the experimental results of [46]. The results showed that even with the use of moderately ductile profiles in the column, the connection's performance was not affected, and no damage to the column or brittle failure mechanisms was evident. They also demonstrated that the use of slender-width-to-thickness ratio columns reduces the stiffness and the strength of the connection, concentrating plastic deformations in the panel and external diaphragms' zone. Nevertheless, these studies did not encompass the use of CFT columns and the influence of the width-to-thickness ratio of CFT columns on reducing damage in the column. Moreover, although in special composite moment frames (SCWBs), the first plastic hinge appears at the base of the column, this can be reinforced by transferring the problem to the area of the node where plastic hinges should not be expected in columns.

In this research, the influence of the local slenderness effect of concrete-filled tubular columns (CFTs) on the seismic performance of moment connections with bidirectional strength is studied. A design methodology is proposed to consider the local slenderness of the column in the connection design. This parametric study is used to assess the impact of the slenderness ratio in connections for CFT columns manufactured with box sections. Numerical models using the finite element method (FEM) through the ANSYS software v2022 [47] were developed. The size of sections in beams and columns, joint configurations of connected beams for different CFT columns, and different width-to-thickness ratios, such as high-ductility, non-compact, and slender ratios, are considered as input variables in this parametric study to evaluate the cyclic performance of the I-beam–CFTT moment connection. A total of 70 high-fidelity FEM models were analyzed and calibrated using experimental data. To obtain a reliable response of numerical models, cyclic loads and joint configurations of bidirectional moment connections were employed. The response of the steel moment connections was evaluated in terms of the moment rotation, stiffness, failure mechanism, evaluation of concrete core damage, and rupture index.

This paper starts by presenting the fundamental considerations of the steel moment connections using CFTs subjected to cyclic load and the performance of such structures in the most destructive earthquakes, along with their theoretical background. The experimental data used in the calibration of the numerical models are described, followed by a section with the considerations of high-fidelity models. Finally, the results, a discussion of the results, and the conclusions are presented.

2. Design Considerations of Joint Configurations

In this study, high-ductility slenderness ratios are assumed for beams according to those established in AISC-341 [14]. Thus, the maximum flexural capacity assumed is $Mpr = CprRyFyZx$ (Cpr is the hardening factor according to AISC-341 [14], Ry is the overstrength factor, Fy is the material yield stress, and Zx is the beam plastic modulus). Likewise, the assumed maximum shear is $Vu = Vg + 2Mpr/Lh$ (Vg is the shear due to gravity forces, and Lh is the distance between plastic hinges). Furthermore, to assess the performance of the connection for IMF and SMF use, the span-to-depth ratio was restricted to 12, complying with the lower limit of 7 imposed by Chapter 6.3 in [14]. Moreover, to comply with stability bracing according to [14], lateral bracing was granted to the beams to limit the laterally unbraced length of the beam and reach the yielding given by AISC-360 [48].

Furthermore, CFT built-up columns were designed by varying the width-to-thickness ratio from the high ductility limit, established in Table D.1.1 in [14], to non-compact and slender columns, according to Table B.4.1.a in [48]. Following the moment connection proposed by the authors of [44], the end plates were linked to the column by horizontal stiffeners. Moreover, a length–depth column ratio of 10.32 was considered for all models to simulate a typical story height of a domestic Chilean construction. In addition, the axial load for the column design was equivalent to 0.25 Py, considering Py as the yield axial load ($Py = FyAg$). Likewise, as a continuation of the research previously conducted by the author, the design of the end plate was executed based on the analytical expression derived from the experimental and numerical research conducted by the authors of [44]. The shear capacity of the column was designed in accordance with the requirements of the Seismic Provisions [14], considering two webs of the CFT column for its verification and the filling concrete strength according to [49].

On the other hand, the external diaphragms were designed to support the maximum factored force on the beam flange (Ffu) considering a yielding failure mechanism, as suggested in [50]. However, a geometrical limitation was added to avoid the joint stress concentration, which led to a column-yielding mechanism. In this sense, the minimum length of the external diaphragm is considered as the half of column width (0.5 hc) and a support length equivalent to the width of the end plate (bp), as shown in Figure 1a with the model configurations of this study (Figure 1b–d).

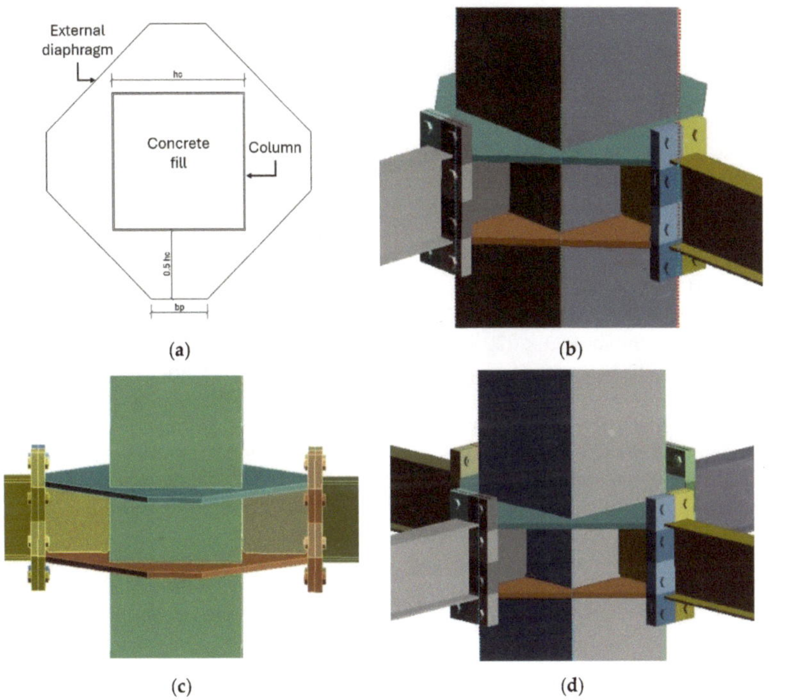

Figure 1. (**a**) Geometrical limitation of external diaphragms, (**b**) 2BC joint configuration, (**c**) 2BI joint configuration, and (**d**) 4B joint configuration.

The strong column–weak beam ratio (SCWB) was computed with the consideration of the moment of attached beams and the axial load of the column, similar to that proposed by the authors of [50]. On this basis, 50% of the flexural strength moment will be in the orthogonal direction to the analysis plane beams (My) according to [14]. The axial load of 0.25 Py was considered as design load because is the typical axial load level of Chilean SMFs. However, the specification of [14] for the estimation of column strength for this verification considers the expected plastic moment of the column as a high-ductility member. In this sense, the proposed design methodology established a new way to compute the maximum available column strength considering the slenderness effect, axial load, and bidirectional effect, modifying the equation specified in [14] to Equation (1).

$$Mpc = Mn * \left[\frac{Fyc - \left(\frac{P}{A} + \frac{My}{Zy} \right)}{Fyc} \right] \quad (1)$$

where Mn is the available column strength estimated according to Chapter H in [48] for composite members. The current standard only considers the use of high-ductility members as columns, whose capacity is only referred to the plastic moment Mp. In a sense, this modification allows to consider the slender effect and the other terms of Equation (1), such as the reduction of strength by the bidirectional effect. Thus, to consider the bidirectional strength effect, the number of connected beams is varied considering three configurations: two beams in the plane (2BI), two beams connected in the corner (2BC), and four beams (4B) in total. The incorporation of this factors allows to obtain designs using elements with a greater width-to-thickness ratio, with the purpose of achieving a smaller amount of steel material in the final design. This goal leads to a reduction of 28% of water footprint according to emission calculations given in [51].

In addition, a variation in the cross-section of the beam is introduced to look for a robust trend in the behavior and evaluate patterns and possible dimensional limits of the connection, such as the connections established in [18]. Finally, the dimensions obtained from these assumptions were used for the other column width-to-thickness ratio examined in this study, with the objective of studying the influence of slenderness in the SCWB ratios considering the modification of the column thickness. An outline of the design results is given in Table 1, considering the column as a slender element and then increasing their thickness to comply with the width-to-thickness ratio of non-compact and high-ductility members. Moreover, in previous research [43] where a high-ductility box column was employed for the same connection configuration, the dimension of the column obtained for an IPE200 beam was 220 × 220 × 14. In this sense, in comparison to the dimension obtained in the current study, a reduction of 41.60% of the steel weight per unit length of the column is obtained.

All previous designs were evaluated for the three connected beam configurations previously mentioned, as well as two different column slenderness conditions for two-beam models and three column slenderness conditions for four-beam models. The purpose of these configurations is to assess the influence of the bidirectional effect and the column slenderness simultaneously, which has not been evaluated in previous studies (e.g., [39,40]). Table 2 summarizes the number of models per configuration evaluated, and a diagram of these configurations is given in Figure 1.

Table 2 shows the simulation matrix with the 70 models considered in the numerical study. In this sense, three types of joint configurations are considered: 2BC, a column with two beams connected as a corner joint; 2BI, a column with two beams connected as an interior joint; and 4BC, a column with four beams connected as an interior joint. The beams 2BC and 2BI were modeled to study the effect of bidirectional load and the shear transfer mechanism to the web panel zone, while the 4BC models were used to compare the cyclic performance at maximum applicable loads and to identify the failure mechanisms of such connections.

In all cases, the beams were connected with bolts through the end plate to the columns. The columns have vertical and horizontal stiffeners that stiffen the node and improve the transfer of the cut from the beams to the column. In all cases, the load was applied at the top end of the column. However, a more detailed description of the numerical models is given in the following sections.

Table 1. Results of the designed models.

Beam		Column				Span between Columns (m)	Bolt Diameter (in)	End Plate Thickness (mm)	Horizontal Stiffener Thickness (mm)	Vertical Stiffener Thickness Vertical (mm)	Ratio WPZS (Panel Zone)
Section	Steel	Section Slender	Section Non-Compact	Section High-Ductility	Steel						
IPE-200	A-36	425 × 425 × 4	425 × 425 × 6	425 × 425 × 14	A-36	2.4	3/4″	20	16	6	0.93
IPE-220	A-36	450 × 450 × 5	450 × 450 × 6	450 × 450 × 14	A-36	2.64	7/8″	22	16	8	0.90
IPE-240	A-36	500 × 500 × 5	500 × 500 × 6	500 × 500 × 16	A-36	2.88	7/8″	25	18	8	0.89
IPE-270	A-36	525 × 525 × 6	525 × 525 × 8	525 × 525 × 16	A-36	3.24	1″	25	20	8	0.88
IPE-300	A-36	575 × 575 × 6	575 × 575 × 8	575 × 575 × 18	A-36	3.6	1″	28	22	8	0.87
IPE-330	A-36	625 × 625 × 6	625 × 625 × 8	625 × 625 × 18	A-36	3.96	1 1/8″	28	25	8	0.87
IPE-360	A-36	700 × 700 × 8	700 × 700 × 10	700 × 700 × 22	A-36	4.32	1 1/4″	38	28	10	0.74
IPE-400	A-36	750 × 750 × 8	750 × 750 × 10	750 × 750 × 22	A-36	4.8	1 1/4″	38	30	10	0.75
IPE-450	A-36	825 × 825 × 8	825 × 825 × 10	825 × 825 × 25	A-36	5.4	1 3/8″	40	38	10	0.72
IPE-500	A-36	875 × 875 × 10	875 × 875 × 12	875 × 875 × 28	A-36	6	1 1/2″	50	38	10	0.68

Table 2. Summary of the studied models.

Beam/Configuration	Column Slender CFT			Column Non-Compact CFT			Column High-Ductility CFT	
	2BC	2BI	4B	2BC	2BI	4B	4B	Total
IPE-200	1	1	1	1	1	1	1	7
IPE-220	1	1	1	1	1	1	1	7
IPE-240	1	1	1	1	1	1	1	7
IPE-270	1	1	1	1	1	1	1	7
IPE-300	1	1	1	1	1	1	1	7
IPE-330	1	1	1	1	1	1	1	7
IPE-360	1	1	1	1	1	1	1	7
IPE-400	1	1	1	1	1	1	1	7
IPE-450	1	1	1	1	1	1	1	7
IPE-500	1	1	1	1	1	1	1	7
Total	**10**	**10**	**10**	**10**	**10**	**10**	**10**	**70**

3. Numerical Models

This section describes the numerical modeling of a steel I-beam and CFT column moment connection employing the finite element method (FEM) within the ANSYS software [47] framework. The aim is to simulate the structural behavior, considering the interaction between the beam, column, end plate, and bolts under cyclic load conditions. The numerical model provides a high-fidelity analysis, incorporating detailed considerations such as accurate mesh, realistic boundary conditions, and the application of bolt pretension. This computational approach serves to simulate the laboratory condition in a virtual environment, according to the established in prequalification code [14]. The results obtained from the numerical simulations contribute to the validation of the proposed design equation and offer insights into the structural integrity and reliability of the proposed moment connection. The model includes the material fundamental law, loading conditions, boundary settings, mesh size, and FE type to solve the nonlinear interaction between the elements when a cyclic load in a nonlinear range is analyzed. Nonlinearities are considered using sub-steps for each load stage employing the Incremental Newton–Raphson procedure [47,52].

3.1. Material Fundamental Law

In this research, the material data employed for steel were obtained from the material experimental test performed in [46]. In this sense, the ASTM A36 material is considered for horizontal and vertical stiffeners, end plates, beams, and columns, whereas ASTM-A325 is assumed for the bolts. For the modeling of these materials, the von Mises yield criterion, along with the nonlinear bilinear kinematic fundamental law, was employed; therefore, the yield surface is constant in magnitude. In addition, the properties to be introduced in the model were converted to true strains and true stress to consider the actual change in the length of a material as it deforms. On the other hand, we used concrete material with a compressive strength of 28 MPa and the *Microplane* model based on the Drucker Prager model calibration as several studies [53–55] suggest for confined concrete with 41 MPa as the confined compressive strength, 46 MPa as the biaxial compressive strength, 3.8 MPa as the uniaxial tensile strength, 1 as the damage parameter (Rt), 40,000 MPa/m as the hardening parameter (D), and −35 MPa as the compression cap location as the model parameters. This consideration allows the capture of the brittle failure and fracture mechanism in the concrete. Table 3 shows the material properties of the steel materials employed.

Table 3. Mechanical characteristics of materials derived from tests [46], transformed into true values.

Material	*Fy* (MPa)	ε_y (mm/mm)	*Fu* (MPa)	ε_u (mm/mm)
ASTM A36	320.33	0.0014017	441.67	0.1036
ASTM A-325	627.45	0.00393	934.32	0.1601

3.2. Loading and Boundary Conditions

In the FEM models, the following considerations were assumed: the length of the column was considered as the distance between the points of zero moment for each case (with the points of zero moment assumed to be at half of the height of the columns). The beam length was determined by the clear span of the beam, denoted as (L). Furthermore, the length of $L/2$ was used to simulate the distance to the point of zero moment for seismic loads. Welds were excluded from the model because inelastic behavior was not anticipated in these components. The diameters of the holes were assumed to be standard in accordance with the requirements specified in Section J3 in [48]. These assumptions were validated and utilized by the authors of [10,44]. An optimized end plate thickness based on an analytical proposal using yield line theory was derived from previous research [44].

The base connection of the column was modeled as being partially restrained, allowing rotation only in the direction of the load application. The beams were modeled as being simply supported, restricted only in the "Z" direction to simulate the load cell boundary condition. Additionally, lateral bracing was applied to the two beams to ensure their stability in accordance with [14,48]. For all configurations, the loading protocol was applied at the top of the column to ensure similar loading in all beams as suggested in [43]. Nevertheless, to prevent out-of-plane instability, lateral support was provided at the top of the column. Figure 2 illustrates the boundary conditions employed in the four-beam (4B) model configuration. The loading protocol involved applying horizontal displacements at the top of the column following the prequalification protocol outlined in [14], with models 2BC and 4B subjected to a 45° load to induce the bidirectional effect. Furthermore, bolt pretension was applied to 70% of the nominal tensile strength of the bolt shank surface, according to [48].

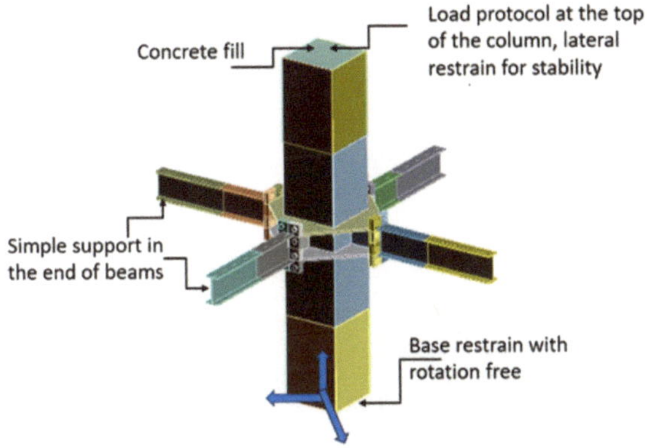

Figure 2. Boundary and loading conditions applied in the numerical model.

3.3. Finite Element Type, Mesh Size, and Contacts

In this study, SOLID185 was used in steel elements as a prism-shaped body, which has a good competence to creep, plasticity, large deflection, and stress stiffening. On the other hand, the finite element "CPT215" was used to fill concrete due to the incorporation of the "microplane" model, which included elasticity, tension rigidity, large deflection, and deformation capabilities to simulate the gradual failure of concrete. Additionally, this element was used because the mesh sensitivity of the "CONCRETE/SOLID65" element does not allow tracking the cracking behavior of concrete when it fails, without having to track individual cracks, as demonstrated by previous research [56,57]. The "microplane" model with a newer CPT215 element returned force–displacement results that did not have significant mesh sensitivity and agreed very well with the experimental data. In this sense, a simple model calibration for the element type was developed and is shown in Figure 3, based on the previous experimental research of [58] and numerical results of [59], which employed the SOLID65 element. The results indicate that the local buckling effects are possible to represent and that the numerical results of the use of the CPT215 element have a good agreement with previous experimental and numerical data.

Various mesh sizes and shapes were employed to establish a rational mesh that delivers precise results, incurring excessive computational costs. For instance, a fine mesh was used to capture the inelastic behavior in plastic hinge zones. Furthermore, the author conducted a sensitivity analysis to validate the model's accuracy and its convergence with mesh refinement, varying the mesh element size from 40 mm to 5 mm. Despite these variations, the structural response remained consistent. Consequently, a maximum element size of 8 mm was chosen for areas with expected nonlinear behavior and 25 mm for areas with linear behavior, based on the conducted tests. Figure 4 illustrates a schematic view of the mesh employed for a four-beam model.

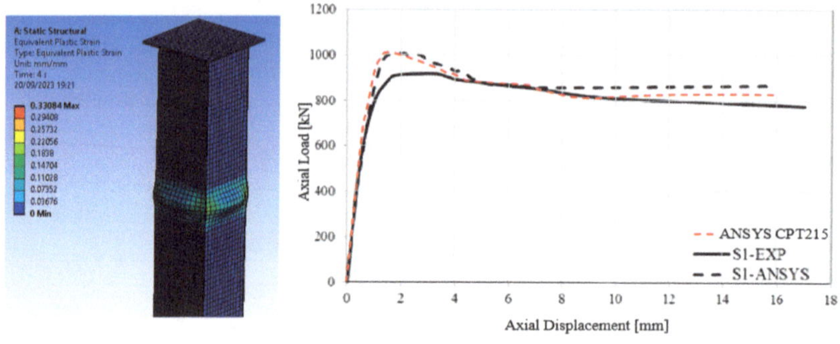

(**a**) (**b**)

Figure 3. Calibration of the concrete model from experimental data. (**a**) Failure mechanism of the calibrated model. (**b**) Calibration curve with experimental data from [58].

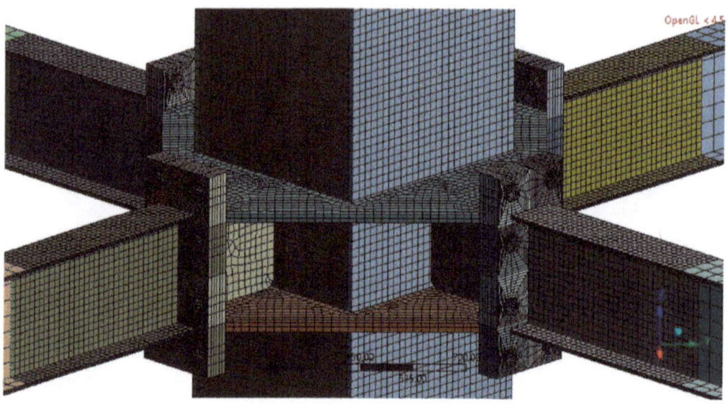

Figure 4. Mesh used in numerical models.

Moreover, to evidence possible p-δ effects and to evade the sub-estimation of the cyclic performance r of the connection during the inelastic deformation, geometrical imperfections were considered longitudinally and transversally in the model [60]. These longitudinal and transversal imperfections were considered according to the maximum limits established in [61,62], which define the imperfections that can appear due to eccentricity boundaries, erection processes, or lack of precision during erection or fabrication. Therefore, in this research, these imperfections were considered using a previous nonlinear case to deform the elements up to the tolerance limits previously mentioned and then running the cyclic load from the previous deformed shape using the "UPGEOM" command. This command updates the geometry of the finite element model according to the displacement results of the previous analysis and creates a revised geometry at the deformed configuration. Thus, when UPGEOM is employed, the current finite element model is overwritten by finite element information from the results file.

On the other hand, in this research, a critical aspect of the numerical modeling process involves replicating welding conditions within the computational framework. To achieve this, a "Bonded" contact, also referred to as a completely restrained contact, was employed. This contact type mimics the welded joints, ensuring that the meshed elements are seamlessly connected to simulate the structural continuity anticipated in a welded moment connection for concrete-filled tubular (CFT) members. In addition, by utilizing the "Bonded" contact, the model accounts for the rigid connection established during the welding process. Likewise, the interplay between different elements within the numerical model was addressed by implementing a "frictional" contact mechanism. This contact type allows for the simulation of potential relative displacements between connected elements, a crucial consideration in the cyclic behavior of moment connection. In particular, a friction coefficient of 0.3 was applied to represent the friction zones between steel elements as end plates, bolts, or nuts, while a coefficient of 0.6 was assigned to concrete–steel interfaces. These values were chosen based on prior research findings (references [10,24]), ensuring a realistic portrayal of material interactions. Therefore, incorporating these coefficients, the simulation not only captures the theoretical aspects of CFT moment connections but also draws on practical insights derived from prior experimental studies. Furthermore, Table 4 shows the summary of the contacts employed in the model.

Table 4. Type of contact interaction by element in the numerical model.

N°	Interaction Contact Zone	Type of Contact
1	End Plate–End Plate	Frictional $\mu = 0.3$
2	Bolt shank–End Plate	Frictional $\mu = 0.3$
3	Bolt shank–Bolt head	Bonded
4	Bolt head–End Plate	Frictional $\mu = 0.3$
5	Bolt shank–Nut	Bonded
6	Nut–End Plate	Frictional $\mu = 0.3$
7	End Plate–Horizontal Diaphragm	Bonded
8	End Plate–Vertical Stiffener	Bonded
9	Horizontal Diaphragm–Vertical Stiffener	Bonded
10	Horizontal Diaphragm–Column	Bonded
11	End Plate–Beam	Bonded
12	Beam–Beam	Bonded
13	Column Plates–Column Plates	Bonded
14	Column Plates–Concrete Core	Frictional $\mu = 0.6$

4. Calibration and Validation of the Numerical Model

Some studies, such as [63,64], included parametric analyses and the examination of different connection configurations through experimental tests, not necessarily from the same research. Numerical models were calibrated based on these tests to accurately replicate the studied physical phenomena. For instance, the authors of [65] performed a numerical study on circular tubular connections, calibrated from a specimen tested in [66], while the authors of [67] analyzed baseplates with chairs using FE numerical models calibrated from a test with ductile bolts developed in [68]. In this sense, it is possible to calibrate numerical models from different experimental test configurations, with the failure mechanism and physical phenomenon that control the model and the test being similar. Consequently, in this study, a previous numerical model was adjusted using experimental data from [46], which used the same connection configuration as that proposed in the present study but using a built-up column without concrete-filled. The results of the calibration, showing an acceptable match between the models' normalized hysteresis curve and the experimental specimens, are illustrated in Figure 5.

Moreover, two additional four-beam (4B) numerical models were developed, considering a plastic model of column like steel column without concrete filled and steel column with concrete filled, denoted as adt-01 and adt-02, respectively. The purpose of these models is to ensure that the calibrated model can reproduce the local buckling and failure mechanism when the connection is not designed with the proposed methodology. In this case, only the 4B model was used because it is the most severe condition for the node. Figure 6 shows the plastic strain distribution for the models adt-01 (Figure 6a) and adt-02 (Figure 6b), and the high-ductility and slender models, designed according to Section 2, are shown in Figure 6c,d, respectively. These results illustrate how the local buckling effects are strongly denoted in the additional models, even in the presence of concrete filling. Furthermore, the failure mechanism manifested in the additional models were local buckling and plastic hinges in the column for filled and non-filled concrete models, respectively. The results obtained validated that the numerical model is capable of reproducing fragile failure mechanisms. On the other hand, the other models designed using a high-ductility column or a slender column with the proposed design methodology exhibit a failure mechanism characterized by plastic hinges in beams, as is desired for connections for SMFs. In addition, no brittle failure or column damage was evident in these configurations.

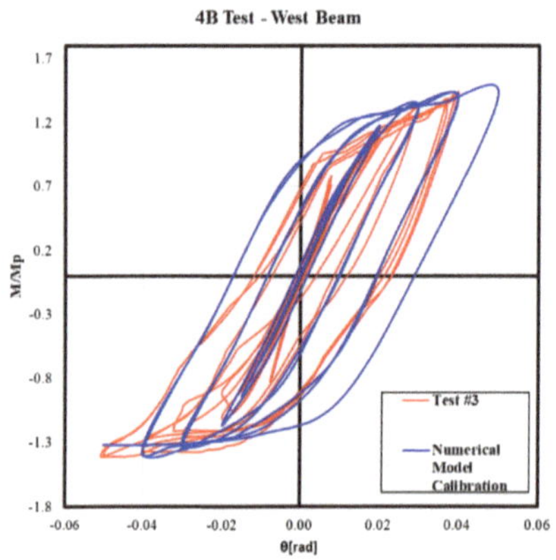

Figure 5. Calibration of the numerical model with Test #3 [46].

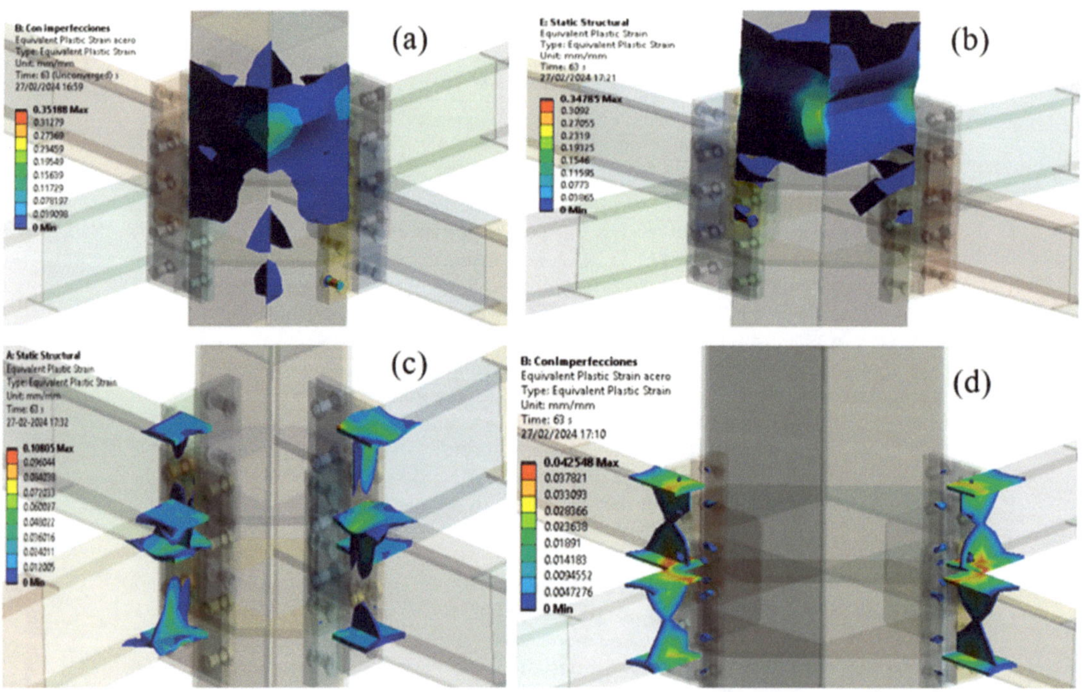

Figure 6. Plastic strain distribution for the (**a**) adt-01, (**b**) adt-02, (**c**) high-ductility CFT, and (**d**) slender CFT proposed.

5. Results of the Parametric Assessment of Cyclic Behavior

The research aims to propose a moment connection with a similar configuration as that in [44], incorporating a CFT slender column for multiple beams. Considering this objective, the results of the bidirectional effect are evaluated from models 2BC and 2BI because they have a similar number of connected beams. In addition, these same models are compared for non-compact and slender columns in terms of moment rotation, secant stiffness, and dissipated energy. To further gauge the effect of the slenderness of the column, the following section shows the results of 4B models for slender, non-compact, and high-ductility CFT columns in terms of moment rotation, secant stiffness, and dissipated energy, supplemented by equivalent stress and plastic strain distribution. Additionally, the assessment of the concrete core damage section shows the equivalent plastic strain and normal stress of the concrete core, while the rupture index assessment section presents this parameter for different configurations concerning slender columns and connected beams. Notably, for all cases, the slender and non-compact columns were designed with a new methodology that avoids brittle failure mechanisms and predominantly showcases plastic strain concentration in the beams.

5.1. Bidirectional Effect

Comparing the performance of the bidirectional effect models 2BC and 2BI for non-compact and slender columns in terms of moment rotation, secant stiffness, and dissipated energy reveals differences in their behavior under cyclic loading. This comparison aims to elucidate how these bidirectional effect models react to cyclic loading, shedding light on their practicality and reliability in structural analysis. Figures 7 and A1 show this comparison in terms of cyclic response, where it is observed that as the beam section increases, the effect of the pinching on the hysteretic curve is evident. Furthermore, for all cases, the maximum resistance of the connection (Mp) is reached and is never lower than the prequalification limits proposed by the authors of [14], i.e., 0.8 Mp at a drift of 0.04 rad, which was reached by all the models. On the other hand, the models that consider the bidirectional effect (2BC) present a reduction in their stiffness compared to the other models (2BI), which have lower resistance. Nevertheless, a greater pinching is denoted in the response of 2BI models, which allows us to infer that the bidirectional effect provides stability to the system. The reduction in strength and stiffness obtained in the 2BC joint models compared to the 2BI joint models is due to the lower resistance reached by the rhombus-shaped column when two beams are connected in the corner joint configuration. In this sense, the shear transfer mechanism in the web panel of the 2BI joint (interior joints) configuration causes the two walls of the web panel to be aligned with the direction of the load; therefore, there is greater effectiveness in resisting the applied cyclic load. Finally, an insignificant difference in the hysteretic behavior is observed comparing the models for slender and non-compact columns. However, for all cases, less throttling and loss of resistance capacity is evidenced for non-compact models.

Moreover, the secant stiffness is a measure of the strength of the connection to deformation at a specific point along its loading history. The stiffness itself is a measure of how much force is required to produce a unit of displacement; this parameter helps to identify the extent of stiffness degradation in the connection as it undergoes cyclic loading. A decreasing secant stiffness over cycles indicates that the connection experiences damage and yielding. In this sense, Figures 8 and A2 illustrate the normalized secant stiffness degradation of the previously analyzed models. This analysis of the secant stiffness allows for more quantitative evidence of the loss of stiffness shown in the hysteresis curves. The results show that the secant stiffness degrades more than its elastic value (0.01 rad) as the beam section increases. On the other hand, for most of the simulated models, greater degradation was evident for the 2BI models than for the 2BC models where the bidirectional effect is present. This is because the degradation of connection resistance was greater in the models with a bidirectional effect. On the other hand, this effect was evident in both slender and non-compact models, with a difference of approximately 5%. However, this gap between slender and non-compact models decreases as the beam section increases, with the behavior being practically the same. In addition, the variation in the secant stiffness allows us to identify a critical point (2% interstory drift ratio) of the cyclic response, from which, the resistance degradation is initiated for all the models studied. However, this phenomenon is more pronounced as the beam increases in size.

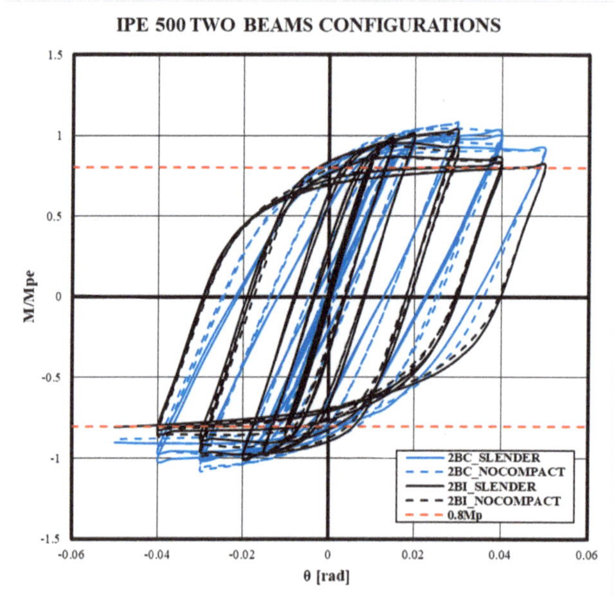

Figure 7. Comparison of hysteresis curves considering the local slenderness of the column for 2B models.

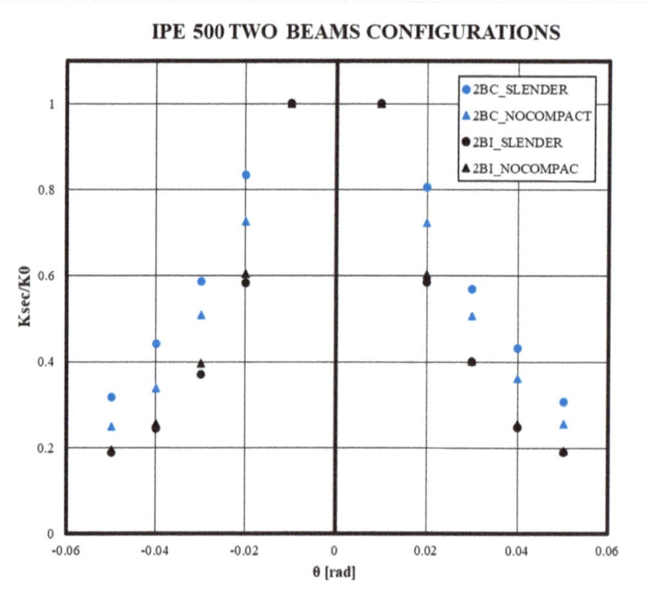

Figure 8. Comparison of secant stiffness considering the local slenderness of the column for 2B models.

Finally, the accumulated dissipated energy of the system was calculated, considering all sources of inelasticity. This energy was determined by calculating the area enclosed in each cycle of the force–displacement curve of the system using the equation proposed in [69]. The results of the analysis shown in Figures 9 and A3 indicate that the bidirectional effect reduces the system's dissipated energy by up to 41% of the original value, similar to the findings demonstrated in previous research [63]. Additionally, models with non-compact columns exhibited a smaller reduction in dissipated energy, although in most cases, the difference with slender models was less than 2%. It could be considered that the bidirectional effect increases with slenderness. However, for models exceeding IPE400, an increase in dissipated energy is evident in slender 2BI models due to the concentration of stresses and deformations in the panel zone.

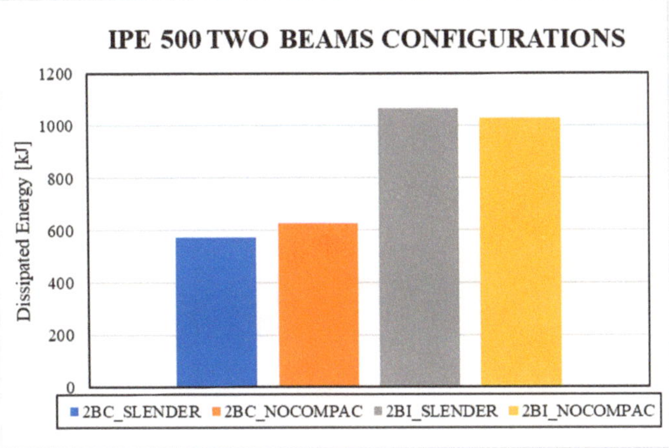

Figure 9. Comparison of dissipated energy considering the local slenderness of the column for 2B models.

Finally, the cyclic behavior of the connection in comparison to other proposals [70,71] is similar in terms of moment rotation and expected failure mechanisms when the bidirectional effect is not considered. However, for the opposite case, with a bidirectional effect, the performance of other models [41,46] with high-ductility columns is only slightly higher in terms of strength and stiffness. Therefore, this similarity obtained in the behavior between the models in the present study and the other proposals also allowed us to validate that the numerical model can adequately reproduce the physical phenomenon for which it was calibrated.

5.2. *Effect of the Slenderness of the CFT Column*

Analyzing 4B models, this study investigates the influence of column slenderness on slender, non-compact, and highly ductile CFT column. The results are presented in terms of moment rotation, dissipated energy, and equivalent stress/plastic strain. This examination aims to reveal the implications of column slenderness on the functionality and efficacy of moment connection systems, contributing to a better understanding of its effects within CFT moment connection configurations and enhancing seismic design codes and guidelines. In this sense, a comparison of slender, non-compact, and highly ductile CFT columns is presented in Figures 10 and A4 in terms of moment rotation curves of the connection. The results reveal that for all configurations, the prequalification requirements are reached for all models, although the configurations are subjected to four-beam forces simultaneously. The effect of slenderness was not predominant in terms of the hysteretic behavior of the connection; for the first models of smaller beams, the models with slender and non-compact CFT columns presented a very similar behavior. On the other hand, models with high-ductility CFT columns showed greater strength and stiffness compared to models without this width-to-thickness ratio. However, for larger beams over IPE270, the behavior was similar for the three width-to-thickness ratios, which indicates that for the present connection configuration, there is no variation in cyclic behavior because of slenderness. Furthermore, a small loss of resistance and throttling was evident for cycles greater than 0.03rad of drift. Nevertheless, despite this reduction, the connection reached its maximum resistance (*Mp*) for all load cycles.

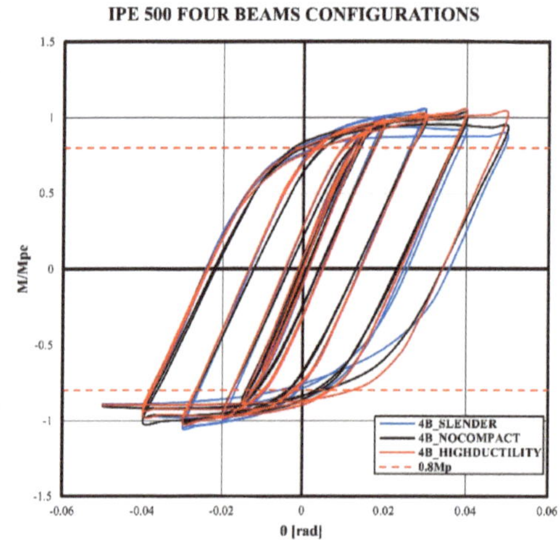

Figure 10. Comparison of hysteresis curves considering the local slenderness of the column for 4B models.

For the failure mechanism characterization, the equivalent plastic strains are shown in Figures 11 and A5. For all models, the failure mechanism was given from the yielding in bending at beams, which is expected for special moment frames. Moreover, plastic strains were not evidenced in the column or the panel zone, avoiding the brittle failure mechanism for all cases. The plastic strains were mainly concentrated in the flanges of the beams, starting near the area where the end plate joins the beam and extending along the flange and, finally, towards the area of the web in yielding in shear. Nevertheless, for models with greater beams, up to IPE270, the inelastic incursion was not uniform in the web zone, which can be an indication of a possible local buckling instead of yielding in shear. On the other hand, models with CFT slender and non-compact columns developed lower levels of plastic strain in beams than the high-ductility models because the greater stiffness of the columns limits deformation in the column area, limiting further deformation in the beams.

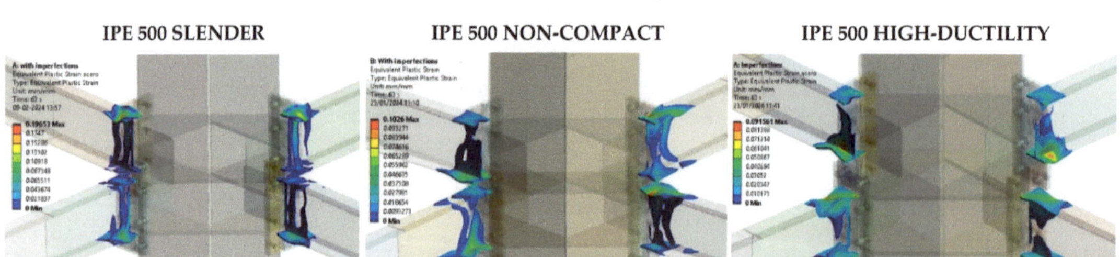

Figure 11. Equivalent plastic strain of 4B models.

In addition, the evolution of plastic deformation through loading cycles showed that inelastic incursion begins after approximately 0.015 rad of drift for beams lower than IPE300. However, for larger beams, the plastic deformations begin around 0.02 rad of drift, mainly concentrated in the contact area between the beam and the end plate. Furthermore, this concentration manifests itself in the upper or lower flange of the beam in an asymmetric manner, evidencing greater plastic deformations and local buckling effects in the flange. Conversely, although plastic deformation offers a perspective on the failure mechanism and energy dissipation of the connection, there may be a stress concentration in certain areas of connection. Therefore, that stress concentration can lead to a modification of the dissipation mechanism under higher loading cycles. In this sense, Figures 12 and A6 show a comparison between equivalent stress distributions to analyze if the slenderness effect could modify the failure mechanism and other possible sources of inelasticity in the connection.

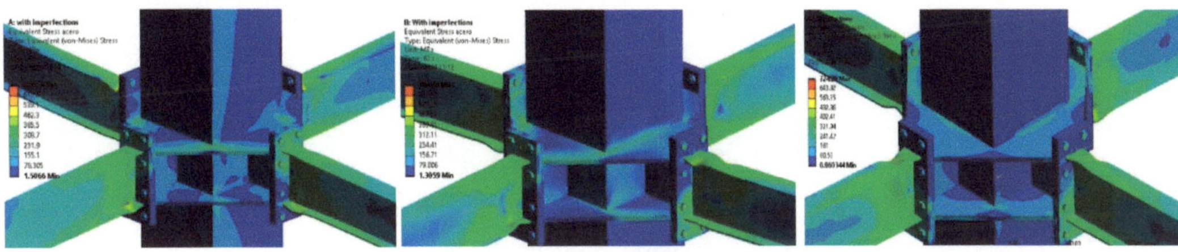

Figure 12. Equivalent von Mises stress for 4B models.

For all cases, a joint stress concentration is shown in the node zone as well as in the flange of beams. This stress concentration in the node zone naturally decreases as the slenderness decreases. In addition, for slender models, the stress concentration along the column denotes that the axial load has a major impact in comparison to the other model where the beam action predominates. Likewise, while the external and vertical stiffeners contribute to avoiding damage in the node zone, stress concentration at the union of the external stiffener with the corners of the column is evidenced. Thus, this suggests that damage in this area could be evident in slender CFT columns. In addition, for all cases, a stress distribution in the external stiffener denotes that it is subjected to flexure. Eventually, plastic deformations could be evidenced in this stiffener, in conjunction with a reduction in the stiffness of the connection.

The seismic capacity assessment of moment connections plays a crucial part in the design and analysis of structures, particularly with CFT columns. In this context, one of the aspects for understanding and improving the behavior of the moment connection is the dissipated energy because provides a measure of the capacity of the connection to resist and absorb seismic energy. Moreover, the dissipated energy of the connections contributes to the mitigation of structural damage and ensures the safety of the structures. In this research, the assessment of the dissipated energy is developed through the integration of system forces over applied displacements according to [69], considering plastic strains and tensions reached during the application of cyclic loads. Figure 13 shows the comparison between the dissipated energy computed for four-beam models (4B), emphasizing that for all cases, the high-ductility models dissipate more energy. Nevertheless, models with slender and non-compact columns have similar dissipated energy, which is slightly greater in non-compact in comparison to slender columns. Furthermore, the difference between high-ductility and other slender and non-compact columns increases with the size of the beam because as the beam increases, the stresses in the external stiffener increase, which contributes to energy dissipation from the connection.

According to the moment obtained for all 4B models, a normal distribution was considered for the slenderness data. The normal distribution is a common choice to model continuous data when it is possible to assume a mean value and a standard deviation, applicable to many scientific and engineering contexts. In addition, Monte Carlo sampling was used to simulate the propagation of uncertainty in the data because it is a useful method when the relationship between variables is not known analytically or when the variables are subject to uncertainty. This approach is based on generating random samples of the input variables (in this case, the moment) and evaluating the objective function (the relationship between slenderness and moment) for each sample. A total of 10,000 samples were obtained for each slenderness value and are shown, in Figure 14, with the 5th and 95th percentiles to establish a confidence interval of the values obtained. The results obtained support the findings of the rotation moment behavior, demonstrating that even for other slenderness values not analyzed in the present study, the moment is always higher than the minimum value proposed in [14].

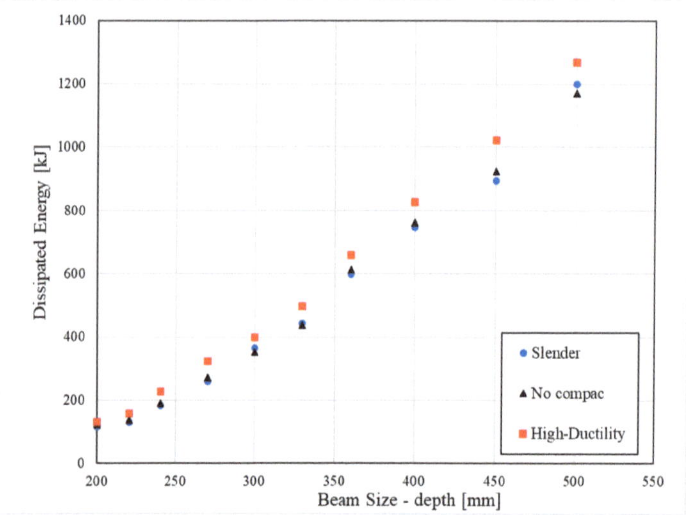

Figure 13. Comparison of the total dissipated energy for 4B models.

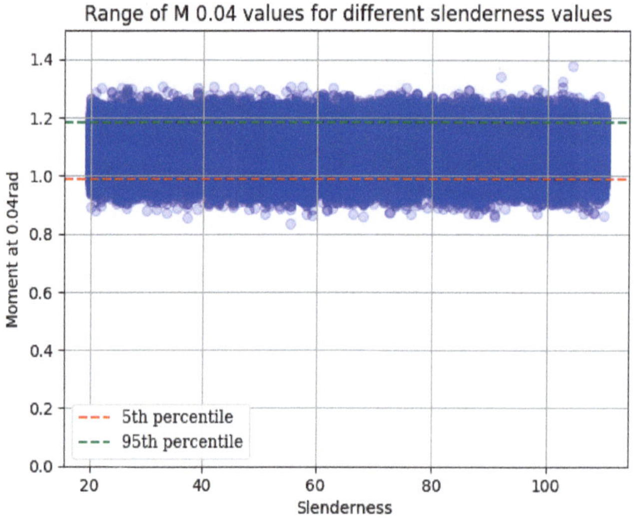

Figure 14. Uncertainty assessment analysis for the moment at 0.04 rad of drift.

5.3. *Assessment of Concrete Core Damage*

The evaluation of concrete core damage is pertinent to evaluate the reliability of concrete after being subjected to cyclic loading. Additionally, identifying and assessing concrete damage allows precautionary measures to be considered to repair or retrofit connections after an earthquake. In this sense, the plastic deformations in the concrete core are compared, as shown in Figures 15 and A7, denoting that there is no damage for high-ductility models, as expected. Moreover, most of the other models exhibit plastic deformation, which is an indication of damage to the concrete. However, as the width-to-thickness ratio is reduced, the plastic deformations in the concrete core decrease, showing that for non-compact sections, there could be damage, but to a lesser extent than a slender section. Conversely, plastic strain concentrations occur near the column's corner, with most of them appearing after a drift of 0.045 rad. Nonetheless, the IPE200SLENDER model shows that plastic strain is concentrated in the external stiffener–flange beam zone, which does not occur in the other models. Furthermore, although the distribution of plastic deformations does not follow a clear trend in its behavior, the damage to the concrete core is concentrated in slender models due to the action of the external stiffener under the column.

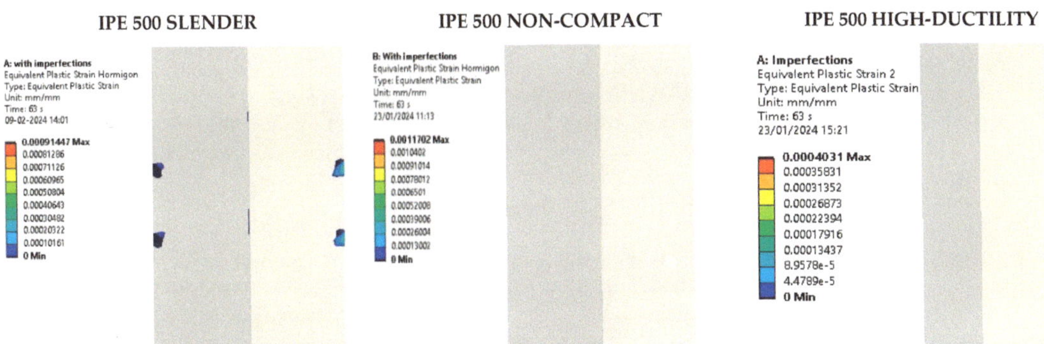

Figure 15. Plastic strains in concrete material for 4B models.

Based on the findings of this study, the concrete damage is also dependent on plastic deformation because when the concrete reaches its tensile strength. In this context, the analysis of the normal stresses in the concrete core is shown in Figures 16 and A8 as a comparison between the different slenderness values analyzed. These values are compared to the ultimate compression strength ($f'c$) of 41 MPa and the ultimate tension strength of 3.8 MPa, estimated according to [49], considering confined concrete. The results indicate that for most scenarios, even in high-ductility models, the tensile strength is exceeded, which means the concrete is cracked. Nevertheless, the stress distribution shows that the concrete is cracked beyond the node zone of external stiffeners, mainly in the corners of the column. On the other hand, the maximum compression stress is evident in the zone of the beam flange subjected to compression by the action of the moment. This shows that the effect of the external stiffener allows these stresses to be dissipated in the node area compared to a scenario where the beams are directly connected to the face of the column.

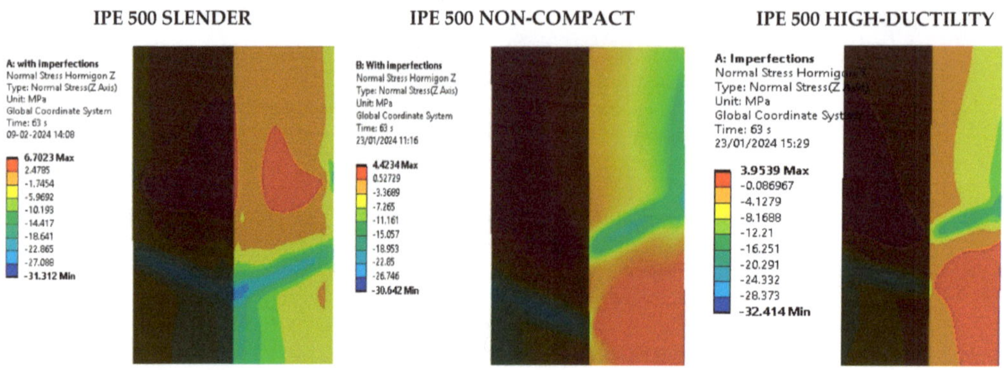

Figure 16. Normal stress (vertical direction) in concrete material for 4B models.

On the other hand, the damage in 2BC models was greater than in 2BI models, independent of the slenderness ratio. Plastic strain was shown in most of the 2BC models around the external stiffeners and beam flange zone, in contrast to 2BI models where they were concentrated only in the beam flange zone and column corners. Moreover, the normal stress distribution of 2BC models shows a major area of the concrete core, which surpasses the rupture modulus and is cracked on the face of the column where the beam is connected. Likewise, based on a similar analysis, the normal stress distribution of 2BI denotes a cracked area on the panel zone without surpassing the normal compression strength in the other zones with compression stresses.

5.4. Rupture Index Assessment

The rupture index provides a measure of the vulnerability of the connection subjected to cyclic loads. In addition, this allows us to identify the critical zones of the connection that have a high potential for rupture and could be weak points in the overall structural design. The identification of these points makes it possible to take preventive action to reinforce or modify the connection as necessary. In this research, given that the explicit fracture has not been simulated, the rupture index is assessed to obtain an additional validation of the results obtained in the previous analysis. This helps to strengthen confidence in the study findings and support the results presented. Thus, the rupture index is computed according to Equation (2) for a drift level of 0.03 rad, as previous studies suggest [72].

$$RI = \frac{PEEQ/\varepsilon_f}{\exp(1.5\frac{p}{q})} \tag{2}$$

where the $PEEQ$ is the equivalent plastic strain, ε_f is the ductile fracture strain, and p and q are equal to the hydrostatic pressure and von Mises stress, respectively. The choice of this level of damage is because damage to the structures begins to occur when rotations are greater than 0.02 rad. Furthermore, according to the Seismic Provisions, it is not expected that there will be repairable damage for rotation levels greater than 0.03 rad. Figure 17 shows the typical rupture index distribution greater than 1 obtained for 2BC, 2BI, and 4B models in the case of beam IPE450 and a slender column, denoting how the rupture index is mainly concentrated in the beams and flanges. However, for 2BI models, there is a similar distribution along the plastic hinge of the beam, with a punctual rupture concentration in the flange of the beam, which is an indication of a large plastic deformation in that zone. In this sense, higher values of the rupture index are due to the large plastic strains that are developed in the flange of the beams, denoting that as the number of beams increases, the rupture index also increases. Furthermore, the stress distribution shows a minimum value of the rupture index in the corners of the column as a product of the stress concentration of this zone. Nevertheless, in general, the results show that the connection behavior is essentially ductile.

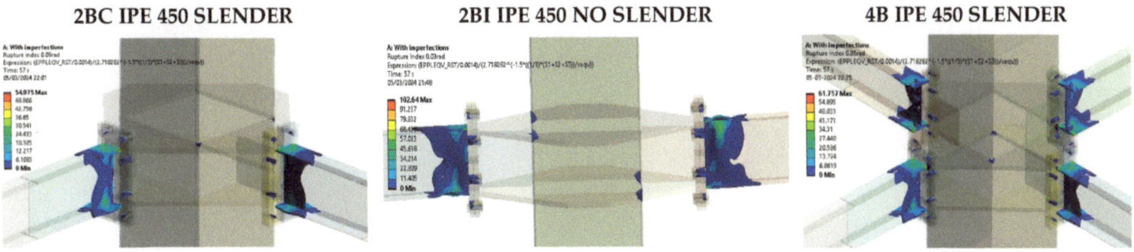

Figure 17. Rupture index for IPE 450 slender models.

On the other hand, although the rupture index distribution and values were similar for all models analyzed, a comparison of the average rupture index values at the flange of the beam for 0.03 rad is shown in Figure 18. This analysis aims to evaluate the impact of the slenderness effect and the moment connection configuration on the connection damage and rupture index. The results show that as the slenderness decreases, the rupture index increases due to the large plastic deformations in the beam by avoiding the deformation of the columns. In addition, the comparison between 2BC and 4B models shows that as the number of beams increases, the rupture index also increases. However, an increase around the double rupture index is evidenced in 2BI models as a product of stress concentration generated in the flange of beams by local buckling. Therefore, these results denote that although slender models present some pinching in comparison to non-compact or high-ductility models, they are less susceptible to fracture phenomena. Moreover, these findings suggest that the fracture potential of the connection reaches the limit of rupture after the material develops its ductility.

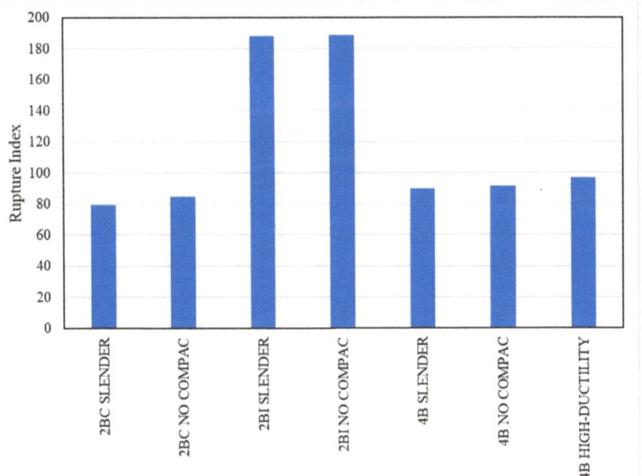

Figure 18. Average rupture index for all models' configurations.

In Table 5, a summary of the key parameters of the studied models is shown.

Table 5. Summary of the key parameters for the studied models.

Beam Size	Joint Configuration	Slenderness Ratio	Max Lateral Load (kN)	$M_{0.04}/M_p$	M_{Max}/M_p	θ_{max} (rad)	Initial Stiffness West Beam (kN·m/rad)	Dissipated Energy (kJ)
IPE-200	2BC	Slender	73,150	1.084	1.109	0.050	6,489,480	67,406
		Non-compact	82,059	1.109	1.109	0.050	8,424,600	59,053
	2BI	Slender	86,817	1.113	1.141	0.050	7,277,120	103,647
		Non-compact	87,396	1.117	1.143	0.050	7,980,777	112,484
	4B	Slender	127,100	1.086	1.111	0.050	6,730,320	131,587
		Non-compact	124,790	1.079	1.104	0.050	6,051,840	115,365
		High-ductility	127,650	1.081	1.106	0.050	6,036,240	122,743
IPE-220	2BC	Slender	69,704	1.073	1.098	0.050	7,239,012	71,640
		Non-compact	70,156	1.075	1.100	0.050	7,401,372	73,453
	2BI	Slender	106,770	1.118	1.133	0.050	9,213,502	132,523
		Non-compact	103,480	1.110	1.138	0.050	9,359,856	135,814
	4B	Slender	147,630	1.069	1.120	0.050	6,710,939	130,802
		Non-compact	148,110	1.072	1.096	0.050	6,943,929	136,508
		High-ductility	151,270	1.078	1.102	0.050	7,785,917	157,828
IPE-240	2BC	Slender	82,090	1.076	1.099	0.050	9,737,136	70,330
		Non-compact	86,549	1.216	1.216	0.040	11,312,208	74,899
	2BI	Slender	120,400	1.100	1.100	0.040	12,247,776	134,874
		Non-compact	120,540	1.100	1.100	0.040	12,377,664	138,819
	4B	Slender	172,430	1.072	1.097	0.050	9,213,840	183,422
		Non-compact	172,520	1.070	1.091	0.050	9,427,824	189,265
		High-ductility	185,470	1.220	1.248	0.050	11,973,168	228,389

Table 5. *Cont.*

Beam Size	Joint Configuration	Slenderness Ratio	Max Lateral Load (kN)	$M_{0.04}/M_p$	M_{Max}/M_p	θ_{max} (rad)	Initial Stiffness West Beam (kN·m/rad)	Dissipated Energy (kJ)
IPE-270	2BC	Slender	104,360	1.062	1.062	0.050	13,470,462	106,096
		Non-compact	105,400	1.064	1.069	0.050	13,974,120	143,525
	2BI	Slender	152,120	1.050	1.055	0.050	16,856,100	242,183
		Non-compact	152,620	1.037	1.057	0.050	17,507,340	251,553
	4B	Slender	217,600	1.059	1.079	0.050	12,725,748	258,978
		Non-compact	219,180	1.061	1.077	0.050	13,335,516	272,294
		High-ductility	229,920	1.184	1.204	0.050	17,126,640	324,394
IPE-300	2BC	Slender	132,130	1.122	1.135	0.050	19,509,660	194,437
		Non-compact	131,730	1.134	1.134	0.050	20,172,600	188,772
	2BI	Slender	184,550	0.939	1.062	0.050	22,271,544	324,591
		Non-compact	183,960	0.947	1.059	0.050	22,813,200	332,266
	4B	Slender	270,040	1.138	1.138	0.050	19,376,010	366,289
		Non-compact	270,490	1.148	1.148	0.050	19,405,549	353,614
		High-ductility	270,160	1.126	1.135	0.050	21,826,800	398,715
IPE-330	2BC	Slender	145,320	1.109	1.112	0.050	25,256,880	235,315
		Non-compact	145,570	1.095	1.107	0.050	25,664,760	238,151
	2BI	Slender	206,300	0.907	1.041	0.050	28,700,100	385,390
		Non-compact	207,370	0.890	1.023	0.050	31,345,380	417,632
	4B	Slender	302,440	1.121	1.121	0.050	24,550,020	444,289
		Non-compact	303,020	1.117	1.117	0.050	24,771,780	436,629
		High-ductility	305,170	1.081	1.110	0.050	27,707,460	497,868

Table 5. *Cont.*

Beam Size	Joint Configuration	Slenderness Ratio	Max Lateral Load (kN)	$M_{0.04}/M_p$	M_{Max}/M_p	θ_{max} (rad)	Initial Stiffness West Beam (kN·m/rad)	Dissipated Energy (kJ)
IPE-360	2BC	Slender	163,260	1.157	1.157	0.050	34,097,040	316,829
		Non-compact	163,330	1.141	1.141	0.050	34,749,360	320,434
	2BI	Slender	234,500	0.931	1.071	0.050	40,690,080	511,352
		Non-compact	234,540	0.920	1.072	0.050	40,992,994	519,173
	4B	Slender	343,290	1.157	1.157	0.050	31,680,720	598,759
		Non-compact	343,550	1.158	1.158	0.050	32,365,440	613,515
		High-ductility	378,330	1.131	1.131	0.050	33,274,800	658,428
IPE-400	2BC	Slender	191,420	1.108	1.108	0.050	43,903,200	390,951
		Non-compact	191,700	1.107	1.107	0.050	44,624,800	396,125
	2BI	Slender	276,380	0.871	1.040	0.050	50,081,720	638,187
		Non-compact	277,300	0.883	1.051	0.050	52,508,800	655,127
	4B	Slender	403,420	1.110	1.110	0.050	41,503,200	746,655
		Non-compact	404,010	1.111	1.111	0.050	42,305,600	761,987
		High-ductility	406,390	1.194	1.194	0.050	50,238,067	826,726
IPE-450	2BC	Slender	227,470	0.979	1.088	0.050	56,189,700	507,024
		Non-compact	218,660	0.981	1.088	0.050	56,891,700	517,100
	2BI	Slender	327,410	0.880	1.027	0.050	64,810,800	827,266
		Non-compact	319,270	0.850	1.008	0.050	68,096,100	519,173
	4B	Slender	476,000	0.997	1.083	0.050	53,592,823	898,216
		Non-compact	462,040	1.027	1.086	0.050	55,668,600	986,388
		High-ductility	480,280	0.976	1.092	0.050	61,230,600	1,071,928

Table 5. *Cont.*

Beam Size	Joint Configuration	Slenderness Ratio	Max Lateral Load (kN)	$M_{0.04}/M_P$	M_{Max}/M_P	θ_{max} (rad)	Initial Stiffness West Beam (kN·m/rad)	Dissipated Energy (kJ)
IPE-500	2BC	Slender	257,850	1.014	1.014	0.050	57,042,000	571,528
		Non-compact	324,950	1.023	1.078	0.050	68,831,840	627,984
	2BI	Slender	388,800	0.847	1.040	0.050	78,791,100	1,066,038
		Non-compact	389,030	0.840	1.063	0.050	80,298,000	1,030,614
	4B	Slender	560,070	0.999	1.052	0.050	57,494,850	1,199,222
		Non-compact	551,830	1.019	1.019	0.050	60,936,000	1,170,716
		High-ductility	556,670	1.040	1.040	0.050	65,319,000	1,269,456

6. Conclusions

In this paper, the influence of the column slenderness ratio on the behavior of the bidirectional moment connection between the I-beam and CFT column was studied numerically through a parametric study. In addition, new considerations in the design of bidirectional moment connection with slender columns are proposed and implemented in the design of the models evaluated in this study. A total of 70 high-fidelity FEM models are developed using ANSYS software calibrated from experimental studies. Furthermore, different width-to-thickness ratios, such as slender, non-compact, and high-ductility ratios, were considered in column members. The main conclusions are described as follows:

- The models exhibited a stable cyclic behavior until 0.03 rad of drift, without the loss of strength and stiffness. After this limit, the strength decreased a 10% of the maximum strength for 0.04 rad of drift and 20% for 0.05 rad of drift. However, the flexural strength of 0.8 Mp was reached for 4% of the drift ratio according to the Seismic Provisions.
- Although the bidirectional effect reduces the stiffness and the strength of moment connections with CFT columns, this phenomenon makes it possible to maintain the stability of the hysteretic behavior of the connection, reducing pinching in comparison to the 2BI models. Therefore, 2BI models may develop out-of-plane instability.
- The slenderness column effect slightly reduces the strength and the dissipated energy of the connection in comparison to columns with a high-ductility ratio. However, the prequalification limitations according to Chapter K-Seismic Provisions are satisfied.
- The concrete damages achieved a fracture in concrete for cyclic loads higher than 3% of the drift. This effect was obtained for all models studied. Furthermore, the maximum compression strength was not reached; therefore, the main use of concrete is to avoid deformation by local buckling.

7. Final Remarks

In the current design, the use of slender columns is not allowed according to the Seismic Provisions; however, the joint configurations designed according to the considerations proposed allow the use of concrete-filled columns with a slender wall of tube, in addition to using bidirectional moment connections, avoiding brittle failure mechanisms or column damage to ensure plastic hinges in the beams. On the other hand, the use of concrete with a strength greater than 25 MPa that meets the criteria in ACI-318 [49] is recommended in order to achieve high levels of confinement and plastic deformation, as shown in the results. In this research, the main use of concrete is to avoid the local buckling of the steel tube.

The proposed seismic design of moment frames in seismic zones should consider the design of the web panel zone shear, the number of connected beams, the axial load level, and the SC-WB requirement. Additionally, the verification of drifts must be performed considering the rigidity of the connection. Thus, by fulfilling these requirements, it is possible to create a design that controls the damage in columns and guarantees the failure mechanism in beams.

Author Contributions: Conceptualization, E.N.; methodology, E.N.; software, R.M.; validation, E.N., R.M.; formal analysis, E.N. and R.M.; investigation, R.M.; resources, E.N.; data curation, E.N. and R.M.; writing—original draft preparation, E.N. and R.M.; writing—review and editing, E.N. and R.M.; visualization, E.N. and R.M.; supervision, E.N.; project administration, E.N.; and funding acquisition, E.N. All authors have read and agreed to the published version of the manuscript.

Funding: This research was funded by the Dirección de postgrado de la Universidad Católica de la Santísima Concepción.

Data Availability Statement: Data are contained within the article.

Conflicts of Interest: The authors declare no conflicts of interest.

Appendix A

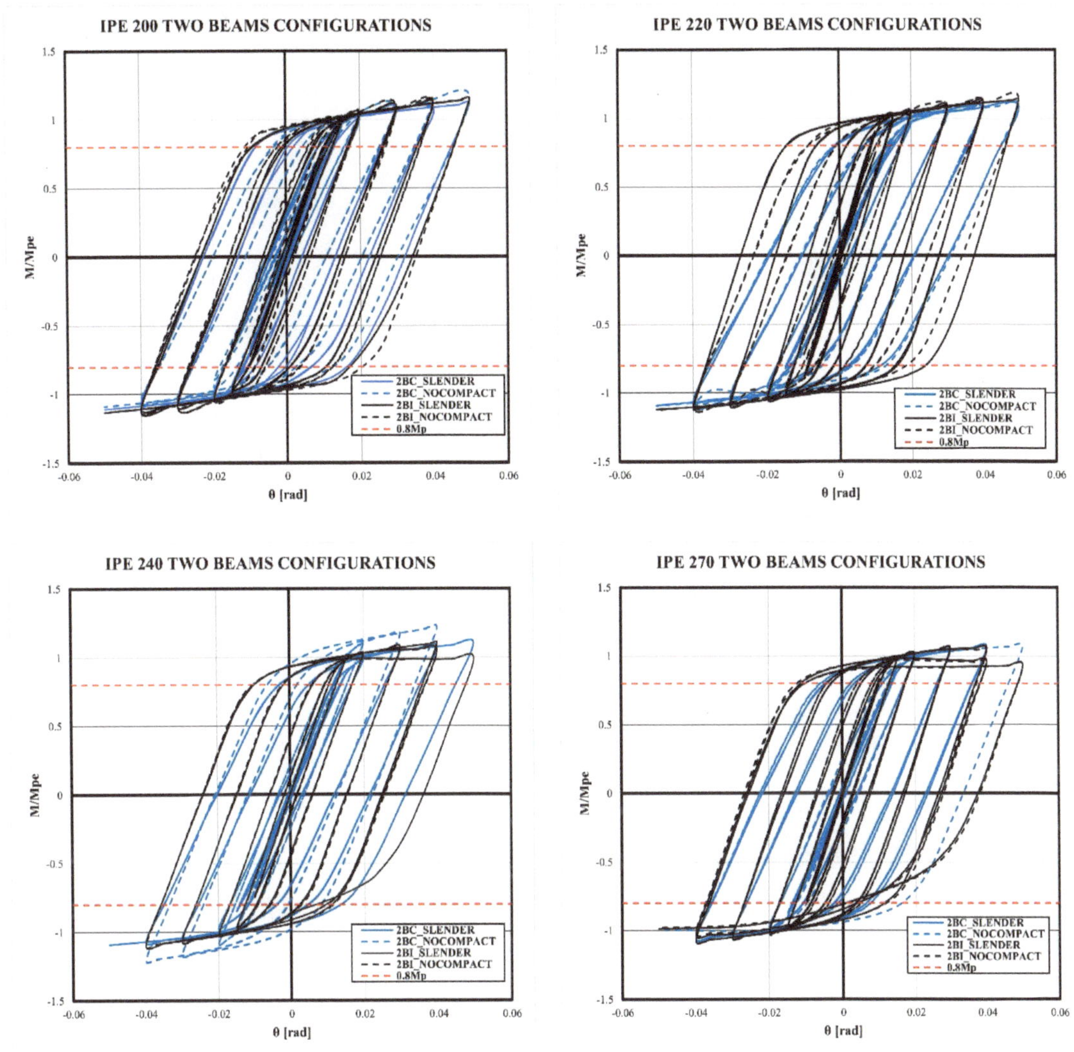

Figure A1. *Cont.*

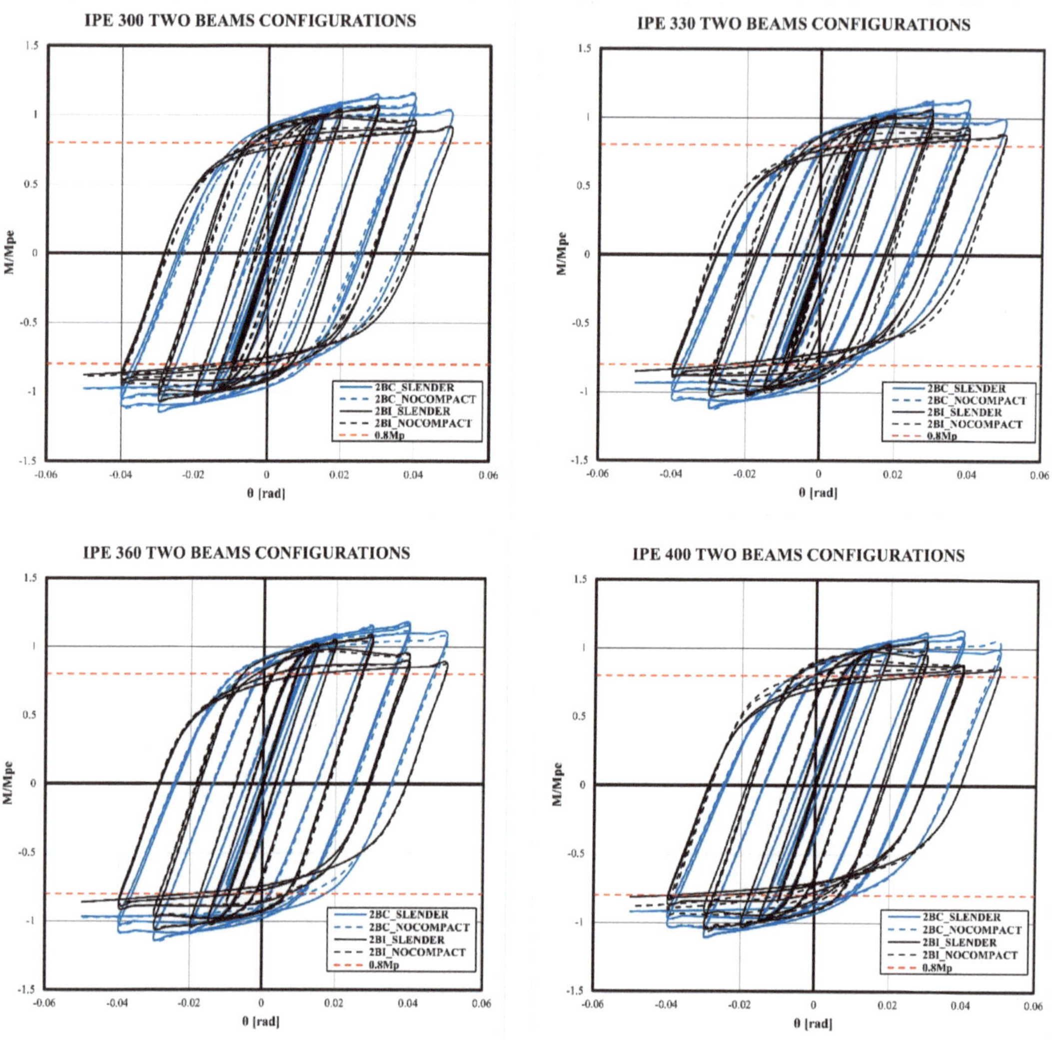

Figure A1. *Cont.*

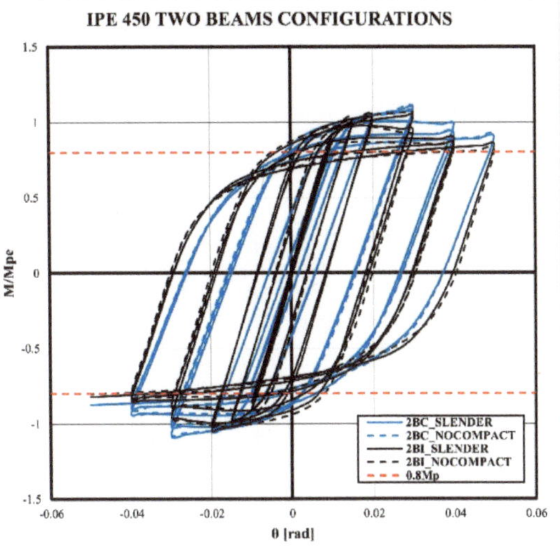

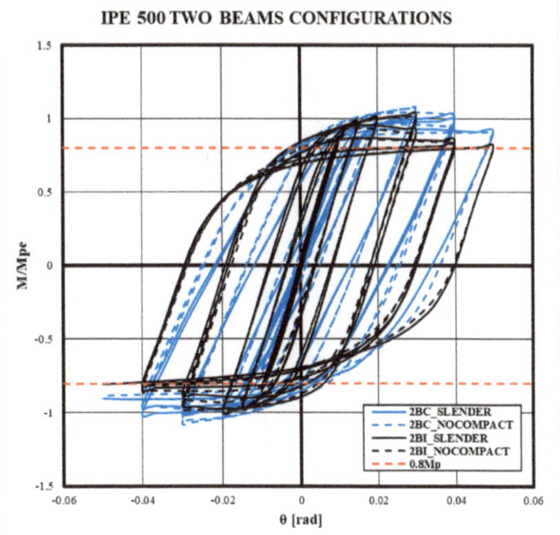

Figure A1. Comparison of hysteresis curves considering the local slenderness of the column for 2B models.

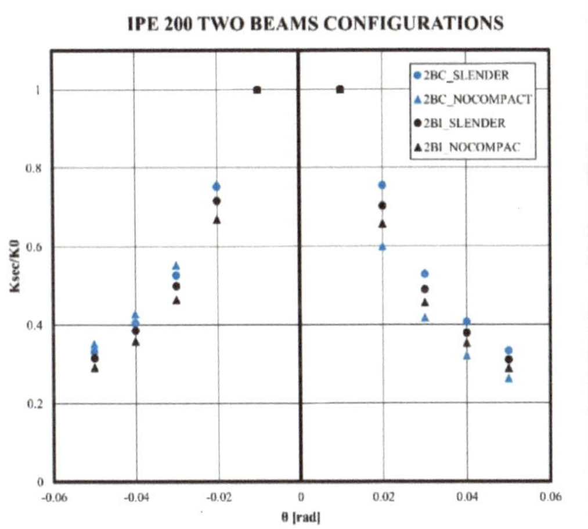

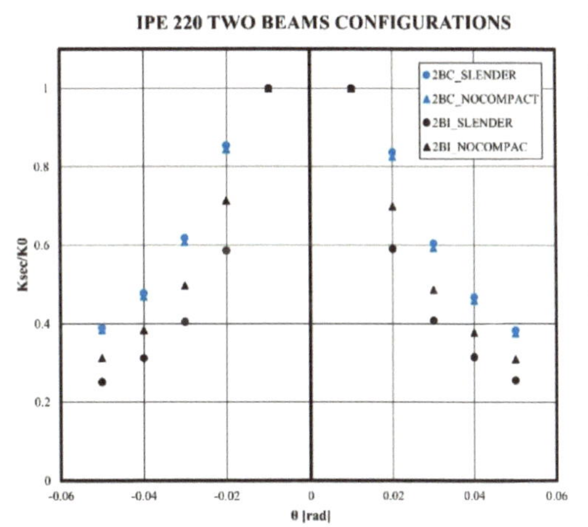

Figure A2. *Cont.*

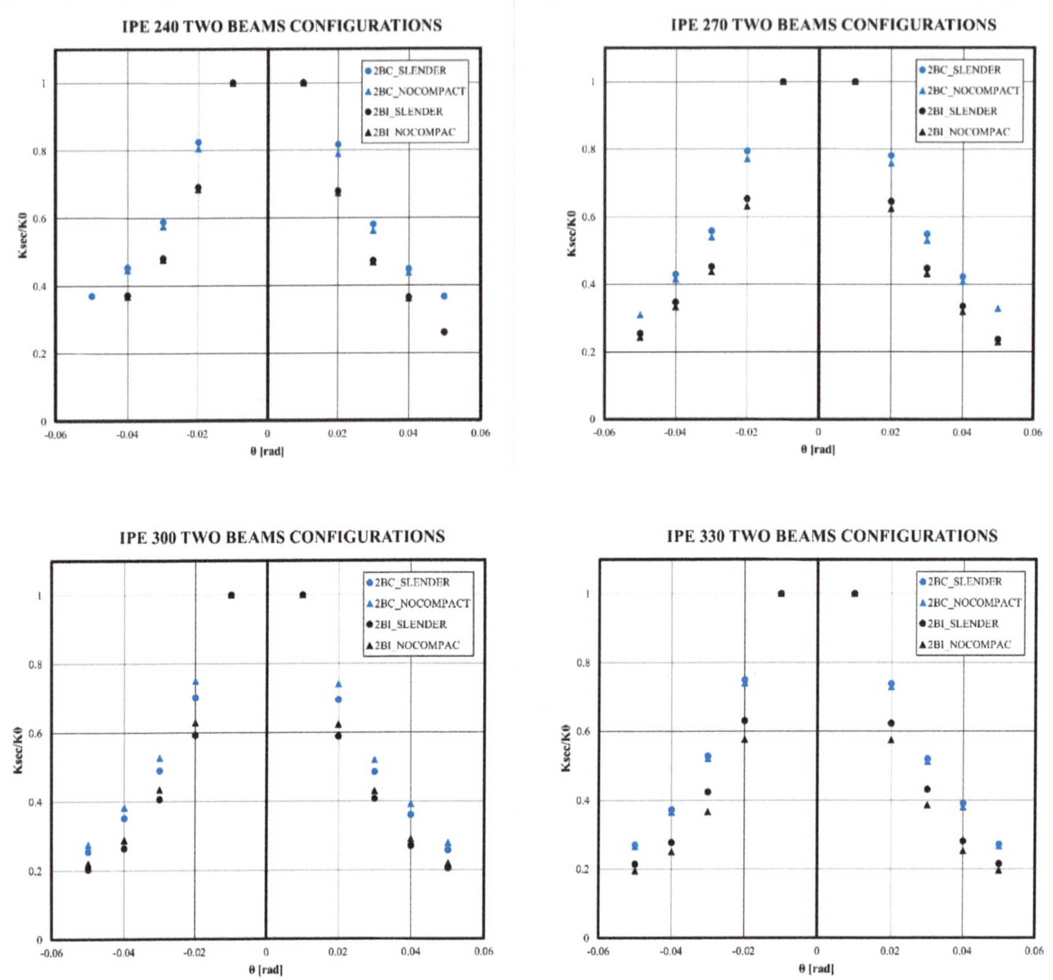

Figure A2. *Cont.*

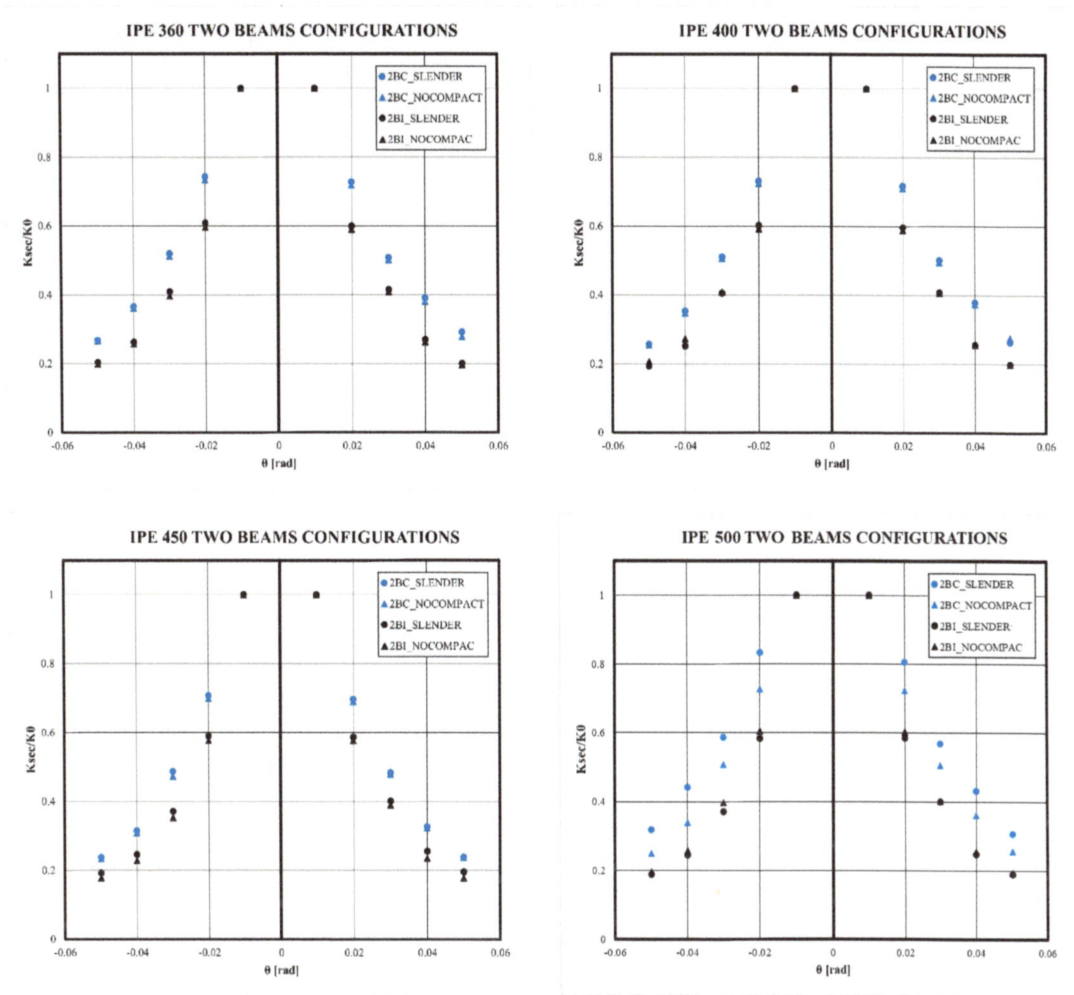

Figure A2. Comparison of secant stiffness considering the local slenderness of the column for 2B models.

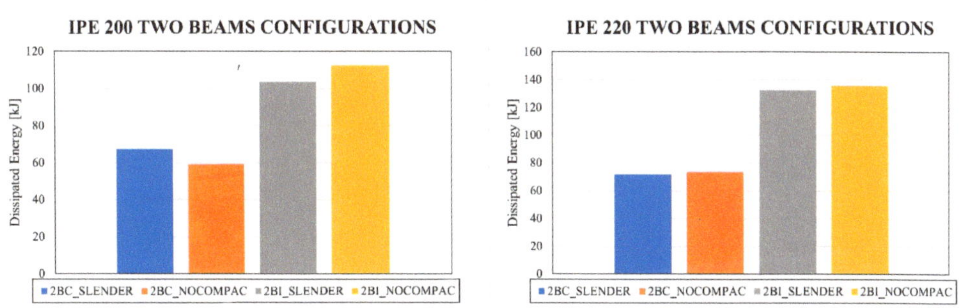

Figure A3. *Cont.*

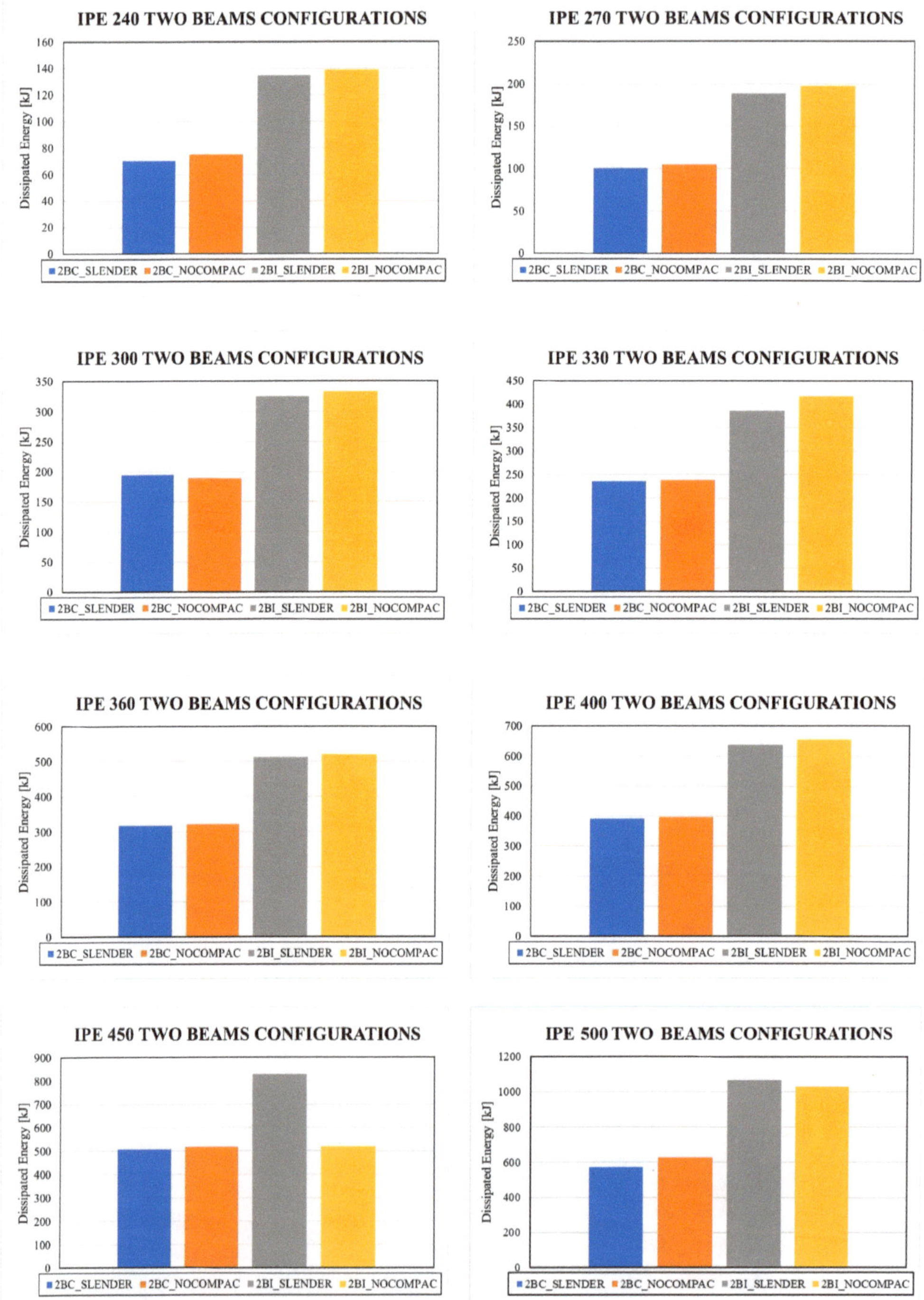

Figure A3. Comparison of dissipated energy considering the local slenderness of the column for 2B models.

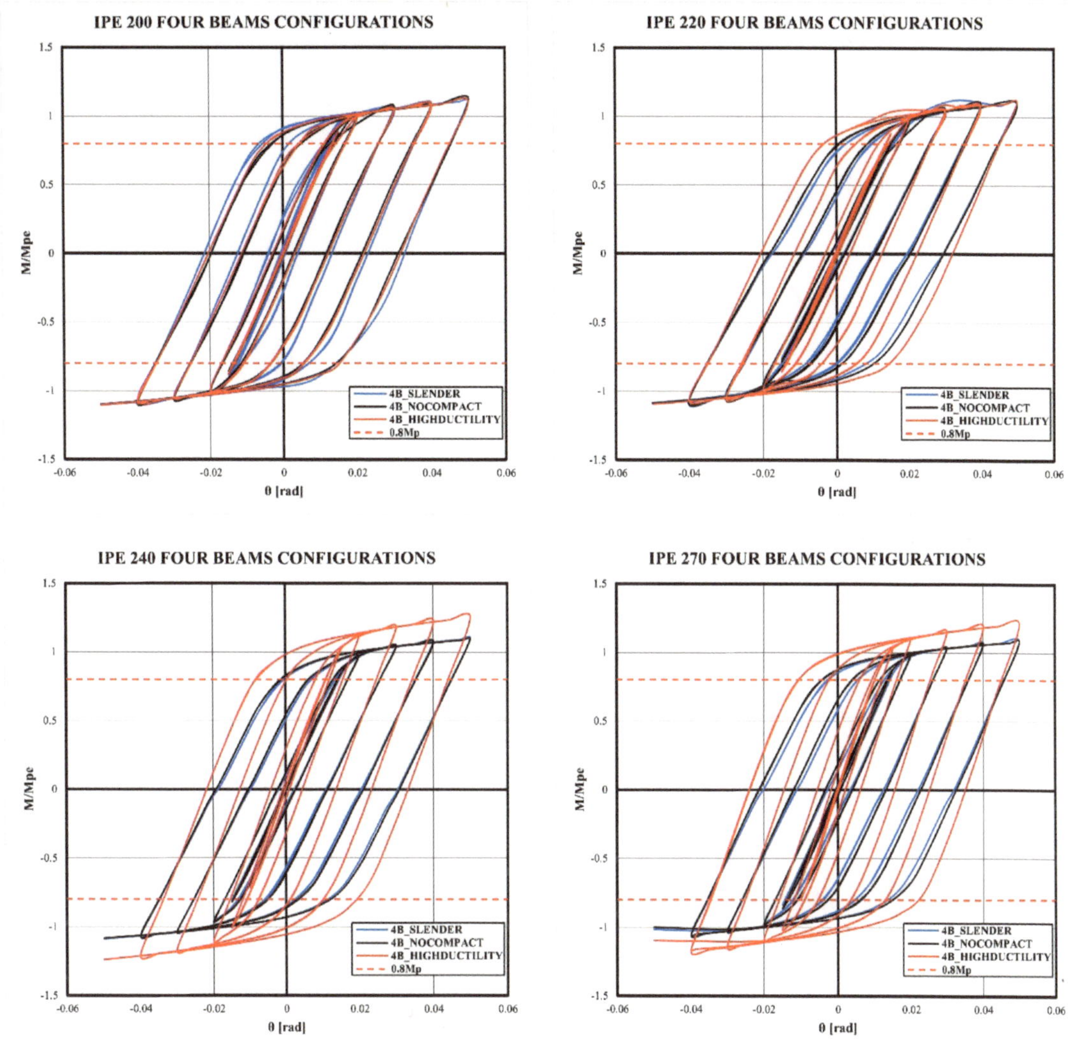

Figure A4. *Cont.*

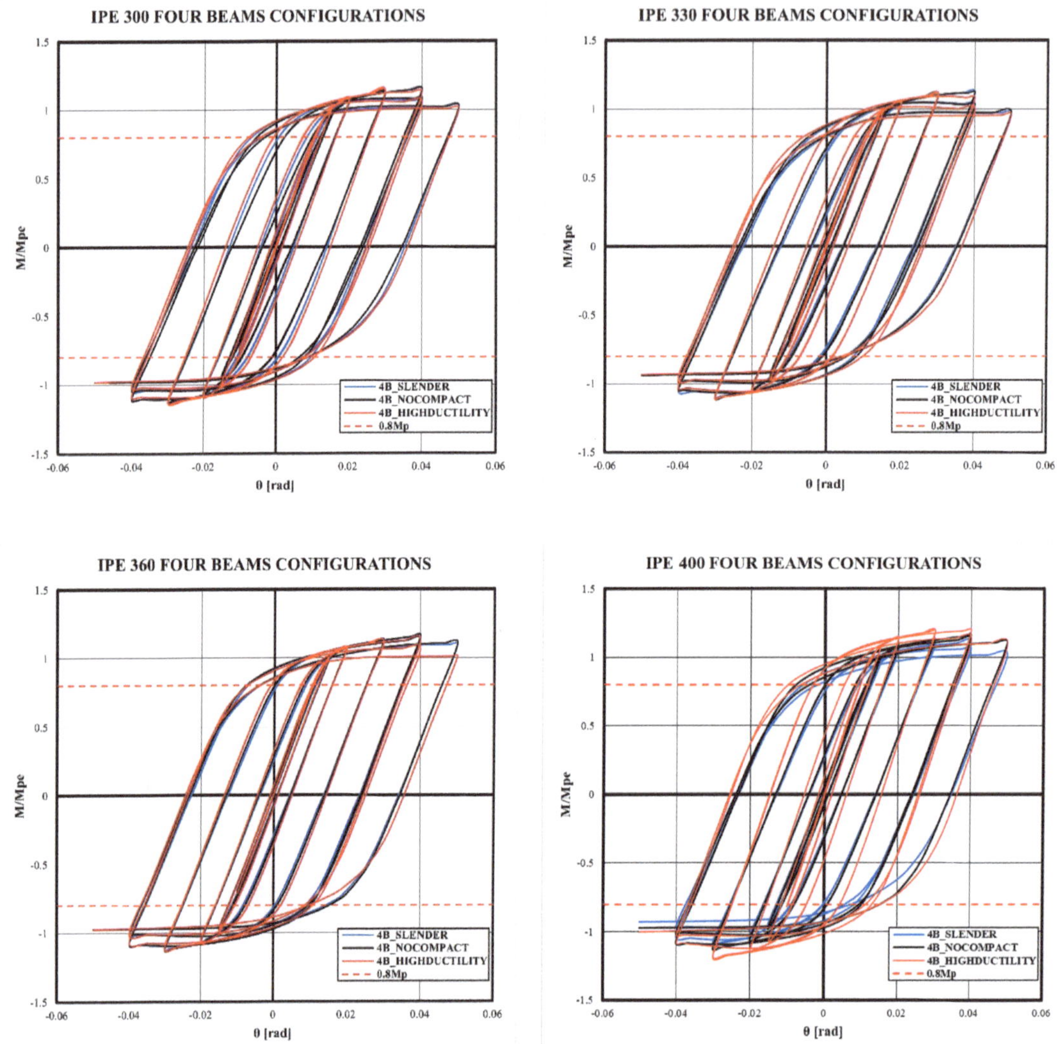

Figure A4. *Cont.*

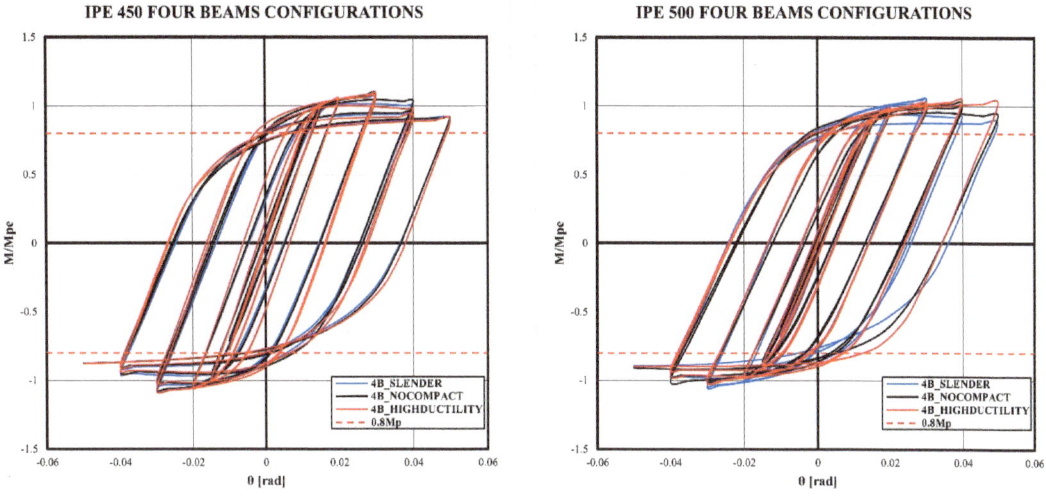

Figure A4. Comparison of hysteresis curves considering the local slenderness of the column for 4B models.

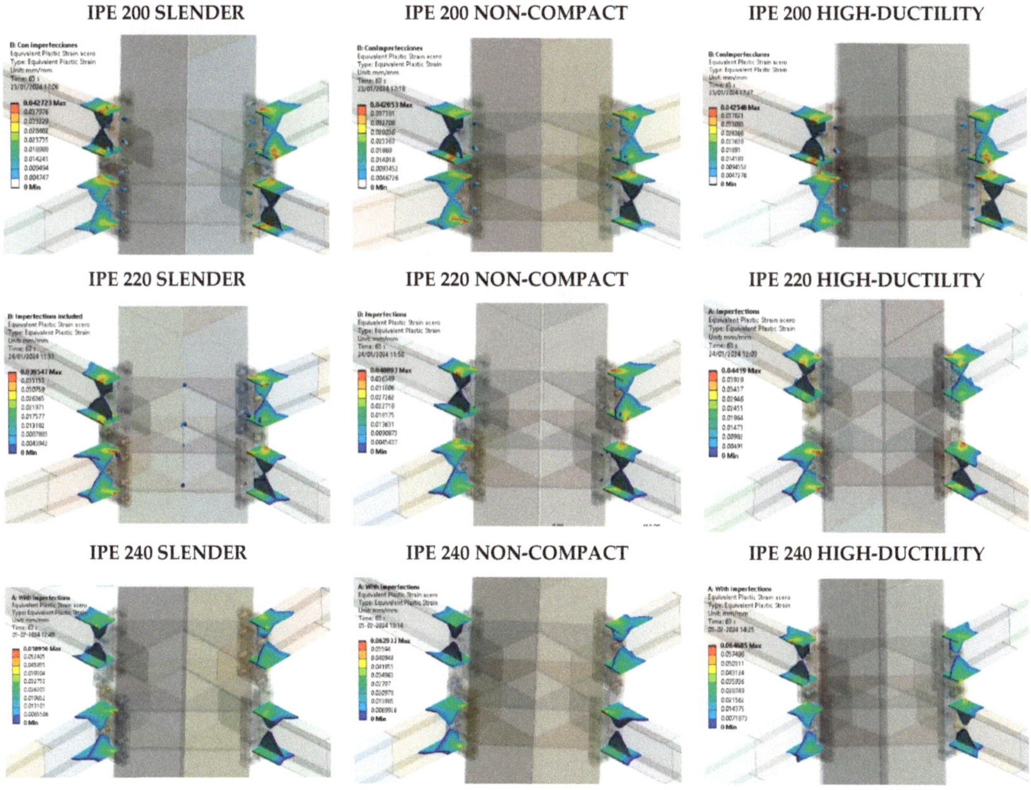

Figure A5. *Cont.*

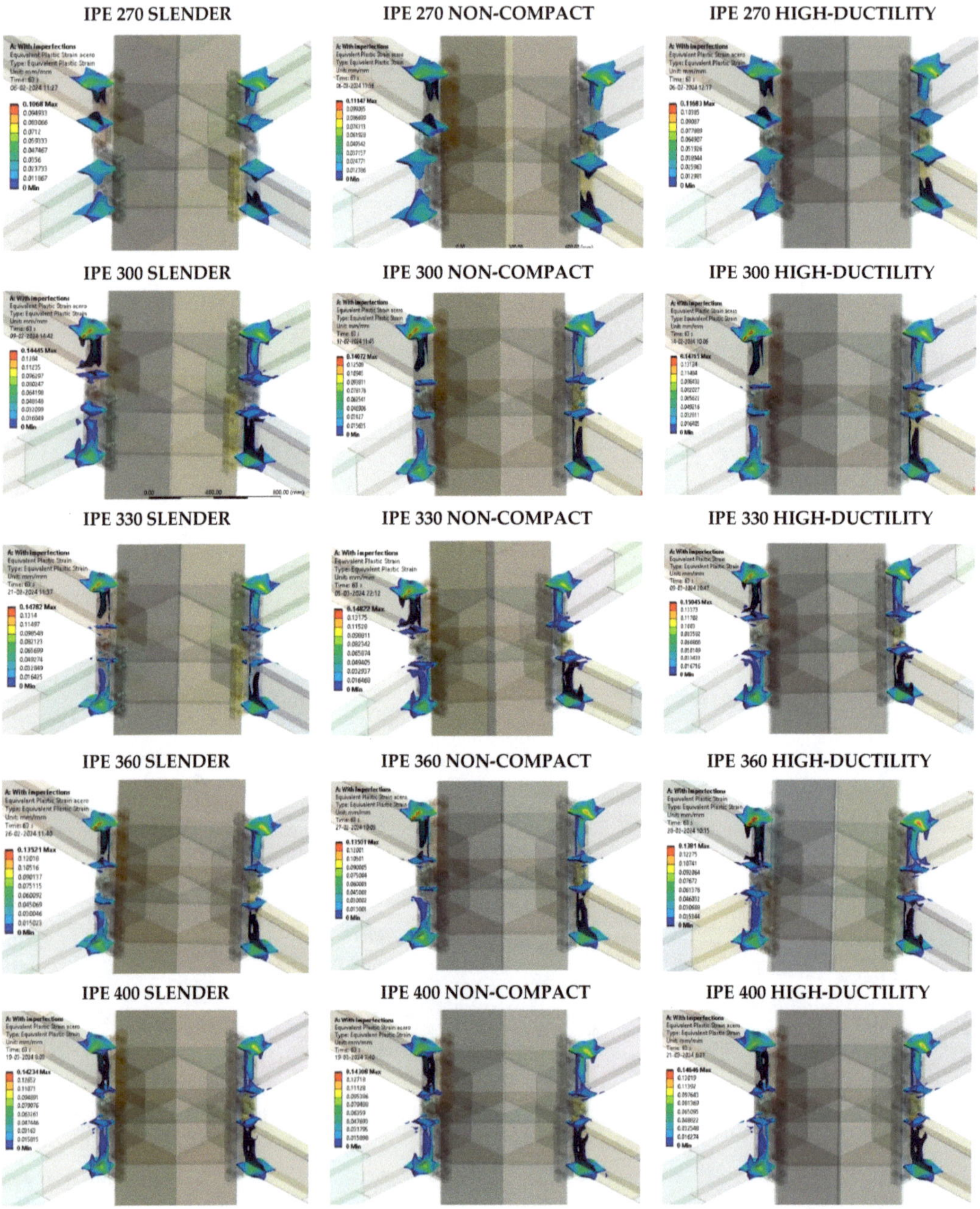

Figure A5. *Cont.*

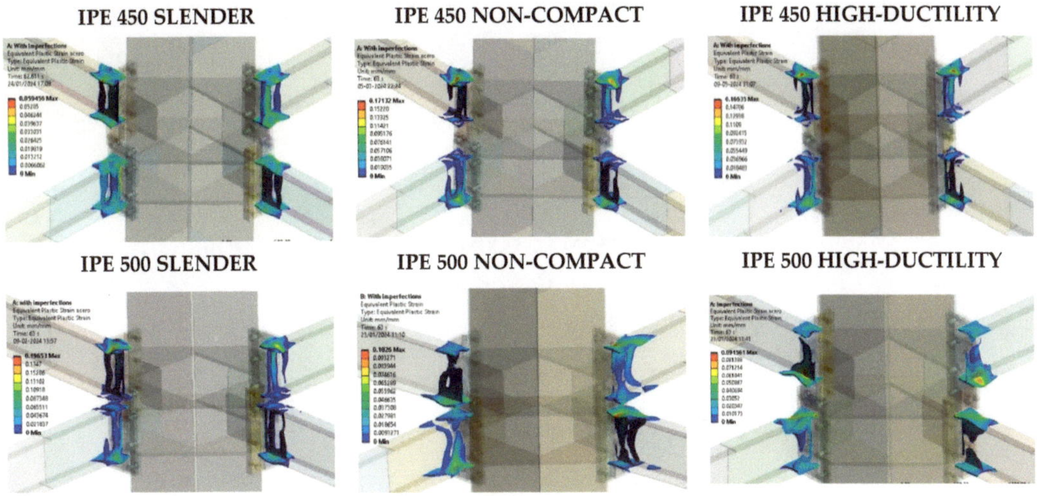

Figure A5. Equivalent plastic strain of 4B models.

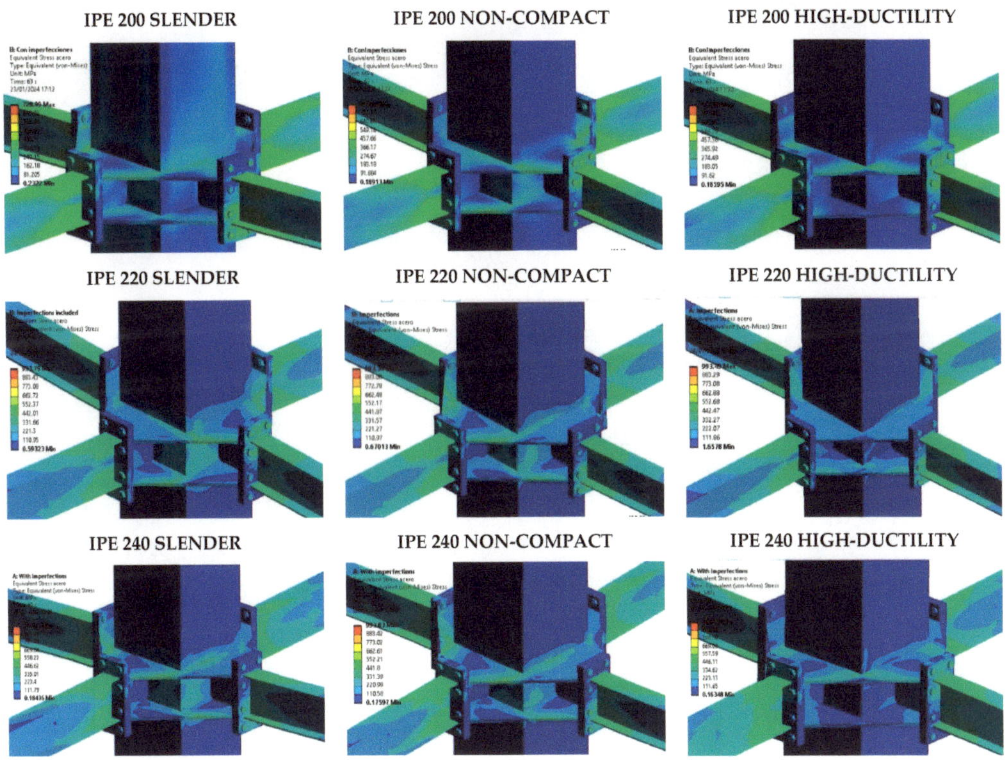

Figure A6. *Cont.*

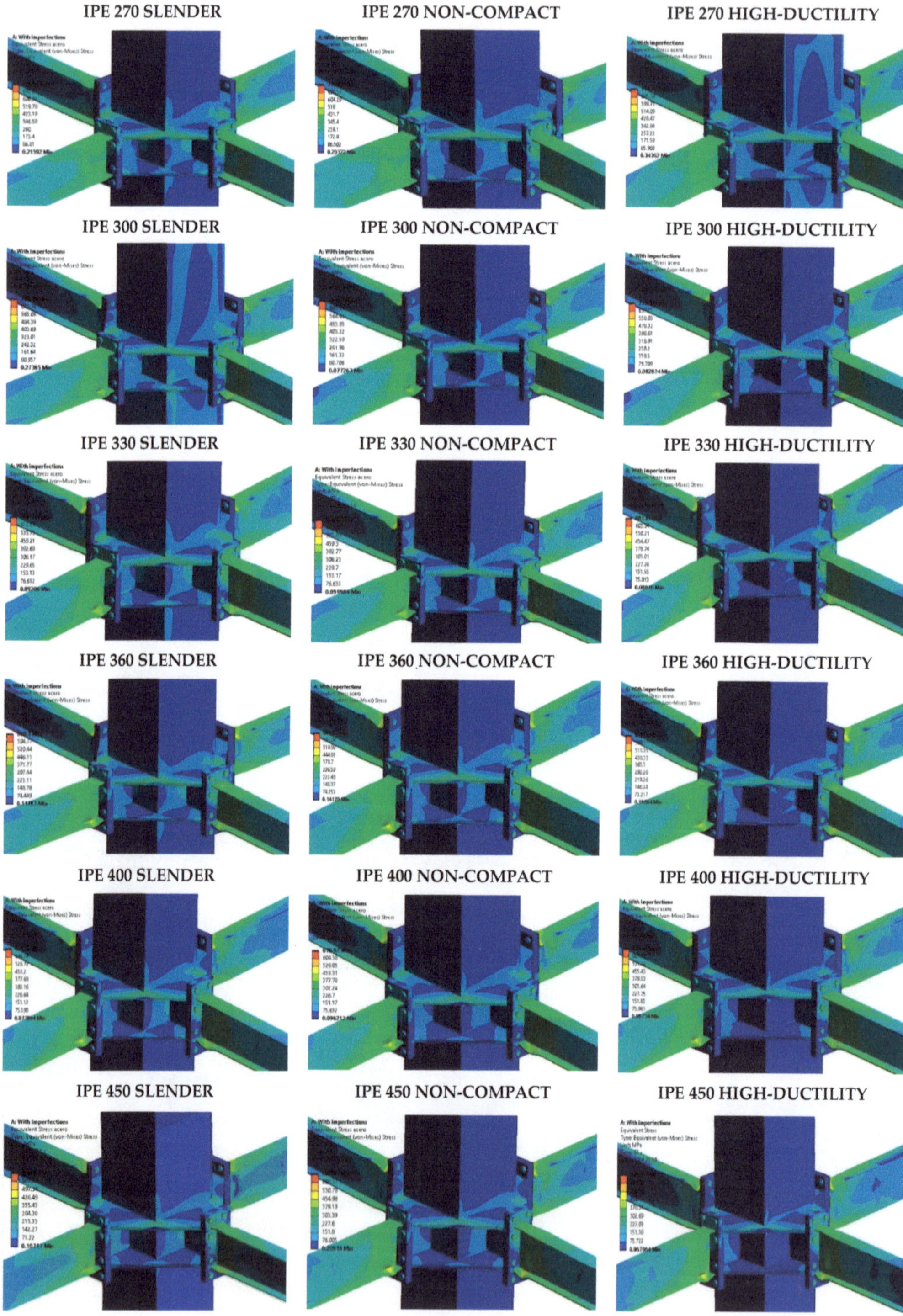

Figure A6. *Cont.*

Figure A6. Equivalent von Mises stress for 4B models.

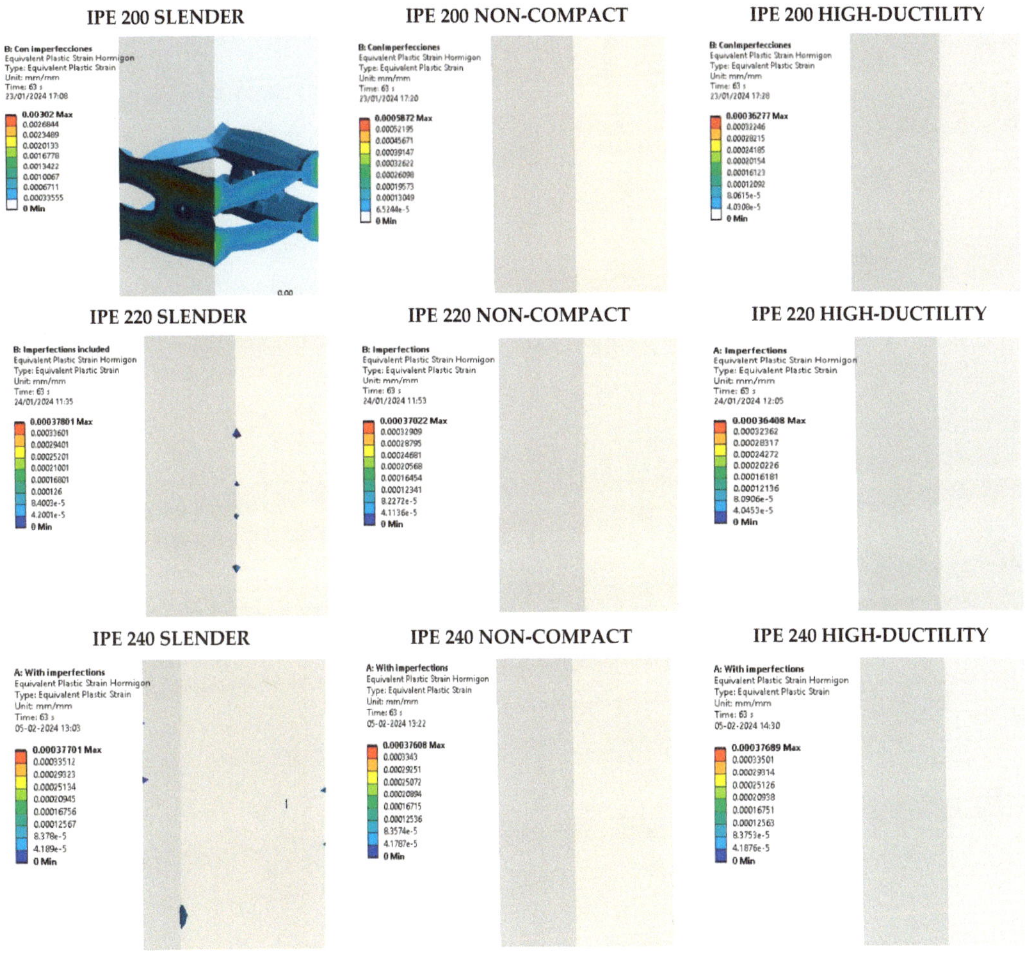

Figure A7. *Cont.*

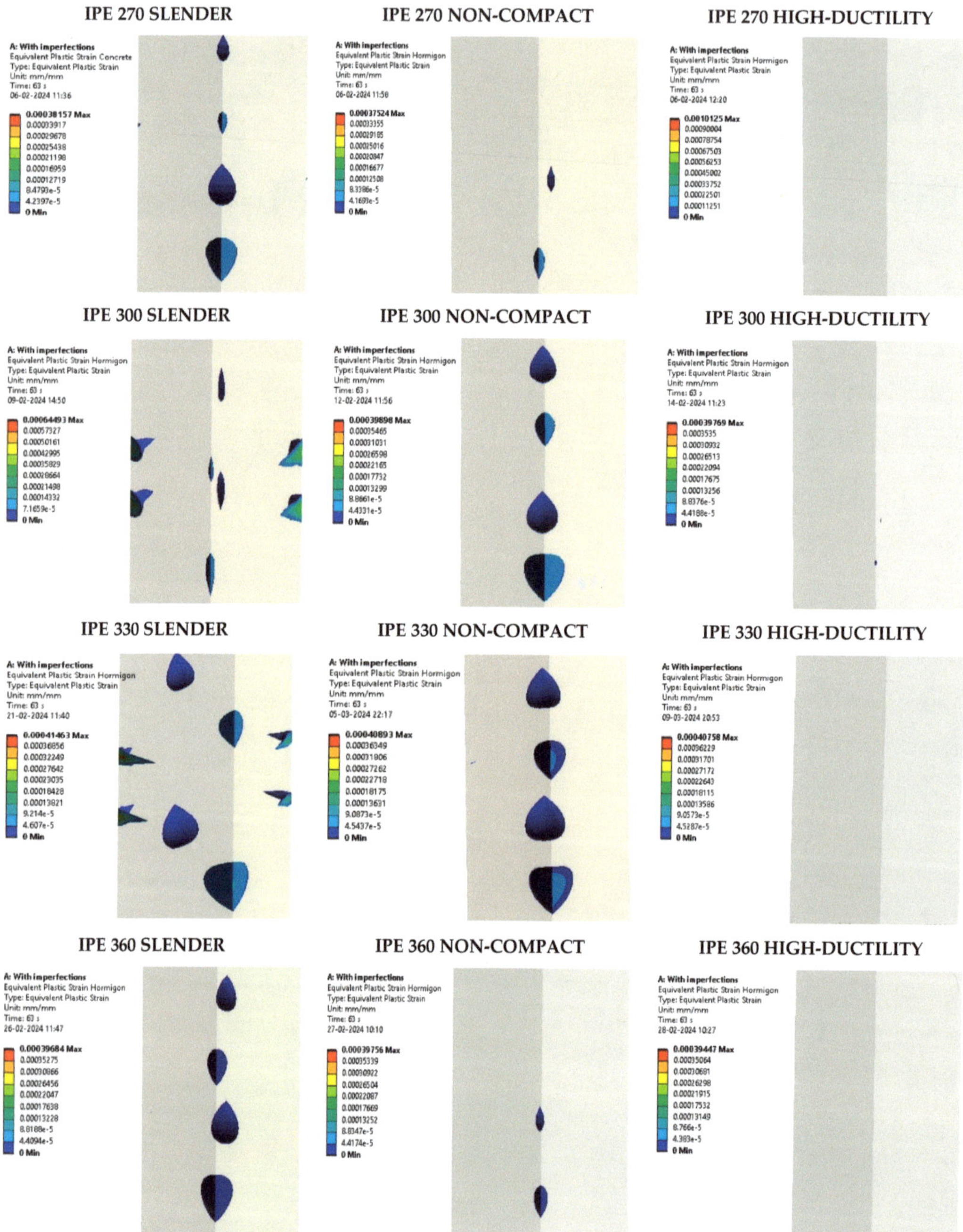

Figure A7. *Cont.*

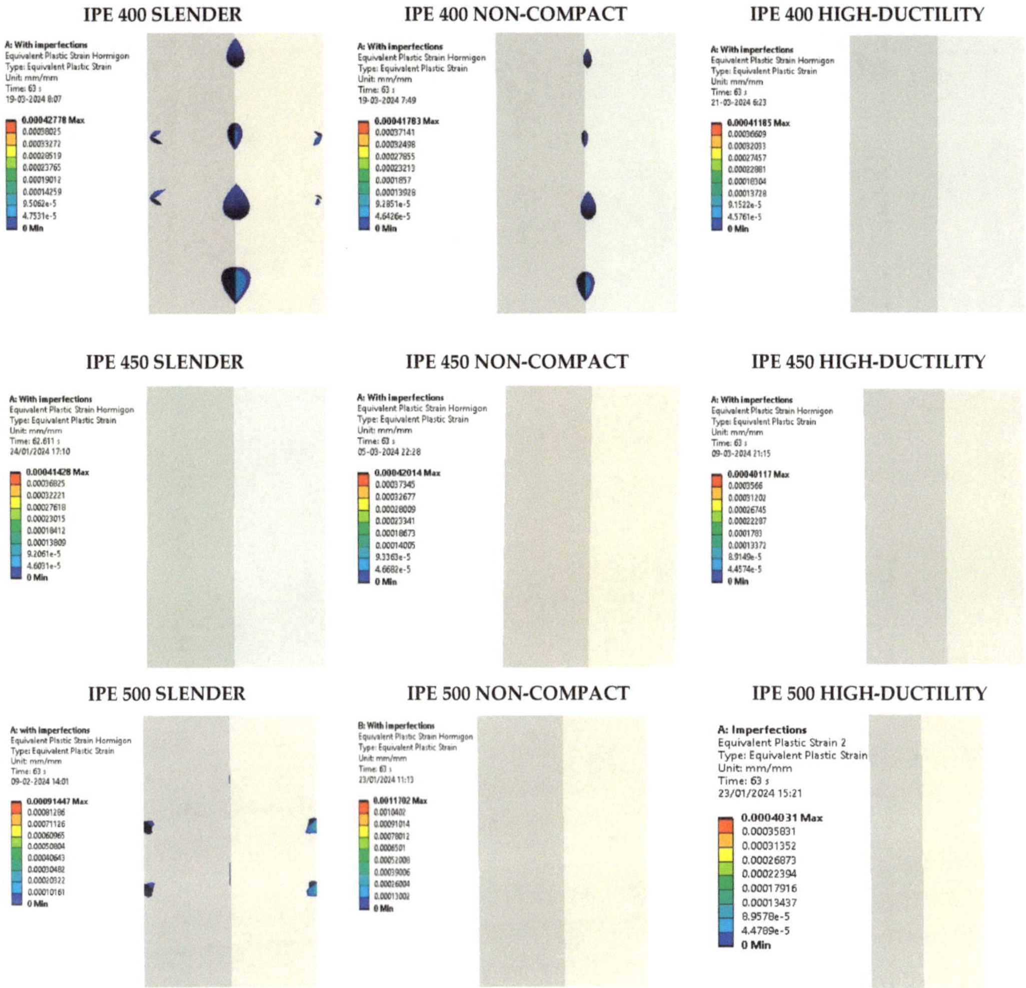

Figure A7. Plastic strains in concrete material for 4B models.

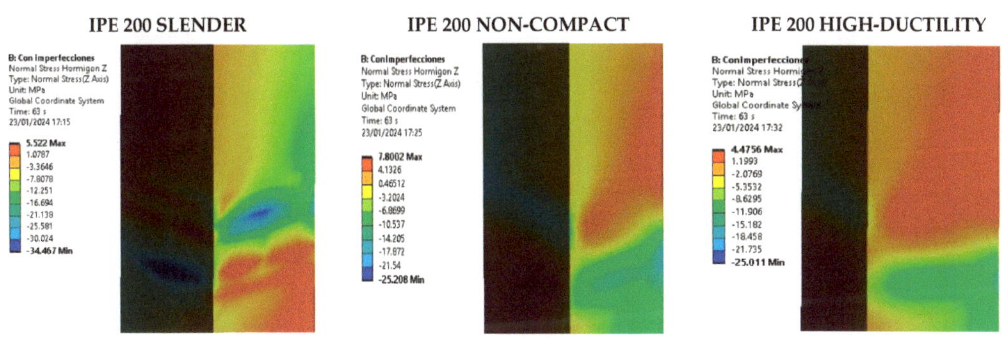

Figure A8. *Cont.*

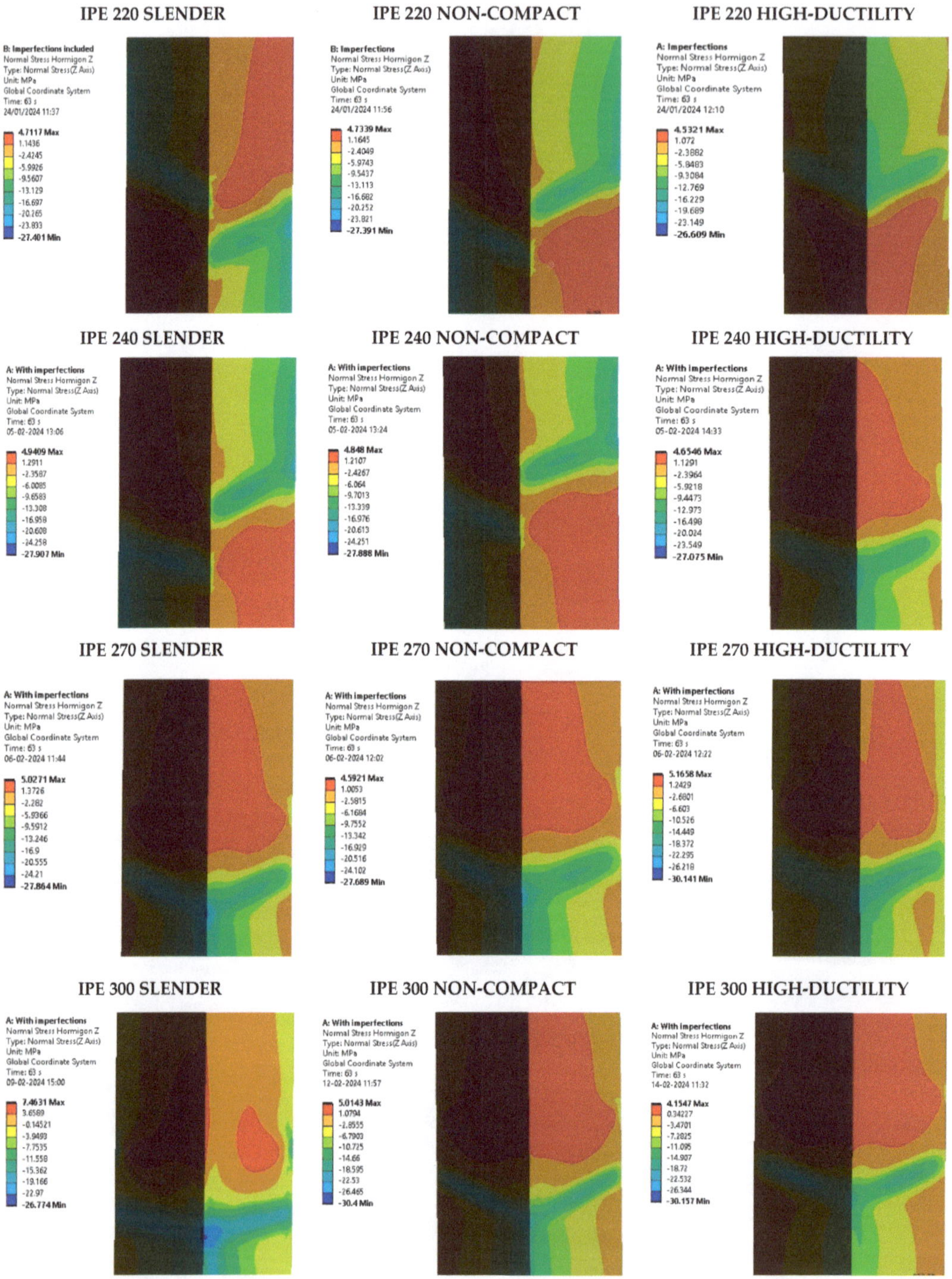

Figure A8. *Cont.*

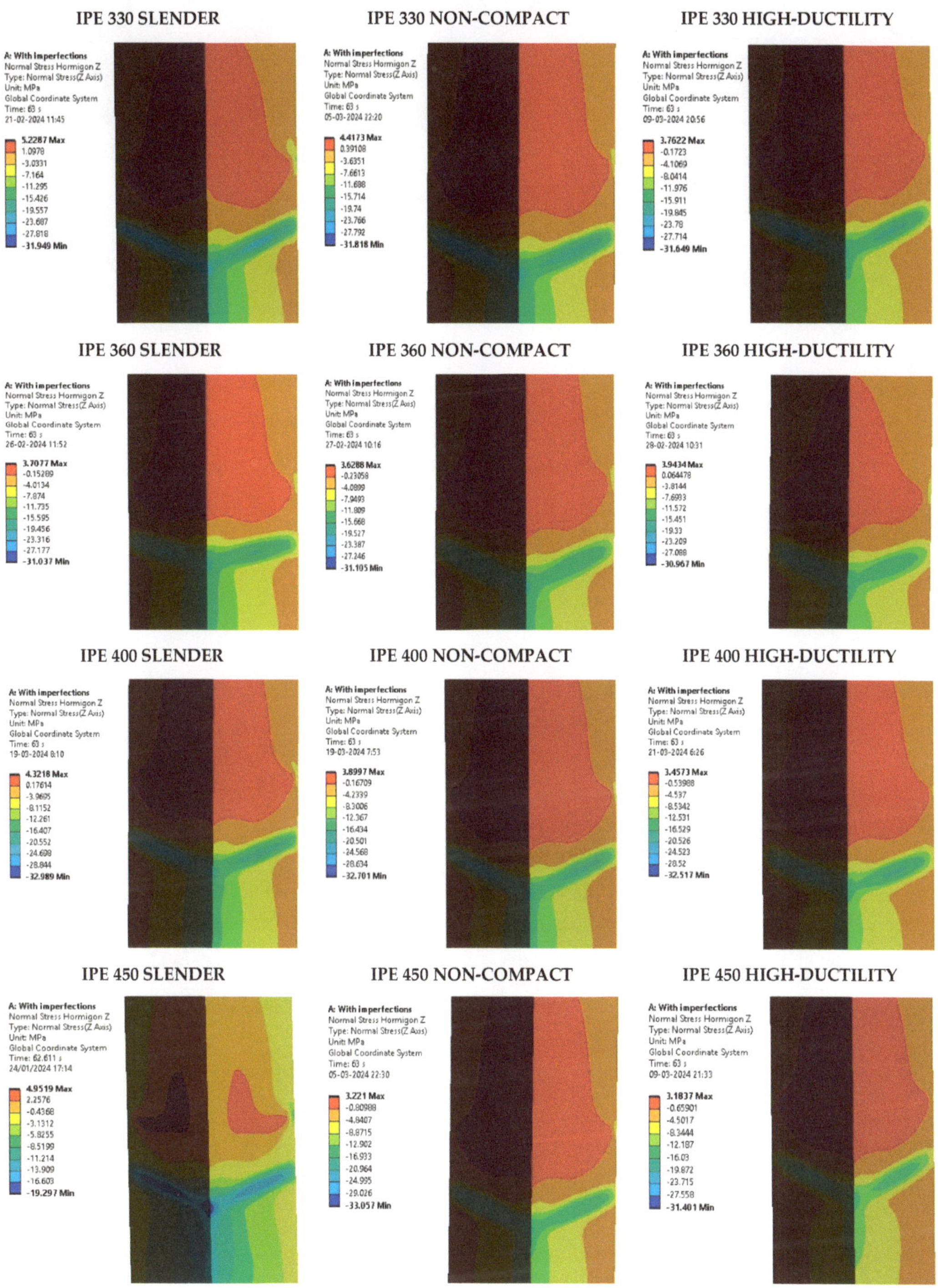

Figure A8. *Cont.*

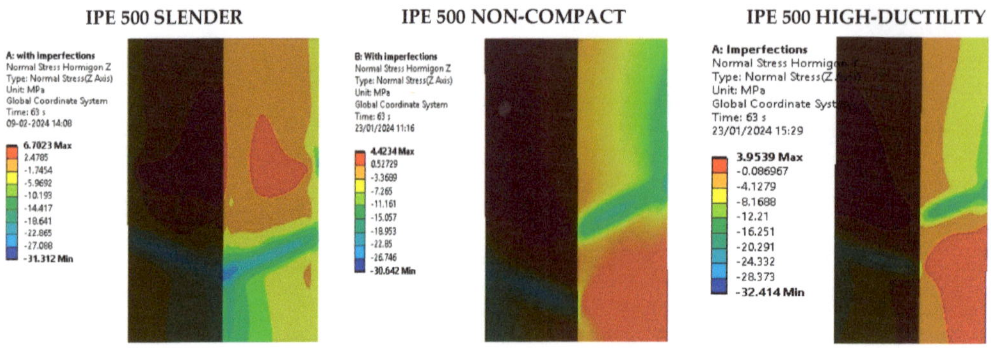

Figure A8. Normal stress (vertical direction) in concrete material for 4B models.

References

1. Okazaki, T.; Lignos, D.G.; Midorikawa, M.; Ricles, J.M.; Love, J. Damage to steel buildings observed after the 2011 Tohoku-oki earthquake. *Earthq. Spectra* **2013**, *29* (Suppl. 1), 219–243. [CrossRef]
2. Midorikawa, M.; Okazaki, T. Earthquake and tsunami damage to steel structures. In Proceedings of the 2012 Structures Congress, Tokyo, Japan, 1–4 March 2012. [CrossRef]
3. Nishiyama, I.; Yamanouchi, H.; Aoyama, H. U.S.-Japan cooperative earthquake engineering research on composite and hybrid structures. *AIJ J. Technol. Des.* **1999**, *5*, 107–110. [CrossRef] [PubMed]
4. Longo, A.; Montuori, R.; Nastri, E.; Piluso, V. On the use of HSS in seismic-resistant structures. *J. Constr. Steel Res.* **2014**, *103*, 1–12. [CrossRef]
5. Horikawa, K.; Sakino, Y. Damages to steel structures caused by the 1995 Kobe earthquake. *Struct. Eng. Int. J. Int. Assoc. Bridge Struct. Eng. (IABSE)* **1996**, *6*, 181–182. [CrossRef]
6. Xiao, Y.; He, W.; Choi, K. Confined Concrete-Filled Tubular Columns. *J. Struct. Eng.* **2005**, *131*, 488–497. [CrossRef]
7. Mao, X.Y.; Xiao, Y. Seismic behavior of confined square CFT columns. *Eng. Struct.* **2006**, *28*, 1378–1386. [CrossRef]
8. Goto, Y.; Kumar, G.P.; Kawanishi, N. Nonlinear Finite-Element Analysis for Hysteretic Behavior of Thin-Walled Circular Steel Columns with In-Filled Concrete. *J. Struct. Eng.* **2010**, *136*, 1413–1422. [CrossRef]
9. Gajalakshmi, P.; Helena, H.J. Behaviour of concrete-filled steel columns subjected to lateral cyclic loading. *J. Constr. Steel Res.* **2012**, *75*, 55–63. [CrossRef]
10. Aydin, F.; Saribiyik, M. Investigation of flexural behaviors of hybrid beams formed with GFRP box section and concrete. *Constr. Build. Mater.* **2013**, *41*, 563–569. [CrossRef]
11. Skalomenos, K.A.; Hatzigeorgiou, G.D.; Beskos, D.E. Parameter identification of three hysteretic models for the simulation of the response of CFT columns to cyclic loading. *Eng. Struct.* **2014**, *61*, 44–60. [CrossRef]
12. Skalomenos, K.A.; Hayashi, K.; Nishi, R.; Inamasu, H.; Nakashima, M. Experimental Behavior of Concrete-Filled Steel Tube Columns Using Ultrahigh-Strength Steel. *J. Struct. Eng.* **2016**, *142*, 04016057. [CrossRef]
13. Wang, Y.T.; Cai, J.; Long, Y.L. Hysteretic behavior of square CFT columns with binding bars. *J. Constr. Steel Res.* **2017**, *131*, 162–175. [CrossRef]
14. *ANSI/AISC 341-22*; Seismic Provisions for Structural Steel Buildings. American Institute of Steel Construction: Chicago, IL, USA, 2022.
15. Araújo, M.; Macedo, L.; Castro, J.M. Evaluation of the rotation capacity limits of steel members defined in EC8-3. *J. Constr. Steel Res.* **2017**, *135*, 11–29. [CrossRef]
16. Ning, K.Y.; Yang, L.; Ban, H.Y.; Sun, Y.N. Experimental and numerical studies on hysteretic behaviour of stainless steel welded box-section columns. *Thin-Walled Struct.* **2019**, *136*, 280–291. [CrossRef]
17. Chou, C.C.; Wu, S.C. Cyclic lateral load test and finite element analysis of high-strength concrete-filled steel box columns under high axial compression. *Eng. Struct.* **2019**, *189*, 89–99. [CrossRef]
18. *ANSI/AISC 358-22*; Prequalified Connections for Special and Intermediate Steel Moment Frames for. American Institute of Steel Construction: Chicago, IL, USA, 2022.
19. Simmons, R.J. Quick-Set, Full-Moment-Lock, Column and Beam Building Frame System and Method. US8453414B2, 4 June 2013.
20. Bock, C.; Bong, W. Moment Resisting Connection Apparatus and Method. 20020124520A1, 12 September 2002.
21. Kang, C.H.; Shin, K.J.; Oh, Y.S.; Moon, T.S. Hysteresis behavior of CFT column to H-beam connections with external T-stiffeners and penetrated elements. *Eng. Struct.* **2001**, *23*, 1194–1201. [CrossRef]
22. Varma, A.H.; Ricles, J.M.; Sause, R.; Lu, L.W. Seismic behavior and modeling of high-strength composite concrete-filled steel tube (CFT) beam-columns. *J. Constr. Steel Res.* **2002**, *58*, 725–758. [CrossRef]
23. Tizani, W.; Wang, Z.Y.; Hajirasouliha, I. Hysteretic performance of a new blind bolted connection to concrete filled columns under cyclic loading: An experimental investigation. *Eng. Struct.* **2013**, *46*, 535–546. [CrossRef]

24. Sheet, I.S.; Gunasekaran, U.; MacRae, G.A. Experimental investigation of CFT column to steel beam connections under cyclic loading. *J. Constr. Steel Res.* **2013**, *86*, 167–182. [CrossRef]
25. Qin, Y.; Chen, Z.; Wang, X. Experimental investigation of new internal-diaphragm connections to CFT columns under cyclic loading. *J. Constr. Steel Res.* **2014**, *98*, 35–44. [CrossRef]
26. Qin, Y.; Chen, Z.; Bai, J.; Li, Z. Test of extended thick-walled through-diaphragm connection to thick-walled CFT column. *Steel Compos. Struct.* **2016**, *20*, 1–12. [CrossRef]
27. Vulcu, C.; Stratan, A.; Ciutina, A.; Dubina, D. Beam-to-CFT High-Strength Joints with External Diaphragm. I: Design and Experimental Validation. *J. Struct. Eng.* **2017**, *143*, 04017001. [CrossRef]
28. Azar, A.S.; Shaghaghi, T.M. Seismic behavior of conxl connections in concrete filled steel Tube Columns (CFT). *J. Fundam. Appl. Sci.* **2017**, *9*, 217. [CrossRef]
29. Koloo, F.A.; Badakhshan, A.; Fallahnejad, H.; Jamkhaneh, M.E.; Ahmadi, M. Investigation of Proposed Concrete Filled Steel Tube Connections under Reversed Cyclic Loading. *Int. J. Steel Struct.* **2018**, *18*, 163–177. [CrossRef]
30. Rezaifar, O.; Younesi, A. Experimental study discussion of the seismic behavior on new types of internal/external stiffeners in rigid beam-to-CFST/HSS column connections. *Constr. Build. Mater.* **2017**, *136*, 574–589. [CrossRef]
31. Lai, Z.; Fischer, E.C.; Varma, A.H. Database and Review of Beam-to-Column Connections for Seismic Design of Composite Special Moment Frames. *J. Struct. Eng.* **2019**, *145*, 04019023. [CrossRef]
32. Fanaie, N.; Moghadam, H.S. Experimental study of rigid connection of drilled beam to CFT column with external stiffeners. *J. Constr. Steel Res.* **2019**, *153*, 209–221. [CrossRef]
33. Parvari, A.; Zahrai, S.M.; Mirhosseini, S.M.; Zeighami, E. Numerical and experimental study on the behavior of drilled flange steel beam to CFT column connections. *Structures* **2020**, *28*, 726–740. [CrossRef]
34. Ahmadi, M.M.; Mirghaderi, S.R. Experimental studies on through-plate moment connection for beam to HSS/CFT column. *J. Constr. Steel Res.* **2019**, *161*, 154–170. [CrossRef]
35. Parvari, A.; Zahrai, S.M.; Mirhosseini, S.M.; Zeighami, E. Comparing cyclic behaviour of RBS, DFC and proposed rigid connections in a steel moment frame with CFT column. *Aust. J. Civ. Eng.* **2021**, *19*, 164–183. [CrossRef]
36. Jeddi, M.Z.; Sulong, N.H.R.; Ghanbari-Ghazijahani, T. Behaviour of double-sleeve TubeBolt moment connections in CFT columns under cyclic loading. *J. Constr. Steel Res.* **2022**, *194*, 107302. [CrossRef]
37. Paul, S.; Deb, S.K. Experimental study on a new V-cut RBS and CFT connections with bidirectional bolts under cyclic loadings. *J. Build. Eng.* **2022**, *46*, 103688. [CrossRef]
38. Du, H.; Zhao, P.; Wang, Y.; Sun, W. Seismic experimental assessment of beam-through beam-column connections for modular prefabricated steel moment frames. *J. Constr. Steel Res.* **2022**, *192*, 107208. [CrossRef]
39. Habibi, A.; Fanaie, N.; Shahbazpanahi, S. Experimental and numerical investigation of I-beam to concrete-filled tube (CFT) column moment connections with pipe-stiffened internal diaphragm. *J. Constr. Steel Res.* **2023**, *200*, 107648. [CrossRef]
40. Razavi, S.A.; Kandi, A.H.; Alimardani, M.; Jovaini, E. Tube-in-tube rigid beam to CFT column connection in moment-resisting frames: An experimental study. *Soil Dyn. Earthq. Eng.* **2023**, *171*, 107901. [CrossRef]
41. Bai, Y.; Wang, S.; Mou, B.; Wang, Y.; Skalomenos, K.A. Bi-directional seismic behavior of steel beam-column connections with outer annular stiffener. *Eng. Struct.* **2021**, *227*, 111443. [CrossRef]
42. Gallegos, M.; Nuñez, E.; Herrera, R. Numerical study on cyclic response of end-plate biaxial moment connection in box columns. *Metals* **2020**, *10*, 523. [CrossRef]
43. Mata, R.; Nuñez, E. Parametric study of 3D steel moment connections with built-up box column subjected to biaxial cyclic loads. *J. Constr. Steel Res.* **2022**, *197*, 107453. [CrossRef]
44. Nuñez, E.; Torres, R.; Herrera, R. Seismic performance of moment connections in steel moment frames with HSS columns. *Steel Compos. Struct.* **2017**, *25*, 271–286. [CrossRef]
45. Mata, R.; Nuñez, E.; Sanhueza, F.; Maureira, N.; Roco, Á. Assessment of Web Panel Zone in Built-up Box Columns Subjected to Bidirectional Cyclic Loads. *Buildings* **2023**, *13*, 71. [CrossRef]
46. Nuñez-Castellanos, E.; Bustos-Figueroa, J.; Mata-Lemus, R.; Sanhueza-Espinoza, F.; Lapeña-Mañero, P. Cyclic behavior of 3D moment connections subjected to bidirectional load: Experimental approach. *Eng. Struct.* **2023**, *291*, 116392. [CrossRef]
47. ANSYS Inc. *ANSYS v2022*; ANSYS Inc.: Canonsburg, PA, USA, 2022.
48. *ANSI/AISC 360-22*; Specification for Structural Steel Buildings. American Institute of Steel Construction: Chicago, IL, USA, 2022.
49. ACI Committee 318. *Building Code Requirements for Structural Concrete (ACI 318-19) and Commentary (ACI 318R-19)*; American Concrete Institute: Indianapolis, IN, USA, 2019.
50. Nuñez, E.; Mata, R. Strong column-weak beam relationship of 3D steel joints with tubular columns: Assessment, Validation and Design proposal. *J. Build. Eng.* **2024**, *82*, 108399. [CrossRef]
51. Gerbens-Leenes, P.W.; Hoekstra, A.Y.; Bosman, R. The blue and grey water footprint of construction materials: Steel, cement and glass. *Water Resour. Ind.* **2018**, *19*, 1–12. [CrossRef]
52. Vilela, P.M.L.; Carvalho, H.; Grilo, L.F.; Montenegro, P.A.; Calçada, R.B. Unitary model for the analysis of bolted connections using the finite element method. *Eng. Fail Anal.* **2019**, *104*, 308–320. [CrossRef]
53. Zreid, I.; Kaliske, M. A gradient enhanced plasticity–damage microplane model for concrete. *Comput. Mech.* **2018**, *62*, 1239–1257. [CrossRef]

54. Jiang, J.F.; Wu, Y.F. Identification of material parameters for Drucker-Prager plasticity model for FRP confined circular concrete columns. *Int. J. Solids Struct.* **2012**, *49*, 445–456. [CrossRef]
55. Kedziora, S.; Anwaar, M.O. Concrete-filled steel tubular (CFTS) columns subjected to eccentric compressive load. *AIP Conf. Proc.* **2019**, *2060*, 020004. [CrossRef]
56. Bayraktar, A.; Şahin, A.; Özcan, D.M.; Yildirim, F. Numerical damage assessment of Haghia Sophia bell tower by nonlinear FE modeling. *Appl. Math. Model.* **2010**, *34*, 92–121. [CrossRef]
57. Avci, O.; Bhargava, A. Finite-Element Analysis of Cantilever Slab Deflections with ANSYS SOLID65 3D Reinforced-Concrete Element with Cracking and Crushing Capabilities. *Pract. Period. Struct. Des. Constr.* **2019**, *24*, 05018007. [CrossRef]
58. Schneider, S.P. Axially Loaded Concrete-Filled Steel Tubes. *J. Struct. Eng.* **1998**, *124*, 1125–1138. [CrossRef]
59. Gupta, P.K.; Khaudhair, Z.; Ahuja, A.K. 3D numerical simulation of concrete filled steel tubular columns using ANSYS. In *UKIERI Concrete Congress—Innovation in Concrete Construction*; Jalandhar, India; 2013; pp. 2262–2271. Available online: https://www.researchgate.net/publication/317380097_3D_NUMERICAL_SIMULATION_OF_CONCRETE_FILLED_STEEL_TUBULAR_COLUMNS_USING_ANSYS (accessed on 25 February 2024).
60. Tartaglia, R.; D'Aniello, M.; Rassati, G.A.; Swanson, J.A.; Landolfo, R. Full strength extended stiffened end-plate joints: AISC vs recent European design criteria. *Eng. Struct.* **2018**, *159*, 155–171. [CrossRef]
61. ASTM Committee. *Standard Specification for General Requirements for Rolled Structural Steel Bars, Plates, Shapes, and Sheet Piling*; American Society for Testing and Materials (ASTM): West Conshohocken, PA, USA, 2001.
62. *A1085/A1085M−15*; Standard Specification for Cold-Formed Welded Carbon Steel Hollow Structural Sections (HSS). American Society for Testing and Materials (ASTM): West Conshohocken, PA, USA, 2015.
63. Behrooz, S.M.; Erfani, S. Parametric study of Stub-Beam Bolted Extended End-Plate connection to box-columns. *J. Constr. Steel Res.* **2020**, *171*, 106155. [CrossRef]
64. Goudarzi, A.; Erfani, S. Seismic performance of beam to box-column connection by a short stub beam. *J. Constr. Steel Res.* **2022**, *190*, 107145. [CrossRef]
65. Wei, J.P.; Tian, L.M.; Guo, Y.; Qiao, H.Y.; Bao, Y.; Jiao, Z.A.; Shi, X.J. Numerical study of the seismic performance of a double-hinge steel frame joint. *J. Constr. Steel Res.* **2021**, *187*, 106963. [CrossRef]
66. Zhang, A.L.; Qiu, P.; Guo, K.; Jiang, Z.Q.; Wu, L.; Liu, S.C. Experimental study of earthquake-resilient end-plate type prefabricated steel frame beam-column joint. *J. Constr. Steel Res.* **2020**, *166*, 105927. [CrossRef]
67. Torres-Rodas, P.; Medalla, M.; Zareian, F.; Lopez-Garcia, D. Cyclic behavior and design methodology of exposed base plates with extended anchor bolts. *Eng. Struct.* **2022**, *260*, 114235. [CrossRef]
68. Trautner, C.A.; Hutchinson, T.; Grosser, P.R.; Silva, J.F. Effects of Detailing on the Cyclic Behavior of Steel Baseplate Connections Designed to Promote Anchor Yielding. *J. Struct. Eng.* **2016**, *142*, 04015117. [CrossRef]
69. Chopra, A.K. *Dynamic of Structures*; Pearsons: London, UK, 2012.
70. Wu, L.Y.; Chung, L.L.; Tsai, S.F.; Lu, C.F.; Huang, G.L. Seismic behavior of bidirectional bolted connections for CFT columns and H-beams. *Eng. Struct.* **2007**, *29*, 395–407. [CrossRef]
71. Li, X.; Xiao, Y.; Wu, Y.T. Seismic behavior of exterior connections with steel beams bolted to CFT columns. *J. Constr. Steel Res.* **2009**, *65*, 1438–1446. [CrossRef]
72. Rahnavard, R.; Hassanipour, A.; Siahpolo, N. Analytical study on new types of reduced beam section moment connections affecting cyclic behavior. *Case Stud. Struct. Eng.* **2015**, *3*, 33–51. [CrossRef]

buildings

Article

A Study on the Bearing Performance of an RC Axial Compression Shear Wall Strengthened by a Replacement Method Using Local Reinforcement with an Unsupported Roof

Yuanwen Liu [1], Jie Deng [1], Yigang Jia [2,*], Guangyu Wu [2,*], Naiwen Ke [1] and Xianglan Wei [1]

1 School of Civil Engineering and Architecture, Nanchang University, Nanchang 330031, China; 402500220023@email.ncu.edu.cn (Y.L.)
2 Design and Research Institute, Nanchang University, Nanchang 330047, China
* Correspondence: jiayigang999@ncu.edu.cn (Y.J.); wuguangyu@ncu.edu.cn (G.W.)

Abstract: When compared with conventional replacement reinforcement methods, the method of replacement using a local reinforcement with an unsupported roof has the advantages of shortening the reinforcement cycle and reducing material loss, and many scholars have carried out useful explorations thereof. At present, the formula for the bearing capacity of reinforcement by replacing concrete in the Code does not consider the effect of stress hysteresis on the parts of the reinforcement; so when the initial stress level is greater than 0.4, the Code's strength utilization coefficient of 0.8 for the new concrete in the replaced area is on the side of insecurity. In this study, we are trying to improve and supplement the formula in the Code through the following work. Firstly, 18 groups of shear wall models were constructed using a VFEAP finite element analysis program to analyze the bearing performances of the shear walls after the replacement. The results showed that the replacement concrete strength, the initial stress level and the size of the replaced area had a significant influence on the bearing capacity of the shear wall after the replacement. Secondly, utilizing the replacement concrete strength, the initial stress level and the size of the replacement area as key parameters, then introducing the strength improvement coefficient considering the constraints of the stirrups, the modified strength utilization coefficient of new concrete in the replaced area was formulated. Finally, based on the modified strength utilization coefficient, the replacement bearing capacity formulas for the one-batch, two-batch, and three-batch replacements were derived, and an N-batch replacement bearing capacity formula was regressed and fitted on the basis of these equations, which are less discrete and more secure than the Code's formula.

Keywords: shear wall; replacement method using local reinforcement with an unsupported roof; stress hysteresis; nonlinear analysis

1. Introduction

The replacement concrete reinforcement method is favored because it does not increase the size of the structure on which it is used and maintains its original form. There are many engineering examples of RC shear walls that have been strengthened using replacement methods at home and abroad [1,2], and a series of research results have been obtained. The current Code makes it clear that this method is suitable for the local reinforcement of concrete with low strength or serious defects in the compression area of the RC-structure-bearing members, and relevant provisions on the strength grades of non-replacement concrete and replacement concrete are described [3]. In practical engineering, compared to the conventional replacement reinforcement method, the method of replacement using local reinforcement with an unsupported roof can reduce the scope of replacement construction, allow full use of the original section to be made, and shorten the reinforcement period. Therefore, our research group performed a useful exploration of the strengthening of RC shear walls using a local strengthening replacement method without a supporting roof [4].

At present, many scholars in China mainly explore the bearing performance of shear walls strengthened using local replacement reinforcement methods [5]. Liu Yuwei [6] took a residential building under construction as the background and used a local strengthening replacement method to carry out a structural reinforcement analysis and research; finally, it was found that the vertical and lateral bearing capacity of a residential building reinforced with this method could meet the original design requirements. Wang Xin [7] conducted a low-cyclic-reciprocating-load comparison test on four shear wall specimens without initial vertical stress (two of which were locally reinforced replacements at both ends), and the results showed that the C50 concrete or modified concrete in the replaced area had good cooperation with the C20 concrete in the non-replaced area; in addition, the seismic bearing capacity, energy dissipation capacity, ductility, cracking load, and peak load of the replacement specimens were significantly improved. Stress hysteresis refers to the phenomenon where the timing of stress on the new material lags behind that of the old material due to an uncoordinated deformation between the old and new materials. Stress hysteresis may reduce the reinforcement effect of the shear wall, and it can even cause local damage in severe cases. However, up to now, there have been few studies on the cooperative work and overall performance of high-performance materials [8,9] in the replaced areas of RC shear walls and the low-strength concrete in non-replaced areas while considering stress hysteresis. In particular, the bearing capacity of RC shear walls when using high-performance materials for local strengthening replacements and the effects of stress hysteresis have yet to be thoroughly investigated.

In order to study the cooperative work performance of the replaced and non-replaced areas after replacement while considering stress hysteresis, in this study, we performed a useful exploration on it through the following work. Firstly, the VFEAP finite element analysis program was utilized to establish 18 groups of shear wall models. Based on these models, the influences of multiple factors on the bearing performance of the reinforced shear walls were analyzed, and it was pointed out that the formula in the Code is unsafe at high stress levels. Secondly, in order to improve or supplement the Code's formula, this study considered the influences of the replacement concrete strength, initial stress level, and the size of the replaced area; in addition, we introduced the strength improvement coefficient considering the constraints of the stirrups, then established a modified strength utilization coefficient of the new concrete in the replaced area. Based on the modified strength utilization coefficient, we deduced the bearing capacity formulas for the one-batch, two-batch, and three-batch unsupported locally reinforced replacements, and conducted a regression fitting for the bearing capacity formulas of the N-batch replacements. Finally, according to the existing data, the formula derived in this study was compared with the formula in the Code; the results showed that the formula derived here had a higher degree of coincidence with the test results, and the dispersion was smaller, giving it a certain reliability.

2. Analysis Method and an Example

In this study, a nonlinear finite element analysis method based on the theory of a three-dimensional solid degenerate virtual laminated unit [10] was adopted, as this allowed a better simulation of the actual force state and deformation of the structure [11]. The feasibility and accuracy of the VFEAP finite element analysis program were verified through an example, which provided a theoretical basis for the subsequent nonlinear analysis of members reinforced through the use of a replacement method using local reinforcement with an unsupported roof.

2.1. Analysis Method

A three-dimensional solid degraded virtual laminate unit was constructed by introducing the relevant assumptions into isoparametric units of a spatial solid to create three-dimensional solid degraded units using the concepts of "virtual nodes" and "virtual regions" [12]. Blocks with different geometric and material parameters were allowed to exist in the same unit and divisions into units and blocks can be performed according to the actual situation, thus improving the modeling efficiency, analysis efficiency, and greatly increasing the accuracy of the analysis and calculation. In addition, the theory of three-dimensional solid degenerate virtual laminated units involves the double nonlinearity of geometry and material, that is, the T.L. [13] method, which was employed for geometric nonlinearity, and the multiaxial strengthening plastic model from Ohtani and Chen was introduced for material nonlinearity. Finally, in the process of modeling, a three-node one-dimensional isoparametric unit was used to describe the relative positions of the steel bars in any section, allowing the true simulation of the stress conditions of the steel bars.

2.2. Example Verification

The test in [14] was selected for the simulation analysis to verify the feasibility and accuracy of the VFEAP finite element analysis program. The sectional dimensions and reinforcements of the members before and after the reinforcement of the enclosure are shown in Figure 1. The detailed parameters of the relevant samples, the preparation process, and the main experiments are all described in [14].

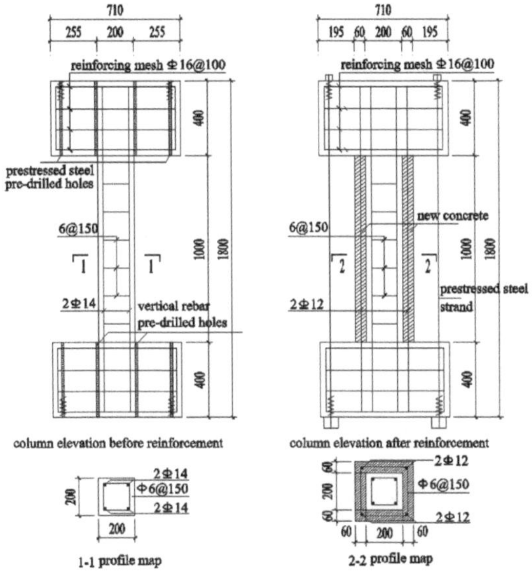

Figure 1. Cross-sectional dimensions and reinforcements of the test specimen.

In the process of a finite element analysis, it is easy for the phenomenon of stress concentration to appear; to prevent stress concentration at the bottom of concrete members in the early stage, which would lead to the termination of the calculation, a rigid pad was established at both ends of the column for force transmission. The models of the concrete and steel bars are shown in Figures 2 and 3.

According to the curve of the axial force–strain relationship of the concrete shown in Figure 4, the failure process and deformation simulated in the finite element analysis program were basically consistent with the test process. In addition, the ultimate value of the axial compression bearing capacity of the BJG specimen simulated with the finite element method was 1175 KN, and the experimental value was 1241 KN, with an error of 5.3%. The ultimate value of the bearing capacity simulated with the finite element method for a JGZ-13 member was 3684 KN, and the test value was 3795 KN, with a difference of 2.92%. The overall error was less than 6%. To sum up, the VFEAP finite element analysis program used in this study was able to simulate the actual failure process and deformation of the structure well, and the finite element analysis results had a high degree of fitting with the test results, which showed that the VFEAP finite element analysis program was completely feasible for a simulation analysis.

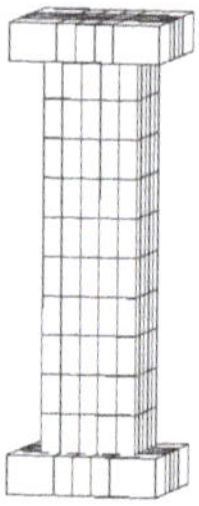

Figure 2. Concrete model.

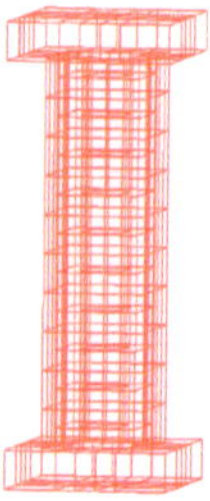

Figure 3. Finite element mesh for steel reinforcement.

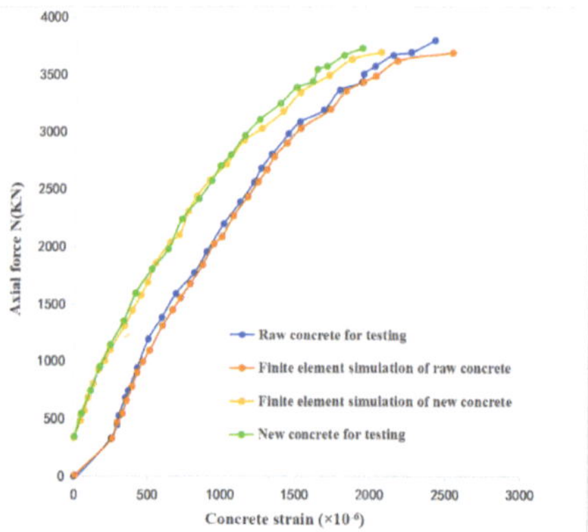

Figure 4. Curve of the relationship between an existing column and newly poured concrete using JGZ-13.

3. Analysis of the Simulation of the Whole Process of Local Strengthening Replacement of an RC Shear Wall without a Supporting Roof under Axial Compression

Based on the verification of the example in the second section, this section takes an RC shear wall reinforced using the local strengthening replacement method without a supporting roof as the research object, and the strengthening mechanism of the RC shear wall is studied using the VFEAP finite element analysis program. In addition, the influences of the initial stress level, replacement concrete strength, and the size of the replacement area on the bearing capacity of the shear wall are comprehensively analyzed. Finally, according to the results of the simulation analysis, it is pointed out that the value of the strength utilization coefficient for the replacement concrete in the Code is unsafe at high stress levels, which provides a theoretical basis for the later derivation of a calculation formula for the bearing capacity of RC shear walls strengthened using the local strengthening replacement method without a supporting roof.

3.1. Analysis Scheme and Model Design

The research object in this simulation analysis was a rectangular-section RC shear wall strengthened using a local one-batch replacement and two-batch replacement without a supporting roof. The section size and reinforcement of the specimen are shown in Figure 5.

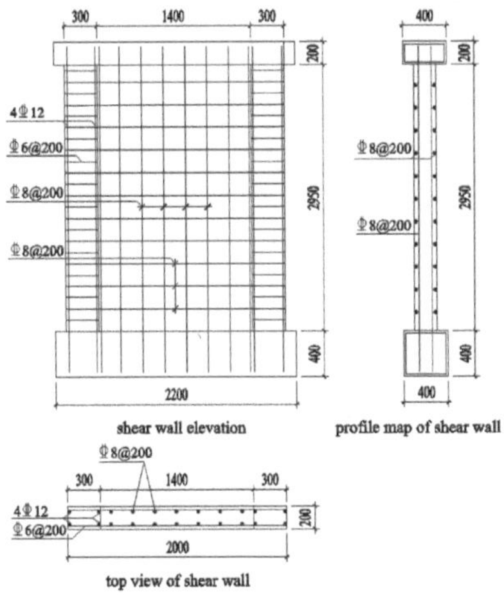

Figure 5. Shear wall section size and reinforcement.

According to the existing research results, this study mainly considered the impacts of the initial stress level, the replacement concrete strength, the size of the replacement area, the replacement batch, and the longitudinal reinforcement ratio for a concealed column; the influencing factors and their levels are listed in Table 1.

Table 1. Experimental influencing factors and their levels.

Influencing Factor	Level					
Initial stress level	0.15	0.25	0.35	0.4	0.5	0.6
Replacement concrete grade	C50		C65		C80	
Replacement section size (mm × mm)	300 × 200		350 × 200		400 × 200	
Reinforcement ratio of a concealed column	0.85%		1.50%		profile steel	

According to the factors and levels selected in this table, experimental schemes with a one-batch replacement and two-batch replacement were designed using the orthogonal test method. The overall test scheme is shown in Table 2.

According to existing research results, during the replacement process of a structure, the stress of each wall segment will be redistributed, and the distribution mainly occurs in the reinforced floor and the two adjacent floors; the stress of the demolished wall will be shared by the undemolished wall, and the stress of the shear wall perimeter and that of the upper and lower floors of the connecting beams, girders, slabs, columns, and other components will be redistributed [15]. This study only considers a single shear wall, so it is assumed that stress redistribution only occurs on this shear wall; the effect of the replacement of shear walls in practical engineering will be better than that of this study.

Table 2. Overall test scheme.

Specimen	Initial Stress Level	Replacement Concrete Grade	Length of Replaced Wall (mm)	Reinforcement Ratio of a Concealed Column	Replacement Batch
YPZH-1	0.15	C50	300 × 2	0.85%	one-batch
YPZH-2	0.15	C65	400 × 2	1.50%	one-batch
YPZH-3	0.15	C80	350 × 2	profile steel	one-batch
YPZH-4	0.25	C50	400 × 2	profile steel	one-batch
YPZH-5	0.25	C65	350 × 2	0.85%	one-batch
YPZH-6	0.25	C80	300 × 2	1.50%	one-batch
YPZH-7	0.35	C50	350 × 2	1.50%	one-batch
YPZH-8	0.35	C65	300 × 2	profile steel	one-batch
YPZH-9	0.35	C80	400 × 2	0.85%	one-batch
EPZH-10	0.4	C50	300 × 2	0.85%	two-batch
EPZH-11	0.4	C65	400 × 2	1.50%	two-batch
EPZH-12	0.4	C80	350 × 2	profile steel	two-batch
EPZH-13	0.5	C50	400 × 2	profile steel	two-batch
EPZH-14	0.5	C65	350 × 2	0.85%	two-batch
EPZH-15	0.5	C80	300 × 2	1.50%	two-batch
EPZH-16	0.6	C50	350 × 2	1.50%	two-batch
EPZH-17	0.6	C65	300 × 2	profile steel	two-batch
EPZH-18	0.6	C80	400 × 2	0.85%	two-batch

3.2. Analysis of the Simulation Results

3.2.1. One-Batch Replacement without a Supporting Roof

Experiments using one-batch replacements were carried out at a low stress level (between 0.15 and 0.35). The one-batch replacements were divided into four working conditions: working condition one represented the action of initial stress; working condition two represented the removal of the wall section to be replaced; working condition three represented the casting of the wall section to be replaced with high-strength concrete; and working condition four represented the action of the ultimate load. This study took two components, YPZH-1 and YPZH-9, as representatives for analysis; their working condition stress diagrams and load–stress curves are shown in Figures 6 and 7.

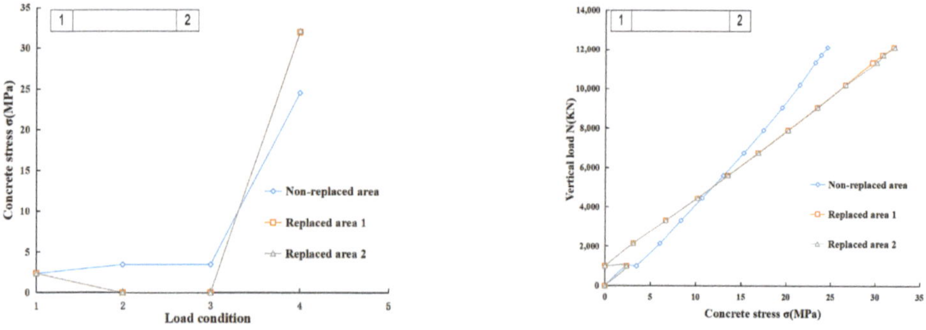

(**a**) (**b**)

Figure 6. (**a**) Stress–strain diagram under different load conditions for YPZH-1; (**b**) load–stress curve for YPZH-1.

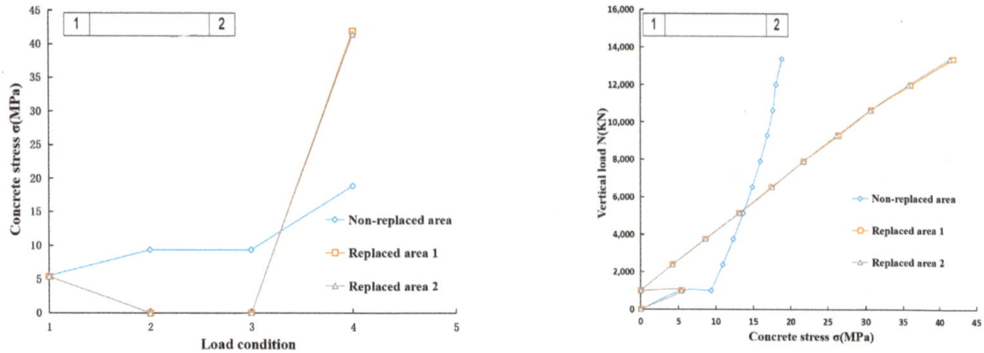

(a) (b)

Figure 7. (**a**) Stress–strain diagram under different load conditions for YPZH-9; (**b**) load–stress curve for YPZH-9.

As shown in Figure 6, at the initial stress level of 0.15, the YPZH-1 component was simultaneously dismantled in replaced areas 1 and 2, the load removed from these two parts was completely borne by the middle area that was not replaced. The stress in replaced area 1 and replaced area 2 decreased from the initial 2.4 MPa to 0, and the stress of the middle wall increased from the initial 2.4 MPa to 3.47 MPa. As shown in Figure 7, at the initial stress level of 0.35, the YPZH-9 component was simultaneously dismantled in replaced areas 1 and 2, the stress in replaced area 1 and replaced area 2 decreased from the initial 5.4 MPa to 0, and the stress of the middle wall increased from the initial 5.4 MPa to 9.4 MPa. After removing the shear wall in the replaced area, the load borne by the replaced area was completely borne by the middle area that was not replaced; however, when the replaced area had just been refilled with high-strength concrete, it did not share the load immediately, and the stress in the non-replaced area and the replaced area remained unchanged at this time; the load was continuously applied until it was destroyed and the stress of the two parts was continuously redistributed. When the replacement had just been completed, the stress in the non-replaced area was greater than that in the replaced area. Later, because the replaced area was replaced by high-strength concrete, the stress in it gradually increased, and finally, the stress in the replaced area was greater than that in the non-replaced area. In addition, after the replacement of the shear wall, the stress difference between the non-replaced and replaced areas increased with the increase in the initial stress level and the area that had been replaced.

3.2.2. Two-Batch Replacement without a Supporting Roof

When using the method of one-batch replacement reinforcement without a supporting roof at high stress levels, the load in the replaced area will all be transferred to the non-replaced area; at this point, the stress difference between the non-replaced area and the replaced area is very large, which may lead to the crushing of the concrete in the non-replaced area. Therefore, at high stress levels, using a multi-batch replacement method can effectively reduce the stress difference between the various regions of the wall and improve the cooperative working ability of each region, thus enhancing the overall performance of the wall after reinforcement.

Two batches of experimental replacement groups were tested at high stress levels; the initial stress level was between 0.4 and 0.6. The two-batch replacement was divided into six working conditions. Working condition one represented the action of initial stress; working condition two represented the removal of the wall section of replaced area 1; working condition three represented the casting of the wall section of replaced area 1; working condition four represented the removal of the wall section of replaced area 2; working condition five represented the casting of the wall section of replaced area 2, and working condition 6 represented the action of the ultimate load. This study took two components, EPZH-10 and EPZH-18, as representatives for analysis; their working condition stress diagrams and load–stress curves are shown in Figures 8 and 9.

(**a**) (**b**)

Figure 8. (**a**) Stress diagram under different load conditions for EPZH-10; (**b**) load–stress curve for YPZH-10.

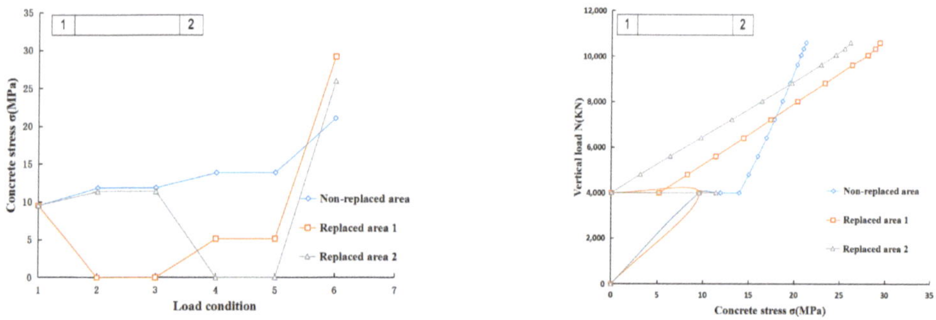

(**a**) (**b**)

Figure 9. (**a**) Stress diagram under different load conditions for EPZH-18; (**b**) load–stress curve for EPZH-18.

As shown in Figure 8, at the initial stress level of 0.4, the EPZH-10 component first demolished replaced area 1, and the stress in area 1 decreased to 0. The original 6.4 MPa stress in area 1 was borne by the middle area and area 2, and their stress increased from the initial 6.4 MPa to 7.7 MPa. After casting area 1, the stress of the middle area and area 2 had little change. In the same way, after demolishing replaced area 2, the stress in area 2 dropped to 0, and the load that was unloaded from area 2 was borne by the middle area and area 1; the stress in area 1 increased from 0 to 2.5 MPa, and the stress in the middle area increased from 7.7 MPa to 8.8 MPa. When the replacement had just been completed, the stresses in area 1, the middle area, and area 2 were 2.5 MPa, 8.8 MPa, and 0.014 MPa, respectively.

As shown in Figure 9, at the initial stress level of 0.6, the EPZH-18 component first demolished replaced area 1, and the stress in area 1 reduced to 0. The original 9.5 MPa stress in area 1 was borne by the middle area and area 2, and their stress increased from the initial 9.5 MPa to 11.4 MPa. After casting area 1, the stress of the middle area and area 2 had little change. Likewise, after demolishing replaced area 2, the stress in area 2 dropped to 0, and the load that was unloaded from area 2 was borne by the middle area and area 1; the stress in area 1 increased from 0 to 5.14 MPa, and the stress in the middle area increased from 11.4 MPa to 13.86 MPa. When the replacement had just been completed, the stresses in area 1, the middle area, and area 2 were 5.14 MPa, 13.86 MPa, and 0.017 MPa, respectively.

To sum up, when the replacement had just been completed, the stress of each wall segment was different; the stress of the shear wall in the non-replaced area was the greatest, while the stress of the shear wall in the replaced area was related to the replacement order; that is, the earlier the shear wall was replaced, the greater the stress was. Therefore, for the local replacement of the shear wall without a roof under a high stress level, the stress difference in each wall section could be reduced by increasing the number of replacement batches and optimizing the replacement sequence so as to improve the utilization rate of the concrete's strength.

3.2.3. Analysis of the Ultimate Bearing Capacity and Its Significance

The ultimate bearing capacity of each component, the stress in each region in the ultimate state, and the concrete strength utilization coefficient are shown in Table 3. According to the data in Table 3, the ultimate bearing capacity of the one-batch and two-batch local reinforcement replacements without a supporting roof was greater than the design value of 9185 KN for the bearing capacity of a shear wall, and the ultimate bearing capacity was increased by 8.8–69.8%; the ultimate bearing capacity of the profile steel reinforcement was increased by 50% on average.

Table 3. Finite element ultimate bearing capacity analysis.

Specimen	Simulation Value of the Ultimate Bearing Capacity in VFEAP (KN)	Stress in the Ultimate State (Mpa)			Concrete Strength Utilization Coefficient		
		Area 1	Non-Replaced Area	Area 2	Area 1	Non-Replaced Area	Area 2
YPZH-1	12,119	32.0	24.5	32.1	0.99	1.05	0.99
YPZH-2	14,302	40.0	24.6	39.4	0.96	1.05	0.95
YPZH-3	15,596	51.7	23.8	51.7	1.03	1.02	1.03
YPZH-4	13,556	32.6	26.3	32.8	1.01	1.12	1.01
YPZH-5	13,214	39.3	24.2	39.8	0.95	1.03	0.96
YPZH-6	14,097	46.1	24.1	46.8	0.92	1.03	0.93
YPZH-7	11,830	28.9	20.0	29.1	0.89	0.86	0.90
YPZH-8	11,561	42.8	21.4	42.7	1.03	0.91	1.03
YPZH-9	14,000	41.9	18.9	41.5	0.84	0.81	0.83
EPZH-10	11,335	28.2	23.0	27.8	0.87	0.98	0.86
EPZH-11	11,456	31.0	21.1	29.7	0.75	0.90	0.71
EPZH-12	12,620	35.9	21.4	34.4	0.71	0.91	0.69
EPZH-13	10,374	24.7	21.2	22.3	0.76	0.91	0.69
EPZH-14	10,860	28.2	21.6	27.3	0.68	0.92	0.66
EPZH-15	12,063	36.1	22.2	35.8	0.72	0.95	0.71
EPZH-16	9995	23.4	22.1	20.9	0.72	0.94	0.65
EPZH-17	11,383	27.6	22.7	26.6	0.67	0.97	0.64
EPZH-18	10,725	29.3	21.1	26.0	0.58	0.90	0.52

According to the current Code for the design of strengthened concrete structures, when the vertical RC components are strengthened using a replacement method and without a supporting roof, the strength utilization coefficient of the new concrete in the replaced area will be 0.8 when using the calculation formula for the bearing capacity of a normal section. However, as can be seen in Table 3, with the increase in the initial stress level, the strength utilization coefficient of concrete gradually decreased. When the initial stress level was less than 0.4, the strength utilization coefficient of the new concrete in the replaced area was greater than 0.8; however, when the initial stress level exceeded 0.4, the concrete strength utilization coefficient in the replaced area was mostly less than 0.8. In particular, in the subsequently replaced area, there was a greater difference between the concrete strength utilization coefficient and 0.8 with the increase in the initial stress level. To sum up, when the initial stress level was less than 0.4, it was conservative to take the strength utilization coefficient of the new concrete in the replaced area as 0.8 according to the Code; however, when the initial stress level was greater than 0.4, it was unsafe to take the value as 0.8 according to the Code.

Using the data in Table 3, an analysis of the results of the orthogonal experiment was conducted, as shown in Table 4. According to k_{ii} in Table 4, the curve of the relationship between different factors and the ultimate bearing capacity was drawn, as shown in Figures 10–12.

Table 4. Analysis of the results of the orthogonal experiment.

Factor		Initial Stress Level	Strength of the Replacement Concrete	The Size of Replaced Area
Orthogonal analysis of one-batch replacement	K1	42,017	37,505	37,777
	K2	40,867	39,077	40,640
	K3	37,391	43,693	41,858
	k_{11}	14,005.7	12,501.7	12,592.3
	k_{22}	13,622.3	13,025.7	13,546.7
	k_{33}	12,463.7	14,564.3	13,952.7
	Range (R)	1542	2062.6	1360.4
Orthogonal analysis of two-batch replacement	K1	35,411	31,704	34,781
	K2	33,297	33,699	33,475
	K3	32,103	35,408	32,555
	k_{11}	11,803.7	10,568	11,593.7
	k_{22}	11,099	11,233	11,158.3
	k_{33}	10,701	11,802.7	10,851.7
	Range (R)	1102.7	1234.7	742

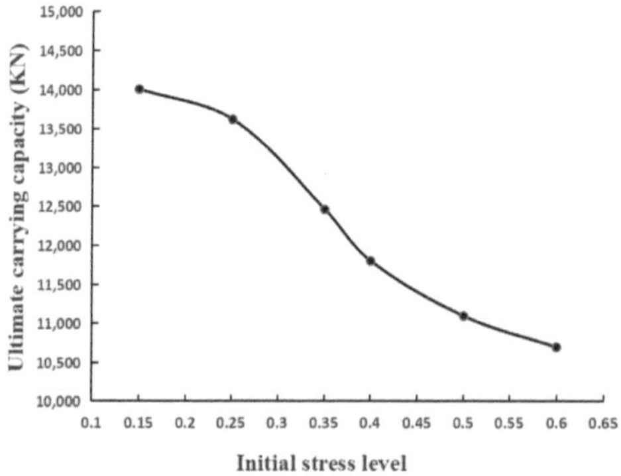

Figure 10. Relationship between the initial stress level and the ultimate bearing capacity.

As can be seen in Figure 10, the ultimate bearing capacity of the strengthened components decreased with the increase in the initial stress level. At a stress level of 0.15, the ultimate bearing capacity was 14,005 KN. At a stress level of 0.6, the ultimate bearing capacity was 10,701 KN, which was a reduction of 23.6%. It can be seen that the initial stress level had a significant impact on the ultimate bearing capacity of the strengthened components. The initial stress level represents the original stress state of the shear wall before replacement. The higher the initial stress level, the more cracks inside the shear wall develop, and the worse the reinforcement effect of the shear wall. Therefore, in order to obtain an ideal reinforcement effect, the replacement of the shear wall should be carried out at a low stress level.

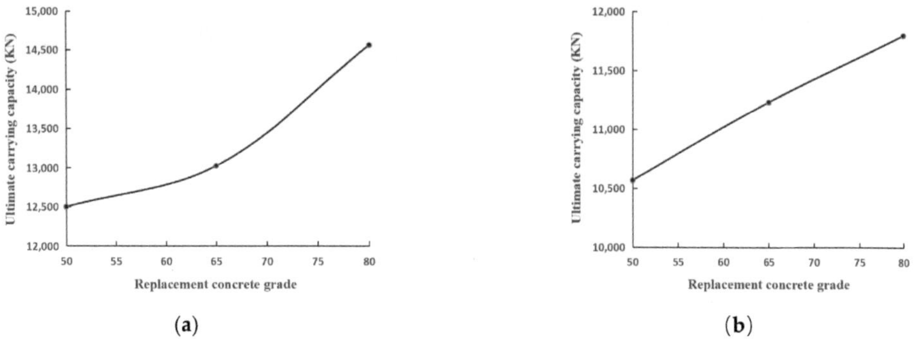

Figure 11. (**a**) For a stress level ranging from 0.15 to 0.35, the relationship between the concrete strength at displacement and the ultimate bearing capacity is shown; (**b**) for a stress level ranging from 0.4 to 0.6, the relationship between the concrete strength at displacement and the ultimate bearing capacity is shown.

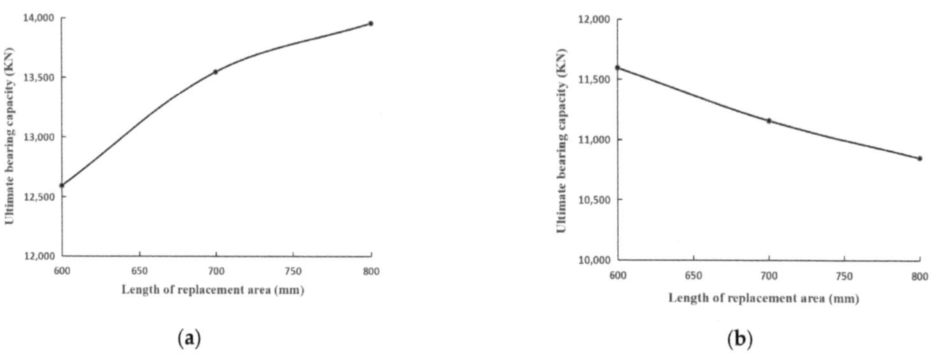

Figure 12. (**a**) For stress levels ranging from 0.15 to 0.35, the relationship between the displacement area and the ultimate bearing capacity is shown; (**b**) for stress levels ranging from 0.4 to 0.6, the relationship between the displacement area and the ultimate bearing capacity.

As can be seen in Figure 11, the ultimate bearing capacity of the strengthened components increased with the increase in the replacement concrete strength. At stress levels from 0.15 to 0.35, the replacement concrete strength of C50 had an ultimate bearing capacity of 12,501 KN, while the replacement concrete strength of C80 had an ultimate bearing capacity of 14,564 KN, which was an enhancement of 16.5%. At stress levels from 0.4 to 0.6, the replacement concrete strength of C50 had an ultimate bearing capacity of 10,568 KN, while the replacement concrete strength of C80 had an ultimate bearing capacity of 11,802 KN, which was an enhancement of 11.7%. It can be seen that the influence of the replacement concrete strength on the ultimate bearing capacity was more significant at low stress levels than at high stress levels.

As can be seen in Figure 12, at low stress levels, the ultimate bearing capacity of the strengthened components increased with the increase in the size of the replacement area; however, at high stress levels, the ultimate bearing capacity decreased with the increase in the size of the replacement area. At stress levels from 0.15 to 0.35, when the total length of the replaced wall on both sides was 600 mm, the ultimate bearing capacity was 12,592 KN; when the total length of the replaced wall on both sides was 800 mm, the ultimate bearing capacity was 13,952 KN, which was an enhancement of 10.8%. At stress levels from 0.4 to 0.6, when the total length of the replaced wall on both sides was 600 mm, the ultimate bearing capacity was 11,593 KN; when the total length of the replaced wall on both sides was 800 mm, the ultimate bearing capacity was 10,851 KN, which was a reduction of 6.4%. At low stress levels, the cracks and deformations in the shear wall are small, and the stress distribution in the shear wall is more uniform. At this time, the bearing capacity of the shear wall increases with the increase in the size of the replaced area. In contrast, at high stress levels, the shear wall may have developed obvious cracks and deformations, resulting in an uneven stress distribution in the shear wall, and cooperative work between the old and new materials becomes more difficult. At this time, the bearing capacity of the shear wall decreases with the increase in the size of the replacement area.

4. Calculation Formula for the Bearing Capacity of a Local Strengthening Replacement of an RC Shear Wall with an Unsupported Roof under Axial Compression

According to the simulation analysis in the previous section, at high stress levels, the strength utilization coefficient of the new concrete in the replaced area was lower than the value of 0.8 in the Code. Therefore, at high stress levels, the Code's value of the strength utilization coefficient of the new concrete in the replaced area is unsafe. In addition, the Code's formula does not take the stress hysteresis of the reinforced components into account but simply discounts the strength of the new concrete in the replaced area and then accumulates the bearing capacity of the new and old concrete. Based on these two points, this study comprehensively considered the influences of stress hysteresis, the initial stress level, and the strength of the replacement concrete to derive the calculation formulas for the bearing capacity of a normal section with the one-batch, two-batch, and three-batch replacements in order to perfect and supplement the formula in the Code.

In this study, the original column concrete adopts the constrained concrete constitutive model [16] considering the constraints of the stirrups in the reinforcement area. The model reflected the improvement of the concrete strength and ductility due to the stirrups in the original column by introducing the strength improvement coefficient K. The calculation formulas of the constrained concrete constitutive model are detailed in [16]. However, the unconstrained concrete constitutive model was adopted for the new concrete. The model did not consider the improvement of the concrete strength or ductility due to the stirrups. The calculation formulas of the unconstrained concrete constitutive model were similar to those of the constrained concrete constitutive model calculation; except, they did not take the strength improvement coefficient K into account. The constitutive curve is shown in Figure 13. The model for Figure 13 can be found in [16]. A bilinear constitutive model that described complete elastoplasticity was adopted for the steel bars, and the constitutive curve is shown in Figure 14.

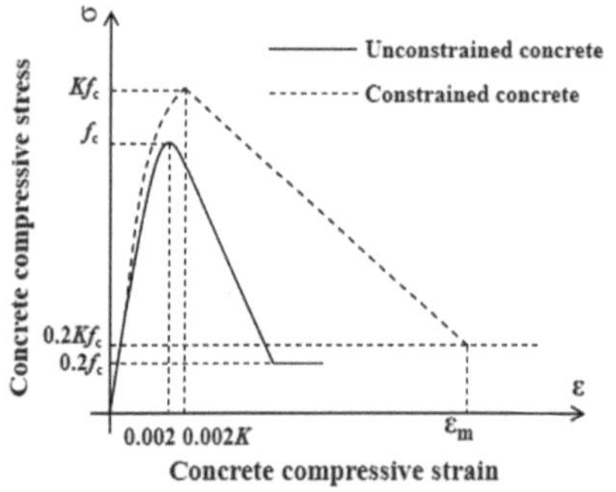

Figure 13. Kent–Park concrete model.

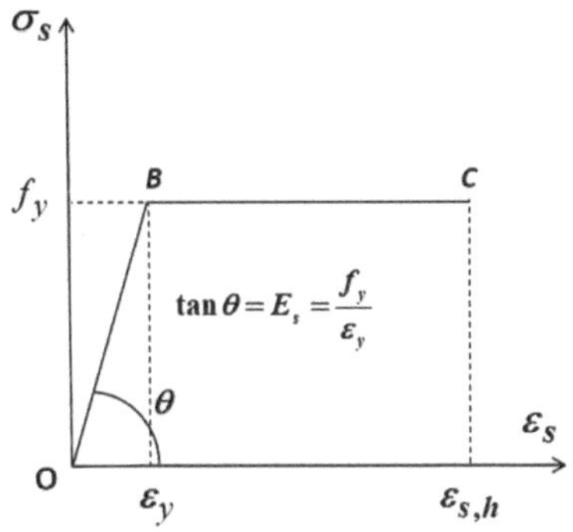

Figure 14. Fully plastic model.

4.1. Theoretical Analysis

In this section, the derivation of the formula for the ultimate bearing capacity of the three-batch RC shear walls under axial compression was taken as an example. (Due to the similar ideas regarding the derivation of the formulas, the derivation processes for the bearing capacities of the one-batch and two-batch replacements are not repeated, and the corresponding formulas will be given directly in the following section.) Finally, based on the formulas for the bearing capacities of the one-batch, two-batch, and three-batch replacements, a formula for the ultimate bearing capacity of the N-batch replacements was summarized.

4.1.1. Example of the Derivation of a Bearing Capacity Formula for a Locally Reinforced Replacement with an Unsupported Roof

The shear wall strengthened using a three-batch replacement was simplified as the theoretical model shown in Figure 15.

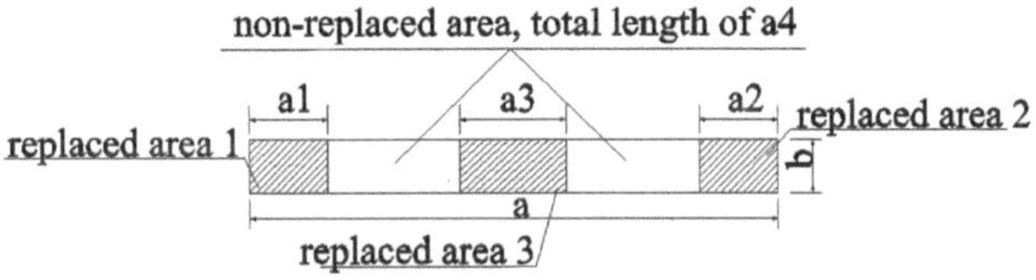

Figure 15. Theoretical modeling of the three-batch replacement reinforcement.

(1) The stress state analysis of the shear wall before replacement.

The axial pressure on the shear wall before the replacement is

$$N_1 = A_{c1}\sigma_c^1 + A_{s1}\sigma_s^1 \quad (1)$$

where σ_c^1 and σ_s^1 are, respectively, the compressive stress of the concrete and the longitudinal reinforcement of the shear wall before replacement; A_{c1} and A_{s1} are, respectively, are the cross-sectional areas of the concrete and the longitudinal reinforcement of the shear wall before replacement; and A_{c1} = a × b.

According to the constitutive relation of constrained concrete, we can obtain

$$\sigma_{c4}^1 = \sigma_c^1 = Kf_c\left[2\left(\frac{\varepsilon_c^1}{\varepsilon_{c0}}\right) - \left(\frac{\varepsilon_c^1}{\varepsilon_{c0}}\right)^2\right] \quad (2)$$

$$\varepsilon_c^1 = \left(1 - \sqrt{1 - \frac{1}{K}\sigma_c^1/f_{c0}}\right)\varepsilon_{c0} = \left(1 - \sqrt{1 - \frac{1}{K}\beta}\right)\varepsilon_{c0} \quad (3)$$

where σ_{c4}^{1} is the concrete stress of the non-replaced area at the first stage, the letter in the subscript represents the concrete and the number represents the area, such as "c2" is the concrete in the replaced area 2; the superscript represents the stage reached, such as "2" means it is in the second stage; and the descriptions of stress and strain in the following text is the same. ε_{c}^{1} is the concrete strain in the shear wall at the first stage. β is the initial stress level of the concrete in the shear wall, whereby $\beta = \sigma_{c}^{1}/f_{c0}$. f_{c0} is the compressive strength of the concrete before the replacement, which is the same later. K is the concrete strength improvement coefficient considering the effect of constraints due to the stirrups in the reinforcement area, $K = 1 + \frac{\rho_{s1} f_{yh}}{f_c}$, and it is the same later. ρ_{s1} and f_{yh} are the stirrup ratio and stirrup yield strength in the reinforcement area; $\varepsilon_{c0} = 0.002K$.

(2) The stress state analysis after removing and re-casting replaced area 1.

After removing replaced area 1, the stress unloaded from it is shared by replaced area 2, replaced area 3, and the non-replaced area; at this time, the stress and strain of replaced area 1 are both zero, and the stress and strain of the other areas are as follows:

$$\sigma_{c2}^{2} = \sigma_{c3}^{2} = \sigma_{c4}^{2} = \sigma_{c}^{1} + \sigma_{c}^{1}\frac{a_1}{a - a_1} = \frac{a}{a - a_1}\sigma_{c}^{1} \tag{4}$$

$$\varepsilon_{c2}^{2} = \varepsilon_{c3}^{2} = \varepsilon_{c4}^{2} = \left(1 - \sqrt{1 - \frac{a}{a - a_1}\frac{1}{K}\beta}\right)\varepsilon_{c0} \tag{5}$$

(3) The stress state analysis after removing and re-casting replaced area 2.

Similarly, after removing replaced area 2, the stress unloaded from it is shared by replaced area 3, the non-replaced area, and the newly added replaced area 1; at this time, the stress and strain of replaced area 2 are both zero, and the stress and strain of the other areas are as follows:

$$\sigma_{c1}^{3} = \sigma_{c2}^{2}\frac{a_2}{a - a_2} = \frac{a \times a_2}{(a - a_1) \times (a - a_2)}\sigma_{c}^{1} \tag{6}$$

$$\sigma_{c3}^{3} = \sigma_{c4}^{3} = \sigma_{c4}^{2} + \sigma_{c2}^{2}\frac{a_2}{a - a_2} = \frac{a^2}{(a - a_1) \times (a - a_2)}\sigma_{c}^{1} \tag{7}$$

$$\varepsilon_{c1}^{3} = \left(1 - \sqrt{1 - \frac{a \times a_2}{(a - a_1) \times (a - a_2)}\frac{1}{K}\beta}\right)\varepsilon_{c0} \tag{8}$$

$$\varepsilon_{c3}^{3} = \varepsilon_{c4}^{3} = \left(1 - \sqrt{1 - \frac{a^2}{(a - a_1) \times (a - a_2)}\frac{1}{K}\beta}\right)\varepsilon_{c0} \tag{9}$$

(4) The stress state analysis after removing and re-casting replaced area 3.

Similarly, after removing replaced area 3, the stress unloaded from it is shared by the non-replaced area and the newly added replaced area 1 and replaced area 2; at this time, the stress and strain of replaced area 3 are both zero, and the stress and strain of the other areas are as follows:

$$\sigma_{c1}^{4} = \sigma_{c1}^{3} + \sigma_{c3}^{3}\frac{a_3}{a - a_3} = \frac{a^2 \times (a_2 + a_3) - a \times a_2 \times a_3}{(a - a_1) \times (a - a_2) \times (a - a_3)}\sigma_{c}^{1} \tag{10}$$

$$\sigma_{c2}^{4} = \sigma_{c3}^{3}\frac{a_3}{a - a_3} = \frac{a^2 \times a_3}{(a - a_1) \times (a - a_2) \times (a - a_3)}\sigma_{c}^{1} \tag{11}$$

$$\sigma_{c4}^{4} = \sigma_{c4}^{3} + \sigma_{c3}^{3}\frac{a_3}{a - a_3} = \frac{a^3}{(a - a_1) \times (a - a_2) \times (a - a_3)}\sigma_{c}^{1} \tag{12}$$

$$\varepsilon_{c1}^{4} = \left(1 - \sqrt{1 - \frac{a^2 \times (a_2 + a_3) - a \times a_2 \times a_3}{(a - a_1) \times (a - a_2) \times (a - a_3)}\frac{1}{K}\beta}\right)\varepsilon_{c0} \tag{13}$$

$$\varepsilon_{c2}^{4}=\left(1-\sqrt{1-\frac{a^2\times a_3}{(a-a_1)\times(a-a_2)\times(a-a_3)}\frac{1}{K}\beta}\right)\varepsilon_{c0} \tag{14}$$

$$\varepsilon_{c4}^{4}=\left(1-\sqrt{1-\frac{a^3}{(a-a_1)\times(a-a_2)\times(a-a_3)}\frac{1}{K}\beta}\right)\varepsilon_{c0} \tag{15}$$

(5) The stress state analysis of the shear wall after damage.

In the ultimate state, when the concrete in the non-replaced area reaches a peak compressive strain, the increase in the strain of the non-replaced area is

$$\Delta\varepsilon=\varepsilon_{c0}-\varepsilon_{c4}^{4}=\varepsilon_{c0}\sqrt{1-\frac{a^3}{(a-a_1)\times(a-a_2)\times(a-a_3)}\frac{1}{K}\beta} \tag{16}$$

According to the assumption of a planar cross-section, after the replacement of replaced areas 1, 2, and 3 is completed, the strain produced by the concrete in the replaced areas and the non-replaced area is equal under the ultimate load. Therefore,

$$\varepsilon_{c1}^{5}=\Delta\varepsilon+\varepsilon_{c1}^{4},\ \varepsilon_{c2}^{5}=\Delta\varepsilon+\varepsilon_{c2}^{4},\ \varepsilon_{c3}^{5}=\Delta\varepsilon \tag{17}$$

After substituting the strain of replaced areas 1, 2, and 3 in the ultimate state into the constitutive relation of unconstrained concrete, the compressive stress of newly added concrete in the ultimate state of the bearing capacity can be obtained as follows:

$$\sigma_1^5=f_c\left[2\left(\frac{\varepsilon_{c1}^{5}}{\varepsilon_0}\right)-\left(\frac{\varepsilon_{c1}^{5}}{\varepsilon_0}\right)^2\right] \tag{18}$$

$$\sigma_2^5=f_c\left[2\left(\frac{\varepsilon_{c2}^{5}}{\varepsilon_0}\right)-\left(\frac{\varepsilon_{c2}^{5}}{\varepsilon_0}\right)^2\right] \tag{19}$$

$$\sigma_3^5=f_c\left[2\left(\frac{\varepsilon_{c3}^{5}}{\varepsilon_0}\right)-\left(\frac{\varepsilon_{c3}^{5}}{\varepsilon_0}\right)^2\right] \tag{20}$$

here $\varepsilon_0=0.002$. The modified concrete strength utilization coefficients in replaced areas 1, 2, and 3 are defined as $\alpha_{c1}=\sigma_1^5/f_c$, $\alpha_{c2}=\sigma_2^5/f_c$, and $\alpha_{c3}=\sigma_3^5/f_c$, respectively. Then, we can obtain the modified strength utilization coefficient of the newly added concrete α_{c1}, α_{c2}, and α_{c3} using the following:

$$\alpha_{c1}=2K\sqrt{\left(1-\frac{C_1}{K}\beta\right)\left(1-\frac{C_3}{K}\beta\right)}+K(C_1+C_3)\beta-K^2 \tag{21}$$

$$\alpha_{c2}=2K\sqrt{\left(1-\frac{C_2}{K}\beta\right)\left(1-\frac{C_3}{K}\beta\right)}+K(C_2+C_3)\beta-K^2 \tag{22}$$

$$\alpha_{c3}=2K\sqrt{1-\frac{C_3}{K}\beta}+KC_3\beta-K^2 \tag{23}$$

To sum up, for the shear wall under axial compression, the ultimate bearing capacity formula for the three-batch local strengthening replacement method with an unsupported roof is as follows:

$$N_u\le\varphi\left[f_{c0}A_{c0}+\alpha_{c1}f_cA_{c1}+\alpha_{c2}f_cA_{c2}+\alpha_{c3}f_cA_{c3}+f_{y0}'A_{s0}'\right] \tag{24}$$

where C_1, C_2, and C_3 are the ratios of the length of the shear wall section, $C_1=\frac{a^2\times(a_2+a_3)-a\times a_2\times a_3}{(a-a_1)\times(a-a_2)\times(a-a_3)}$, $C_2=\frac{a^2\times a_3}{(a-a_1)\times(a-a_2)\times(a-a_3)}$, and $C_3=\frac{a^3}{(a-a_1)\times(a-a_2)\times(a-a_3)}$, where a is the total length of the shear wall, and a_1, a_2, and a_3 are, respectively, the lengths of replaced areas 1, 2, and 3; φ is the stability factor for the compression members.

For the shear wall under axial compression, the ultimate bearing capacity formula for the two-batch local strengthening replacement method with an unsupported roof is as follows:

$$N_u \leq \varphi\left[f_{c0}A_{c0} + \alpha_{c1}f_cA_{c1} + \alpha_{c2}f_cA_{c2} + f'_{y0}A'_{s0}\right] \tag{25}$$

where α_{c1} is the modified strength utilization coefficient of the newly added concrete in replaced area 1, $\alpha_{c1} = 2K\sqrt{\left(1 - \frac{C_1}{K}\beta\right)\left(1 - \frac{C_2}{K}\beta\right)} + K(C_1 + C_2)\beta - K^2, 0 < \alpha_{c1} \leq 1$; α_{c2} is the modified strength utilization coefficient of the newly added concrete in replaced area 2, $\alpha_{c2} = 2K\sqrt{1 - \frac{C_2}{K}\beta} + KC_2\beta - K^2, 0 < \alpha_{c2} \leq 1$; C_1, C_2 are the ratios of the length of the shear wall section, $C_1 = \frac{a \times a_2}{(a-a_1)\times(a-a_2)}$, $C_2 = \frac{a^2}{(a-a_1)\times(a-a_2)}$, where a is the total length of the shear wall, a_1 is the length of replaced area 1, and a_2 is the length of replaced area 2.

For the shear wall under axial compression, the ultimate bearing capacity formula for the one-batch local strengthening replacement method with an unsupported roof is as follows:

$$N_u \leq \varphi\left[f_{c0}A_{c0} + \alpha_{c1}f_cA_c + f'_{y0}A'_{s0}\right] \tag{26}$$

where α_{c1} is the modified strength utilization coefficient of the newly added concrete in the replaced area, $\alpha_{c1} = 2K\sqrt{1 - \frac{C}{K}\beta} + CK\beta - K^2$; C is the ratio of the length of the shear wall section, $C = a/a_3$, where a_3 is the length of the non-replaced area, and a is the total length of the shear wall.

4.1.2. Regression Analysis of the Formula for Bearing Capacity of N-Batch Replacements

Based on the bearing capacity formulas for the one-batch, two-batch, and three-batch locally reinforced replacements mentioned above, the ultimate bearing capacity formula for the N-batch locally reinforced replacements with an unsupported roof can be summarized as follows:

$$N_u \leq \varphi\left[f_{c0}A_{c0} + \alpha_{c1}f_cA_{c1} + \alpha_{c2}f_cA_{c2} + \ldots\ldots + \alpha_{cn}f_cA_{cn} + f'_{y0}A'_{s0}\right] \tag{27}$$

where α_{c1} is the modified strength utilization coefficient of the newly added concrete in replaced area 1, $\alpha_{c1} = 2K\sqrt{(1 - \frac{C_1}{K}\beta)(1 - \frac{C_n}{K}\beta)} + K(C_1 + C_n)\beta - K^2, 0 < \alpha_{c1} \leq 1$; $\alpha_{c(n-1)}$ is the modified strength utilization coefficient of the newly added concrete in replaced area (N−1), $\alpha_{c(n-1)} = 2K\sqrt{(1 - \frac{C_{(n-1)}}{K}\beta)(1 - \frac{C_n}{K}\beta)} + K(C_{(n-1)} + C_n)\beta - K^2, 0 < \alpha_{c(n-1)} \leq 1$; α_{cn} is the modified strength utilization coefficient of the newly added concrete in replaced area N, $\alpha_{cn} = 2K\sqrt{1 - \frac{C_n}{K}\beta} + KC_n\beta - K^2, 0 < \alpha_{cn} \leq 1$; and C_n is the ratio of the length of the wall section in replaced area N, $C_n = \frac{a^n}{(a-a_1)(a-a_2)\cdots(a-a_n)}$.

For Equation (27) to have a reliable probability, a regression analysis was used to establish a practical calculation formula for a 95% guarantee. At the same time, considering randomness, the ultimate bearing capacity of the RC components strengthened using the local strengthening replacement method with an unsupported roof was regarded as a random variable. In order to clarify its probability distribution, the distribution law was verified with a probability density function in combination with mathematical statistics software. The results showed that the bearing capacity of the strengthened components conformed to the normal distribution law, so the following model was adopted:

$$\hat{N} = \sigma N$$

According to the results of the nonlinear analysis in the third section, the numerical values of random variables corresponding to the bearing capacity of each shear wall were obtained by using the above formula. Figure 16 shows a statistical histogram of the random variables, and it can be intuitively seen that they approximately obey the normal distribution. According to the mathematical analysis, the average value is 1.03, and the standard deviation is 0.03.

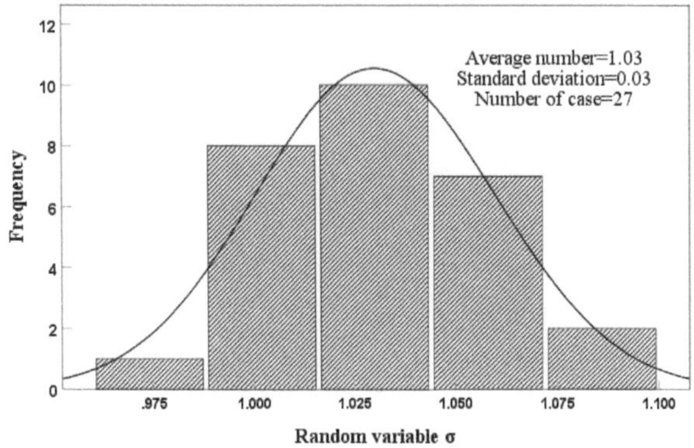

Figure 16. Histogram of the random variable.

Using the K–S method, the probability distribution of the hypothetical random variables was tested for goodness of fit. The sample size n was 27, the maximum observed $D_{27} = 0.03 < D_{27}^{0.05} = 0.26$ indicated that the test was accepted, and the random variable followed a normal distribution, with an average value of 1.03 and a standard deviation of 0.03. Therefore, the final formula for strengthening the bearing capacity of the RC members using the local strengthening replacement method without a supporting roof is

$$\hat{N} = (1.03 - 0.03 \times 4.325)\, N = 0.9N$$

Finally, after finishing, the bearing capacity formula for the RC members strengthened using the unsupported local strengthening replacement method is as follows:

$$N_u \leq 0.9\varphi\left[f_{c0}A_{c0} + \alpha_{c1}f_cA_{c1} + \alpha_{c2}f_cA_{c2} + \ldots\ldots + \alpha_{cn}f_cA_{cn} + f'_{y0}A'_{s0}\right] \quad (28)$$

4.2. Formula Reliability Test

According to the results of the simulation analysis and the collected test data, the ratios of the values calculated using the formula in this study to the values calculated using the formula in the Code were obtained by substituting the parameters of each specimen into each formula, as shown in Table 5. According to the data in Table 5, a comparison between this formula and the Code's formula could be drawn, as shown in Figures 17 and 18. As can be seen in the figures, the values calculated with this formula were all less than the experimental simulation values. When compared with the Code's formula, the formula in this study was more consistent with the experimental simulation results, and the dispersion is smaller. To sum up, the formula in this study is more reliable than the Code's formula.

Table 5. Correlated experimental data.

Data Source	Specimen	Finite Element Simulation Value N_1 (KN)	The Calculated Value N_2 of the Formula in this Paper (KN)	Ratio of N_1 to N_2	The Calculated Value N_3 of the Code's Formula (KN)	Ratio of N_1 to N_3
VFEAP simulation in this study	YPZH-1	12,119	10,980	1.104	10,441	1.161
	YPZH-2	14,302	13,065	1.095	10,758	1.329
	YPZH-3	15,596	13,615	1.146	11,754	1.327
	YPZH-4	13,556	12,170	1.114	9578	1.415
	YPZH-5	13,214	12,190	1.084	10,846	1.218
	YPZH-6	14,097	13,263	1.063	12,529	1.125
	YPZH-7	11,830	11,121	1.064	10,353	1.143
	YPZH-8	11,561	11,552	1.001	11,387	1.015
	YPZH-9	14,000	13,184	1.062	11,213	1.249
	EPZH-10	11,335	10,660	1.063	10,441	1.086
	EPZH-11	11,456	11,353	1.009	10,758	1.065
	EPZH-12	12,620	11,978	1.054	11,754	1.074
	EPZH-13	10,374	10,218	1.015	9578	1.083
	EPZH-14	10,860	10,699	1.015	10,876	0.999
	EPZH-15	12,063	11,869	1.016	12,529	0.963
	EPZH-16	9995	9886	1.011	10,353	0.965
	EPZH-17	11,383	10,657	1.068	11,397	0.999
	EPZH-18	10,725	10,289	1.042	11,213	0.956
Yao Zexin [17]	ZH-1	9065	5922	1.531	5550	1.633
	ZH-2	9647	8348	1.156	7734	1.247
	ZH-3	10,055	10,010	1.004	9903	1.015
	ZH-4	8664	7471	1.160	6924	1.251
	ZH-5	8988	8483	1.060	8029	1.119
	ZH-6	9459	8522	1.110	8212	1.152
	ZH-7	8481	7399	1.146	7224	1.174
	ZH-8	8861	7125	1.244	7123	1.244
	ZH-9	9319	8914	1.045	8812	1.058

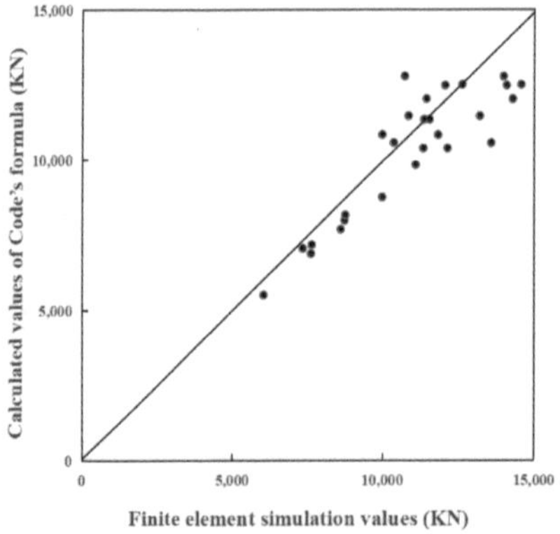

Figure 17. Comparison of the values calculated using the Code's formulas with the experimental simulation values.

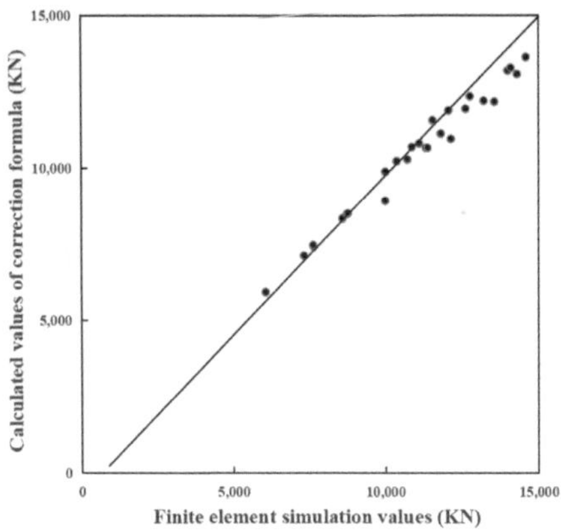

Figure 18. Comparison of the values calculated using the formulas in this study with the experimental simulation values.

5. Conclusions

(1) The three-dimensional solid degraded virtual laminated nonlinear finite element program VFEAP can simulate the actual failure processes and deformation of structures well. The simulated results were in good agreement with the test results, which indicated that the program is suitable for simulating the bearing performance of the shear walls after replacement.

(2) After shear wall replacement, the stress difference between the non-replaced area and replaced area increased with an increase in the initial stress level and the size of the replaced area; the larger the stress difference was, the more pronounced the phenomenon of stress hysteresis became. When the replacement had just been completed, the stress of the shear wall was not the same; the stress of shear wall in the non-replaced area was the greatest, while the stress of the shear wall in the replaced area was related to the replacement order, that is, the earlier the shear wall was replaced, the greater the stress was.

(3) The replacement concrete strength, the initial stress level, and the size of the replaced area had a significant influence on the bearing capacity of the shear wall after the replacement. The ultimate bearing capacity increased with an increase in the replacement concrete strength; it decreased with an increase in the initial stress level. At low stress levels, it increased with an increase in the size of the replaced area, while at high stress levels, it decreased with an increase in the size of the replaced area.

(4) Considering the influence of the replacement concrete strength, the initial stress level, and the size of the replaced area, then introducing a strength improvement coefficient considering the constraints of the stirrups, the modified strength utilization coefficient of the new concrete in the replaced area was established. Finally, a bearing capacity formula for a shear wall reinforced using a multi-batch local reinforcement replacement method was derived, and it had a higher degree of coincidence with the test simulation results, as well as a smaller degree of dispersion, and a higher degree of reliability than the Code's formula.

Author Contributions: Conceptualization, Y.J., G.W. and Y.L.; methodology, G.W. and Y.L.; investigation, Y.L.; formal analysis, Y.L., X.W., J.D. and N.K.; validation, Y.J. and G.W.; date curation: Y.L.; writing—original draft, Y.L.; writing—review and editing, Y.L., Y.J. and G.W.; supervision, Y.J. and G.W.; project administration, Y.J. All authors have read and agreed to the published version of the manuscript.

Funding: The authors gratefully acknowledge the research grant provided by the National Natural Science Foundation of China (No. 52268050 (January 2023–December 2026)).

Data Availability Statement: All data are contained within the article.

Conflicts of Interest: The authors confirm that they have no conflicts of interest.

References

1. Lei, T.; Liu, Y.; Dong, J. Replacement reinforcement and construction monitoring of high-rise concrete shear wall. *Sci. Technol. Eng.* **2020**, *20*, 11295–11301.
2. Che, J.; Song, X. Strengthening construction technology of concrete shear wall members in high-rise buildings. *Dev. Orientat. Build. Mater.* **2023**, *21*, 142–144.
3. *GB 50367-2013*; Ministry of Housing and Urban-Rural Development of the People's Republic of China. Code for Design of Reinforcement of Concrete Structures; China Building Industry Press: Beijing, China, 2013.
4. Jia, Y.; Wei, X.; Wu, G.; Yao, Z.; Deng, J.; Li, A.; Yuan, Z. Simulation analysis of RC shear wall by unsupported local strengthening replacement method. *J. Nanchang Univ. Sci. Ed.* **2023**, *47*, 69–77.
5. Wang, F.; Luo, X.; Wu, M.; Li, H.; Wan, J.; Yu, C. Analysis and calculation of vertical bearing capacity for RC shear walls with unsupported local concrete replacement. *J. Basic Sci. Eng.* 2024, *in press*.
6. Liu, Y. Nonlinear Simulation Analysis of Local Reinforced Replacement Method for Shear Wall Structure Considering Construction Process. Master's Thesis, Nanchang University, Nanchang, China, 2021.
7. Wang, X.; Li, J.; Hu, S.; Li, S.; Xu, Y. Experimental study on seismic performance of local concrete replacement of shear walls. *Build. Struct.* **2023**, *53*, 55–62.
8. Ma, L.; Dong, H. Experimental analysis of seismic performance of shear walls strengthened by local replacement of highly ductile concrete. *China Cem.* **2024**, *8*, 91–93.
9. Song, Y. Performance analysis of high ductility concrete for shear wall replacement applications. *New Technol. New Prod. China* **2024**, *12*, 92–94.
10. Ling, D.; Xu, X. *Nonlinear Finite Element Method and Program*, 1st ed.; Zhejiang University Press: Hangzhou, China, 2004.
11. Wei, X.; Lu, L.; Ke, N.; Jia, Y.; Wu, G. A study on the effects of punch-to-span ratio and longitudinal reinforcement eigenvalues on the bearing capacity of RC Slab–Column connections. *Buildings* **2024**, *14*, 208. [CrossRef]
12. Ling, D.; Zhang, J.; Xiang, Y. Virtual laminated element method and its application in bridge engineering. *J. Civ. Eng.* **1998**, *3*, 22–29.
13. Chen, H.; Saripu, A.F.; Yu, T. *Constitutive Equations of Concrete and Soil*; Yu, T.; Wang, X.; Liu, X., Translators; China Building Industry Press: Beijing, China, 2004.
14. Zhang, Z. The Compressive Bearing Capacity of Normal Section of RC Short Column Reinforced by Section-Increased Method Considering Secondary Loading. Master's Thesis, Fuzhou University, Fuzhou, China, 2016.
15. Lei, T.; Liu, Y.; Li, K.; Jiang, D.; Dong, J. Stress redistribution of shear wall structure replacement reinforcement. *J. Build. Sci. Eng.* **2020**, *37*, 132–141.
16. Scott, B.D.; Park, R.; Priestley, M.J.N. Stress-strain behavior of concrete confined by overlapping hoops at low and high strain rates. *J. Proc.* **1982**, *79*, 13–27.
17. Yao, Z. Nonlinear Simulation Study of RC Shear Wall Structure Strengthened by Unsupported Local Strengthening Replacement Method Considering Different Factors. Master's Thesis, Nanchang University, Nanchang, China, 2022.

buildings

MDPI

Article

Analysis of Factors Affecting the Seismic Performance of Widened Flange Connections in Mid-Flange H-Beams and Box Columns

Tigist Demssice Gemechu and Linfeng Lu *

School of Civil Engineering, Chang'an University, 75 Chang'an Middle Rd, Xi'an 710062, China; 2021028906@chd.edu.cn
* Correspondence: lulinfeng@chd.edu.cn

Abstract: Following the Northridge and Kobe earthquakes, research has increasingly focused on achieving high ductility in beam-to-column connections. This study investigates the seismic performance of connections featuring widened beam-end flanges in mid-flange H-beams and box columns, an area with limited prior research compared to I-section columns and narrow-flange H-beams. Detailed finite element modeling using ABAQUS 6.1.4 demonstrates that widened beam-end flanges significantly improve bending capacity and ductility by relocating the plastic hinge away from the connection, thereby enhancing seismic resilience. Key findings include the identification of optimal design parameters: flange length ranging from 0.55 to 0.75 times the beam flange width, beam flange cutting length between 0.36 and 0.39 times the beam depth, and flange cutting depth from 0.19 to 0.23 times the beam flange width. These parameters ensure effective plastic hinge development and improved structural performance. This study introduces a novel approach that emphasizes geometric optimization over material-based enhancements, offering a cost-effective and practical solution for improving seismic performance and extending previous research insights.

Keywords: steel moment frame; seismic performance; connection; box column; widened beam-end flange

1. Introduction

As the nation progresses, the push for energy-saving and emission reduction has driven the move away from traditional building materials high in energy consumption and pollution. Steel structures, with their lightweight, strong, seismically resilient, highly industrialized, and environmentally friendly advantages, are anticipated to become mainstream in the construction industry [1]. The beam–column connection, vital to a steel frame's seismic performance, must effectively balance deformation and energy dissipation. Design principles emphasize "strong columns, weak beams" and "strong joints, weak members". In China's steel frames, rigid connections, including bolt, weld, and hybrid types, are commonly used [2]. However, the brittle failure of traditional welded joints during the Northridge (1994) and Kobe (1995) earthquakes has led to a shift toward high-ductility connections [3]. Researchers have since focused on enhancing beam–column connections by either reinforcing the beam end to shift the plastic hinge away from the joint or weakening the beam to ensure plastic deformation occurs before the beam ends, thus improving ductility and preventing brittle failure [4]. Developing high-ductility beam–column connections has become a critical area of study [5].

Recent studies emphasized the importance of joint ductility in steel frames for better seismic performance. Han et al. [6] analyzed the behavior of steel frames under cyclic loads and developed a high-performance joint design that enhances energy dissipation. Their findings align with earlier studies and further highlight the need for improved ductility in beam–column connections to mitigate the impact of seismic forces.

Yang et al. [7] also conducted an experimental investigation into the cyclic performance of UHPSFRC-filled thin-walled steel tubular columns subjected to combined axial compression and cyclic lateral displacement loading. Their study meticulously analyzed failure modes, hysteretic behaviors, and energy dissipation capacities. The research revealed that UHPSFRC significantly enhances the cyclic behavior of thin-walled steel tubular columns compared to high-strength concrete (HSC) and normal-strength concrete (NSC). Notably, UHPSFRC-filled columns demonstrated superior ductility, stiffness, and energy dissipation, making them more resilient to seismic activities. The investigation also highlighted the influence of various design parameters, such as steel tube thickness and steel fiber volume ratio, on the seismic performance of these columns.

Further enhancing this understanding, a study by Wei et al. [8] focused on concrete-filled steel tubular (CFST) composite columns with ultra-high-performance concrete (UHPC) plates. Their research revealed that UHPC improved the specimen's initial stiffness by 13.7% and increased hysteretic energy dissipation by 41.2%. Additionally, the specimens demonstrated excellent seismic performance, even under mainshock–aftershock sequences, indicating the structure's ability to withstand multiple earthquakes. This highlights the critical role of innovative materials in enhancing the seismic performance of steel and composite structures.

Two main types of improved connections have been proposed to shift the plastic hinge outward, as illustrated in Figure 1. The first type is the beam-end reinforced connection, designed to increase joint stiffness. By enhancing the bending stiffness at the beam–column connection, the beam flange can plastically deform away from the column flange surface, forming a plastic hinge that reduces stress concentration at the beam–column groove weld. The second type, shown in Figure 2, is a weakened connection. This approach aims to induce plastic deformation in the weakened part of the beam before the beam ends, enhancing joint ductility and protecting the joint from brittle failure.

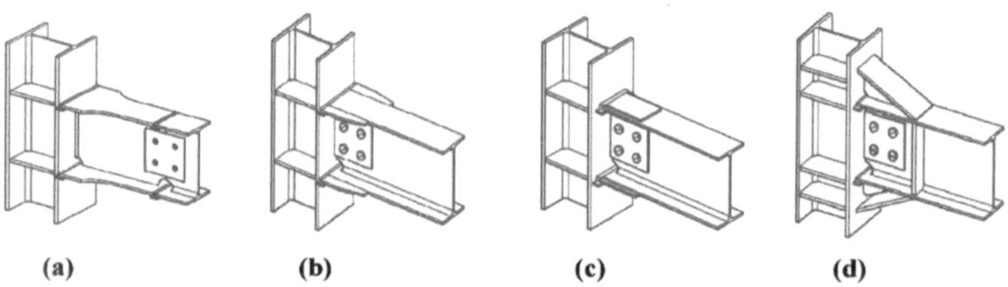

Figure 1. Enhanced connections. (**a**) Widened flange; (**b**) Side plate reinforced; (**c**) Cover reinforced; (**d**) Haunch reinforcement.

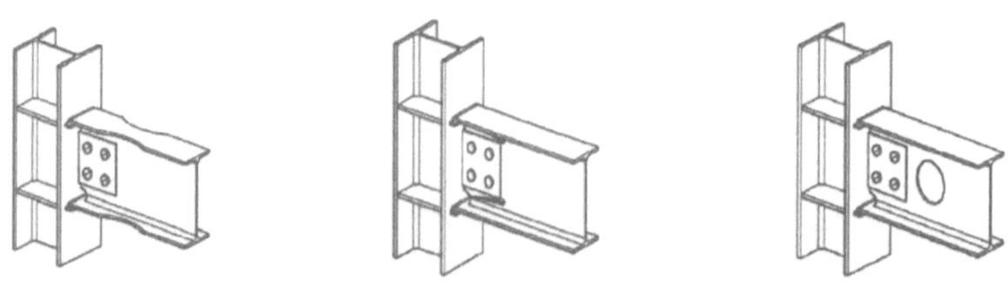

Figure 2. Weakened connections. (**a**) RBS; (**b**) Beam web slit; (**c**) Beam web opening.

Various studies have explored the seismic performance of beam–column connections with expanded beam flanges. Wang [9] found that flange expansion in large-scale joints significantly reduces stress concentration and meets FEMA standards. Nia et al. [10] showed that joints with expanded flanges and shear cover plates improve seismic performance, achieving a plastic angle of 0.06 rad. Lu et al. [11] emphasized the importance of weld quality in hybrid connections. Ma et al. [12] tested the impact of flange size on cyclic behavior, finding that while flange expansion slightly affects stiffness and energy dissipation, it influences plastic hinge positioning. Wang [13] further analyzed the effects of beam flange widening, concluding that appropriate weakening and flange size adjustments enhance seismic performance, although excessive widening yields diminishing returns.

Recent research further advanced the understanding of beam–column connections. Zohdi et al. [14] investigated the seismic performance of enlarged beam section connections, highlighting joint design and performance improvements. Xiao et al. [15] conducted finite element analysis on prefabricated semi-rigid concrete beam–column joints with steel connections, providing new insights into their behavior under seismic loads. Han et al. [6] explored the seismic behavior of high-performance steel weakened beam–column joints through experimental tests and numerical analysis, offering valuable data on the effectiveness of various connection designs.

Despite extensive research on the seismic performance of beam–column connections with I-section columns and narrow-flange H-beams, the behavior of mid-flange H-beams, which are increasingly used in modern engineering, remains underexplored. The unique characteristics of widened beam-end flanges in mid-flange H-beams and box columns, particularly their influence on plastic hinge formation and stress distribution, require further investigation. While the outward shift of the plastic hinge reduces weld stress, it introduces new challenges, such as increased deformation in the beam web and potential local buckling. This study addresses these gaps by conducting detailed finite element simulations to evaluate the seismic performance of these connections. The findings will provide insights into improving connection design and enhancing steel structure stability and resilience in modern construction practices.

2. Design Specification for Widened Flange Connection

This study examines the expanded flange joint with a weld-bolt hybrid connection, as shown in Figure 3.

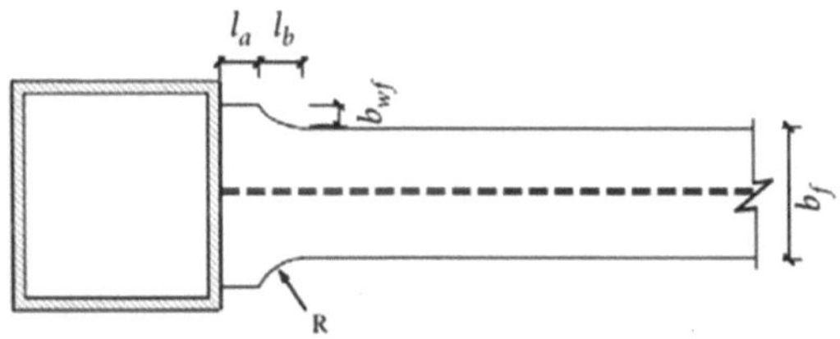

Figure 3. Shapes and sizes of beam.

A widened beam flange connection was proposed to shift the plastic hinge on the beam away from the column [16]. In this design, the beam flanges are enlarged at the beam ends near the column, reinforcing the beam-to-column connection. This reinforcement aims to increase the beam's flexural capacity at the interface, as illustrated in Figure 3. With the widened beam flange at the joint, the plastic hinge is expected to form on the beam section beyond the widened flange. When the auxiliary expansion type connects the flange at the beam's end, the length of the added auxiliary section should follow these dimensions: l_a = 0.50–0.75 b_f, l_b = 0.30–0.45 h_b, and b_{wf} = 0.15–0.25 b_f, where l_a and l_b are the widened flange length, beam flange cutting length, and b_{wf} is the cutting depth of the widened flange [17]:

$$R = \frac{l_b^2 + b_{wf}^2}{2b_{wf}} \tag{1}$$

3. Numerical Analysis Methods

3.1. Numerical Modeling

To fully account for the effects of boundary conditions and joint performance on the upper and lower columns and beam ends, this simulation uses an external-column joint in a steel frame between beam–column inflection points under horizontal load. Typically, a column's inflection point is located at its midpoint under horizontal load, while a beam's inflection point depends on the load position and geometry of the column and beam. For a horizontal load, the reverse bending point is at mid-span; for a vertical load, it is approximately 1/3 to 1/4 of the span from the beam end. This study's beam and column dimensions are 3000 mm and 3600 mm (measured from the hinge center).

The ABAQUS finite element analysis technology is mature and reliable. In this analysis, the beam–column connection is expected to deform locally. To improve accuracy, the C3D8I element was chosen for its ability to handle shear locking and contact-heavy models, requiring minimal elements for bending analysis. The mesh is refined at critical areas like plastic hinges and contact points, while a coarser mesh is used at distant locations to save computation time, as shown in Figure 4. Sweep meshing is applied to screw holes, and structured meshing is used elsewhere to maintain mesh quality.

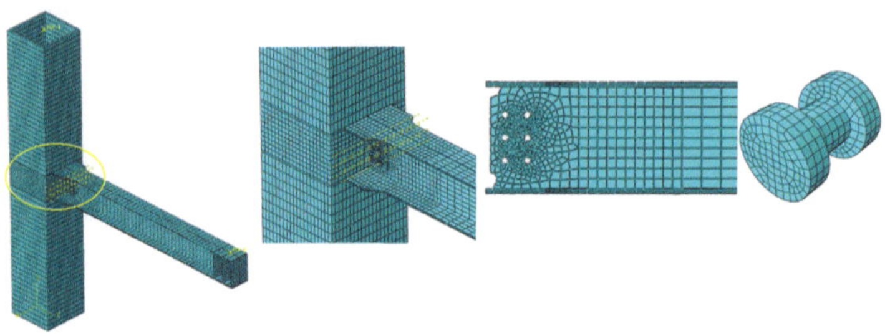

(**a**) (**b**) (**c**)

Figure 4. Mesh of a specimen. (**a**) Overall mesh of a specimen and joint; (**b**) Mesh of screw holes and beam; (**c**) Mesh of a bolt.

Contact is nonlinear, making analysis and convergence challenging. With pre-tension applied to high-strength bolts, various contact relationships form. To simplify modeling, bolts are represented with circular heads, smooth surfaces, and no threads. Key contacts include plate-to-beam web, bolt head-to-web, and screw-to-hole wall, simulated using the ABAQUS "surface-to-surface" method. The higher stiffness surface is the master, and the lower is the slave. Normal contact is defined as "hard", with surfaces touching under pressure and separating otherwise, while tangential contact is modeled as "Coulomb friction", using the actual anti-slip coefficient 0.44 [18] of the friction surface of the bolt connection. The design pre-tension value is given according to JGJ82-2010 [19], and the "bolt load" of ABAQUS is used to simulate. The "TIE" command to simulate the welding; the local schematic diagram of contact and binding constraints of the finite element model is shown in Figure 5. Reference coupling points are placed at the top, bottom, and beam end. The column bottom restricts all translational movement, the column top restricts out-of-plane movement, and the beam end restricts out-of-plane and vertical movement. Lateral support is simulated by restricting out-of-plane movement at the beam flange. Vertical axial force and horizontal cyclic displacement are applied to the column top.

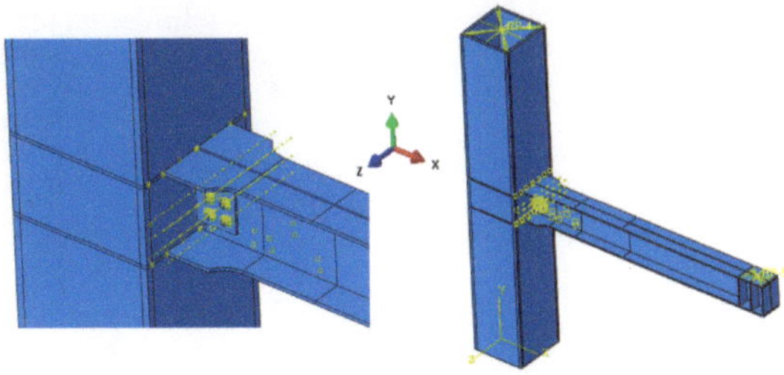

(a) (b)

Figure 5. Constraints and boundary conditions. (**a**) Contact and binding constraints; (**b**) Boundary conditions.

3.2. Material Property Assumptions and Parameters

As shown in Figure 6, assuming the steel and bolt materials follow a three-line isotropic hardening model with a corresponding stress–strain curve, σ_y, σ_u, and σ_{st} represent the yield, maximum, and failure stresses, while ε_y, ε_u, and ε_{st} are the associated strains. Q235 steel's properties are averaged from the stress–strain data of Nie et al. [20] and 8.8 grade M20 bolt properties based on Kontolati et al. [21], as listed in Table 1.

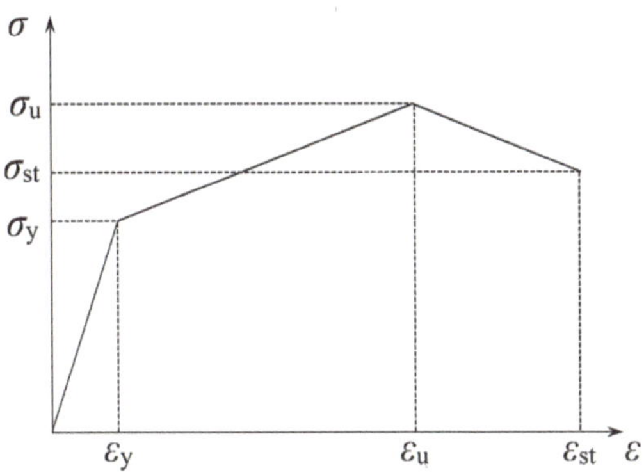

Figure 6. Three-line stress–strain relationship curve.

Table 1. Material properties of Q235 steel and 8.8 grade bolt.

Material Samples	E/MPa	σ_y/MPa	ε_y	σ_u/MPa	ε_u	σ_{st}/MPa	ε_{st}
Q235 steel	197,000	320	0.01748	451.667	0.14361	360.643	0.19653
8.8 grade M20 bolt	161,000	645.0	0.004	793.5	0.025	636.9	0.121

3.3. Validation

The SPS-2 specimen by Gao [22] is selected with specific dimensions, and the finite element model is shown in Figure 7. The beam and column specimens were selected as hot-rolled steel sections HN300 × 150 × 6.5 × 9 and HW250 × 250 × 9 × 14, respectively. The steel material is Q235, with a measured yield strength of 299.2 MPa and tensile strength of 420.6 MPa. The specimen loading protocol followed ANSI/AISC 341-05 [23], with six cycles at inter-story drift angles of 0.375% rad, 0.5% rad, 0.75% rad, and four cycles at 1% rad. For drift angles of 1.5% rad, 2% rad, 3% rad, and 4% rad, two cycles were applied for each. Afterwards, for every additional 1% rad, two cycles are applied. Loading was stopped when the bearing capacity dropped to 85% of the maximum load. Following the test loading protocol, displacement-controlled cyclic loading is applied at the beam end [22].

During the test, the column top force is 850 kN (the corresponding axial compression ratio is 0.4), the bolt pre-tension is 155 kN, and the friction coefficient between the connection plate, web, and bolts is 0.35.

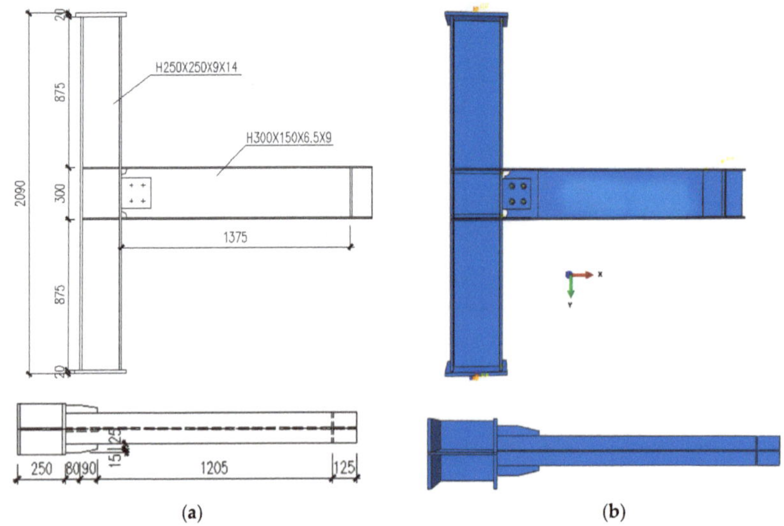

Figure 7. SPS-2 Specimen. (**a**) Schematic diagram of dimensions; (**b**) FE Model.

The P–Δ hysteresis curve obtained from the experiment for the SPS-2 specimen is shown in Figure 8a, while the P–Δ hysteresis curve obtained from the finite element analysis is shown in Figure 8b. The failure modes of the SPS-2 specimen from the experiment and the finite element analysis are shown in Figures 9a and 9b, respectively.

The failure modes of the SPS-2 specimen from the experiment and the finite element model are generally consistent. Although there are some differences between the experimental and finite element hysteresis curves, their shapes are similar, showing a spindle shape with near-symmetry and no significant pinching. The maximum load obtained from the finite element model was 164.23 kN, while the experimental maximum load was 158.97 kN, with a 3.3% error. This indicates that the numerical analysis method used in this study is highly reliable.

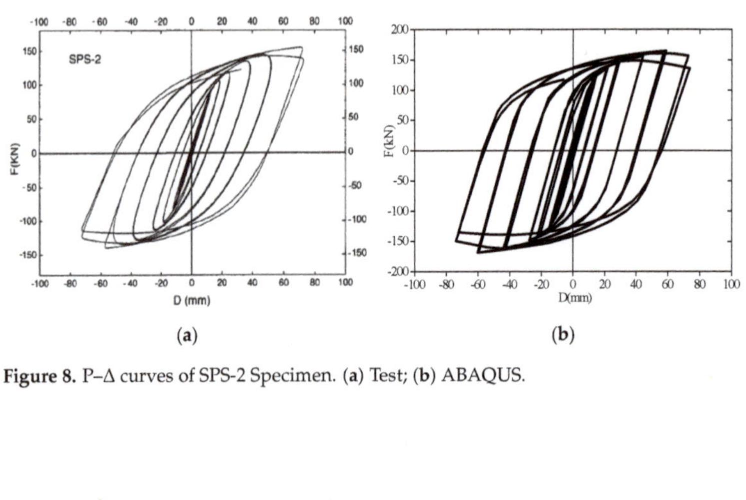

Figure 8. P–Δ curves of SPS-2 Specimen. (**a**) Test; (**b**) ABAQUS.

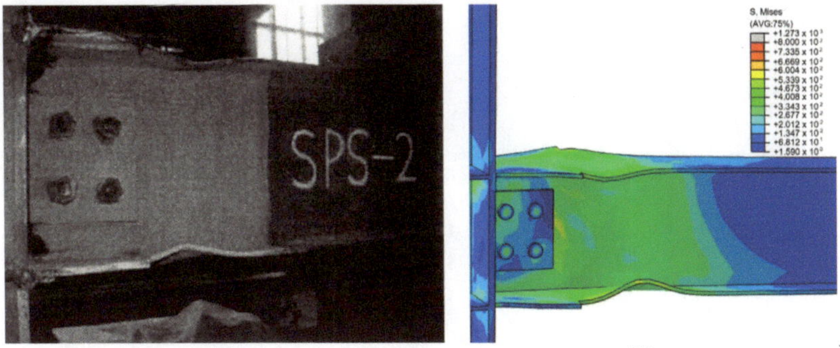

(**a**) (**b**)

Figure 9. Failure modes of SPS-2 Specimen. (**a**) Test; (**b**) ABAQUS.

4. Design of Analysis Specimens

Three key design parameters were investigated for the widened flange connection: l_a, l_b, and b_{wf}. Three series were analyzed: WFS-250A, WFS-250B, and WFS-250C. Six joint configurations were modeled for each series for finite element analysis (see Table 2, and R is calculated by Equation (1)). A connection with no widened flange is termed "UW" for comparison. The column was box-section 500 × 500 × 16, and the H-shape beam was HM340 × 250 × 9 × 14. The steel was grade 235, and the material property data were in reference [11]. The parameters all met design specification JGJ99-2015 [17].

Table 2. Properties of WFS-250 specimens.

Specimens			l_a/mm	l_a/b_f	l_b/mm	l_b/h_b	b_{wf}/mm	b_{wf}/b_f	R/mm
WFS-250	A	A-1	125	0.50	102	0.30	62.5	0.25	114.5
		A-2	137.5	0.55	102	0.30	62.5	0.25	114.5
		A-3	150	0.60	102	0.30	62.5	0.25	114.5
		A-4	162.5	0.65	102	0.30	62.5	0.25	114.5
		A-5	175	0.70	102	0.30	62.5	0.25	114.5
		A-6	187.5	0.75	102	0.30	62.5	0.25	114.5
	B	B-1	125	0.50	102	0.30	62.5	0.25	114.5
		B-2	125	0.50	112.2	0.33	62.5	0.25	132.0
		B-3	125	0.50	122.4	0.36	62.5	0.25	151.1
		B-4	125	0.50	132.6	0.39	62.5	0.25	171.9
		B-5	125	0.50	142.8	0.42	62.5	0.25	194.4
		B-6	125	0.50	153	0.45	62.5	0.25	218.5
	C	C-1	125	0.50	102	0.30	37.5	0.15	157.5
		C-2	125	0.50	102	0.30	42.5	0.17	143.7
		C-3	125	0.50	102	0.30	47.5	0.19	133.3
		C-4	125	0.50	102	0.30	52.5	0.21	125.3
		C-5	125	0.50	102	0.30	57.5	0.23	119.2
		C-6	125	0.50	102	0.30	62.5	0.25	114.5

5. Analysis Results and Discussing

5.1. WFS-250-A Series Joints

5.1.1. Joint Failure Mode

Figure 10 shows the PEEQ cloud diagram for the WFS-250-A series joints at failure. Figure 11 presents the curve of the maximum PEEQ value at the plastic hinge of WFS-250-A series joints as a function of loading displacement.

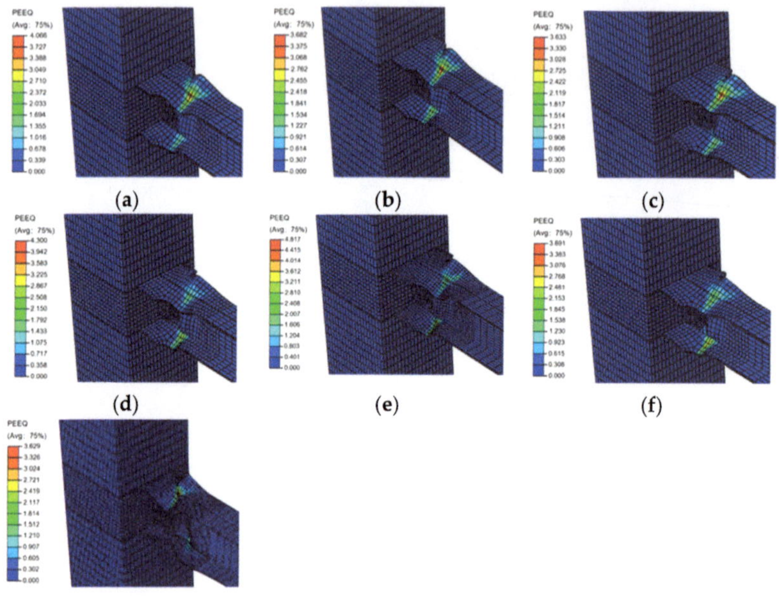

Figure 10. PEEQ cloud chart of WFS-250-A series joints at failure time. (**a**) WFS-250-A-1; (**b**) WFS-250-A-2; (**c**) WFS-250-A-3; (**d**) WFS-250-A-4; (**e**) WFS-250-A-5; (**f**) WFS-250-A-6; (**g**) UW-250.

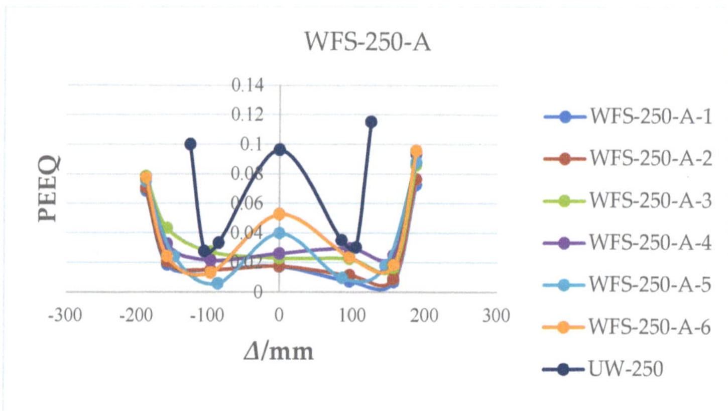

Figure 11. PEEQ maximum at the plastic hinge of WFS-250-A series and UW-250 joints.

Figure 10 shows that the equivalent plastic deformation distribution for the WFS-250-A series joints is uniform at failure. The upper and lower beam flanges expand and deform at the widened sections and their ends, forming distinct plastic hinges. As the reinforced section l_a length increases, the plastic hinge moves outward, while the PEEQ value at the beam-to-column flange joint decreases, indicating reduced cumulative plastic deformation in the weld. The maximum PEEQ value is found on the beam web where the weld passes through, suggesting a reduced risk of brittle fracture in the weld and achieving the goal of weld protection [24,25]. The joint domain remains intact, with the column flange largely elastic and with minimal deformation, demonstrating a "strong joint" characteristic.

In the WFS-250-A series, the failure is predominantly influenced by the plastic deformation in the widened beam flanges. High levels of equivalent plastic strain (PEEQ) are observed at the ends of the reinforced sections, leading to local buckling of the flanges. This buckling is initiated due to the cyclic nature of seismic loading, where plastic deformation progressively accumulates.

A plastic hinge forms near the ends of the widened flange sections, which helps reduce stress concentration at the weld between the beam and column flanges. The outward shift of the plastic hinge protects the weld from brittle fracture. However, the widening of the flanges, while effective in energy dissipation, creates instability in these regions, causing flange buckling. This buckling, in turn, undermines the overall integrity of the joint, leading to failure.

The key reason for the failure in the WFS-250-A series is the inability of the widened flange sections to sustain plastic deformation without local instability. As the flanges buckle, the joint loses its capacity to continue absorbing seismic energy, resulting in structural failure.

Figure 11 shows the peak PEEQ values observed at the extreme edges of the beam flange groove weld. When the widened flange was used in WFS-250-A-6, the maximum PEEQ decreased by 16.9% (from 0.115 rad to 0.095 rad) and by 36.9% for WFS-250-A-1 (from 0.115 rad to 0.073 rad) compared to UW-250. A wider flange at the beam–column interface results in a lower PEEQ index. The WFS-250-A-6 joint showed the highest maximum PEEQ, indicating the most accumulated plastic deformation at the plastic hinge, while WFS-250-A-1 had the least. As the length l_a increased, the outward displacement of the plastic hinge decreased, and the maximum PEEQ value first was reduced and then increased, reflecting a similar trend in the cumulative plastic deformation at the plastic hinge.

5.1.2. M–θ Hysteresis Curve and Skeleton Curve

Figure 12 shows the moment-angle M–θ hysteretic and skeleton curves for the WFS-250-A series joints. In Figure 12a, the M–θ hysteretic curves for all joints display a similar, full shuttle shape. With a transition section length l_b of 102 mm and a reinforced section width b_{wf} of 62.5 mm, the seismic performance of the WFS-250-A joints is favorable when l_a is 0.50–0.75 times the beam width. At an inter-story displacement angle of 0.06 rad, the corresponding bending capacity exceeds 0.85 M_p (the plastic moment of the H beam).

Additionally, the M–θ hysteresis curve (Figure 12a) demonstrates the cyclic response of the connection, highlighting its ability to dissipate energy under repeated loading conditions. *X*-axis (Rotation, θ): the horizontal axis represents rotation (θ) in radians, ranging from −0.1 to +0.1 radians. *Y*-axis (Moment, M): the vertical axis represents moment (M) in kNm, with values ranging from −600 kNm to +600 kNm.

The shape of the hysteresis loop illustrates the connection's energy dissipation and ductility under cyclic lateral loads. The broad and stable loops indicate excellent energy absorption with minimal stiffness degradation or slippage. The curves for WFS-250-A-1, WFS-250-A-2, and others show consistency across tests, indicating strong repeatability and reliability of the seismic performance of the connections.

The lack of significant pinching further suggests the connection retains its stiffness, providing robust resistance to seismic forces. The gradual, consistent spread of the loops confirms good ductility, a critical factor in seismic design, allowing the connection to accommodate large rotations without a notable loss of strength.

In Figure 12b, the skeleton curves for all six joints nearly coincide during the elastic stage, where the bending moment and rotation angle increase linearly. As the joints enter the elastic–plastic stage, the curves begin to diverge. The moment at the beam end increases with the length l_a of the reinforced section, placing the WFS-250-A-6 joint skeleton curve on the outermost side. After reaching the peak moment, the skeleton curve shows a decline, with the rate of decline increasing as l_a lengthens, indicating that a longer expanded flange accelerates the reduction in joint bearing capacity.

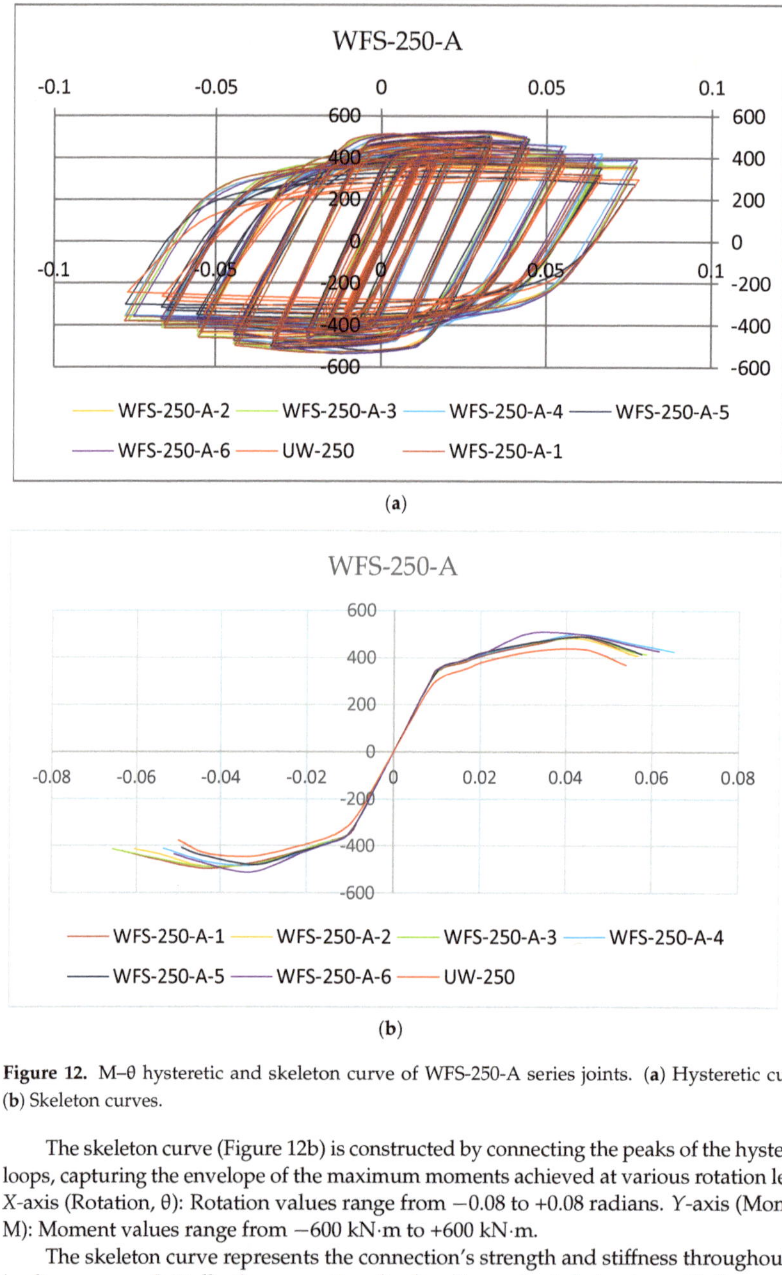

Figure 12. M–θ hysteretic and skeleton curve of WFS-250-A series joints. (**a**) Hysteretic curves; (**b**) Skeleton curves.

The skeleton curve (Figure 12b) is constructed by connecting the peaks of the hysteresis loops, capturing the envelope of the maximum moments achieved at various rotation levels. *X*-axis (Rotation, θ): Rotation values range from −0.08 to +0.08 radians. *Y*-axis (Moment, M): Moment values range from −600 kN·m to +600 kN·m.

The skeleton curve represents the connection's strength and stiffness throughout the loading process. Initially, the connection displays linear elastic behavior, with the moment increasing proportionally to the rotation up to approximately ±0.02 radians. Beyond this point, yielding occurs, indicated by the curve flattening, which signifies plastic deformation. After the maximum moment was reached, a slight reduction in stiffness was observed, signaling some degradation due to cyclic loading.

According to the skeleton curves of the WFS-250-A joint in Figure 12b and the calculation method in reference [11], the joint's mechanical property index under pushing load is summarized in Table 3. R_{ki} is the initial rotation stiffness. M_y, M_{max}, and M_u are the yield, peak, and ultimate moments, and θ_y, θ_{max}, and θ_u correspond to the rotation displacement.

Table 3. Mechanical performance indexes of WFS-250-A series joints.

Joint Number	R_{ki} /kN·m·rad^{-1}	M_y /kN·m	θ_y /rad	M_{max} /kN·m	θ_{max} /rad	M_u /kN·m	θ_u /rad	θ_p /rad	μ
WFS-250-A-1	35,768.45	389	0.017	489.741	0.044	416.28	0.055	0.037	3.163
WFS-250-A-2	35,882.74	382	0.015	481.171	0.044	409.00	0.056	0.041	3.649
WFS-250-A-3	35,991.44	388	0.016	488.030	0.044	414.83	0.058	0.042	3.661
WFS-250-A-4	36,107.55	399	0.017	500.944	0.043	425.80	0.065	0.047	3.694
WFS-250-A-5	35,951.63	385	0.015	488.460	0.045	415.20	0.057	0.042	3.755
WFS-250-A-6	36,316.66	409	0.019	503.702	0.031	428.15	0.061	0.042	3.237
UW-250	33,086.98	356	0.017	434.696	0.044	369.49	0.054	0.036	3.167

θ_p is the rotation capacity according to FEMA 350 [26]. μ is the ductility factor, $\mu = \theta_u/\theta_y$.

From Table 3, increasing l_a strengthens the expanded flange at the beam end, slightly raising joint rotational stiffness by up to 1.5%. However, this increase in stiffness weakens rotational capacity, with yield and ultimate angles differing by 19.5% and 15.8%, respectively. The yield and maximum bending moments initially decrease and then increase, with maximum increases of 6.6% and 4.5%. The angle ductility shows minimal change, following an initial rise and then a decrease.

By comparing the performance indices of the UW-250 specimen, it is evident that when l_a/b_f is between 0.55 and 0.75 (the corresponding specimens WFS-250-A-2 to WFS-250-A-6.), the joint exhibits excellent plastic rotation capacity and seismic performance.

5.2. WFS-250-B Series Joints

5.2.1. Joint Failure Mode

Figure 13 illustrates the PEEQ distribution for WFS-250-B series joints at failure. Figure 14 shows the curve of the maximum PEEQ at the plastic hinge versus loading displacement for the same series.

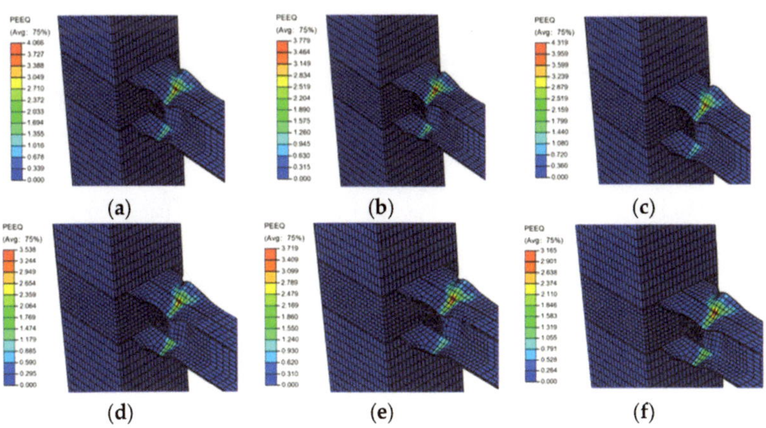

Figure 13. PEEQ cloud chart of WFS-250-B series joints at failure time. (**a**) WFS-250-B-1; (**b**) WFS-250-B-2; (**c**) WFS-250-B-3; (**d**) WFS-250-B-4; (**e**) WFS-250-B-5; (**f**) WFS-250-B-6.

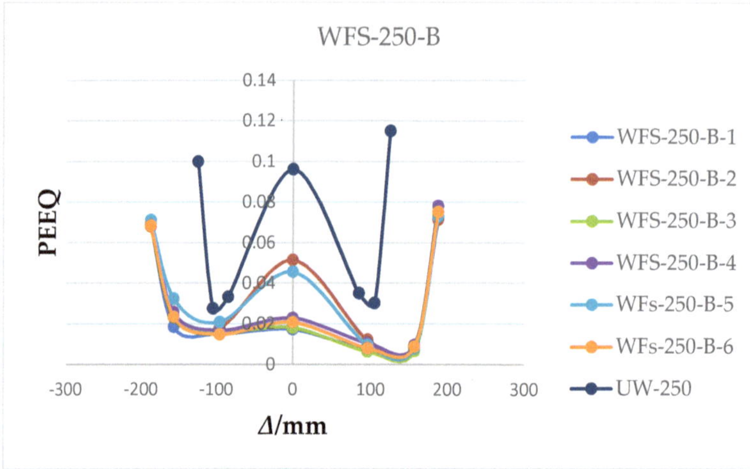

Figure 14. PEEQ maximum at the plastic hinge of WFS-250-B series and UW-250 joints.

As shown in Figures 10 and 13, the WFS-250-B and WFS-250-A series joints exhibit similar failure modes. The upper and lower flanges of the beam show significant buckling deformation, and a plastic hinge forms at the end of the expanded flange. As the length l_b increases, the PEEQ value at the weld between the beam and column flanges gradually decreases, causing the plastic deformation zone to move away from the beam end. The equivalent plastic strain in the beam web outside the connecting plate significantly increases, and the plastic hinge gradually shifts outward. The column flange and joint area experience minimal overall plastic deformation, maintaining their shape and exhibiting the clear characteristics of a "strong joint".

In the WFS-250-B series, the longer transition section (l_b) between the widened flange and the regular beam geometry plays a critical role in the failure mechanism. This longer transition length shifts the plastic hinge further from the beam–column weld, thus reducing the likelihood of weld failure due to brittle fracture. However, this design feature introduces new vulnerabilities.

The outward movement of the plastic hinge leads to increased plastic deformation in the beam web, making it the next critical region of concern. This concentration of deformation in the web may eventually compromise the overall performance of the connection, as the web experiences higher stress levels and potential instability under cyclic loading conditions. As the plastic strain accumulates in the web, it becomes susceptible to significant deformation and eventual buckling. The interaction between the beam flanges and the web, combined with the cyclic nature of seismic loading, leads to the progressive weakening of the web, eventually compromising the joint's integrity.

Thus, while the WFS-250-B series mitigates the risk of weld failure, the plastic strain concentration in the beam web is a significant cause of failure. The beam web's inability to withstand this accumulated deformation ultimately leads to joint failure.

Figure 14 shows that for model UW-250, peak PEEQ values were observed at the extreme edges of the beam flange groove weld. When a widened flange was used, the maximum PEEQ index decreased by 32% for WFS-250-B-4 (from 0.115 to 0.078) and 37.8% for WFS-250-B-2 (from 0.115 to 0.071) compared to UW-250. A wider flange at the beam–column interface reduces the PEEQ index. Figure 13 shows that WFS-250-B-4 has the highest maximum PEEQ at the plastic hinge, indicating the greatest accumulated plastic deformation, while WFS-250-B-2 has the least.

5.2.2. M–θ Hysteresis Curve and Skeleton Curve

Figure 15 shows the moment-angle M–θ hysteretic and skeleton curves for the WFS-250-B series joints. Figure 15a shows that the moment-angle M–θ hysteretic curves for the WFS-250-B series joints are consistently full shuttle-shaped, exhibiting bearing capacity and energy dissipation. When the inter-story drift angle reaches 0.06 rad, the corresponding bending capacity exceeds 0.85 Mp. This indicates that with a reinforced section length l_a of 125 mm and width b_{wf} of 62.5 mm, increasing the transition section length l_b from 0.3 to 0.45 times, the beam depth maintains good seismic performance, similar to WFS-250-A series joints.

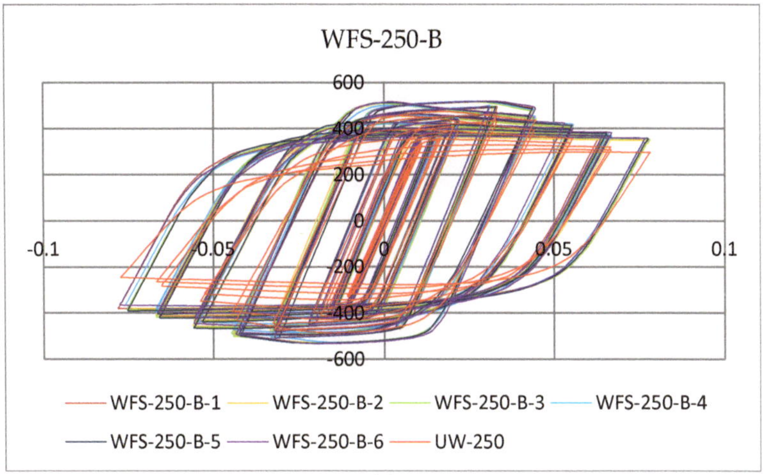

(**a**)

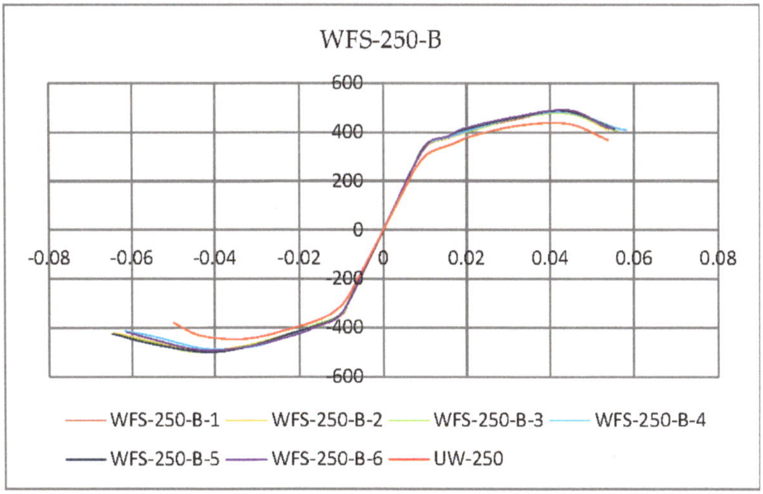

(**b**)

Figure 15. M–θ hysteretic and skeleton curve of WFS-250-B series joints. (**a**) Hysteretic curves; (**b**) Skeleton curves.

The M–θ hysteresis curves for the WFS-250-B series joints, shown in Figure 15a, display the cyclic response of the connections under repeated loading, highlighting their energy dissipation, ductility, and stiffness degradation. The *X*-axis represents rotation (θ) in radians, ranging from −0.1 to +0.1, while the *Y*-axis represents the moment (M) in kNm, ranging from −600 kNm to +600 kNm. The broad loops indicate significant energy dissipation, essential for seismic performance, while the stable shape of the curves demonstrates good ductility, allowing for large rotations without significant stiffness loss. Minimal pinching in the curves suggests the connections maintain their stiffness well during cyclic loading. Although slight stiffness degradation is observed with increasing cycles, it occurs slowly, indicating strong post-yield performance. The consistent behavior across different specimens (WFS-250-B-1–WFS-250-B-6) confirms the repeatability and reliability of these connections for seismic applications.

The skeleton curve in Figure 15b reveals that all six joints show a similar trend during displacement loading, with bending capacity increasing as l_b increases, and the skeleton curve of WFS-250-B-6 being the outermost. Based on the skeleton curve, Table 4 presents the mechanical performance indices of WFS-250-B series joints. As l_b increases, the initial rotational stiffness and flexural capacity also gradually increase, with maximum increases of 0.7% in initial stiffness and 4.5% in yield moment, while the maximum difference in bending capacity is 2.7%. The yield angle changes by 13%, and the ultimate angle decreases, with a maximum reduction of 7.6%. Angle ductility initially increases and then decreases as bending stiffness increases.

Table 4. Mechanical performance indexes of WFS-250-B series joints.

Joint Number	R_{ki} /kN·m·rad^{-1}	M_y /kN·m	θ_y /rad	M_{max} /kN·m	θ_{max} /rad	M_u /kN·m	θ_u /rad	θ_p /rad	μ
WFS-250-B-1	35,768.45	389	0.017	489.740	0.044	416.28	0.055	0.037	3.164
WFS-250-B-2	35,826.66	378	0.015	484.583	0.044	411.89	0.054	0.038	3.456
WFS-250-B-3	35,875.24	381	0.015	477.993	0.044	406.30	0.056	0.041	3.751
WFS-250-B-4	35,923.59	379	0.015	482.707	0.044	410.29	0.058	0.043	3.891
WFS-250-B-5	35,973.79	393	0.017	487.599	0.044	414.46	0.055	0.038	3.255
WFS-250-B-6	36,022.17	396	0.017	491.550	0.044	417.82	0.055	0.038	3.235
UW-250	33,086.98	356	0.017	434.696	0.044	369.49	0.054	0.036	3.167

The skeleton curve, with rotation (θ) in radians on the *X*-axis and moment (M) in kilonewton-meters (kN·m) on the *Y*-axis, illustrates the bending behavior of the joints under cyclic loading. Initially, the curves are linear, reflecting the elastic stage where the moment increases proportionally with rotation. As loading progresses, the joints transition to a plastic stage, with nonlinear growth eventually reaching a peak moment, representing maximum bending capacity. After the peak, the moment stabilizes or decreases, indicating stiffness degradation as the joints approach failure. The different specimens (WFS-250-B-1, WFS-250-B-2, etc.) show consistent performance, with minor variations in peak moments and stiffness likely due to differences in configuration or testing conditions. Overall, the WFS-250-B joints exhibit good ductility, plastic rotation capacity, and energy dissipation under cyclic loading, with small divergences in their ability to resist repeated loading and seismic performance.

Comparing the performance indices of the UW-250 specimen, it is clear that when l_b/h_b ranges from 0.36 to 0.39 (corresponding to specimens WFS-250-B-3 and WFS-250-B-4), the joint demonstrates excellent plastic rotation capacity and seismic performance.

5.3. WFS-250-C Series Joints

5.3.1. Joint Failure Mode

Figure 16 depicts the PEEQ distribution at failure for the WFS-250-B series joints. Figure 17 presents the curve of the maximum PEEQ at the plastic hinge as a function of loading displacement for the same series.

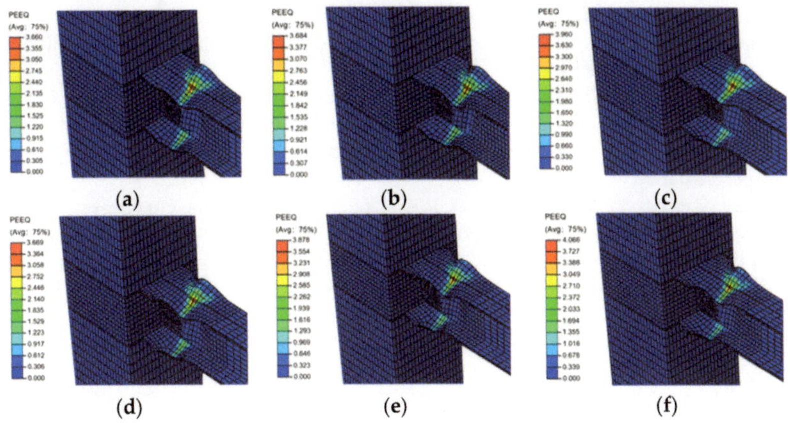

Figure 16. PEEQ cloud chart of WFS-250-C series joints at failure time. (**a**) WFS-250-C-1; (**b**) WFS-250-C-2; (**c**) WFS-250-C-3; (**d**) WFS-250-C-4; (**e**) WFS-250-C-5; (**f**) WFS-250-C-6.

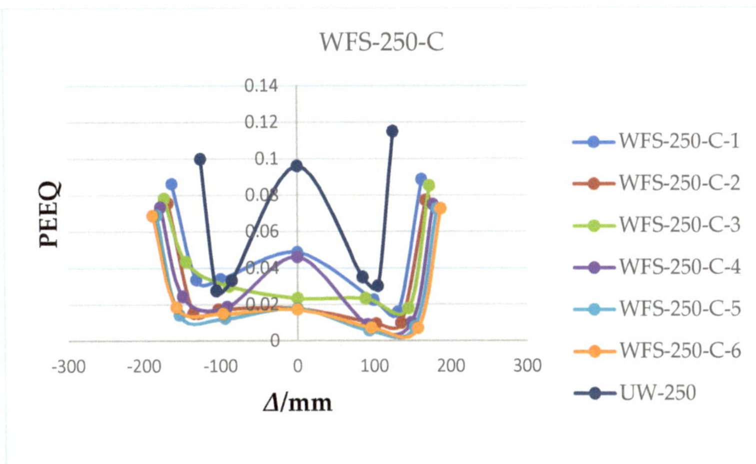

Figure 17. PEEQ maximum at the plastic hinge of WFS-250-C series and UW-250 joints.

Compared to Figures 10 and 13, Figure 16 shows that the failure mode of the WFS-250-C series joints is consistent with that of the WFS-250-A and WFS-250-B series. The upper and lower flanges of the beam buckle form a plastic hinge at the end of the expanded flange. The PEEQ value of the bolt weld between the beam and column flanges gradually decreases while the joint domain remains in the elastic stage with no significant deformation, characteristic of a 'strong joint'. As b_{wf} increases, the plastic deformation decreases, the plastic hinge shifts outward, and the equivalent plastic strain in the beam web outside the connecting plate increases significantly.

The WFS-250-C series introduces an even wider flange section (b_{wf}), greatly influencing the joint's failure mode. The increased width allows for better distribution of plastic deformation across the flanges, further shifting the plastic hinge away from the critical weld zone. This design feature helps reduce the stress concentration at the weld, improving the joint's performance in terms of energy dissipation.

However, the wider flanges experience higher strain at their outer edges, leading to localized buckling as cyclic deformation progresses. The increased flange width also places additional stress on the beam web, making it vulnerable to buckling and other forms of local deformation. Under continued cyclic loading, the interaction between the wider flanges and the beam web leads to a complex failure mode where both the flanges and the web contribute to the overall failure.

The primary cause of failure in the WFS-250-C series is the combined effect of flange buckling and web deformation. While the widened flange section enhances energy dissipation, it also introduces new failure mechanisms that compromise the joint's long-term seismic performance.

Figure 17 illustrates that for model UW-250, peak PEEQ values were observed at the extreme edges of the beam flange groove weld. When a widened flange was used, the maximum PEEQ index decreased by 23% for WFS-250-C-1 (from 0.115 to 0.088) and 36.9% for WFS-250-C-5 (from 0.115 to 0.073) compared to UW-250. A wider flange at the beam–column interface reduces the PEEQ index. Additionally, Figure 16 shows that WFS-250-C-1 has the highest maximum PEEQ at the plastic hinge, indicating the greatest accumulated plastic deformation, while WFS-250-C-5 has the least. As b_{wf} increases, the plastic hinge's outward displacement increases, and the maximum PEEQ value first increases and then decreases, indicating that the scalar value of cumulative plastic deformation at the plastic hinge follows the same trend.

5.3.2. M–θ Hysteresis Curve and Skeleton Curve

Figure 18 presents the moment-angle $M-\theta$ hysteretic and skeleton curves for the WFS-250-C series joints.

Figure 18a shows that the moment-angle $M-\theta$ hysteretic curves for the WFS-250-C series joints are all fully shuttle-shaped, exhibiting bearing capacity and energy dissipation. Increasing the width of the reinforced section b_{wf} from 0.15 to 0.25 times the beam width maintains good seismic performance. When the inter-story drift angle reaches 0.06 rad, the joint's bending capacity exceeds 0.85 Mp, similar to the WFS-250-A and WFS-250-B series joints.

The M–θ hysteresis curve shows the relationship between moment (M) in kilonewton-meters (kN·m) and rotation (θ) in radians, reflecting the angular deformation experienced by the structural elements under cyclic loading. The loops in the curve represent energy dissipation, with larger and more spread-out loops indicating greater resilience to repeated loading cycles. Different labels (WFS-250-C-1, WFS-250-C-2, etc.) correspond to various test specimens or conditions. The curve demonstrates a gradual increase in rotation as the moment increases during the loading cycle, followed by a return during unloading, forming hysteresis loops. The overlapping nature of the curves suggests similar performance between the specimens, though some divergence at larger rotations indicates varying degrees of energy dissipation, ductility, or stiffness degradation among the different specimens.

Figure 18b shows that the skeleton curves for all six joints follow the same trend during displacement loading. In the elastic stage, the skeleton curves are nearly straight and coincide with the bending moment, increasing linearly with the rotation angle. After the elastic–plastic stage, nonlinear growth begins, and once the bending capacity peaks, the curves gradually diverge as the joints approach failure, with WFS-250-C-6 exhibiting the fastest decline.

The skeleton curve, with rotation (θ) in radians on the *X*-axis and moment (M) in kilonewton-meters (kN·m) on the *Y*-axis, is derived by connecting the peak points of each cycle from the M–θ hysteresis loops, providing insights into the system's strength, stiffness, and ductility. After a certain rotation, the moment peaks and either stabilizes or decreases slightly, indicating stiffness degradation. The curves for the different specimens show similar trends, indicating consistent performance with only minor variations in peak moments or stiffness, likely due to testing conditions or structural configurations. The hysteresis and skeleton curves reveal crucial information about energy dissipation and stiffness degradation, with some divergence at higher loads, suggesting differences in the ability of the specimens to sustain loads and resist damage under repeated cycles.

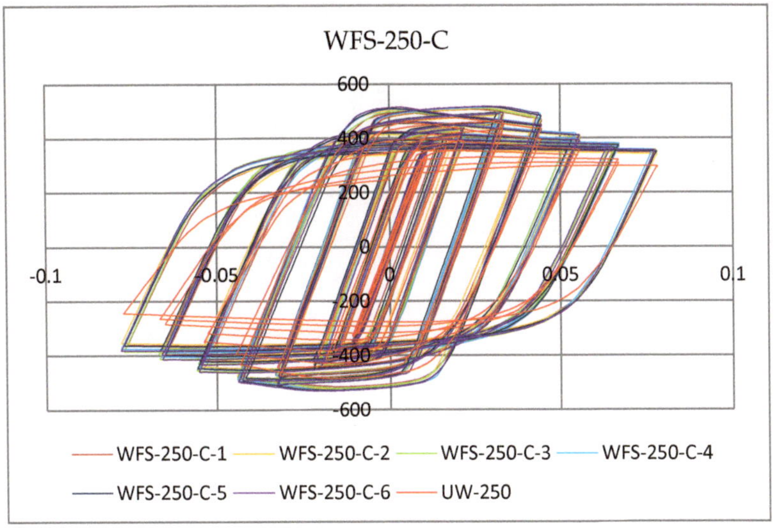

(**a**)

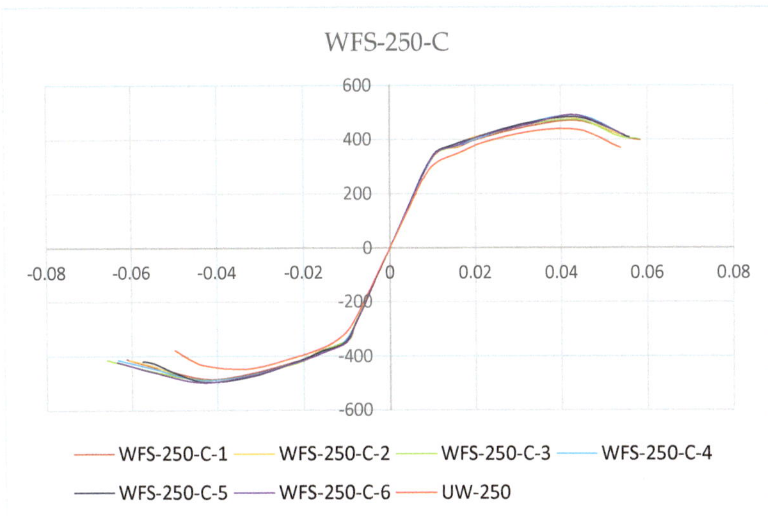

(**b**)

Figure 18. M–θ hysteretic and skeleton curve of WFS-250-B series joints. (**a**) Hysteretic curves; (**b**) Skeleton curves.

Additionally, according to the M–θ skeleton curve of the WFS-250-C series joint in Figure 18b, the joint's mechanical property index when the loading direction is pushing is obtained, as shown in Table 5. The data indicate that increasing the width b_{wf} of the reinforced section increases the yield moment and maximum moment of the WFS-250-C series joints by up to 5.4% and 4.4%, respectively. This suggests that increasing the width of the expanded flange's reinforced section enhances the joint's bending capacity. The yield angle of each joint decreases slightly, while changes in the ultimate angle are minimal, and corner ductility remains similar across the joints.

Table 5. Mechanical performance indexes of WFS-250-C series joints.

Joint Number	R_{ki} /kN·m·rad^{-1}	M_y /kN·m	θ_y /rad	M_{max} /kN·m	θ_{max} /rad	M_u /kN·m	θ_u /rad	θ_p /rad	μ
WFS-250-C-1	34,915.13	370	0.0155	468.151	0.044	397.93	0.055	0.037	3.548
WFS-250-C-2	35,118.17	387	0.0163	479.200	0.043	407.32	0.056	0.038	3.436
WFS-250-C-3	35,300.88	373	0.0150	471.675	0.044	400.92	0.058	0.043	3.881
WFS-250-C-4	35,469.80	391	0.0152	489.144	0.044	415.77	0.055	0.042	3.618
WFS-250-C-5	35,618.35	382	0.0150	480.783	0.044	408.67	0.056	0.041	3.720
WFS-250-C-6	35,768.45	389	0.0173	489.740	0.044	416.28	0.055	0.037	3.164
UW-250	33,086.98	356	0.017	434.696	0.044	369.49	0.054	0.036	3.167

By comparing the performance indices of the UW-250 specimen, it is evident that when b_{wf}/b_f ranges from 0.19 to 0.23 (corresponding to specimens WFS-250-C-3 to WFS-250-C-5), the joint exhibits excellent plastic rotation capacity and seismic performance. This finding aligns with studies on beam–column connections, such as those by Xiao et al. [15] and Han et al. [6], who observed that optimized flange dimensions contribute significantly to energy dissipation and plastic hinge relocation. However, unlike previous research primarily focused on I-section columns and narrow-flange beams, this study extends the understanding to mid-flange H-beams and box columns, revealing that the widened flange connection can further reduce stress concentration and improve cyclic performance under seismic loads.

Furthermore, this research introduces a novel approach to balancing stiffness and ductility in joint design, contrasting with material-based improvements emphasized in prior studies like Yang et al. [7]. While earlier efforts, such as those by Wang et al. [9], focused on enhancing materials like ultra-high-performance concrete, this study prioritizes geometric optimization as a more cost-effective and practical solution.

5.4. Discussion

In summary, the WFS-250 series shifts the focus from material improvements to geometric optimization, offering a streamlined and economical approach to enhancing the seismic performance of steel structures.

These findings align with previous studies by Xiao et al. [15] and Han et al. [6], demonstrating that optimized flange dimensions enhance energy dissipation and aid in plastic hinge relocation. However, this study extends these insights to mid-flange H-beams and box columns, revealing that widened flange connections can reduce stress concentrations and improve cyclic performance under seismic loads.

Additionally, this research introduces a novel approach to balancing stiffness and ductility through geometric optimization, contrasting with earlier studies that emphasized material-based improvements, such as those by Yang et al. [7] and Wang et al. [9], who focused on enhancements like ultra-high-performance concrete. By prioritizing geometric adjustments over material enhancements, this work provides a more cost-effective and practical solution for improving seismic resilience.

6. Conclusions

This study bridges a significant gap in understanding the seismic performance of beam-to-column connections, specifically those involving box columns and mid-flange H-beams. The research demonstrates that incorporating widened beam-end flanges can substantially improve bending capacity and ductility by effectively relocating the plastic hinge away from the connection—an essential factor for enhancing seismic resilience. Key findings from the detailed analysis of failure mechanisms, hysteretic behavior, and mechanical properties are as follows:

1. Performance Comparison: Connections featuring widened flanges outperform traditional H-beam connections with unmodified ends and box columns. The widened

flanges exhibit superior bearing capacity, enhanced seismic performance, and increased ductility, highlighting their effectiveness in improving joint resilience under seismic loads.
2. Flange Length Recommendations: The optimal length for widened beam-end flanges (l_a) ranges from 0.55 to 0.75 times the beam flange width (b_f). This range ensures that the connection benefits from increased moment capacity and energy dissipation without compromising joint performance.
3. Beam Flange Cutting Length: For effective performance, the beam flange cutting length (l_b) should be between 0.36 and 0.39 times the beam depth (h_b). This range balances the need for adequate plastic hinge formation with the structural integrity of the connection.
4. Flange Cutting Depth: The recommended depth for beam flange cutting (b_{wf}) is between 0.19 and 0.23 times the beam flange width (b_f). Adhering to this range optimizes the balance between enhancing ductility and maintaining the structural strength of the connection.

Author Contributions: Conceptualization, L.L.; software, T.D.G.; validation, T.D.G.; resources, L.L.; writing—original draft preparation, T.D.G. and L.L.; writing—review and editing, L.L.; project administration, L.L.; funding acquisition, L.L. All authors have read and agreed to the published version of the manuscript.

Funding: This research was funded by the Nature Science Foundation of China (NSFC), grant number 51278061.

Data Availability Statement: The original contributions presented in the study are included in the article, further inquiries can be directed to the corresponding author.

Conflicts of Interest: The authors declare no conflict of interest.

References

1. Peng, Y. The application and research progress of steel structures in construction industrialization. *IOP Conf. Ser. Earth Environ. Sci.* **2019**, *330*, 022003. [CrossRef]
2. Song, Z.; Gu, Q.; Guo, B. Experimental research on the connection between beam and column of rigid steel frame. *J. Build. Struct.* **2001**, *22*, 53–57.
3. Nakashima, M.; Inoue, K.; Tada, M. Classification of damage to steel buildings observed in the 1995 Hyogoken-Nanbu earthquake. *Eng. Struct.* **1998**, *20*, 271–281. [CrossRef]
4. Mahin, S.A. Lessons from damage to steel buildings during the Northridge earthquake. *Eng. Struct.* **1998**, *20*, 261–270. [CrossRef]
5. Gates, W.E.; Morden, M. Professional structural engineering experience related to welded steel moment frames following the Northridge earthquake. *Struct. Des. Tall Build.* **1996**, *5*, 29–44. [CrossRef]
6. Han, W.; Nie, S.; Chen, Z.; Yang, B.; Deng, Y.; Elchalakani, M. Seismic behavior of high-performance steel (HPS) weakened beam-column joints: Experimental tests and numerical analysis. *J. Build. Eng.* **2024**, *92*, 109807. [CrossRef]
7. Yang, Z.M.; Chen, J.; Wang, F.; Wang, J. Seismic performance of circular concrete-filled steel tube columns reinforced with inner latticed steel angles. *J. Constr. Steel Res.* **2023**, *205*, 107908. [CrossRef]
8. Wei, J.; Ying, H.; Yang, Y.; Zhang, W.; Yuan, H.; Zhou, J. Seismic performance of concrete-filled steel tubular composite columns with ultra high performance concrete plates. *Eng. Struct.* **2023**, *278*, 115500. [CrossRef]
9. Wang, W.Z.; Chen, F.Y.; Huang, Y.Q.; Wang, X.T. Fracture mechanism of joint of H-style steel beam-square steel tubular column and study on joints with horizontal haunch and slot holes at beam flanges. *Eng. Mech.* **2013**, *30*, 197–204.
10. Saneei Nia, Z.; Mazroi, A.; Ghassemieh, M.; Pezeshki, H. Seismic performance and comparison of three different I beam to box column joints. *Earthq. Eng. Eng. Vib.* **2014**, *13*, 717–729. [CrossRef]
11. Lu, L.; Xu, Y.; Zheng, H. Investigation of composite action on seismic performance of weak-axis column bending connections. *J. Constr. Steel Res.* **2017**, *129*, 286–300. [CrossRef]
12. Ma, H.; Wang, J.; Lui, E.M.; Wan, Z.; Wang, K. Experimental study of the behavior of beam-column connections with expanded beam flanges. *Steel Compos. Struct.* **2019**, *31*, 319–327.
13. Wang, M.; Ke, X. Seismic design of widening flange connection with fuses based on energy dissipation. *J. Constr. Steel Res.* **2020**, *170*, 106076. [CrossRef]
14. Zohdi, A.; Matinpour, M.H.; Rafati Bonab, R.; Kiumarsi Oskuei, F. Seismic performance of the enlarged beam section connection. *Steel Constr.* **2024**, *17*, 29–43. [CrossRef]
15. Xiao, Y.; Yu, M.; Liu, W. Finite Element Analysis of Prefabricated Semi-Rigid Concrete Beam–Column Joint with Steel Connections. *Appl. Sci.* **2024**, *14*, 5070. [CrossRef]

16. Wang, Y.; Gao, P. Low-cycle Cyclic Load Test Research on Expanded Joints of Beam End Flanges in Steel Frames. *Build. Struct.* **2010**, *31*, 94–101.
17. *JGJ99-2015*; Technical Specification for Steel Structures of Tall Buildings. China Architecture & Building Press: Beijing, China, 2015.
18. Lu, L.; Zhu, P.; Ding, S.; Ma, Z.; Li, R.; Nie, S.; Wang, W.; Raftery, G.M. Impact of anti-corrosion coatings and maintenance on high-strength bolt friction connections in C4 marine environment. *Structures* **2024**, *68*, 107098. [CrossRef]
19. *JGJ 82-2011*; Technical Specification for High Strength Bolt Connections of Steel Structures. China Architecture & Building Press: Beijing, China, 2011.
20. Nie, W.; Lu, L.; Li, R.; Hao, H.; Luo, T. Global buckling investigation of T-shaped built-up columns composed of two hot-rolled H-shapes. *Structures* **2023**, *58*, 105485. [CrossRef]
21. Kontolati, K.; Apostolos, K.; Olympia, P. Numerical investigation of weak axis I profile connections. In Proceedings of the 9th Hellenic National Conference on Steel Structures, Larissa, Thessaly, Greece, 5–7 October 2017.
22. Gao, P. Experimental Evaluation on the Behavor of Side-Plate Reinforced Section and Widen Flange Section of Steel Frame. Master's Thesis, Qingdao University of Technology, Qingdao, China, 2009.
23. *ANSI/AISC 341-05*; Seismic Provisions for Structural Steel Buildings. American Institute of Steel Construction: Chicago, IL, USA, 2005.
24. Chen, C.C.; Lin, C.C.; Lin, C.H. Ductile moment connections used in steel column-tree moment-resisting frames. *J. Constr. Steel Res.* **2006**, *62*, 793–801. [CrossRef]
25. Chen, C.C.; Lin, C.C.; Tsai, C.L. Evaluation of reinforced connections between steel beams and box columns. *Eng. Struct.* **2004**, *26*, 1889–1904. [CrossRef]
26. *FEMA-350*; Recommended Seismic Design Criteria for New Steel Moment-Frame Buildings. FEMA: Washington, DC, USA, 2000.

buildings

MDPI

Article

Fatigue Life Prediction for Stud Shear Connectors Based on a Machine Learning Model

Dong-Hyun Kang [1], Gi-Tae Roh [1], Chang-Su Shim [1,*] and Kyoung-Chan Lee [2]

[1] Department of Civil and Environmental Engineering, Chung-Ang University, Seoul 06974, Republic of Korea; rkdehdgus@cau.ac.kr (D.-H.K.); shrlxo2001@cau.ac.kr (G.-T.R.)
[2] Department of Civil and Railroad Engineering, Pai Chai University, Daejeon 35345, Republic of Korea; kclee@pcu.ac.kr
* Correspondence: csshim@cau.ac.kr

Abstract: The shear connector of a steel composite bridge is designed by predicting fatigue life using the fatigue strength curves (S-N curve) based on push-out test results. The fatigue strength curves of the current design codes present only a linear relationship between the stress range and fatigue life on a log scale based on push-out experiment results. However, an alternative to the current empirical formula is necessary for the fatigue design of shear connections involving many detailed variations or high strength steel materials. This study collected and reanalyzed data from push-out fatigue tests to determine the factors influencing fatigue life and propose a machine learning-based fatigue life prediction model. The proposed machine learning model demonstrated an improvement in predictive performance of approximately 2 to 8 times compared to the existing design codes when evaluated against experimental data. Feature importance analysis based on the proposed model revealed that the stress range most significantly influenced fatigue life prediction. Model validation results indicated that the proposed model provided reliable predictions with accuracy and generalization performance. Moreover, it effectively accounted for uncertainty by incorporating features previously overlooked in existing design codes. Plans for fine-tuning pretrained models were also discussed.

Keywords: machine learning; stud shear connection; fatigue life prediction; composite bridge; gaussian process regression; pretrained model

1. Introduction

Welded stud shear connectors are critical components in the design of a steel concrete composite bridge, demonstrating reliable performance for over 60 years. These connectors are typically installed using welding guns. The fatigue life of the stud connectors primarily governs the design of shear connections in composite bridges. However, current design codes are based on a relatively limited dataset, often involving standard details. For example, the AASHTO LRFD Bridge Design Specifications were derived from only 44 push-out tests [1]. According to the AASHTO [2], the fatigue shear resistance of an individual stud shear connector, Z_r, is given by Equation (1). Notably, the fatigue life of stud shear connectors is predominantly influenced by their stress ranges.

$$Z_r = \alpha d^2 \geq \frac{38.0}{2} d^2 \tag{1}$$

where $\alpha = 238 - 29.5$ logN; d = stud diameter (mm), and N = number of fatigue cycles.

Oehlers and Foley [3] developed a peak load model that considers the maximum shear force, P_p, based on 280 push tests. However, as Johnson [4] highlighted, discrepancies in fatigue life predictions among different researchers are primarily attributed to the scatter in the test data and differences in the regression analyses. These inconsistencies underscore the challenges in achieving a consensus on fatigue life prediction. In addition, Eurocode 4 [5] provisions for the fatigue of stud shear connectors are constrained by limitations in the underlying design parameters, influenced by the specific set of background data used in their development. The magnitude of the peak load, P_{max}, of the cyclic loading exhibited a significant effect. Consequently, Hanswille et al. [6,7] proposed a modified damage accumulation hypothesis.

$$N_f \left(\frac{\Delta P}{P_R} \right)^m = 10^k \tag{2}$$

ΔP is the range of shear force;
P_R is the static shear strength of the stud;
$N_f = N_e / \left[1 - \left(\frac{P_p}{P_R} \right) \right]$;
N_e is the endurance of uni-directionally loaded specimen;
P_p is peak shear force per stud;
k is characteristic (2.3% fractile for a normal distribution).

Variations in stud shear connection practices primarily include the use of large-diameter studs [8,9], high-performance concrete [10], specific details of shear pockets in precast decks [11], and the group arrangement of studs [12–14]. These variations are illustrated in Figure 1. These design variables, falling outside the scope of current design codes, significantly impact static strength and fatigue life predictions. The diversity of materials and configurations used in stud shear connections for steel concrete composite bridges poses challenges for effectively applying existing design codes in practice.

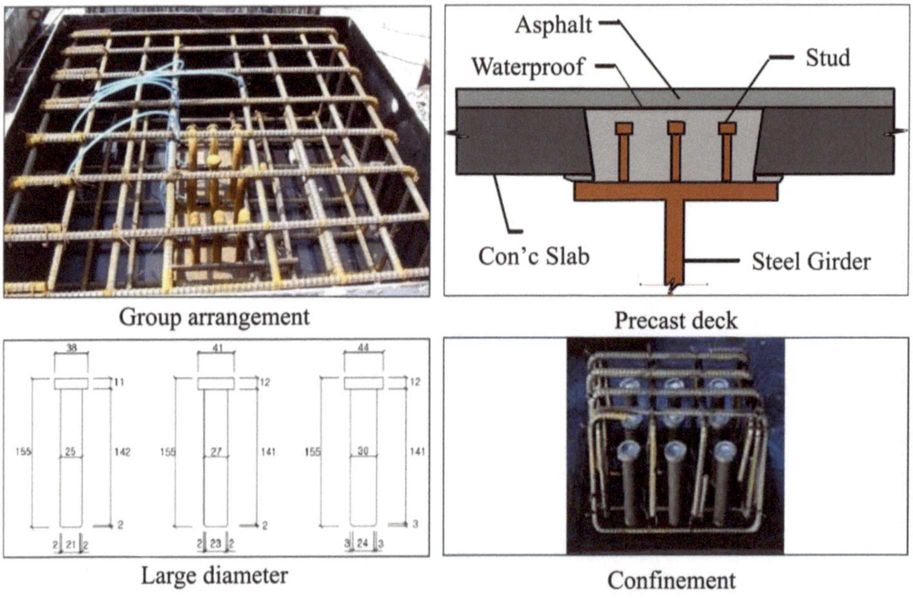

Figure 1. Variation in stud shear connector [12].

Current design criteria for evaluating fatigue life primarily rely on empirical formulas derived from previous experimental data. These formulas are largely based on tests using 19 and 22 mm diameters stud connectors, often with varying experimental setups, including single and double-sided configurations [15]. Although this empirical approach provides a conservative estimate of stud fatigue life, it does not account for critical factors such as the effects of maximum stress, influence of mortar fillers without high-strength concrete or coarse aggregate, and impact of constraints. Consequently, current criteria may not fully capture the complexities of modern materials and design conditions.

To address the limitations of existing design codes, some researchers have explored the replacement of conventional empirical formulas with machine learning models. Using a data-driven approach, machine learning models can overcome challenges, such as limited data availability, uncertainty quantification, and the evaluation and consideration of various influencing not included in current design codes. Avci-Karatas [16] developed advanced machine learning models to predict the shear capacity of headed shear studs using minimax probability machine regression (MPMR) and extreme machine learning (EML). These models were compared and validated against the static strength equations of existing design codes. Rho et al. [17] proposed a machine learning model that integrates finite element method data with experimental data. They introduced the concept of a digital twin model for the static design of stud shear connectors with push-out specimens to replace empirical design codes. Roshanfar et al. [18] compared the performance of six machine learning models in predicting the fatigue life of shear connectors and compared the results with current design codes (Table 1).

Table 1. Research on the replacement of design codes using machine learning.

Reference	Objective	Data and Features	Methodology
Avci-Karatas [16]	Shear capacity of headed stud	Number of data: 215	Minimax probability machine regression (MPMR)
		Features: D, f_{cm}, f_u	Extreme Machine Learning (EML)
Roh et al. [17]	Shear capacity of headed stud	Number of data (Strength): 431 Number of data (Slip): 194	Catboost Xgboost Random forest LightGBM ExtraTrees
		Features: Dimension for stud (D_{sc}, h_{sc}), Confinements (confinement ratio, diameter, f_y) slab (b (mm), d (mm), t (mm)), Concrete material (E_{cm}, f_{cm}) Steel material (E_s, f_u)	
Roshanfar et al. [18]	Fatigue life of shear connector	Number of data:	Linear regression
		Features: Shear stress range of shear connector Compressive strength, Shear strength of shear connector Diameter Type of test Type of shear connector	Decision tree
			Support vector Regressor
			Gaussian process regression (GPR) Bagged tree (BT) Artificial neural network

Machine learning models heavily rely on analyzing the relationship between input and output data using algorithms that maximize outcomes. However, this approach is highly dependent on data quality, posing substantial challenges, particularly when experimental data are limited. In cases where data availability is constrained, such as in regression analysis or artificial intelligence (AI)-based models, reliance on conventional models can lead to difficulties in inferring clear causal relationships. This may reduce the reliability of outcomes and lead to issues such as overfitting.

To address these challenges, an increasing demand for explainable AI models has emerged, gaining widespread applications across various fields. These models validate outcomes by comparing input and output values, and investigating variable correlations through rational processes (e.g., SHapley Addictive exPlanations (SHAP)). In addition, they incorporate methods such as physics-informed machine learning, which replaces the data required for conventional training with physics-based equations (e.g., differential equations) commonly used in engineering fields.

This study proposes a machine learning-based model for predicting the fatigue life of stud shear connectors, as illustrated in Figure 2. To enhance the model performance, we performed feature engineering by analyzing the correlations and causal relationships among the factors influencing fatigue behavior, as identified in previous research. Subsequently, we eliminated features that impaired accuracy owing to linear interdependence among input data and fatigue life with high sensitivity for training. To derive the model outcomes, we further analyzed the significance and trends of the selected variables within the model through feature importance analysis using individual conditional expectation (ICE) and SHAP. Ultimately, this study proposes a pretrained machine learning model, trained on global data from referenced literature, as an alternative to the existing design codes. In addition, it presents the foundation for a fine-tuning model that can be further trained with local data in future.

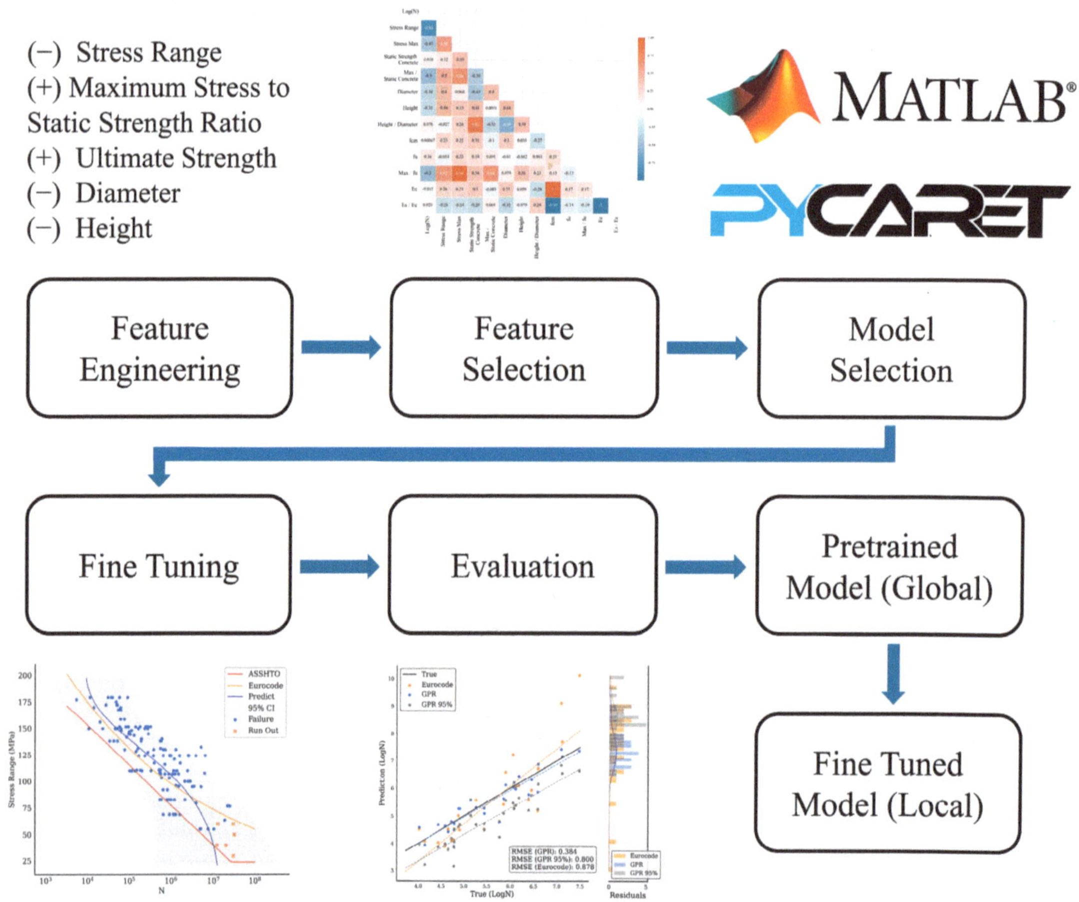

Figure 2. Fatigue life prediction based on a machine learning model.

2. Dataset

2.1. Data Preprocessing

Push-out test data collected from 1966 to 2016 were analyzed to predict the fatigue life of shear connectors. Through feature engineering, factors affect to fatigue life were identified, and relevant physical features were selected. Subsequently, feature selection was conducted to analyze the linearity and select the most reasonable features. Ultimately, a dataset for predicting fatigue life was developed.

2.2. Feature Engineering

To determine the features affecting fatigue behavior and analyze causality, fatigue behavior was categorized into shear transfer under fatigue loading, crack initiation, crack propagation, and fatigue failure.

2.2.1. Mechanism of Shear Transfer by Dowel Action

Figure 3 illustrates the mechanism of shear transfer by dowel action when a cyclic load is applied to a push-out specimen. Under cyclic load, the resultant force, F, in the bearing zone occurs with eccentricity, e, and induces high tensile stresses in the steel failure zone. Oehler and Bradford [19] stated that "in the concrete adjacent to the bearing zone has to withstand compressive stresses of about seven times the cylinder strength of the concrete f_c, and this can only be achieved by the triaxial restraint imposed on the region by the streel element, the dowel and the surrounding concrete".

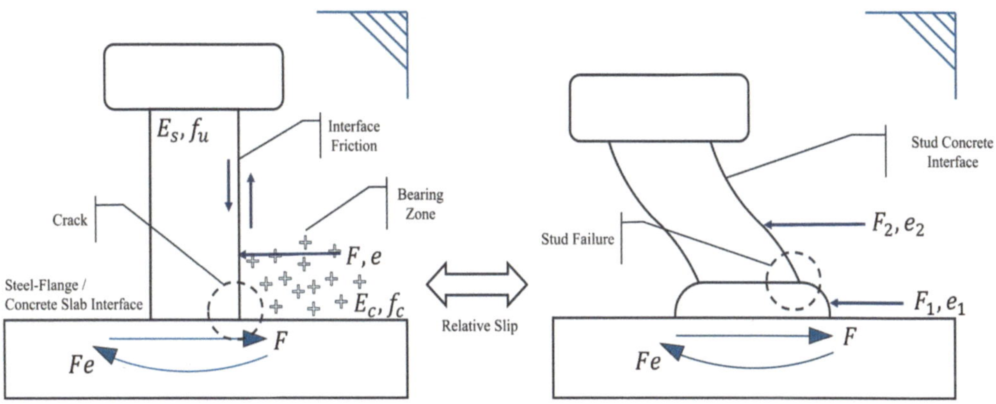

Figure 3. Idealized model of shear transfer by dowel action.

The stress range and maximum stress on the shear connectors induced by cyclic loading, compressive strength of the concrete in the bearing zone, and ultimate strength of the stud were identified as the key features influencing fatigue life in shear transfer.

2.2.2. Crack Initiation

Cracks in welded shear studs typically initiate at the welded part. In smooth-surfaced parts subjected to cyclic loading, dislocation and deformations of crystallographic planes result in surface roughening. Consequently, numerous micro-cracks form along the slip planes. Continued cyclic loading causes some of these micro-cracks to merge and form dominant cracks. The stress at the tips of these larger cracks is extremely high, causing them to propagate more rapidly [20].

Therefore, the peak load of the cyclic load and welding specifications and quality were determined as factors influencing crack initiation. However, for the welding quality, which cannot be quantified, Lee et al. [9] stated that "As the large stud has a lager cross-sectional area, welding required a power source with higher amperage. It affects to the lower of fatigue life in push out test which lead to an unfavorable welding collar". Thus, stud diameter was also identified as an influencing feature indirectly reflecting the effect of welding quality on crack initiation.

2.2.3. Crack Propagation

During cyclic loading, the crack propagation rate in the steel failure zone of the stud is proportional to the stress intensity factor K according to Paris's power law [21]. This factor is a function of the stress at the crack tip, which is proportional to the energy release rate, G. Based on this relationship, the behavior of the stud can be distinguished into cantilever behavior and steel beam supported by an elastic concrete medium [19].

If the stud behaves as a cantilever under cyclic loading, the uncracked area decreases as the crack propagates. This reduction in the uncracked area increases the stress at the crack tip. This increases the energy release rate, accelerating crack propagation. Conversely, if the stud acts as a steel beam supported by an elastic concrete medium, the concrete resists the cyclic load, reducing the eccentricity. This reduces the stress at the crack tip. Consequently, the energy release rate decreases, leading to a reduced crack propagation rate. However, in actual behavior, reduced eccentricity, reduces the area of concrete resisting the load also decreases. This increases the stress on the concrete, causing concrete crushing and affecting the energy release rate. In addition, friction at the stud prevents the crack tip from opening and reduces the stress at the crack tip.

In existing design codes, fatigue life is defined under the assumption that the ultimate and fatigue limit states are independent. However, in push-out tests, fatigue life is typically determined based on the number of cycles at which significant deformation is observed; therefore, the assumption that static strength is preserved until cyclic loading is not desirable. Consequently, researchers have explained that the repeated loading reduces static strength of shear studs [3,6]. Based on this assumption, Hanswille et al. [6,7] performed 71 push-out tests to determine the reduction in static strength after cyclic loading and proposed a fatigue life model that considers the ratio between the peak load and static strength.

Therefore, during the crack propagation process, the stress range, maximum stress under peak load, concrete compressive strength, stud height for reflecting friction, static strength, and their ratio between peak load and static strength derived from existing experiments and analyses were selected as the influencing features.

2.2.4. Failure

As the crack propagates, the dowel shear mechanism can be represented by an interaction model of an equivalent steel beam and concrete subjected to force, F, with eccentricity, e. In this model, an increase in the ratio of the elastic modulus of the steel beam to that of the concrete, $\frac{E_s}{E_c}$, results in greater eccentricity, reducing the dowel strength.

In addition, if the concrete is crushed during crack propagation, its elastic modulus, E_c, decreases, increasing eccentricity. This increases the flexural stress in the steel failure zone, accelerating crack propagation. Consequently, increasing $\frac{E_s}{E_c}$ further increases eccentricity and the bearing pressure on the concrete, causing concrete crushing. This process repeats, ultimately leading to failure. Therefore, to account for the interaction between the stud and concrete after crack propagation leading to failure, the ratio of the elastic modulus of the stud to that of the concrete was selected as a feature.

2.2.5. Feature Definition

The analysis of the fatigue behavior of shear connections and the factors influencing fatigue life, along with the relationships between each feature, is illustrated in the fishbone diagram in Figure 4.

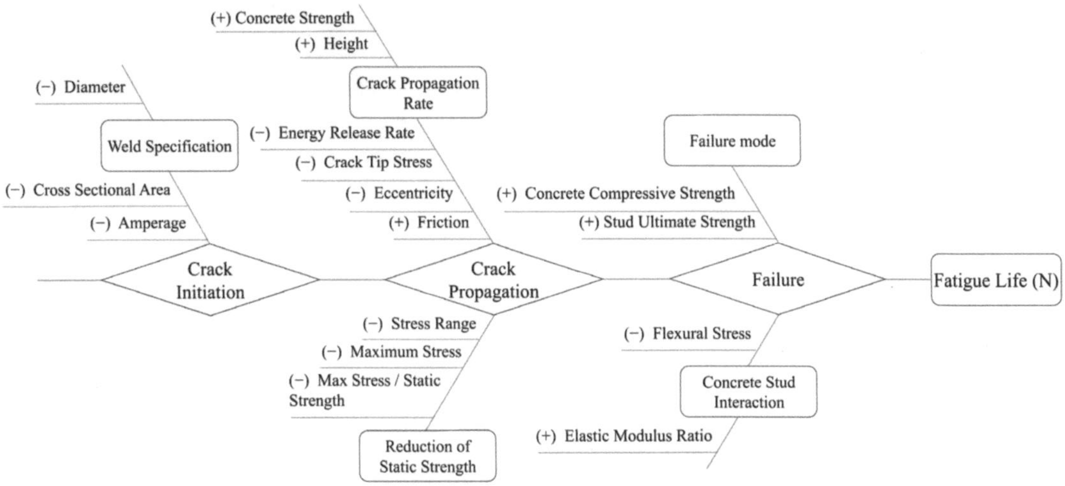

Figure 4. Causality diagram for fatigue life prediction.

The features influencing fatigue life are defined as follows: The N value, representing fatigue life, refers to the number of load cycles at which failure or significant displacement occurs. However, for run-out data in which failure did not occur, the last observed number of cycles is designated as the fatigue life. The stress range and maximum stress values are defined as the stress per stud during cyclic loading. The compressive strength of concrete is determined by the mean cylindrical compressive strength of concrete in the bearing zone, where interaction with the studs occurs. The average value from the direct compressive strength test is used. If compressive strength tests are not performed and an f_{ck} value is provided, it is converted to f_{cm} using Equation (3) from the Eurocode.

$$f_{cm} = f_{ck} + 8 \quad (\text{MPa}) \tag{3}$$

$$f_{cylinder} = 0.85 f_{cube} \tag{4}$$

In cases where the $f_{ck,cube}$ value of cubic specimens is provided, it is converted to the cylinder value, f_{ck}, according to the relevant code, and then reflected as f_{cm}. Because most experiments do not specify the elastic modulus of concrete, it is calculated using Equation (5) from the Eurocode. For static strength, the material properties from each test are converted using Equation (6) from the Eurocode.

$$E_{cm} = 22{,}000 * \left[\frac{f_{cm}}{10}\right]^{0.3} \quad (\text{MPa}) \tag{5}$$

If the test program does not include a static strength experiment, the static strength equation from Equation (6), according to Eurocode 4 [5], is adopted to reflect each failure mode.

$$P_{Rd} = \min\left(\frac{0.29\alpha d_s^2\sqrt{f_{ck}E_c}}{r_v}, \frac{0.8 f_u\left(\frac{\pi d_s^2}{4}\right)}{r_v}\right), \text{ where } r_v = 1.25$$
$$\left(\alpha = 0.2\left(\frac{h_{sc}}{d} + 1\right) for\, 3 \leq \frac{h_{sc}}{d} \leq 4\right) \tag{6}$$

To account for concrete crushing, the static strength by the combined damage from bearing concrete and stud connector, $Q_{u,\ concrete}$, and its ratio to the maximum stress, $\sigma_{max}/Q_{u,\ concrete}$, are selected as features. In addition, to reflect the ultimate failure mode of the stud, the ultimate strength of the stud f_u and its ratio with the maximum stress σ_{max}/f_u are selected as features (Table 2).

$$Q_{u,concrete} = 0.29\alpha d_s^2\sqrt{f_{ck}E_c}, \text{ where } \alpha = 0.2\left(\frac{h_{sc}}{d}+1\right) \quad (7)$$

Table 2. Features for fatigue life prediction by the feature engineering.

Name	Symbol	Definition	Unit
Stress range	σ_{range}	Range of stress per stud under cyclic load	mm
Stress max	σ_{max}	Stress per stud under peak load	MPa
Concrete static strength	$Q_{u,\ concrete}$	Static Strength by local concrete pressure according to Eurocode 4	MPa
Maximum stress concrete ratio	$\sigma_{max}/Q_{u,concete}$	Ratio of maximum stress to static strength by local concrete pressure	None
Diameter	d	Nominal diameter of shear stud	MPa
Height	h	Height of shear stud	MPa
Height diameter ratio	h/d	Height to diameter ratio of shear stud	None
Concrete compressive strength	f_{cm}	Mean value of concrete cylinder compressive strength	MPa
Ultimate strength	f_u	Ultimate strength of shear stud	MPa
Maximum stress ultimate ratio	σ_{max}/f_u	Ratio of maximum stress to ultimate strength of shear stud	None
Concrete elastic modulus	E_c	Concrete elastic modulus from f_{cm} according to Eurocode 2	MPa
Elastic modulus ratio	$\frac{E_s}{E_c}$	Ratio of stud elastic modulus (assumed as 200,000 MPa) to Concrete Elastic Modulus	None
Name	**Symbol**	**Definition**	**Unit**
LogN	N	Log scale cycles	Cycle

2.3. Feature Selection

Among the previously identified features, the final selection of features for constructing the dataset for stud fatigue life prediction was made through data filtering, which involved identifying and removing data that could hinder model training. In addition, a correlation coefficient matrix was analyzed to assess the relationships between the features.

2.3.1. Data Filtering

During the filtering of the collected data, some experiments were identified as outliers that negatively impacted the performance of the machine learning model and were excluded from the dataset. For example, the experiments performed by Mainstone and Menzies [22] were excluded because they significantly underestimated fatigue life. This underestimation was attributed to the high maximum stress ratio ranging from 0.6 to 0.87. This leads to earlier concrete crushing with a lack of reinforcement, resulting in insufficient geometric constraints. Consequently, 145 records [1,6,9,20,22–31] were selected as the final dataset for fatigue life prediction, Figure 5 shows scatter plots and histograms of each feature.

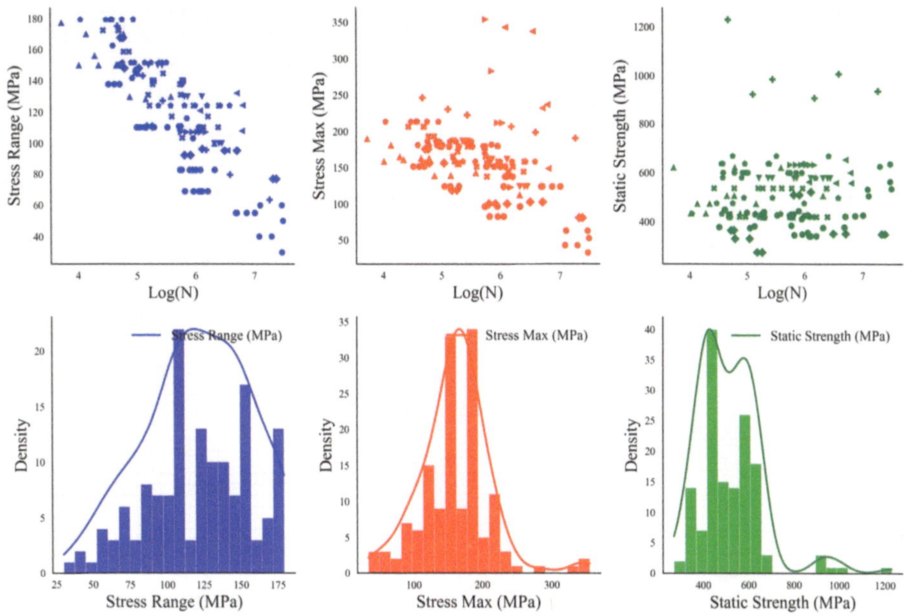

(a)

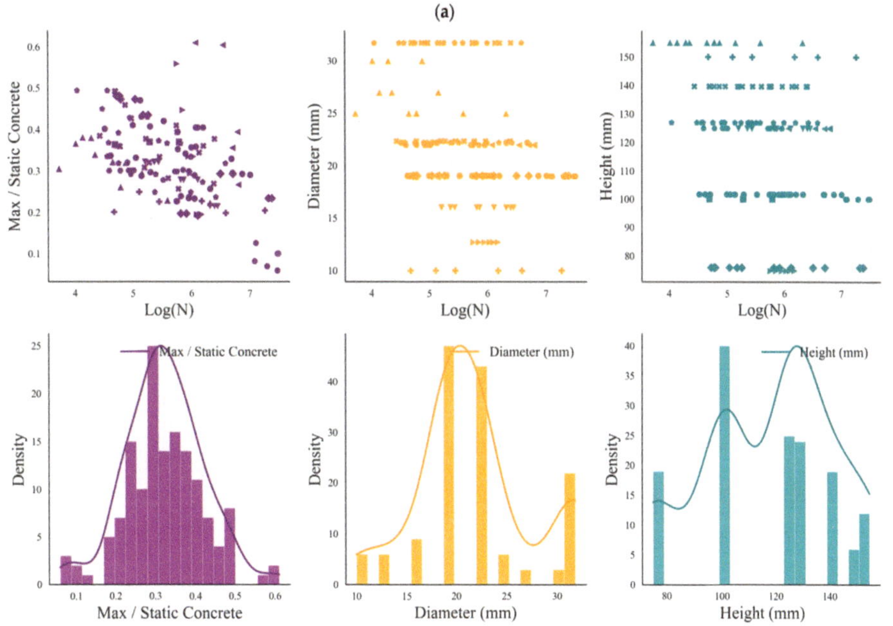

(b)

Figure 5. *Cont.*

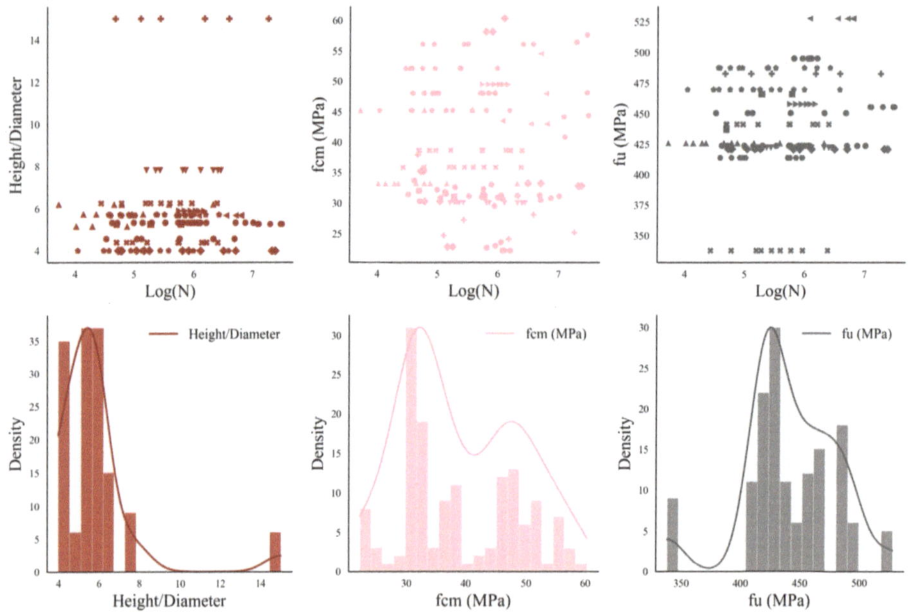

(c)

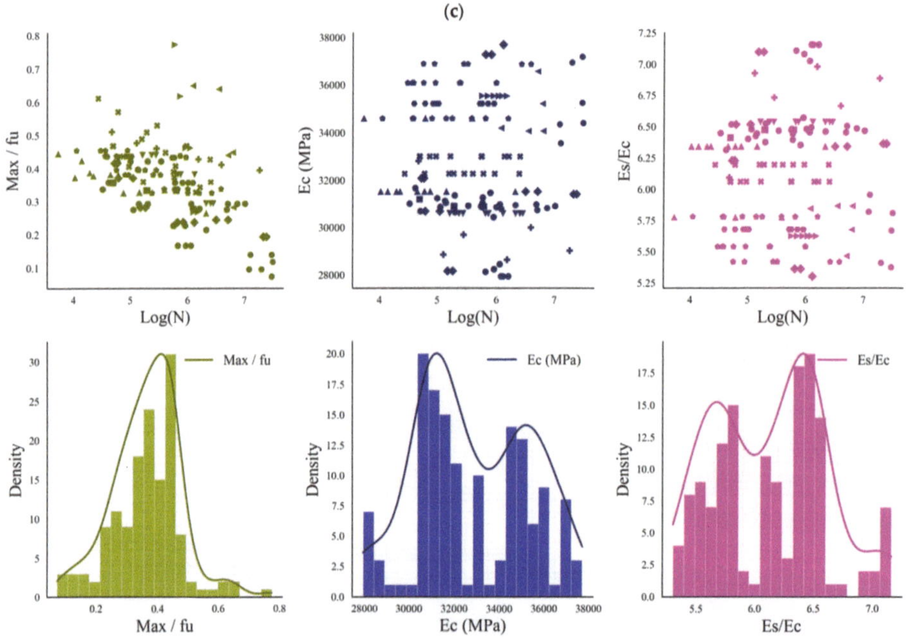

(d)

Figure 5. Scatter plot and histogram of the selected features: (**a**) stress range (blue), maximum stress (red), and static strength (green); (**b**) maximum stress-to-static strength ratio (purple), diameter (orange), and height (dark cyan); (**c**) height-to-diameter ratio (brown), compressive strength (pink), and ultimate strength (gray); (**d**) maximum stress-to-ultimate strength ratio (olive), elastic modulus of concrete (navy), and ratio of the elastic modulus of steel and concrete (magenta).

2.3.2. Pearson Correlation Coefficient Analysis

Pearson correlation coefficient analysis was conducted to evaluate the linearity between fatigue life and the previously selected features, ultimately selecting the most appropriate features for the machine learning models. The relationship between each feature and fatigue life revealed that the stress range, used in existing design codes, exhibited a strong negative correlation with fatigue life, with a coefficient of −0.83. In addition, the maximum stress-to-static strength ratio and maximum stress-to-ultimate strength ratio demonstrated a significant negative correlation of −0.5. This indicates that the failure modes defined during the feature engineering process were well reflected in the data. However, because both features involve the same variable (maximum stress), multicollinearity may occur in regression problems. Therefore, the maximum stress-to-static strength ratio and ultimate strength were selected as the final features. Finally, diameter and height also exhibited a negative correlation, of approximately −0.3, with fatigue life (Figure 6).

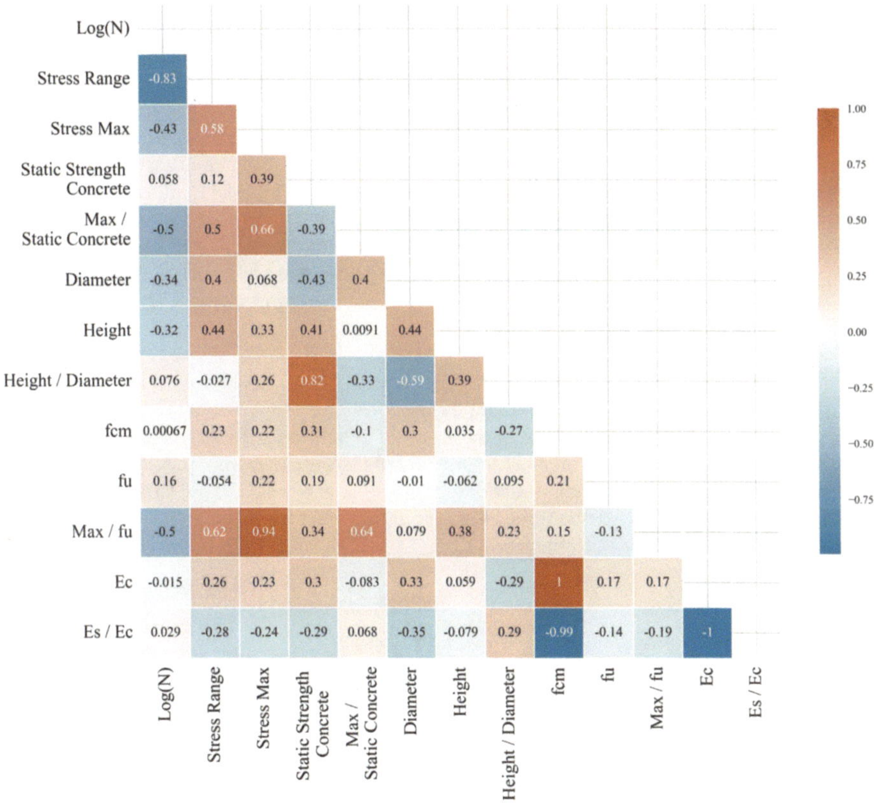

Figure 6. Pearson correlation coefficient analysis of the selected features.

2.4. *Dataset for Fatigue Life Prediction*

The final dataset for fatigue life prediction comprised of the following features: stress range, maximum stress-to-static strength ratio, ultimate strength, diameter and height, as listed in Table 3.

Table 3. Dataset for fatigue life prediction.

Input Feature	Mean	Min.	Max.	Std.	Unit
Stress range	120.798390	30.000000	179.270000	35.338377	MPa
Max stress/static strength	0.323189	0.059405	0.610717	0.097370	None
Ultimate strength	441.546752	337.855000	527.881200	39.703144	MPa
Diameter of stud	21.710124	10.000000	31.750000	5.602554	mm
Height of stud	117.742759	74.900000	155.000000	23.579766	mm
Output Feature	**Mean**	**Min.**	**Max.**	**Std.**	**Unit**
Log(N)	5.628602	3.725912	7.496943	0.816981	Cycle

3. Mahine Learning Models

3.1. Model Selection

3.1.1. AutoML Process

Based on the final data set defined in this study, various machine learning models were analyzed and compared to predict the fatigue life of headed studs using AutoML-based Pycaret [32] in Python and MATLAB Regression Learner [33]. The MATLAB Regression Learner automates the training, validation, and testing processes of the regression models. It allows for comparison across various regression models, including tree-based models and Bayesian-based models, such as Gaussian Process Regression (GPR) models. Pycaret, an open-source, low-code machine learning library in Python, automates the entire machine learning workflow from preprocessing to hyper parameter tuning and model evaluation. Pycaret leverages a wide range of modern machine learning models, such as XGBoost [34], CatBoost [35] and those from scikit-learn [36], providing a comprehensive comparison of the results across multiple models.

3.1.2. Metrics for Model Performance Evaluation

Four metrics were used to evaluate the performance of various models. The mean absolute error (MAE) is the average of the absolute differences between the predicted and actual values. The MAE provides a straightforward measure of error because it quantifies errors without considering their directions. This metric is robust against outliers, because it does not penalize large deviations. However, it may offer a conservative assessment of models when significant deviations occur.

$$MAE = \frac{1}{n}\sum_{i=1}^{n}|y_i - \hat{y}_i| \tag{8}$$

The mean squared error (MSE) is the average of the squared differences between the predicted and actual values. Unlike the MAE, the MSE squares the errors, assigning a larger penalty to larger errors, making it more sensitive to outliers. The square of errors has squared unites of the target variable. Therefore, the root mean squared error (RMSE), which is simply the square of the MSE, is also used for comparing model performances.

$$MSE = \frac{1}{n}\sum_{i=1}^{n}(y_i - \hat{y}_i)^2 \tag{9}$$

$$RMSE = \sqrt{\frac{1}{n}\sum_{i=1}^{n}(y_i - \hat{y}_i)^2} \tag{10}$$

R-squared is the coefficient of determination used to measure the proportion of variance in the target value that can be predicted from the independent variables. It is calculated as shown in Equation (11). R-squared ranges from 0 to 1, with higher values indicating that the model explains a larger proportion of the variance. This makes it a useful and intuitive metric for comparing model performances. However, R-squared does not directly account for prediction errors. Therefore, it should be used in conjunction with other evaluation metrics for a more comprehensive assessment of model performance.

$$R^2 = 1 - \frac{\sum_{i=1}^{n}(y_i - \hat{y}_i)^2}{\sum_{i=1}^{n}(y_i - \overline{y})^2} \quad (11)$$

where

y_i = Actual value
$\hat{y}_i$ = Predicted value
$\hat{y}$ = Mean of actual values
n = Number of data

3.1.3. Model Selection

The performance metrics of various models were compared to select an appropriate machine learning model. The dataset was split into 80% training data and 20% test data, and cross-validation was performed using the K-fold method with five folds. To ensure balanced learning and numerical stability across various model types, all features were scaled to a range between 0 and 1 using a standard scaler. Machine learning models were categorized into tree-based models, ridge regressor, support vector machines, and Gaussian process regressions. Twelve individual models were evaluated based on different sub-models or hyperparameters, with their performance metrics for the test data are summarized in Table 4.

Table 4. Comparison of machine learning models.

Category	Model	MAE	MSE	RMSE	R-Squared
Tree-based model	Gradient boosting regressor	0.2596	0.1338	0.3591	0.7859
	AdaBoost regressor	0.3203	0.1722	0.4092	0.7278
	Light gradient boosting	0.3411	0.1870	0.4236	0.7115
	Extreme gradient boosting	0.3110	0.1848	0.4221	0.7039
	Random forest regressor	0.2992	0.1585	0.3919	0.7464
Ridge regressor	Ridge regression	0.3740	0.1988	0.4439	0.6861
	Kernel ridge regression	0.36947	0.21092	0.45926	0.73712
Support vector machine	Medium gaussian	0.34547	0.2027	0.45023	0.74736
	Quadratic	0.42196	0.27556	0.52494	0.65656
Gaussian process regression	Rational quadratic	0.32	0.15677	0.39594	0.80461
	Matern 5/2	0.32515	0.16166	0.40207	0.79851
	Squared exponential	0.32667	0.16615	0.40761	0.79292
	Exponential	0.29939	0.14345	0.37875	0.82121

Each metrics was calculated based on the test data.

Overall, the best-performing machine learning model was a Gaussian process regression, with the rational quadratic kernel achieving an MAE of 0.34, MSE of 0.15677, RMSE of 0.39594, and R-squared of 0.80461. The next best model was the gradient-boosting regressor (GBR), a tree-based model, achieving an MAE of 0.2596, MSE of 0.1338, RMSE 0.3591, and an R-Squared of 0.7859. Consequently, the GPR model was selected as the final model, and a fine-tuning process was conducted to optimize and further improve the prediction accuracy.

3.2. Fine-Tuning Models

The models and model parameters that were fine-tuned for the GBR and GPR, selected through an earlier model selection process, are as follows.

3.2.1. Gradient-Boosting Regressor (GBR)

Boosting algorithms aim to "boost" the small advantage that a hypothesis produced by a weak learner can achieve over random guessing, by repeatedly applying the weak learning procedure on a sequence of carefully constructed distributions [37]. The GBR is a boosting algorithm that combines decision trees to effectively model nonlinear relationships. Each tree model is sequentially trained with subsequent models to reduce the residual errors of previous models. During the training process new models are added to minimize residuals within each tree, and the final prediction is derived by aggregating the weighted outputs of all tree models. In this process the predictions are adjusted by following the gradient of the loss function from the objective function and fitting new models to the residual errors. This approach results in a model with high predictive performance.

The random grid search method was used to fine-tune the GBR. Table 5 summarizes the optimized model parameters obtained through this fine-tuning process.

Table 5. Hyperparameters of gradient-boosting regressor (GBR).

Model	Parameter	Value
Gradient boosting regressor	Loss	squared_error
	Learning rate	0.4
	N estimators	200
	Subsample	0.9
	Criterion	Friedman_mse
	Min. sample split	9
	Min. sample leafs	3
	Max. depth	1
	Min. impurity decrease	0.05
	Max. features	1.0
	Alpha	0.9
	Tol	10^{-4}
	Ccp alpha	0.0
	Max. features	1.0

3.2.2. Gaussian Process Regression (GPR)

GPR is a probabilistic supervised machine learning method that assumes the data follow a Gaussian distribution. The model is defined for all pairs of data points using a covariance function (kernel function), that reflects prior knowledge and measures the similarity between data points. This kernel function enables the prediction of new data and provides uncertainty quantification for these predictions. The parameters of the GPR model were optimized using the limited-memory Broyden–Fletcher–Goldfarb–Shanno-bound (L-BFGS-B) algorithm. Table 6 lists the optimized model parameters.

Table 6. Hyperparameters of Gaussian process regression (GPR).

Model	Parameter	Value
Gaussian process regression	Kernel	Constant × Rational Quadratic + White Kernel
	Alpha	10^{-10}
	Optimizer	fmin_l_bfgs_b
	N restart optimizer	50
	Normalize y	True
	Constant value	1.0
	Constant value bounds	0.001, 1000
	Kernel length scale	1.0
	Kernel length scale bounds	0.001, 1000
	Kernel alpha	0.1
	Kernel alpha bounds	10^{-10}, 1000.0
	Noise level	1.0

4. Results and Discussion

4.1. Prediction Accuracy of Machine Learning Models

Figure 7 compares the performances of the two proposed machine learning models in predicting stress range based on the test data. The fine-tuned GBR achieved an RMSE of 0.336 and R-squared value of 0.844 for the test data. The GPR model achieved an RMSE of 0.277 and R-squared value of 0.862. Both models demonstrated high accuracy in the training and testing phases, indicating that they were well-generalized and effectively predicted the fatigue life of shear studs on new data.

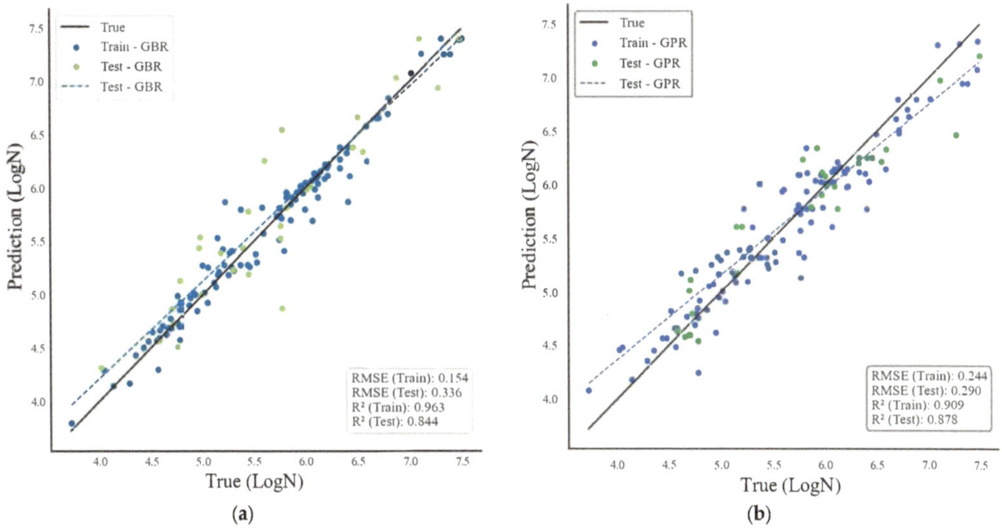

Figure 7. Comparison of model accuracy between machine learning models: (**a**) gradient-boosting regressor; (**b**) Gaussian process regression.

Figure 8 compares the predictions of existing design codes and those of the GPR model with respect to the stress range. The comparison was conducted under the same conditions with the parameters set to $d = 22$ mm, $h = 150$ mm, $f_u = 500$ MPa, $f_c = 35$ MPa. For the GPR model, the maximum stress to static strength ratio was fixed at 0.3, close to the dataset average. The results in Figure 8 indicate that the design codes, based on past experimental data, tend to be more conservative in their fatigue life predictions across the stress range compared with the proposed model. However, when uncertainty for the GPR model was quantified using a 95% confidence interval (CI), a higher level of uncertainty was observed at lower stress ranges. This increased uncertainty was attributed to the limited data available at a lower stress range and the presence of run-out data when fatigue failure did not occur. This uncertainty and prediction are expected to improve as the model is further trained using additional experimental data, enhancing the robustness and reliability of the predictions.

Figure 9 illustrates the accuracy and the distribution of the residual error from test dataset for the GPR model and its 95% CI, as well as for the AASHTO and Eurocode models. The figure shows that the AASHTO model demonstrates a conservative design approach, significantly underestimating fatigue life compared with the actual data. Specifically, the AASHTO model exhibits a high RMSE of 3.224, which is 8.4 times higher than the GPR model's RMSE of 0.384. By contrast, the Eurocode model provided predictions closer to the test data but tended to overestimate fatigue life at lower stress ranges, resulting in an RMSE of 0.878, which is still more than twice as high as RMSE of the GPR model. In addition, the GPR model and its 95% CI exhibited greater predictive accuracy than existing design codes by effectively training on the dataset. And the residual errors of the GPR model and its 95% CI were evenly distributed, indicating a low prediction bias in both models when compared with the test data.

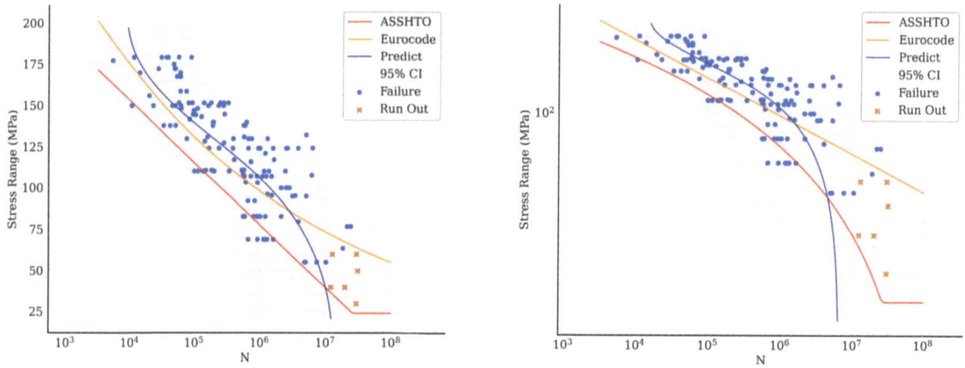

(a) (b)

Figure 8. Comparison between machine learning models and existing design codes by stress range with $d = 22$ mm, $h = 150$ mm, $f_u = 450$ MPa, $f_c = 35$ MPa, and $\sigma_{max}/Q_{u,concrete} = 0.3$: (**a**) semi log scale; (**b**) log scale on stress range.

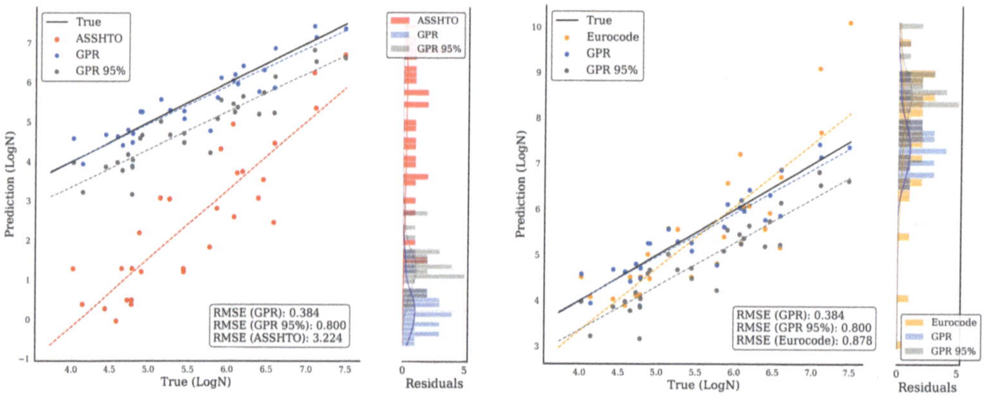

(a) (b)

Figure 9. Comparison of model accuracy and residual errors between machine learning models and current design codes: (**a**) GPR and ASSHTO; (**b**) GPR and Eurocode.

Therefore, the proposed GPR model for predicting fatigue life provided reliable predictions with high accuracy and generalization performance and effectively accounting for the uncertainty by considering features previously overlooked by existing design codes.

4.2. Feature Importance Analysis

4.2.1. Individual Conditional Expectation (ICE)

The proposed machine learning models demonstrated high accuracy in predicting fatigue life, both the training data and on new data from test datasets. However, unlike simple empirical formulas, many machine learning models are complex, multidimensional regression models, which can be difficult for engineers to interpret, limiting their applicability in for decision-making. Therefore, this study used ICE [38] and SHAP [39] values to examine how the predicted fatigue life changes with variations in each selected feature and analyze feature importance.

ICE is a visualization technique that shows how the target value changes at individual data points for each feature. An ICE plot illustrates how the model's predictions vary with changes in a specific input variable, allowing the assessment of the sensitivity of the model to that input variable. For each feature, the predicted changes for the individual data points in the existing dataset are represented by individual lines. In addition, the plot includes lines representing the mean, as well as the 5th and 95th percentiles, to analyze the overall trend of each feature while excluding excessive variations or noise.

Figure 10, the ICE plot used to understand the relationship between each feature and fatigue life in proposed machine learning model, was analyzed according to the range of the dataset as specified in Table 3. For the stress range, a strong negative linearity with fatigue life was observed within the range of 35 MPa to 180 MPa, with a relatively uniform data distribution resulting in lower uncertainty compared with other features. The maximum stress to static strength ratio exhibited an overall negative linearity within the range of 0.05 to 0.6. This indicates that the maximum stress of each stud by cyclic load tend to reduce fatigue life. By contrast, the concrete static strength tended to increase fatigue life, as reflected by the model. The ultimate strength of the stud generally exhibited a positive correlation, particularly for using high-strength studs above 450 MPa. Indicating significant impact on fatigue life. For diameter and height, a slightly nonlinear relationship was observed within the ranges of 10 mm to 30 mm, and 60 mm to 160 mm, respectively. However, owing to the limited number of large-diameter stud experiments within the collected data, higher uncertainty was observed. This indicates that need for additional experimental data on various diameters to improve the reliability of the model.

4.2.2. SHapley Addictive exPlanations (SHAP) Value

SHAP calculates the contribution of each feature to the prediction, providing an overall understanding of the extent to which the model relies on each feature. While ICE visualizes the trend for each data point, SHAP shows how the relative value of each feature contributes to the prediction, allowing for the assessment of the overall feature importance based on mean SHAP values. Figure 11 shows the SHAP values and their averages for the GBR and GPR models. Both models identified stress range as the most significant feature, particularly in higher stress ranges (represented by red data points), where it demonstrated a substantial negative impact on the predicted values.

For the GBR, the next most influential features were, height, maximum stress to static strength ratio, ultimate strength, and diameter. By contrast, the GPR model identified the maximum stress to static strength ratio, ultimate strength, height, and diameter, as most important features following the stress range. In both models, stress range is considered the most important feature. However, in the GBR model, the effect of stress range is observed across a wider range (−1.5 to 1.5), whereas in the GPR model, it constrained to a smaller range (−1.0 to 1.0). This suggests that the influence of Stress Range is relatively lower in the GPR model compared to the GBR model, allowing other features to be considered more significant in the GPR model. In addition, while the GBR model considers height, which exhibits a relatively discrete distribution, as a key feature, the GPR model identifies the maximum stress to static strength ratio as the next most important variable, with a more evenly distributed SHAP value. Moreover, the GBR model as the interactions between features based on tree structure with a structural and hierarchical method, leading to more discontinuous patterns in SHAP values compared to the GPR model. As a result, the GPR model displayed relatively more distinct feature importance than the GBR model.

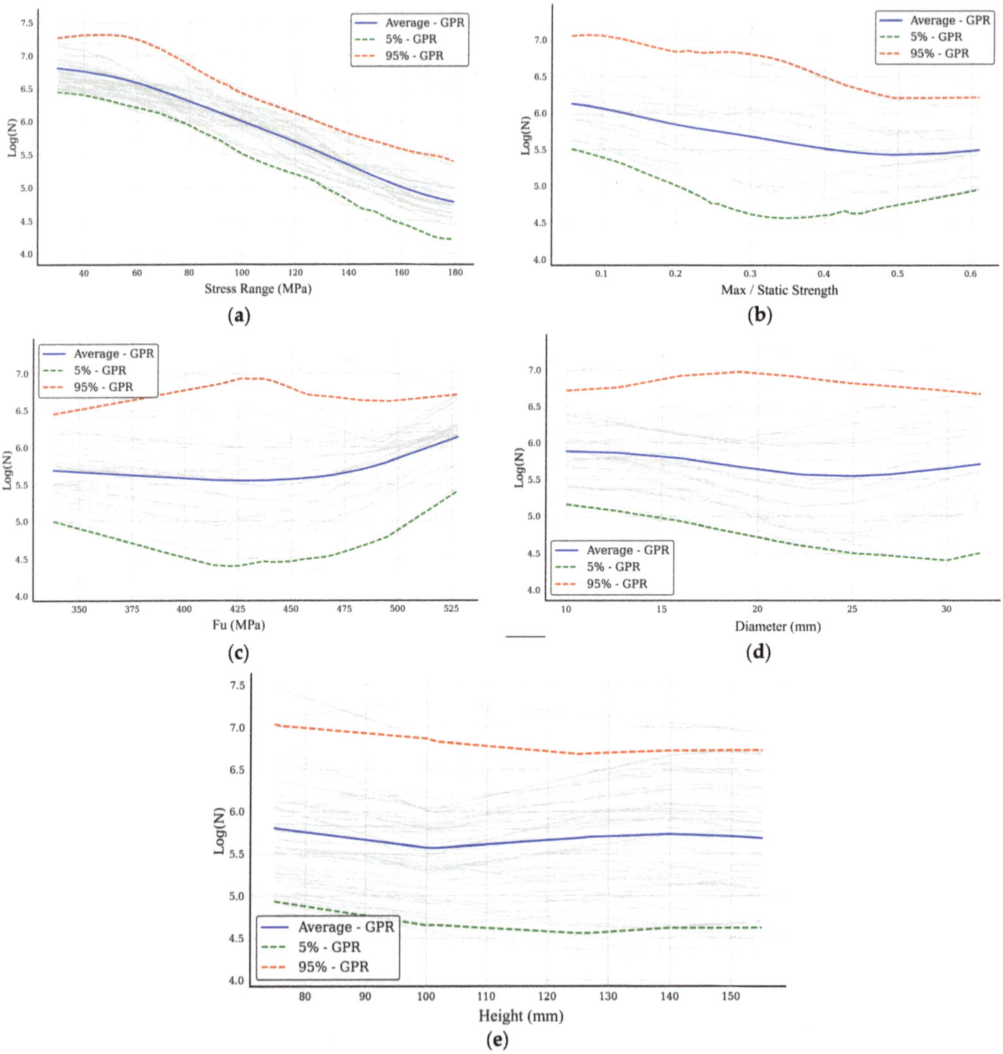

Figure 10. Individual conditional expectation (ICE) plot of Gaussian process regression: (**a**) stress range; (**b**) max./static strength; (**c**) ultimate strength; (**d**) diameter; (**e**) height.

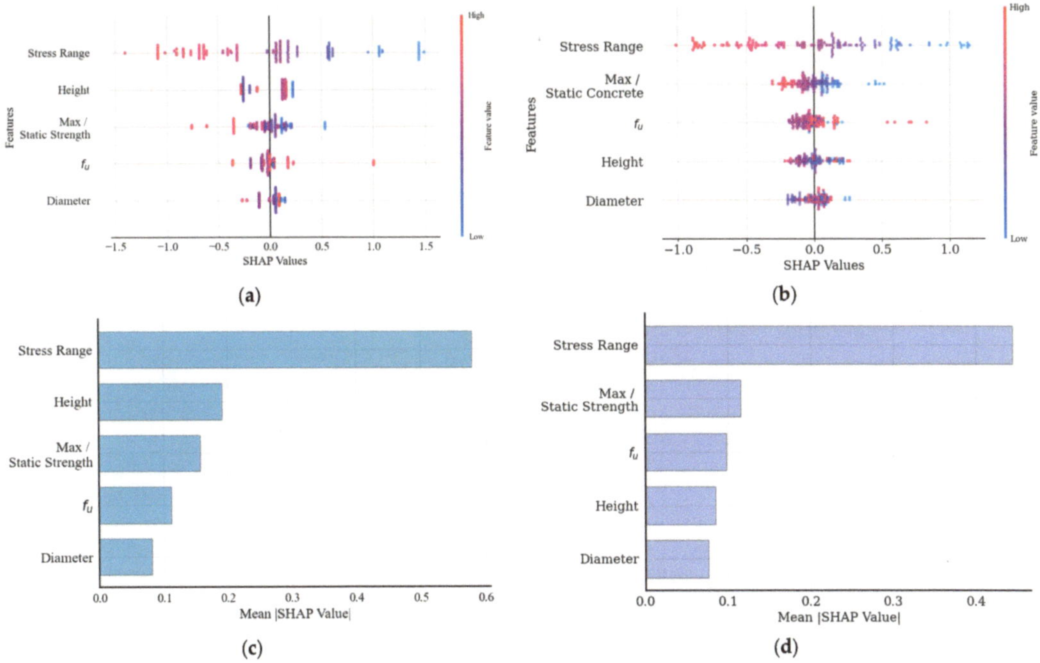

Figure 11. Comparison of SHAP values between machine learning models and current design codes: (**a**) SHAP values of GBR; (**b**) SHAP value of GPR; (**c**) mean SHAP values of GBR; (**d**) mean SHAP values of GPR.

5. Conclusions

Current methods for predicting fatigue life in existing design codes heavily rely on historical experimental data. However, these methods often fail to address the diverse conditions encountered in actual steel concrete composite bridges, such as the use of high-strength materials and precast decks. In addition, the inherent eccentricity in single-sided push-out tests increases crack propagation, leading to overly conservative fatigue life predictions in design codes. Furthermore, the conventional employed statistical technique, the least-squares regression approach, fails to adequately account for uncertainties in confidence levels. This limitation reduces the generalization performance and complicates the assessment of the reliability of predictive outcomes.

To overcome the limitations of existing design codes, this study proposed a machine learning model for predicting the fatigue life of stud shear connectors. The model employs feature engineering to capture critical aspects of fatigue behavior and uses feature selection to evaluate the linearity between variables. The model aims to significantly improve the accuracy of fatigue life predictions by identifying the key factors contributing to fatigue failure and leveraging a comprehensive dataset of push-out test results. The performance and uncertainty of the proposed model were rigorously validated by comparing its predictions with those of existing design codes, using test data excluded from the training phase.

In addition, the impact of each selected feature of fatigue life was analyzed through ICE plots, illustrating the localized relationship between the fatigue life and each feature, as well as SHAP values, which assess the overall trends and feature importances.

The following conclusions were drawn from this process:

1. Through the feature engineering process, the overall fatigue behavior was categorized into four factors: shear transfer, crack initiation, crack propagation, and failure. Twelve

features based on these four factors were selected. Based on their linearity with fatigue life, key features such as stress range, maximum stress-to-static strength ratio, ultimate strength, diameter, and height were identified. These key features were applied to a dataset of approximately 146 push-out tests.

2. To select the most appropriate machine learning model for predicting fatigue life, 12 different models were compared using Pycaret and MATLAB AutoML Process. The GPR model was identified as the most appropriate model exhibiting the highest performance metrics. This model, along with its associated uncertainties, was compared to existing design codes using test data to, analyzing the prediction accuracy and the distribution of residual errors. The results demonstrated that the proposed GPR model provided reliable predictions with high accuracy and generalization performance while effectively accounting for uncertainty by incorporating features previously overlooked in existing design codes.
3. Feature importance analysis based on the proposed GPR model revealed that the stress range most significantly influenced fatigue life prediction. The maximum stress-to-static strength ratio and ultimate strength also exhibited a significant influence. Although the diameter and height, related to stud geometry, exhibited relatively lower importance, they affected fatigue life prediction through certain nonlinear relationships. In addition, the prediction results of the proposed GPR model, along with the feature importance analysis, indicate that each feature exhibits high uncertainty when predicting outside the dataset range specified in Table 3. Therefore, it is recommended to make predictions within this range, and these restrictions can be mitigated with the acquisition of additional data in the future.
4. For the further study, the proposed model, which can replace the empirical equations in existing design codes, can be used as follows: a pretrained model is developed using global data, which includes experimental results sourced from the literature. These data are preprocessed, and the machine learning model is trained to generalize across diverse conditions. Once the pretrained model is established, the second step involves fine-tuning the model using local data specific to the country of interest. Local experimental data, reflective of regional conditions such as materials and practices, are used to adjust the parameter of pretrained model, ensuring accurate representation of the local situation. The fine-tuned model is then integrated into local design codes, replacing empirical equations with a data-driven approach. Continuous validation and monitoring are essential to ensure reliability of the model in practical applications.

Author Contributions: Conceptualization, C.-S.S.; methodology, D.-H.K.; validation, D.-H.K.; investigation, D.-H.K.; resources, D.-H.K., C.-S.S. and K.-C.L.; data curation, D.-H.K. and G.-T.R.; writing—original draft preparation, D.-H.K.; writing—review and editing, C.-S.S., K.-C.L. and G.-T.R.; visualization, D.-H.K.; supervision, C.-S.S.; project administration, C.-S.S.; funding acquisition, C.-S.S. All authors have read and agreed to the published version of the manuscript.

Funding: This research was conducted with the support of the "National R&D Project for Smart Construction Technology (RS-2020-KA156007)", funded by the Korea Agency for Infrastructure Technology Advancement under the Ministry of Land, Infrastructure and Transport, and managed by the Korea Expressway Corporation. Moreover, this research was supported by the Chung-Ang University Research Scholarship Grants in 2023.

Data Availability Statement: The data presented in this study are available at https://github.com/easternwise98/fatigue-life-prediction.git (accessed on 8 October 2024).

Conflicts of Interest: The authors declare no conflicts of interest.

References

1. Slutter, R.G.; Fisher, J.W. Fatigue Strength of Shear Connectors. *Highw. Res. Rec.* **1966**, *147*, 65–88.
2. AASHTO American Association of State Highway and Transportation Officials. *LRFD Bridge Design Specifications*, 9th ed.; AASHTO American Association of State Highway and Transportation Officials: Washington, DC, USA, 2020.

3. Oehlers, D.; Foley, L. The Fatigue Strength of Stud Shear Connections in Composite Beams. *Proc. Inst. Civ. Eng.* **1985**, *79*, 349–364. [CrossRef]
4. Johnson, R.P. Resistance of Stud Shear Connectors to Fatigue. *J. Constr. Steel Res.* **2000**, *56*, 101–116. [CrossRef]
5. *1994-2 Eurocode 4*; Design of Composite Steel and Concrete Structures, General Rules and Rules for Bridges. European Commission: Brussels, Belgium, 2005.
6. Hanswille, G.; Porsch, M.; Ustundag, C. Resistance of Headed Studs Subjected to Fatigue Loading. *J. Constr. Steel Res.* **2007**, *63*, 475–484. [CrossRef]
7. Hanswille, G.; Porsch, M.; Ustundag, C. Resistance of Headed Studs Subjected to Fatigue Loading Part II: Analytical Study. *J. Constr. Steel Res.* **2007**, *63*, 485–493. [CrossRef]
8. Badie, S.S.; Tadros, M.K.; Kakish, H.F.; Splittgerber, D.L.; Baishya, M.C. Large Shear Studs for Composite Action in Steel Bridge Girders. *J. Bridge Eng.* **2002**, *7*, 195–203. [CrossRef]
9. Lee, P.-G.; Shim, C.-S.; Chang, S.-P. Static and Fatigue Behavior of Large Stud Shear Connectors for Steel–Concrete Composite Bridges. *J. Constr. Steel Res.* **2005**, *61*, 1270–1285. [CrossRef]
10. Liu, Y.; Zhang, Q.; Bao, Y.; Bu, Y. Static and Fatigue Push-out Tests of Short Headed Shear Studs Embedded in Engineered Cementitious Composites (ECC). *Eng. Struct.* **2019**, *182*, 29–38. [CrossRef]
11. Shim, C.-S.; Lee, P.-G.; Chang, S.-P. Design of Shear Connection in Composite Steel and Concrete Bridges with Precast Decks. *J. Constr. Steel Res.* **2001**, *57*, 203–219. [CrossRef]
12. Shim, C.-S.; Lee, P.-G.; Kim, D.-W.; Chung, C.-H. Effects of Group Arrangement on the Ultimate Strength of Stud Shear Connection. In *Composite Construction in Steel and Concrete VI*; American Society of Civil Engineers: Tabernash, CO, USA, 2011; pp. 92–101.
13. Okada, J.; Yoda, T.; Lebet, J.-P. A study of the grouped arrangements of stud connectors on shear strength behavior. *Struct. Eng. Earthq. Eng.* **2006**, *23*, 75s–89s. [CrossRef]
14. Shim, C.-S.; Kim, D.-W. Structural Performance of Composite Joints Using Bent Studs. *Int. J. Steel Struct.* **2010**, *10*, 1–13. [CrossRef]
15. Lee, K.-C.; Abbas, H.H.; Ramey, G.E. Review of Current AASHTO Fatigue Design Specifications for Stud Shear Connectors. In *Structures Congress 2010*; American Society of Civil Engineers: Orlando, FL, USA, 2010; pp. 310–321.
16. Avci-Karatas, C. Application of Machine Learning in Prediction of Shear Capacity of Headed Steel Studs in Steel–Concrete Composite Structures. *Int. J. Steel Struct.* **2022**, *22*, 539–556. [CrossRef]
17. Roh, G.-T.; Vu, N.; Jeon, C.-H.; Shim, C.-S. Augmented Data-Driven Machine Learning for Digital Twin of Stud Shear Connections. *Buildings* **2024**, *14*, 328. [CrossRef]
18. Roshanfar, M.; Ghiami Azad, A.R.; Forouzanfar, M. Predicting Fatigue Life of Shear Connectors in Steel-concrete Composite Bridges Using Artificial Intelligence Techniques. *Fatigue Fract. Eng. Mater. Struct.* **2024**, *47*, 818–832. [CrossRef]
19. Oehlers, D.J.; Bradford, M.A. *Composite Steel and Concrete Structural Members: Fundamental Behaviour*, 1st ed.; Pergamon: Oxford, UK; Tarrytown, NY, USA, 1995; ISBN 978-0-08-041919-0.
20. Bro, M.; Westberg, M. Influence of Fatigue on Headed Stud Connectors in Composite Bridges. Master's Thesis, Luleå Tekniska Universitet, Luleå, Sweden, 2004.
21. Paris, P.; Erdogan, F. A Critical Analysis of Crack Propagation Laws. *J. Basic Eng.* **1963**, *85*, 528–533. [CrossRef]
22. Mainstone, R.J.; Menzies, J.B. Shear connectors in steel-concrete composite beams for bridges. I. static and fatigue tests on push-out specimens. *Concrete* **1967**, *1*, 291.
23. National Research Council (U.S.), Highway Research Board. *Highway Research Record*; National Research Council, Highway Research Board: Washington, DC, USA, 1968.
24. Oehlers, D.J. Deterioration in Strength of Stud Connectors in Composite Bridge Beams. *J. Struct. Eng.* **1990**, *116*, 3417–3431. [CrossRef]
25. Zandonini, R.; Bursi, O.S. Cyclic Behavior of Headed Stud Shear Connectors. In *Composite Construction in Steel and Concrete IV*; American Society of Civil Engineers: Banff, AB, Canada, 2002; pp. 470–482.
26. Feldmann, M.; Hechler, O.; Hegger, J.; Rauscher, S. Fatigue Behavior of Shear Connectors in High Performance Concrete. In *Composite Construction in Steel and Concrete VI*; American Society of Civil Engineers: Tabernash, CO, USA, 2011; pp. 39–51.
27. Mundie, D.L. Fatigue Testing and Design of Large Diameter Shear Studs Used in Highway Bridges. Master's Thesis, Auburn University, Auburn, AL, USA, 2011.
28. Xie, E.; Valente, M.I.B. *Research Report Fatigue Strength of Shear Connectors*; University of Minho: Guimarães, Portugal, 2011.
29. Dogan, O.; Roberts, T.M. Fatigue Performance and Stiffness Variation of Stud Connectors in Steel–Concrete–Steel Sandwich Systems. *J. Constr. Steel Res.* **2012**, *70*, 86–92. [CrossRef]
30. Ovuoba, B.; Prinz, G.S. Fatigue Capacity of Headed Shear Studs in Composite Bridge Girders. *J. Bridge Eng.* **2016**, *21*, 04016094. [CrossRef]
31. Wang, W.; Chen, J.; Jin, W. Experimental Investigation of Corroded Stud Shear Connectors Subjected to Fatigue Loading. *J. Mater. Civ. Eng.* **2017**, *29*, 04016175. [CrossRef]
32. Ali, M. PyCaret: An Open-Source, Low-Code Machine Learning Library in Python. Version 3.4. Available online: https://github.com/pycaret/pycaret (accessed on 8 October 2024).
33. The MathWorks, Inc. Statistics and Machine Learning Toolbox Documentation. Available online: https://www.mathworks.com/help/stats/index.html (accessed on 19 August 2024).

34. Chen, T.; Guestrin, C. XGBoost: A Scalable Tree Boosting System. In Proceedings of the 22nd ACM SIGKDD International Conference on Knowledge Discovery and Data Mining, San Francisco, CA, USA, 13–17 August 2016; pp. 785–794.
35. Friedman, J.H. Greedy Function Approximation: A Gradient Boosting Machine. *Ann. Statist.* **2001**, *29*, 1189–1232. [CrossRef]
36. Pedregosa, F.; Varoquaux, G.; Gramfort, A.; Michel, V.; Thirion, B.; Grisel, O.; Blondel, M.; Prettenhofer, P.; Weiss, R.; Dubourg, V.; et al. Scikit-Learn: Machine Learning in Python. *J. Mach. Learn. Res.* **2011**, *12*, 2825–2830.
37. Zemel, R.S.; Pitassi, T. A Gradient-Based Boosting Algorithm for Regression Problems. In Proceedings of the 13th International Conference on Neural Information Processing Systems (NIPS 2000), Denver, CO, USA, 27 November–2 December 2000; MIT Press: Cambridge, MA, USA, 2000.
38. Goldstein, A.; Kapelner, A.; Bleich, J.; Pitkin, E. Peeking Inside the Black Box: Visualizing Statistical Learning with Plots of Individual Conditional Expectation. *J. Comput. Graph. Stat.* **2015**, *24*, 44–65. [CrossRef]
39. Lundberg, S.M.; Lee, S.-I. A Unified Approach to Interpreting Model Predictions. In Proceedings of the 31st International Conference on Neural Information Processing Systems, Long Beach, CA, USA, 4–9 December 2017; Curran Associates, Inc.: Red Hook, NY, USA, 2017.

buildings

MDPI

Article

A New Single-Step Bolt Tightening Method for Multi-Bolt Flange Structures

Jiacheng Zhong [1,2], Jianwen Bao [1], Junyi Wang [1,*], Zhigang Xu [1], Mingyi Yang [1] and Yanjie Shi [3]

1 Shenyang Institute of Automation, Chinese Academy of Sciences, Shenyang 110016, China; zhongjiacheng@sia.cn (J.Z.); baojianwen1@sia.cn (J.B.); zgxu@sia.cn (Z.X.); myyang@sia.cn (M.Y.)
2 School of Computer Science and Technology, University of Chinese Academy of Sciences, Beijing 101408, China
3 College of Science, Inner Mongolia University of Technology, Baotou 014010, China; shi_yanjie2021@163.com
* Correspondence: jywang@sia.cn

Abstract: Single-step tightening is simple and fast compared to multi-step tightening, and it is therefore frequently used during the tightening of bolts in flange structures. However, single-step tightening to maintain uniformity is more difficult to achieve. To address this problem, existing methods such as the Elastic Interaction Coefficient Method (EICM) and Tetraparametric Assembly Method (TAM) have been investigated for load uniformity in single-step tightening. In order to improve the computational efficiency and accuracy of the traditional methods, a new single-step tightening method is proposed in this paper. This method can realize the design of the initial preload force simply by measuring the change in bolt load in a specific sequence. It is verified by numerical simulation that this method can realize the uniform distribution of bolt load. In addition, this paper will provide suggestions for the optimal tightening sequence for the single-step tightening method.

Keywords: multi-bolted joint; bolt tightening sequence; preload; assembly

1. Introduction

Flange connections represent a standard method for joining mechanical systems. They are frequently employed in modern engineering equipment [1]. However, the elastic interaction between bolts influenced by the tightening sequence gives rise to bolt relaxation phenomena [2–4]. The above process hinders the successful implementation of bolt preload [5]. Consequently, the presence of irregular stress distributions, a reduction in fatigue strength [6], and unfavorable impacts on other mechanical properties of the joint system can be observed. These factors ultimately lead to the occurrence of safety accidents. Furthermore, the elastic interaction between bolts is subject to a number of influencing factors, including the geometry and materials of the components being joined, the magnitude of the load, the distance between bolts, and the configuration of the assembly [7–9]. As a result, it is a challenge to mitigate these interactions and achieve uniform load distribution.

Single-step tightening is more convenient and faster than multi-step tightening, so it is often used in flange connections. However, the elastic interaction between bolts is greater in single-step tightening compared to multi-step tightening, which leads to a more dispersed distribution of bolt loads in single-step tightening [10]. Therefore, unlike in multi-step tightening where the preload size for each bolt is the same when designing the tightening sequence, in single-step tightening [11], it is more important to design the preload size for each bolt under different tightening sequences to achieve a uniform distribution of bolts.

Among them, the Elastic Interaction Coefficient Method (EICM) is a method based on the principle of elastic interaction, which assumes that there is a linear relationship between the initial preload and residual preload when the tightening sequence is determined and expresses this relationship by means of an elastic coefficient matrix [12,13]. Once the target preload is set, the initial preload for each bolt can be obtained by solving the inverse of the elastic coefficient matrix and multiplying it with the target preload. Subsequent studies have validated the effectiveness of this method using finite element analysis (FEA) and physical experiments [14–16]. In addition, some studies have attempted to find the optimal tightening sequence by constructing a neural network agent model to achieve a more uniform preload distribution and further improve the effectiveness of the EICM [17,18]. These attempts have not significantly changed the consistency of the optimization results of EICM under different tightening sequences [15]. At the same time, the spring node model provides a theoretical basis for the EICM [19–21]. Even in the case of nonlinear materials such as flexible graphite gaskets, the EICM still ensures a uniform distribution of bolt loads [22]. However, the EICM requires measuring the impact on each already tightened bolt during the tightening process, making the computational cost high for multi-bolt flange structures [16].

To solve the problem of high computational cost of the EICM, a simplified method has been proposed—the Tetraparametric Assembly Method (TAM) [16]. The method neglects elastic interactions between distant bolts, thus greatly simplifying the elastic coefficient matrix. While the TAM simplifies the design process of the initial preload by requiring only a few parameters to be measured, its consistency in optimization results across different tightening sequences is not as robust as that of EICM. In addition, the TAM mainly focuses on the case of unilateral bolts, while it is inadequate for the treatment of bilateral bolts. Therefore, to address the limitations of the EICM and TAM, this paper presents a new method designed to simplify the EICM by considering bilateral bolts by simulating their operation in real situations. Compared with the EICM, this method only needs to measure the variation in individual bolt preload at a specific sequence, which significantly reduces the computational cost. Recommendations for the tightening sequence are also given to minimize the influence of the tightening sequence on the optimization results.

The new method proposed in this paper is dedicated to reducing the computational complexity and improving the efficiency while giving suggestions for the tightening sequence. Specifically, Section 1 demonstrates the EICM using a spring-node model and discusses the effect of the tightening sequence on the load distribution. Further, in Section 2, the simplification approach to the EICM is presented, and a new single-step tightening method is proposed. In order to validate the proposed method, an FE model for numerical simulation is presented in Section 3. Based on FEA, the optimization effects of the TAM and the new method under three different tightening sequences are further compared and analyzed in Section 5, and finally the recommended tightening sequence for single-step tightening is given.

2. The Influence of Tightening Sequence on Elastic Interaction

2.1. Validation of EICM Using the Spring-Node Model

Figure 1 illustrates the tightening process. Despite the nonlinear behavior observed in the contact between the flange and gasket [23], the elastic interaction can be equated to a spring if the deformation of both the bolt and the connector is within the range of the elastic deformation. Therefore, the process of bolt tightening can be described by a spring-node model [19–21]. Elastic deformation is required so that the deformation of each bolt is consistent with the deformation of its associated node. As a result, a set of equilibrium equations describing the elastic deformation behavior is generated during bolt tightening.

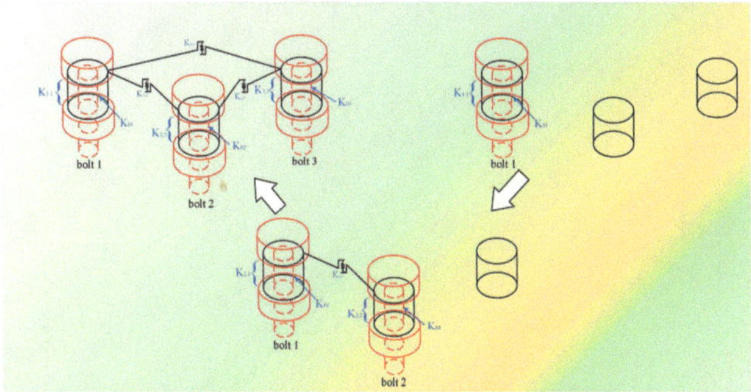

Figure 1. Spring-node system example when bolts are tightened sequentially.

Tightening clockwise produces the following system of equilibrium equations for elastic deformation as we tighten sequentially from the first bolt to the n-th bolt:

$$\begin{bmatrix} K_{1,1}^{-1}+K_{b1}^{-1} & K_{2,1}^{-1} & K_{1,3}^{-1} & \cdots & K_{1,n-1}^{-1} \\ K_{2,1}^{-1} & K_{2,2}^{-1}+K_{b2}^{-1} & K_{2,3}^{-1} & \cdots & K_{2,n-1}^{-1} \\ K_{3,1}^{-1} & K_{3,2}^{-1} & K_{3,3}^{-1}+K_{b3}^{-1} & \cdots & K_{3,n-1}^{-1} \\ \vdots & \vdots & \vdots & \ddots & \vdots \\ K_{n-1,1}^{-1} & K_{n-1,2}^{-1} & K_{n-1,3}^{-1} & \cdots & K_{n-1,n-1}^{-1}+K_{b(n-1)}^{-1} \end{bmatrix} \begin{Bmatrix} \Delta S_{1,n} \\ \Delta S_{2,n} \\ \Delta S_{3,n} \\ \vdots \\ \Delta S_{n-1,n} \end{Bmatrix} = -\begin{Bmatrix} K_{1,n}^{-1}S_n \\ K_{2,n}^{-1}S_n \\ K_{3,n}^{-1}S_n \\ \vdots \\ K_{n-1,n}^{-1}S_n \end{Bmatrix} \quad (1)$$

where $\Delta S_{i,n}$ is the value of the change in preload force of the i-th bolt after preloading the n-th bolt; S_n is the preload force applied to the n-th bolt; $S_{i,j}$ is the elastic interaction stiffness between the i-th and j-th bolts; $S_{i,i}$ is the axial stiffness of the i-th bolt hole, and S_{bi} is the i-th bolt stiffness. The $\Delta S_{i,n}$ is equal to

$$\Delta S_{i,n} = -\frac{\begin{vmatrix} K_{1,1}^{-1}+K_{b1}^{-1} & \cdots & K_{1,n}^{-1} & \cdots & K_{1,n-1}^{-1} \\ \vdots & \ddots & \vdots & \cdots & \vdots \\ K_{i,1}^{-1} & \cdots & K_{i,n}^{-1} & \cdots & K_{i,n-1}^{-1} \\ \vdots & \vdots & \vdots & \ddots & \vdots \\ K_{n-1,1}^{-1} & \cdots & K_{n-1,n}^{-1} & \cdots & K_{n-1,n-1}^{-1}+K_{b(n-1)}^{-1} \end{vmatrix}}{\begin{vmatrix} K_{1,1}^{-1}+K_{b1}^{-1} & \cdots & K_{1,i}^{-1} & \cdots & K_{1,n-1}^{-1} \\ \vdots & \ddots & \vdots & \cdots & \vdots \\ K_{i,1}^{-1} & \cdots & K_{i,i}^{-1}+K_{bi}^{-1} & \cdots & K_{i,n-1}^{-1} \\ \vdots & \vdots & \vdots & \ddots & \vdots \\ K_{n-1,1}^{-1} & \cdots & K_{n-1,i}^{-1} & \cdots & K_{n-1,n-1}^{-1}+K_{b(n-1)}^{-1} \end{vmatrix}} \cdot S_n \quad (2)$$

where the stiffness is determined by the system itself, and the elements contained within the determinant are determined by the tightening sequence. Therefore, as the tightening sequence is determined, $\Delta S_{i,n}$ and F_n are linearly related and can be expressed as follows:

$$\Delta F_{i,n} = A_{i,n} \cdot S_n \quad (3)$$

where $A_{i,n}$ is the elastic interaction coefficient between the i-th bolt and the nth bolt. Assuming a total of m bolts in the flange system, the initial and residual preloads of the bolts can be expressed as the following relationship [12,13]:

$$\begin{Bmatrix} S_{f1} \\ S_{f2} \\ \vdots \\ S_{fn-1} \\ S_{fn} \end{Bmatrix} = \begin{bmatrix} A_{1,1} & A_{1,2} & \cdots & A_{1,n-1} & A_{1,n} \\ 0 & A_{2,2} & \cdots & A_{2,n-1} & A_{2,n} \\ \vdots & \cdots & \ddots & \vdots & \vdots \\ 0 & 0 & \cdots & A_{n-1,n-1} & A_{n-1,n} \\ 0 & 0 & \cdots & 0 & A_{n,n} \end{bmatrix} \begin{Bmatrix} S_1 \\ S_2 \\ \vdots \\ S_{n-1} \\ S_n \end{Bmatrix} \tag{4}$$

where Equation (4) can also be expressed in the following form:

$$S_f = A \cdot S_i \tag{5}$$

where S_f, S_i, and A represent the residual load, the initial load, and the elastic coefficient matrix, respectively. Once the elastic coefficient matrix A has been determined, optimization for bolt tightening can be achieved based on the target residual preload. The initial preload required to achieve the target residual preload can be obtained by inverting elastic coefficient matrix A:

$$S_i = A^{-1} \cdot S_f \tag{6}$$

The method described above is known as the EICM. Although the EICM can achieve the target preload distribution, it is necessary to measure the change in load of the remaining bolts as each bolt is tightened and recalculate the corresponding elastic coefficient matrix when the tightening sequence is changed.

2.2. The Impact of Tightening Sequence on Load Distribution

Figure 2 illustrates the division of the flange into two and five zones. Divide the flange into two zones. Zone I bolts the 1st to the 30-th and zone II bolts the 31-st to the 60-th. Tighten the first bolt of each zone in turn, i.e., 1→31, then tighten the second bolt of each zone, i.e., 2→32, and so on for all bolts. If the flange is divided into five zones, tighten them in a similar manner.

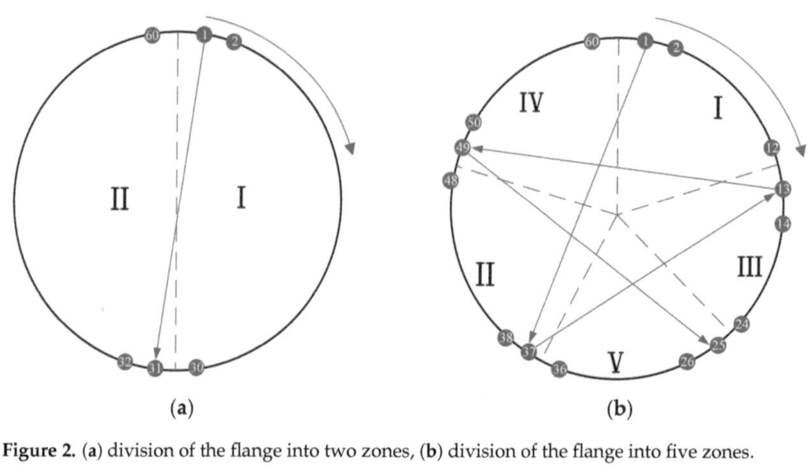

Figure 2. (**a**) division of the flange into two zones, (**b**) division of the flange into five zones.

For dividing the flange into two zones, Equation (1) when tightening to the second zone can be expressed as follows:

$$\begin{bmatrix} K_{11,11}^{-1}+K_{b11}^{-1} & \cdots & K_{11,1n}^{-1} & K_{11,21}^{-1} & \cdots & K_{11,2(n-1)}^{-1} \\ \vdots & \ddots & \vdots & \cdots & \vdots & \vdots \\ K_{1n,11}^{-1} & \cdots & K_{1n,11}^{-1}+K_{b1n}^{-1} & K_{1n,21}^{-1} & \cdots & K_{1n,2(n-1)}^{-1} \\ K_{21,11}^{-1} & \cdots & K_{21,n1}^{-1} & K_{21,21}^{-1}+K_{b21}^{-1} & \cdots & K_{21,2(n-1)}^{-1} \\ \vdots & \vdots & \vdots & \cdots & \ddots & \vdots \\ K_{2(n-1),11}^{-1} & \cdots & K_{2(n-1),1n}^{-1} & K_{2(n-1),21}^{-1} & \cdots & K_{2(n-1),2(n-1)}^{-1}+K_{b2(n-1)}^{-1} \end{bmatrix} \begin{Bmatrix} A_{11,2n} \\ \vdots \\ A_{1n,2n} \\ A_{21,2n} \\ \vdots \\ A_{2(n-1),2n} \end{Bmatrix} = -\begin{Bmatrix} K_{11,2n}^{-1} \\ \vdots \\ K_{1n,2n}^{-1} \\ K_{21,2n}^{-1} \\ \vdots \\ K_{2(n-1),2n}^{-1} \end{Bmatrix} \quad (7)$$

where $A_{ij,nm}$ is the stiffness between the j-th bolt in the i-th zone and the m-th bolt in the n-th zone. The elastic interaction between the bolts decreases with increasing distance [19]. Therefore, it is possible to consider only the stiffness of neighboring bolts. The system of equations combining the symmetry of the flange structure can be simplified to the following form:

$$\begin{bmatrix} K_h^{-1}+K_b^{-1} & K^{-1} & & & & & & \\ & K^{-1} & K_h^{-1}+K_b^{-1} & K^{-1} & & & & \\ & & & \ddots & & & & \\ & & & K^{-1} & K_h^{-1}+K_b^{-1} & & & \\ & & & & & K_h^{-1}+K_b^{-1} & K^{-1} & \\ & & & & & & \ddots & \\ & & & & & & K^{-1} & K_h^{-1}+K_b^{-1} \end{bmatrix} \begin{Bmatrix} A_{11,2n} \\ A_{12,2n} \\ \vdots \\ A_{1n,2n} \\ A_{21,2n} \\ \vdots \\ A_{2(n-1),2n} \end{Bmatrix} = -\begin{Bmatrix} 0 \\ 0 \\ \vdots \\ 0 \\ 0 \\ \vdots \\ K^{-1} \end{Bmatrix} \quad (8)$$

where K is the elastic interaction stiffness of neighboring bolts; K_h is the axial stiffness of the bolt hole and K_b is the bolt stiffness. Clearly, the elastic interaction coefficients beyond the coefficient between the neighboring bolts $A_{2(n-1),2n}$ are zero. $A_{2(n-1),2n}$ is constant.

A similar approach can be used to reach the same conclusion if the flange structure is divided into a different number of zones. Consequently, the resulting residual preloads of the bolts in the same zone k are, respectively,

$$\begin{cases} S_{fkf} = S_{kf} + A \cdot S_{k(f+1)} + A \cdot S_{(k-1)e} \\ S_{fkc} = S_{kc} + A \cdot S_{k(c+1)} \\ S_{fkl} = S_{kl} \end{cases} \quad (9)$$

where S_{fkf} is the residual preload of the first tightened bolt in zone k; S_{kf} is the initial preload of the first tightened bolt in zone k; A is the elastic interaction coefficient between the neighboring bolts; $S_{k(f+1)}$ and $S_{(k-1)e}$ are the initial preloads applied to bolts adjacent to the first tightened bolt of zone k; S_{fkc} is the residual preload of bolts in zone k except for the first and last tightened bolts; S_{kc} is the initial preload of bolts in zone k except for the first and last tightened bolts; $S_{k(c+1)}$ is the initial preload applied to bolt adjacent to the bolts in zone k except for the first and last tightened bolts; S_{fkl} is the residual preload of the last tightened bolt in zone k; S_{kl} is the initial preload of the first tightened bolt in zone k.

Thus, when the same initial preload is applied, the average residual preload of the bolt for different numbers of zones is

$$\overline{S_f} = \frac{\sum^n \sum^{m-2} S_{fkc}}{n(m-2)} + \frac{\sum^n \left(S_{fkf} + S_{fkl}\right)}{2n} = S_{fkc} \quad (10)$$

where m is the number of bolts in zone k; n is the number of zones. Consequently, the average residual preloads of different tightening sequences are equal.

The tightness of the connection is influenced by the uniformity of the residual preload of the bolts [11]. The uniformity of preload is measure by the standard variance (SD). When the same initial preload is

$$\sigma = \sqrt{\frac{\Sigma^n \Sigma^{m-2}\left(S_{fkc} - \overline{S}_f\right)^2 + \Sigma^n\left(S_{fkf} - \overline{S}_f\right)^2 + \Sigma^n\left(S_{fkl} - \overline{S}_f\right)^2}{m \times n}} = \sqrt{\frac{2}{m}} \cdot A \cdot \overline{S}_f \quad (11)$$

Because the number of bolts $m \times n$ in the flange structure is fixed, the number of bolts in zone k is inversely proportional to the number of zones. Hence, as the number of zones increases, the degree of preload dispersion increases.

3. New Methodology for Bolt Tightening Sequence Optimization

3.1. Tetraparametric Assembly Method

To reduce the computational cost of the EICM, the TAM was developed [16]. As shown in Figure 3, the closer the relative distance between bolts, the greater the elastic interaction [19]. Therefore, the TAM focuses on the elastic interactions between adjacent bolts or bolts separated by one bolt. Elastic interactions between bolts that are further apart are not considered, and the elastic interaction coefficients are zero. By measuring the possibilities of all elastic interaction coefficients under simplified conditions, appropriate elastic interaction coefficients are input into the elastic coefficient matrix based on different tightening sequences. It is worth noting that since the TAM is a simplification for the ECIM, the parameters of the TAM and EICM are of the same nature. Therefore, for the same structure, the magnitude of the load does not affect the parameters of the TAM. This selective consideration effectively reduces the computational cost of the EICM. Finite element analysis and physical experiments for different tightening sequences indicate that the TAM can achieve a uniform distribution of loads effectively.

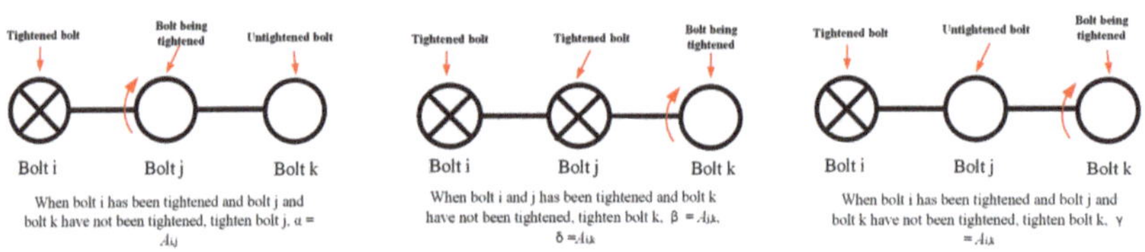

Figure 3. The diagram illustrates the computational methodology for determining parameters in the TAM.

3.2. A Novel Single-Step Bolt Tightening Method for Multi-Bolt Flange Structures

Like the TAM, the proposed method also aims to reduce the EICM by reasonably simplifying the elastic interaction coefficients. Unlike the TAM, the simplification in the proposed method is achieved by simulating the actual flange bolt tightening process through a local bolt tightening. This simplification approach is more in line with the actual situation of flange bolt tightening. The schematic diagram of the proposed methodology is shown in Figure 4. After processing the parameters obtained for a specific tightening sequence, the optimization of the bolt load can be achieved given only the tightening sequence and the target residual preload.

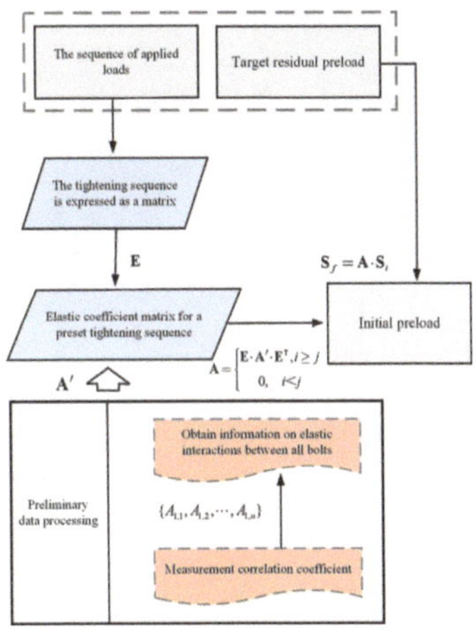

Figure 4. The flowchart of the proposed method for a system with n bolts.

The range of elastic interactions considered in the proposed method is the eight bolts adjacent to each other on both sides. Therefore, considering the flange assembly with twelve bolts illustrated in Figure 5, the preset tightening sequence is 1→2→3→4→5→9→10→11→12.

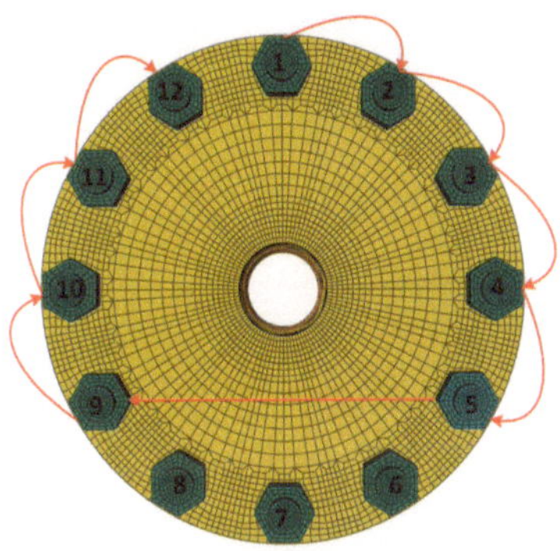

Figure 5. Schematic illustration of the preset tightening sequence employed by the proposed method.

The effects of distant bolts 6, 7, and 8 on bolt 1 are neglected with consideration limited to the impact of bolts closer to bolt 1:

$$\begin{cases} A_{1,1} = 1 & j = 1 \\ A_{1,j} = \frac{S_{1,j} - S_{1,j-1}}{S_j}, & 2 \le j \le 5 \cup 9 \le j \le 12 \\ A_{1,j} = 0, & j = 6,7,8 \end{cases} \tag{12}$$

where $S_{1,j}$ and $S_{1,j-1}$ are the preload on the 1st bolt when the j-th bolt and $(j-1)$-th bolt are tightened, respectively. The coefficient $A_{1,j}$ is the influence of tightening the j-th bolt on the 1st bolt. This gives a complete set of elastic interaction matrix coefficients affecting the 1st bolt. Both $A_{1,2}$ and $A_{1,12}$ denote the coefficients of bolts adjacent to the 1st bolt. Notably, the coefficient $A_{1,12}$ depicts a more complex coupling condition compared to $A_{1,2}$. So, these two coefficients represent predetermined values at different coupling complexities. As shown in Equation (1), in a flange structure with consistent bolt spacing, the elastic interaction matrix coefficients are uniform for each bolt when the bolt spacing and coupling are identical. Therefore, a recursive strategy can logically start with the elastic interaction coefficient of bolt 1:

$$\begin{cases} A_{i,1} = A_{i-1,12}, & 1 < i \leq 12 \\ A_{i,j} = A_{i-1,j-1}, & 1 < i,j \leq 12 \end{cases} \tag{13}$$

The elastic interaction coefficients for each bolt can be determined by Equation (8), which are then used to build the matrix A'. Matrix A' can be represented as follows:

$$A' = \begin{bmatrix} A_{1,1} & A_{1,2} & \cdots & A_{1,12} \\ A_{2,1} & A_{2,2} & \cdots & A_{2,12} \\ \vdots & \vdots & \ddots & \vdots \\ A_{12,1} & A_{12,2} & \cdots & A_{12,12} \end{bmatrix} \tag{14}$$

The matrix A' contains the simplified elastic interaction coefficients between all bolts. With the target residual preload S_f and the specific tightening sequence 1→7→2→8→3→9→4→10→5→11→6→12, the focus is on solving the initial preload S_i. As described in Equation (4), when the bolt tightening sequence is the same as the ordering of the ranks in the matrix, the matrix is an upper triangular matrix. Using this property, the elastic interaction coefficients in the specific tightening sequence can be retained, and the transformation matrix E can be obtained by performing row operations on the identity matrix in sequence according to the specific tightening sequence. Matrix A' suffers simultaneous row and column alterations using E. Hence, by preserving solely the upper triangular constituent of the correction matrix, the matrix A is obtained:

$$A = \begin{cases} E \cdot A' \cdot E^T, i \geq j \\ 0 \quad , i < j \end{cases} \tag{15}$$

$$A = \begin{bmatrix} A_{1,1} & A_{1,7} & \cdots & A_{1,6} & A_{1,12} \\ 0 & A_{7,7} & \cdots & A_{7,6} & A_{7,12} \\ 0 & 0 & \ddots & \vdots & \vdots \\ 0 & 0 & \cdots & A_{6,6} & A_{6,12} \\ 0 & 0 & \cdots & 0 & A_{12,12} \end{bmatrix} \tag{16}$$

Once matrix A is constructed, the initial preload can be calculated by Equation (6).

The parameters in the EICM, the TAM and the proposed method are expressed in different symbols, but the physical meaning and the method of operation are similar. The number of bolts to be tightened and the number of parameters to be measured in the three methods for an O-ring with n ($n \geq 9$) bolts are given in Table 1.

Table 1. Comparison of different methods of calculating costs.

	EICM	TAM	The Proposed Method
the number of bolts	n	5	9
the number of parameters	n(n − 1)/2	4	8

4. Experimental Validation

To validate the effectiveness of the proposed method, a finite element model is created in ABAQUS 2022. As demonstrated in Figure 6, the model is the area where the spacecraft combustion chamber and nozzle are connected. The model has 60 M16 bolts with all materials defined as steel (E = 210 GPa, ν = 0.3, σ_s = 355 MPa). The friction coefficient μ is 0.2. An initial preload of 0.1 N is applied to the untightened bolts, whereas for tightened bolts, a fixed length is maintained post-preload application. The C3D8R element type is adopted with all meshes consisting of hexahedral elements. The final assembly has 6,884,250 nodes and 5,530,800 elements.

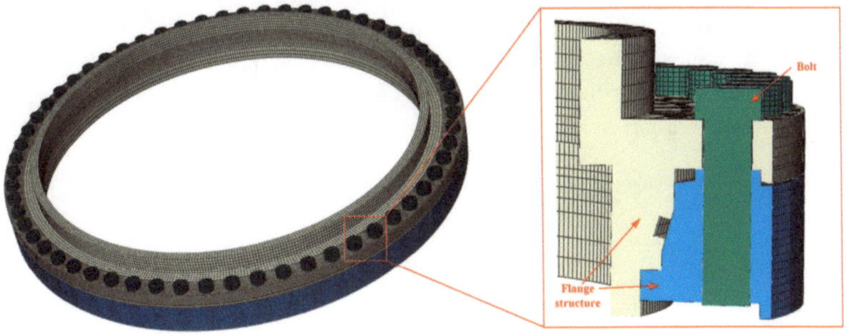

Figure 6. FE model.

To validate that the FE model effectively predicts load change in bolts during joint assembly, a test bench with identical physical properties to the FE model is developed, as shown in Figure 7. We compared FEA and experimental errors for three tightening sequences: clockwise, diagonal and star pattern tightening [24,25]. In the case of clockwise sequential tightening, the bolts are tightened in the sequence $1 \to 2 \to 3 \to \cdots \to 60$. The diagonal pattern tightening sequence and the five-points star pattern tightening sequence are shown in Figures 2a and 2b, respectively. The diagonal tightening sequence divides the bolts into two zones, zone I bolts 1 to 30 and zone II bolts 31 to 60. Tighten the first bolt of each zone in turn, i.e., $1 \to 31$; then, tighten the second bolt of each zone, i.e., $2 \to 32$, and so on for all bolts.

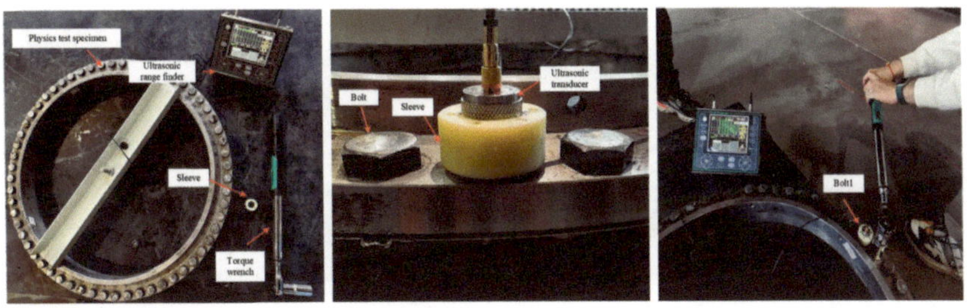

Figure 7. Test bench.

A torque of 147 N·m is applied to the bolt using a torque wrench. The changes of bolt length in the loading process are measured by an ultrasonic rangefinder with a precision of 0.001 mm. It should be noted that the torque wrench has an application torque accuracy of ±3%. The sleeve indicated in Figure 7 guarantees the consistent positioning of the ultrasonic transducer above the bolt throughout measurement. Combined with the bolt size and material, the calculated bolt stiffness is 237,233 N/mm [26]. So, according to Hooke's law, the preload can be calculated from the change in bolt length. Given the linear relationship between both torque and preload [27], an applied torque of 147 N·m is equivalent to applying an initial preload of 54,652 N. The amount of preload ensures that the resulting stress does not exceed 80% of the yield strength.

The experimental steps are as follows:

1. Measure the initial length of each bolt.

2. Tighten the 1st bolt and fix the probe on the 1st bolt, and then tighten the other bolts in sequence to obtain the change in preload force of the 1st bolt.
3. After all the bolts are tightened, measure the change in the length of each bolt to obtain the residual preload force.

The FEA results are compared with the experimental results. Figure 8 illustrates the average residual preload and maximum relative error of all bolts, and Figure 9 shows the variation in the 1st bolt for different tightening sequences. The FEA results and experimental results exhibit close agreement.

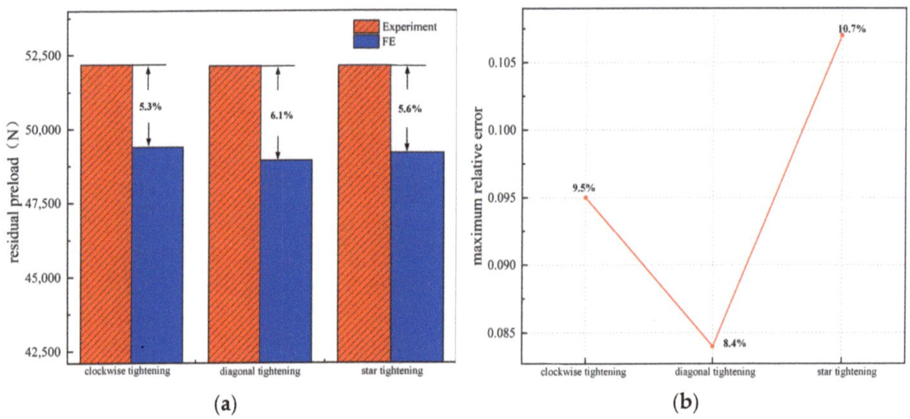

Figure 8. Comparison of overall preload between FE and experimental: (**a**) average residual preload of all bolts, (**b**) maximum relative error of all bolts.

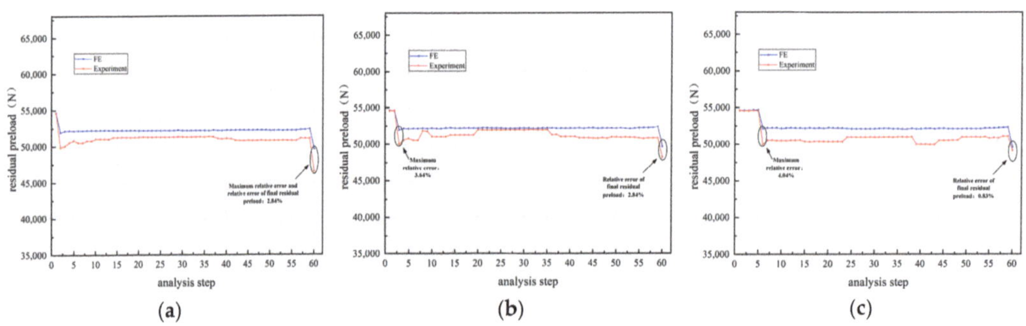

Figure 9. Variation in preload force of the 1st bolt in different sequences: (**a**) clockwise tightening, (**b**) diagonal tightening, (**c**) five-point star pattern tightening sequence.

The discrepancies between the experimental outcomes and the FEA results can be attributed to four primary sources of error: the precision of the torque applied via the wrench, the accuracy of the ultrasonic measurement techniques, the discrepancies between the FEA parameters and the actual frictional coefficients, and the thread damage incurred due to the repeated disassembly of the bolts [28].

The experimental results for the change in preload of the 1st bolt are closer to the FEA results than the residual preload of the individual bolts. The improvement in accuracy is related to the measurement method of the 1st bolt. The ultrasonic sensor is fixed to the head of the 1st bolt, thereby reducing the possibility of error that can arise from fluctuations in environmental variables.

In conclusion, these results demonstrate that the FE model can accurately obtain the preload force of the bolts.

5. Numerical Examples

In order to verify the effectiveness of the proposed method, the numerical examples of three typical tightening sequences are optimized to achieve a uniform distribution of 54,652 N residual preload force by the proposed method and the TAM. Concurrently, a preload force of 54,652 N is uniformly applied to each bolt across all tightening sequences, establishing a benchmark for comparative optimization analysis. The bolt tightening sequence includes the three tightening sequences in Section 4: clockwise sequence tightening sequence, diagonal pattern tightening sequence and five-point star pattern tightening sequence. The specific results of the TAM parameters and the proposed methodological parameters are shown in Tables 2 and 3, respectively.

Table 2. The specific results of the TAM parameters.

α	β	γ	δ
−0.0499	−0.0503	0.0032	0.0039

Table 3. The specific results of the proposed methodological parameters.

$A_{1,2}$	$A_{1,3}$	$A_{1,4}$	$A_{1,5}$	$A_{1,57}$	$A_{1,58}$	$A_{1,59}$	$A_{1,60}$
−0.0499	0.0039	0.0007	0.0001	0.0003	0.0012	0.0007	−0.0499

Figures 10–12 show the FE results for the three tightening sequences. In the radar diagram, the axis is the bolt serial number, and the concentric cycle is the contour of the residual preload. The mean relative error (MRE) is used to evaluate the closeness of optimization results from different methods to the target value. The uniformity of preload is measure by the SD. The comparison results are shown in Table 4. The numerical examples manifest the efficacy of the proposed method in bolt tightening optimization.

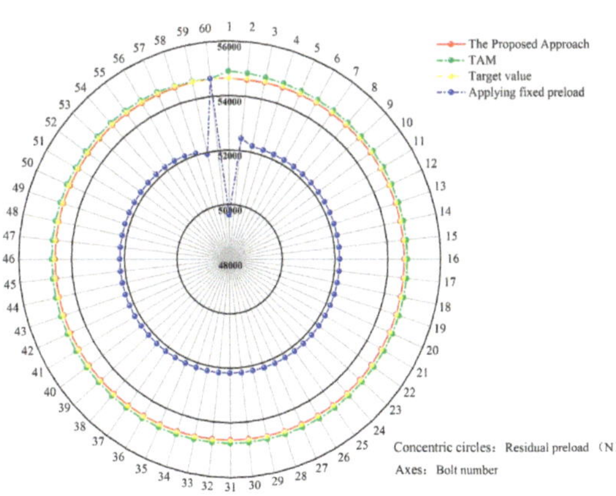

Figure 10. Clockwise sequential tightening optimization comparison (target residual preload 54,652 N).

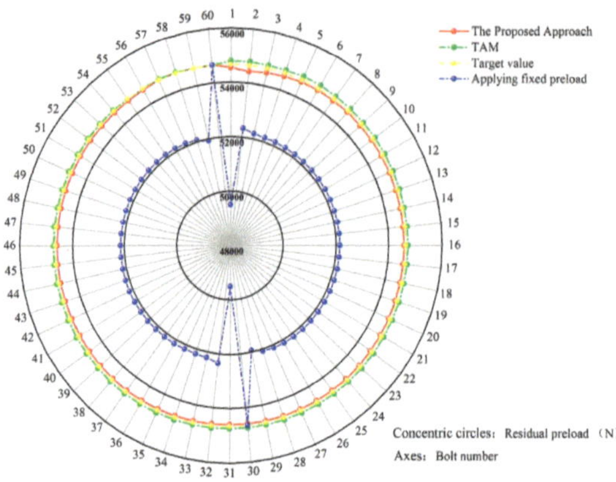

Figure 11. Diagonal pattern tightening optimization comparison (target residual preload 54,652 N).

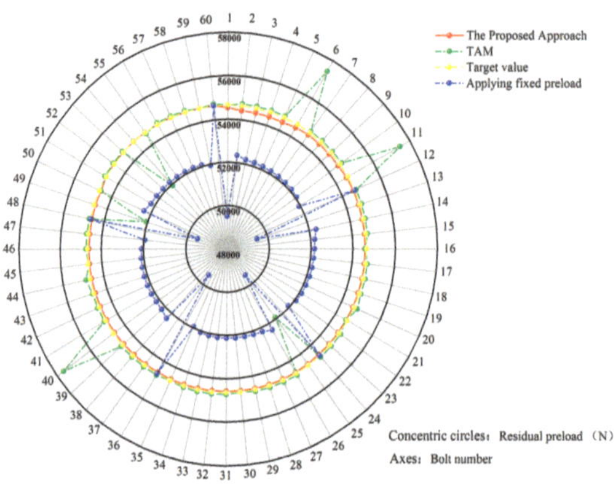

Figure 12. Five-point star pattern tightening optimization comparison (target residual preload 54,652 N).

Table 4. Comparison results of the proposed method with TAM.

	Clockwise Sequential Tightening			Diagonal Pattern Tightening			Five-Points Star Pattern Tightening		
	MRE	**SD**	**Average Residual Preload**	**MRE**	**SD**	**Average Residual Preload**	**MRE**	**SD**	**Average Residual Preload**
applying fixed preload	4.53%	466.07	52,174.6	4.60%	672.14	52,179.3	4.62%	1077.55	52,182.5
TAM	0.168%	32.73	54,743.9	0.136%	29.13	54,726.4	0.64%	876.82	54,718.2
the proposed method	0.037%	7.54	54,622.7	0.152%	31.69	54,584.2	0.132%	55.30	54,581.0

Numerical simulation results indicate that compared to the fixed application of preload, the proposed method and TAM can achieve a uniform distribution of bolt loads with the error from the target load being within an acceptable range. There is no significant difference between the two methods when tightening in a clockwise sequence or a diagonal sequence. However, when using the five-point star tightening sequence, the preload distribution obtained by the proposed method is more uniform than that obtained by TAM. This phenomenon occurs because both methods simplify the elastic coefficient matrix, and the five-point star tightening sequence results in a more discrete distribution of loads. The proposed method more closely approximates the actual tightening conditions, thus providing a more accurate approximation of the elastic coefficient matrix for this sequence.

It is noteworthy that the conclusions drawn from the applying fixed preload are consistent with those derived in Section 2.1. As the number of partitions in the flange structure increases (the clockwise sequence can be considered as dividing the bolts into one zone, diagonal tightening and five-points star tightening can be considered as dividing the flange into two zones and five zones, respectively), the average residual preload of the bolt is approximated and the load distribution is more discrete when the same initial preload is applied. Combining the optimization results of the two methods, it can be observed that as the number of zones increases, the accuracy of the approximation of the elastic coefficient matrix decreases. Therefore, when using a single tightening method that simplifies the elastic coefficient matrix, it is best to use the clockwise tightening sequence. If considering the issue of bolt relaxation caused by coaxiality in actual production [29], diagonal tightening can also be adopted.

6. Conclusions

This paper studies the impact of tightening sequence on bolt load distribution using the spring-node model. It is observed that when only the elastic interaction between adjacent bolts is considered, the average residual preload of bolts is equal under different tightening sequences when the same initial preload is applied. The degree of load distribution dispersion increases as the complexity of the tightening sequence increases. The paper introduces a new single-step tightening method for flanges. This method can be achieved by tightening only local bolts, reducing computational costs compared to EICM. Finite element analysis results confirm the effectiveness of the proposed method in optimizing bolt tightening. When using such single-step tightening methods that simplify the elastic coefficient matrix, it is preferable to use either clockwise tightening or diagonal tightening.

Author Contributions: Methodology, J.Z.; Software, J.B.; Resources, J.W.; Writing—original draft, J.Z.; Writing—review & editing, M.Y. and Y.S.; Project administration, Z.X.; Funding acquisition, J.W. All authors have read and agreed to the published version of the manuscript.

Funding: This research was supported by the LiaoNing Revitalization Talents Program for the Project titled "Digital Twin Assembly Technology for High-end Equipment and Robot Manufacturing System", Project No. YS2023004, organized by the Organization Department of Liaoning Provincial Committee of Communist Party of China and recommended by Shenyang Institute of Automation.

Data Availability Statement: The original contributions presented in the study are included in the article, further inquiries can be directed to the corresponding author.

Conflicts of Interest: The authors declare no conflicts of interest.

References

1. Li, Z.; Zhao, Z.; Zhang, J.; Liu, C. A novel experimental method for bolt stress distribution during the assembly process. *Structures* **2024**, *69*, 107413. [CrossRef]
2. Leray, D.; Vadean, A.; Daidie, A. Fast computation of preloaded bolted circular joint aiming at fatigue bolt sizing: Hypotheses and modeling. *Eng. Comput.* **2008**, *25*, 42–54. [CrossRef]
3. Coria, I.; Martín, I.; Bouzid, A.H.; Heras, I.; Abasolo, M. Efficient assembly of bolted joints under external loads using numerical FEM. *Int. J. Mech. Sci.* **2018**, *142*, 575–582. [CrossRef]
4. Li, H.; Wu, H.; Zhao, J.; Sun, H.; Sun, W.; Wang, Q.; Wang, X. Theoretical and experimental investigations of vibration and damping behaviors of carbon fiber-reinforced composite thin shells with partial bolt looseness constraints. *Eur. J. Mech.-A/Solids* **2023**, *97*, 104839. [CrossRef]
5. Kumakura, S.; Saito, K. Tightening sequence for bolted flange joint assembly. In Proceedings of the ASME Pressure Vessels and Piping Conference, Cleveland, OH, USA, 20–24 July 2003; Volume 16982, pp. 9–16.

6. Rakotondrainibe, L.; Allaire, G.; Orval, P. Topological sensitivity analysis with respect to a small idealized bolt. *Eng. Comput.* **2022**, *39*, 115–146. [CrossRef]
7. Jiang, K.; Liang, Y.; Zhao, O. Machine-learning-based design of high strength steel bolted connections. *Thin-Walled Struct.* **2022**, *179*, 109575. [CrossRef]
8. Mao, J.; Fang, K.; Jia, H.; Wang, Q.; Li, S.; Qian, W.; Xiong, F.; Peng, W. Mechanism and control of preload force loss of precast structural joint bolts under freeze–Thaw environments. *Cold Reg. Sci. Technol.* **2024**, *221*, 104176. [CrossRef]
9. Zheng, M.; Li, Y.; Wang, C.; Bai, J.; Wang, L.; Liu, Z.; Wahab, M.A. Prediction of contact stress in bolted joints using the Polynomial Chaos-Kriging model. *Eng. Fail. Anal.* **2023**, *154*, 107646. [CrossRef]
10. Coria, I.; Abasolo, M.; Aguirrebeitia, J.; Heras, I. Study of bolt load scatter due to tightening sequence. *Int. J. Press. Vessel. Pip.* **2020**, *182*, 104054. [CrossRef]
11. Bausman, A.R.; Waterland, A.F.J. A Process to Consider Practical Applications of Alternative Assembly Procedures of ASME PCC-1. In Proceedings of the Pressure Vessels and Piping Conference, Bellevue, WA, USA, 18–22 July 2010; Volume 49217, pp. 13–20.
12. Bibel, G.D.; Ezell, R.M. An improved flange bolt-up procedure using experimentally determined elastic interaction coefficients. *J. Press. Vessel. Technol.* **1992**, *114*, 405–410. [CrossRef]
13. Bickford, J.H. *Handbook of Bolts and Bolted Joints*; Marcel Dekker: New York, NY, USA, 1998.
14. Coria, I.; Martín, I.; Bouzid, H.; Heras, I.; Aguirrebeitia, J. Optimized bolt tightening sequences in bolted joints using superelement FE modeling technique. In Proceedings of the International Conference on Offshore Mechanics and Arctic Engineering, Madrid, Spain, 17–22 June 2018; American Society of Mechanical Engineers: New York, NY, USA, 2018. V001T01A007.
15. Pan, W.; Li, X.; Sun, L.; Song, H.; Wang, M. General prediction model of residual pretightening force of bolt group with one-step and multi-step pretightening and analysis of influencing factors from the perspective of elastic interaction. *Eng. Comput.* **2023**, *40*, 921–956. [CrossRef]
16. Coria, I.; Abasolo, M.; Olaskoaga, I.; Etxezarreta, A.; Aguirrebeitia, J. A new methodology for the optimization of bolt tightening sequences for ring type joints. *Ocean. Eng.* **2017**, *129*, 441–450. [CrossRef]
17. Liu, S.; Chu, J.; Wang, Y. Research on Prediction Method of Bolt Tightening for Aviation Components Based on Neural Network. *Appl. Sci.* **2023**, *13*, 6771. [CrossRef]
18. Wang, Y.; Liu, Y.; Wang, J.; Zhang, J.; Zhu, X.; Xu, Z. Research on process planning method of aerospace engine bolt tightening based on digital twin. *Machines* **2022**, *10*, 1048. [CrossRef]
19. Wang, Y.-Q.; Wu, J.K.; Liu, H.-B.; Kuang, K.; Cui, X.-W.; Han, L.-S. Analysis of elastic interaction stiffness and its effect on bolt preloading. *Int. J. Mech. Sci.* **2017**, *130*, 307–314. [CrossRef]
20. Jiang, K.; Liu, Z.; Yang, C.; Zhang, C.; Tian, Y.; Zhang, T. Effects of the joint surface considering asperity interaction on the bolted joint performance in the bolt tightening process. *Tribol. Int.* **2022**, *167*, 107408. [CrossRef]
21. Li, Y.; Liu, Z.; Wang, Y.; Cai, L.; Xu, W. Research on preload of bolted joints tightening sequence-related relaxation considering elastic interaction between bolts. *J. Constr. Steel Res.* **2019**, *160*, 45–53. [CrossRef]
22. Nassar, S.A.; Wu, Z.; Yang, X. Achieving uniform clamp load in gasketed bolted joints using a nonlinear finite element model. *J. Press. Vessel. Technol.* **2010**, *132*, 041205. [CrossRef]
23. Krishna, M.M.; Shunmugam, M.S.; Prasad, N.S. A study on the sealing performance of bolted flange joints with gaskets using finite element analysis. *Int. J. Press. Vessel. Pip.* **2007**, *84*, 349–357. [CrossRef]
24. Nassar, S.A.; Alkelani, A.A. Clamp load loss due to elastic interaction and gasket creep relaxation in bolted joints. *J. Press. Vessel. Technol.* **2006**, *128*, 394–401. [CrossRef]
25. Abasolo, M.; Aguirrebeitia, J.; Avilés, R.; Fernández de Bustos, I. A tetraparametric metamodel for the analysis and design of bolting sequences for wind generator flanges. *J. Press. Vessel. Technol.* **2011**, *133*, 041202. [CrossRef]
26. Pedersen, N.L.; Pedersen, P. Stiffness analysis and improvement of bolt-plate contact assemblies. *Mech. Based Des. Struct. Mach.* **2008**, *36*, 47–66. [CrossRef]
27. Luo, W.; Feng, P.; Zhang, J.; Ren, Y.; Yu, D.; Wu, Z. Research on a radiation model for bolt-tightening force attenuation of wheel-type parts. *IEEE Access* **2019**, *7*, 157459–157470. [CrossRef]
28. Zhu, L.; Hong, J.; Jiang, X. On controlling preload and estimating anti-loosening performance in threaded fasteners based on accurate contact modeling. *Tribol. Int.* **2016**, *95*, 181–191. [CrossRef]
29. Yang, H.; Xia, H.; Zhu, L.; Li, X.; Du, C.; Liu, J.; Ao, X.; Li, D. Measurement and assembly methods for bolted flange joints using three-dimensional digital image correlation. *Int. J. Press. Vessel. Pip.* **2024**, *208*, 105136. [CrossRef]

buildings

MDPI

Article

Experimental and Numerical Study of Self-Centering Brace with Arc Steel Plate

Mahdi Malek Mohammad Pour [1], Soheil Monajemi Nejad [1,*] and AbdolReza Sarvgad Moghadam [2]

[1] Civil Engineering Faculty, Islamic Azad University, Central Tehran Branche, Tehran 1955847781, Iran; mahdi_malek_z@yahoo.com
[2] International Institute of Earthquake Engineering and Seismology, Tehran 1953714453, Iran; moghadam@iiees.ac.ir
* Correspondence: soh.monajeminejad@iauctb.ac.ir

Abstract: Self-centering braces are structural systems that help reduce structural drift when the structure is under the influence of seismic forces. This paper introduces a novel self-centering brace designed to enhance the seismic performance of structures, while also minimizing permanent deflection after an earthquake. The advantages of this brace include easy construction, compatibility with construction practices in Iran, affordability, high capacity for axial force, efficient energy dissipation, adaptability for development, and applicability in various structures. The proposed brace components are presented, and the brace's behavior under cyclic loading is analyzed. The results highlight the significant impact of arched steel plates on the damping properties of the proposed brace systems. Various parameters, such as thickness, curvature radius, and arched spring width, are considered. Moreover, experimental studies are conducted to explore the behavior of the proposed brace. A comparison between the experimental and numerical modeling results demonstrates the accuracy of the numerical models in relation to the tests.

Keywords: self-centering; hysteretic behavior; arched spring; residual drifts; damping; flag history hysteretic

1. Introduction

The incorporation of braces into a structure's design is a method for significantly increasing its lateral stiffness. This approach establishes a conventional lateral resistance system that effectively mitigates lateral loads. Both conventional steel braces (CBs) and buckling-restrained braces (BRBs) leverage the energy dissipation capabilities inherent in metal through a process of elastic–plastic deformation. This mechanism provides a robust solution for enhancing the overall structural stability. McCormick's study [1] identified a critical threshold: building maintenance costs become excessive when residual drift caused by earthquakes surpasses 0.5%. To address this challenge, the integration of self-centering mechanisms with force-dissipative brace elements presents a promising solution. These self-centering energy dissipation braces (SCED braces) leverage the superior re-centering capabilities of self-centering devices alongside the exceptional hysteretic behavior of metallic force-dissipative brace elements. This combined approach offers a significant advancement in seismic resilience for structures. Research efforts related to SCED braces explore three primary energy dissipation mechanisms: friction, metal yielding, and viscous damping. While each mechanism demonstrates effective energy dissipation, they are accompanied with distinct limitations. Additionally, investigations into self-centering components are ongoing, with a focus on materials such as memory alloys, springs with disc layouts, and basalt-fiber-reinforced polymers (BFRPs). Building upon the established body of research, Alam [2] explored the potential of shape memory alloy (SMA) rods as a combined energy dissipation and self-centering mechanism. The experimental findings demonstrated encouraging results with regard to both hysteretic behavior, signifying effective energy absorption, and the rod's inherent ability to self-center. However, it is crucial to acknowledge certain limitations that necessitate further investigation. Notably, the efficacy of SMA rods is demonstrably susceptible to variations in temperature. Additionally, their current cost-effectiveness presents a significant obstacle for broader implementation within seismic design practices. Zhou et al. [3] explored the application of basalt-fiber-reinforced polymer (BFRP) tendons within self-centering energy dissipation braces. Their research yielded promising results, demonstrating BFRP's potential as a viable material due to its exceptional elongation, robust centering features, and significant yield strain. However, further research is necessary to definitively validate the effectiveness of the anchorage system, a critical component for ensuring the functionality of BFRP tendons. Additionally, cost reduction strategies for the fiber anchorage system are crucial to facilitate broader adoption within the engineering community. Xu et al. [4] investigated the use of disc springs as a self-centering mechanism within energy dissipation braces. Their research identified several key advantages associated with disc springs, including cost-effectiveness, ease of installation, and consistent self-centering performance. These findings suggest that disc springs represent a highly practical and promising solution for self-centering applications within seismic design. This positive evaluation is reflected in the selection of disc springs to furnish centering capabilities for the brace under consideration. Dolce et al. [5] proposed a novel self-centering energy dissipative (SCED) bracing system employing martensitic shape memory alloy (SMA) wires for energy dissipation and austenitic SMA wires for re-centering capability. Xu et al. [6] further extended the SCED concept by introducing a pre-compressed disc-spring-based SCED brace. This design leverages friction mechanisms for energy dissipation, complementing the re-centering capabilities offered by the pre-stressed disc springs. Christopoulos et al. [7] advanced the development of self-centering energy dissipation (SCED) bracing systems by incorporating post-tensioned aramid-fiber tendons and friction dampers within double-tube braces. This innovative SCED brace design makes use of the inherent self-centering capability of the tendons and the energy dissipation properties of the friction mechanisms to improve seismic performance. Miller et al. [8] presented a novel solution for the residual deformation in buckling-restrained braces (BRBs). Their design employs a tri-tube self-centering BRB (SC-BRB) concept incorporating TPT-SC technology. SMA rods serve as post-tensioning (PT) tendons within the SC-BRB, enabling effective mitigation of residual deformation and enhanced energy dissipation capabilities. Zhou et al. [9] addressed a critical limitation in self-centering buckling-restrained braces (SC-BRBs) by introducing a novel dual-tube design that employs basalt-fiber-reinforced polymer (BFRP) as the post-tensioning (PT) tendon. This material selection capitalizes on the advantageous properties of BFRP, including low elastic elongation and high ultimate elongation capacity. The experimental program yielded compelling results, with the SC-BRB exhibiting a distinct flag-shaped hysteretic response. Watanabe et al. [10], in a study on buckling-restrained braces (BRBs), proposed a novel design that effectively confines the core tube. This innovative BRB utilizes concrete encased within a steel tube to achieve efficient core restraint. Experimental investigations confirmed the BRB's superior performance, characterized by an expectable elastic behavior and stable plastic response under cyclic loading. Notably, the design successfully prevented both overall buckling and fracture during testing. Christopoulos et al. [11] introduced an innovative SCED bracing component which incorporates pre-stressed aramid fibers for centering purposes and utilizes a friction cushion for force absorption. Eatherton et al. [12] introduced a new SC-BRB seismic bracing system.

This design combines a BRB core for energy dissipation with an SMA for self-centering. Hu et al. [13] proposed self-centering friction damper braces (SFDBs) as a novel seismic mitigation strategy. These SFDBs employ friction pads for efficient energy dissipation and shape memory alloys (SMAs) to provide the self-centering restoring force. This design capitalizes on the complementary properties of friction for energy absorption and SMAs for re-centering capability. Chou et al. [14] significantly enhanced the deformation capacity of self-centering braces by proposing a novel dual-core design. This innovative concept effectively doubles the deformation capacity compared to traditional single-core braces. Their research investigated the performance of various pre-stressed elements, including tendons made of steel, fiber tendons (made out of E-glass), and carbon-fiber-based tendons, to identify the optimal material for the self-centering mechanism within the dual-core brace. Xu et al. [15] introduced the pre-pressed spring self-centering energy dissipation (PS-SCED) brace. This device utilizes a pre-stressed spring with disc-like shape for its self-centering capability and integrates a frictional component to efficiently dissipate energy. The PS-SCED brace leverages these distinct mechanisms to achieve enhanced seismic performance. Chou et al. [16] proposed the SC-SBRB, a novel seismic brace combining a BRB core for energy dissipation with pre-tensioned steel for self-centering. This design addresses the trade-off between these functionalities, offering superior energy absorption, self-centering, and deformation capacity as demonstrated by prototype testing. Zhu and Zhang [17] introduced the reusable hysteretic damping brace (RHDB). The proposed brace makes use of highly elastic materials to dissipate seismic forces. Dong et al. [18] further expanded the design landscape of self-centering buckling-restrained braces (SC-BRBs) by proposing a novel concept. Their innovative SC-BRB employs a traditional BRB core for efficient energy dissipation, while incorporating disc springs as the self-centering mechanism. In their publication, Zhou et al. [19] proposed a novel double-tube SC-BRB design. Quasi-static tests confirmed the element's effectiveness, as demonstrated by stable flag-shaped hysteretic loops. Minimal residual deformation in both specimens indicates successful self-centering with suitable energy dissipation. Tremblay and Christopoulos [20] introduced self-centering energy dissipative (SCED) brace. It integrates steel bracing components for structural stability and employs friction dampers to efficiently dissipate energy. Notably, the design incorporates a self-centering mechanism achieved through pre-stressed steel tendons. These tendons demonstrate controlled elongation under both tensile and compressive loads, resulting in symmetrical flag-shaped hysteretic response with suitable self-centering capabilities. Miller et al. [8] introduced a novel self-centering BRB (SC-BRB) concept to address residual deformation in BRBs. The SC-BRB couples a BRB for energy dissipation with pre-stressed SMA rods for re-centering, connected via innovative concentric tubes for efficient elongation. This design exhibits a bi-linear flag-shaped hysteresis diagram, indicating effective self-centering. Zhu and Zhang [21] proposed a novel seismic mitigation strategy for concentrically braced frames (CBFs) with the introduction of the self-centering friction damping brace (SFDB). The SFDB functions as an inactive dissipator, incorporating a self-centering mechanism comprising highly elastic nitinol strands and a frictional component to absorb energy. Xu et al. [22] presented an alternative iteration of self-centering braces employing SMA bolts and post-tensioned tendons. This device implements a pre-stressing mechanism on both the pre-stressed elements and SMA-based bolts, thereby ensuring sustained interaction between the brace and the adjoining structural element. A self-centering brace concept was proposed by Christopoulos et al. [23] that utilizes pre-tensioned (PT) strands as the restoring mechanism and integrates friction plates for energy dissipation. Experimental investigations validated the effectiveness of this design in mitigating residual deformations within the system post-seismic excitation. Tapia et al. [24] introduced the amplified added stiffness and damping (AASD) device, which integrates kinematic amplification capability and a self-centering damper that makes use of frictional forces to dissipate forces. Building upon this concept, Choi et al. [25] investigated the magnetorheological (MR) damper, utilizing a combination of magnets and rubber springs. Manfredi et al. introduced an innovative retrofitting technique for the seismic upgrading of existing structures, utilizing the high-performance dissipating frame (HPDF) system. This system is installed in existing buildings and features shear dampers to dissipate seismic loads. Moreover, the authors proposed a novel design methodology based on displacement for structures. This study also unveiled a self-centering braced member, which underwent rigorous experimental testing. To examine the behavior of the proposed member, physical samples were constructed and put through cyclic loading tests. The results from the experimental analysis were then contrasted with those obtained from simulations. Furthermore, parametric studies were conducted to scrutinize the influence of geometrical parameters on the behavior of the proposed brace.

2. Proposed Braced Member

Based on the previous investigations discussed in Section 1, the self-centering brace presented in this study considers the availability of materials and construction capabilities in Iran. Previous brace designs have had several drawbacks, including construction difficulties, high costs, limited energy dissipation, and restricted axial force capacity. Figure 1 shows a schematic representation of the proposed self-centering brace. This brace consists of three components: (1) an energy dissipator, (2) a self-centering section, and (3) a confined part. The dissipator is composed of arched plates that dissipate energy through flexural deflection. The self-centering capability is achieved through the use of a spring, while the confined part includes loading plates and connection plates. A bar is essential for connecting the spring to the plates. Additionally, the calibration of the bar's connection to the profile, located at the center of the proposed member, is optimized for energy dissipation. The self-centering mechanism of the member works as follows: after the arched plates deform, the spring reverses the displacement caused by load reversal, resetting the position of the sliding plate to complete the self-centering mechanism. This capability is clearly influenced by whether or not the arched plates have yielded.

Figure 1. Schematic figure of proposed braced member.

Figure 1 provides a visual representation of the brace member's components, while Figure 2 outlines the assembly process of the brace member.

1. The first step involves inserting the dowel bar into the hole in the endplate.
2. Next, position the springs in the center of the dowel and apply compression using two bolts for pre-tensioning.
3. Pass the loading plates through the dowel and secure bolt no. 2 on the opposite side of the loading plate. The springs are pre-tensioned using bolt no. 1, as described in the previous step.
4. Connect the arched plates to the loading plates, end plates, and confined parts, ensuring each brace component is attached to the dowel and connecting plates.

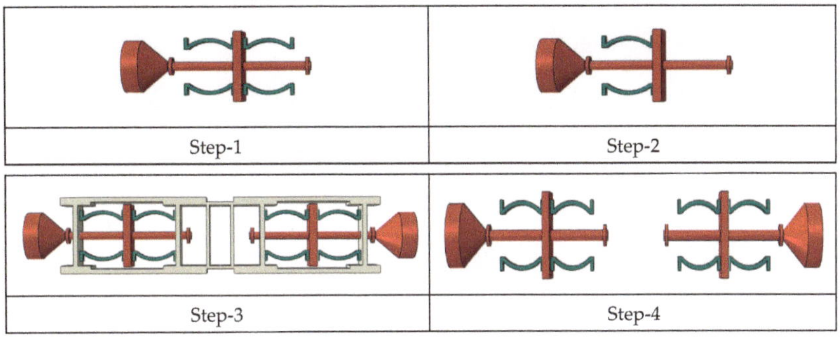

Figure 2. The process of assembling of proposed braced member.

The fabrication, assembly, and installation of the proposed brace are straightforward. Furthermore, the self-centering capability of the member can be adjusted by modifying the configuration and spring models.

The proposed brace exhibits symmetric behavior in both compression and tension, resulting in similar stress conditions. As depicted in Figure 3, it is evident that when the brace is under tension, the plates slide over both sides of the loaded plates. Additionally, while the arched plates situated in sections A1 and B1 are under compression, the arched plates in sections A2 and B2 are under tension. The arched plates effectively dissipate the energy generated by earthquakes. The compression springs in sections A1 and B1 are compressed, generating restoring forces opposing the compression. Similarly, when in compression, the brace performs similarly to the tension condition. When subjected to external forces, the brace exhibits both resilience and energy dissipation within specific regions.

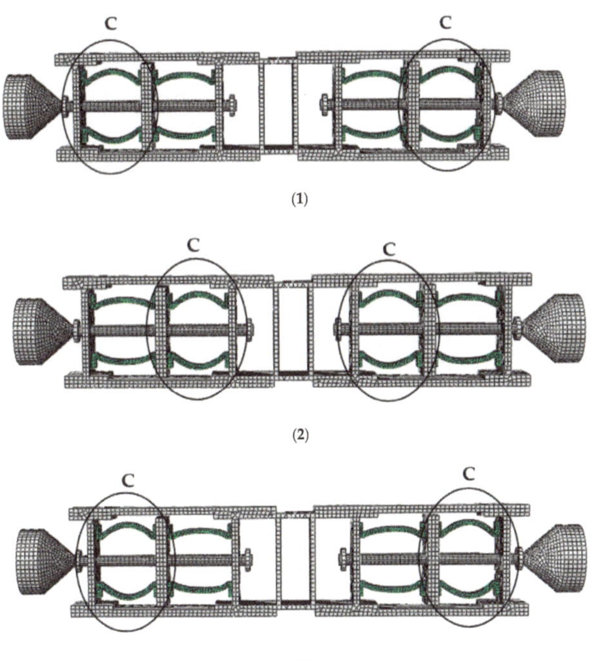

Figure 3. *Cont.*

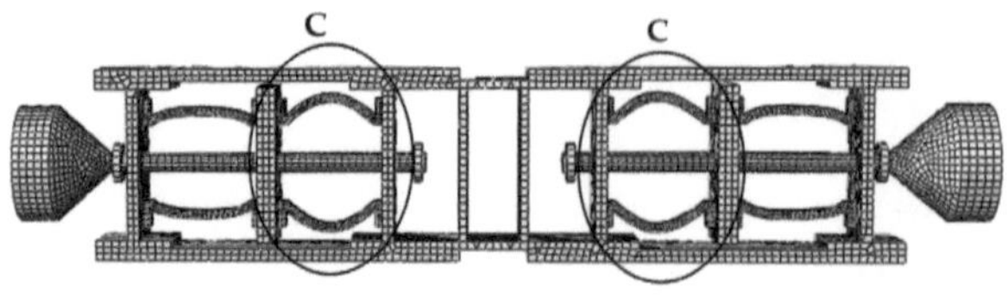

(4)

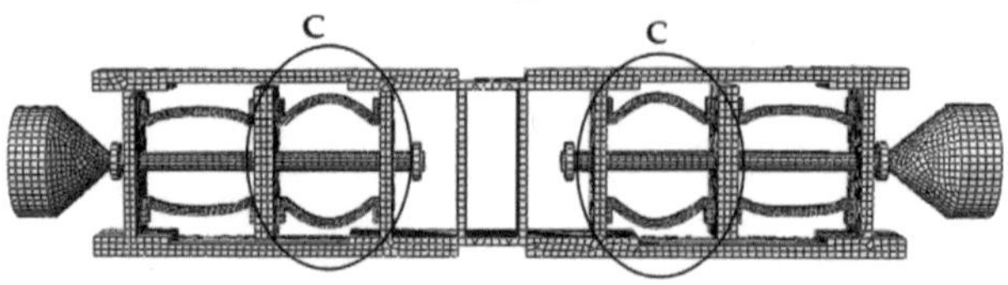

(5)

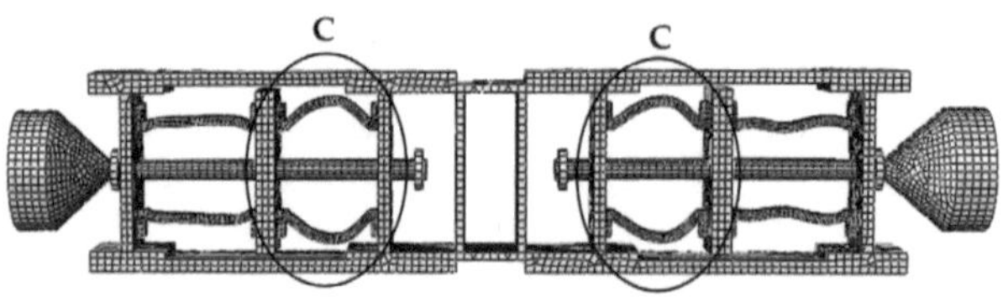

(6)

Figure 3. (**1–6**) The mechanism of proposed self-centering brace.

3. Mechanical Model

3.1. Calculating the Internal Forces of the Arched Steel Plates

In this section, the mechanical model of the proposed brace is examined. For this purpose, the various components that comprise the brace are analyzed. In the mechanical model, the thickness and width of the arched plates are simplified to have a fixed point at one end and a free point at the other end that can move freely. The displacement and rotation angle of the system along the X_1 and X_2 directions are described by Equation (1).

$$\delta_{11}X_1 + \delta_{12}X_2 + \Delta_{1P} = 0 \;\; \& \;\; \delta_{21}X_1 + \delta_{22}X_2 + \Delta_{2P} = 0 \tag{1}$$

Assuming the displacement of the arched plates to be $ds = Rd\theta$, we would have:

$$\delta_{11} = \int \frac{\overline{M_1^2}}{EI} ds = \int_{\varphi}^{\pi-\varphi} \frac{R^3(cos\varphi - cos\theta)^2}{EI} d\theta = \frac{R^3}{EI}\left[(\pi - 2\varphi)cos^2\varphi + \frac{\pi - 2\varphi - sin2\varphi}{2}\right] \tag{2}$$

$$\delta_{21} = \delta_{12} = \int \frac{\overline{M_1}}{EI} ds = \int_{\varphi}^{\pi-\varphi} \frac{R^3(cos\varphi - cos\theta)^2}{EI} d\theta = \frac{R^3}{EI}\left[(\pi - 2\varphi)cos^2\varphi + \frac{\pi - 2\varphi - sin2\varphi}{2}\right] \tag{3}$$

$$\delta_{22} = \int \frac{\overline{M_2}}{EI} ds = \frac{R}{EI}(\pi - 2\varphi) \tag{4}$$

$$\Delta_{1p} = \int \frac{\overline{M_1 M_p}}{EI} ds = \int_{\varphi}^{\pi-\varphi} \frac{PR^3(cos\varphi - cos\theta)(sin\theta - sin\varphi)}{EI} d\theta = \frac{PR^3}{EI}\left[2cos^2\varphi - (\pi - 2\varphi)sin\varphi cos\varphi\right] \tag{5}$$

$$\Delta_{2p} = \int \frac{\overline{M_2 M_p}}{EI} ds = \int_{\varphi}^{\pi-\varphi} -\frac{PR^2(sin\theta - sin\varphi)}{EI} d\theta = \frac{PR^2}{EI}[(\pi - 2\varphi)sin\varphi - 2cos\varphi] \tag{6}$$

Through the above relationships, the following equation is achieved:

$$X_1 = 0;\ X_2 = \frac{PR[2cos\varphi - (\pi - 2\varphi)sin\varphi]}{\pi - 2\varphi} \tag{7}$$

Finally, the equation describing the internal force of the arched plates is derived as follows:

$$M = \overline{M_1}\,X_1 + \overline{M_2}\,X_2 + M_P = PR.\left[sin\theta - \frac{2cos\varphi}{\pi - 2\varphi}\right] \tag{8}$$

3.2. Calculating the Stiffness of the Arched Steel Plates

The stiffness of the arched steel plates is calculated using the following relationship:

$$\overline{M_p} = R(\ sin\theta - sin\varphi); M = \overline{M_1}X_1 + \overline{M_2}X_2 + M_p = PR.[sin\theta - \frac{2cos\varphi}{\pi - 2\varphi}] \tag{9}$$

3.3. The Displacement on the Left-Hand Side Is Equal to

$$\Delta = \int \frac{M.\overline{M}_p}{EI}ds = \frac{PR^3}{EI}\int_{\varphi}^{\pi-\varphi}(sin\theta - sin\varphi)\left(sin\theta - \frac{2cos\varphi}{\pi - 2\varphi}\right)d\theta = \frac{PR^3}{EI}\left(\frac{\pi - 2\varphi + sin2\varphi}{2} - \frac{2cos^2\varphi}{\pi - 2\varphi}\right) \tag{10}$$

The stiffness of the arched plates is equal to:

$$K_{steel} = \frac{1}{f} = \frac{P}{\Delta} \tag{11}$$

When the steel arched plates go through yielding, their hardness experiences a decrease. The modulus of elasticity after yielding is 5% of the pre-yield value. Therefore, the hardness after yielding of the steel arched plates is $E_y = 0.05E$.

3.4. Yield Displacement and Force

The normal stress at the outer edge of the arched plates is equal to σ_y, and the elastic limit moment M_e of the arched plates at the elastic limit at the edge is equal to:

$$M_e = \frac{bh^2\sigma_y}{6} \tag{12}$$

Therefore, the elastic limit bending moment M_p is equal to:

$$M_p = \xi M_e = \frac{\xi bh^2\sigma_y}{6} \tag{13}$$

where ξ is the shape factor of the section. Since the cross-section of the steel arched plates is rectangular, the shape factor is equal to 1.5. b is the width of the steel arched plates, and h is the thickness of the steel arched plates. The bending moment, by substituting $\theta = \frac{\pi}{2}$ and $\theta = \frac{\pi}{2} - \varphi$, is equal to:

$$M_{\theta=\frac{\pi}{2}} = \overline{M_1}X_1 + \overline{M_2}X_2 + M_P = PR.\left[1 - \frac{2cos\varphi}{\pi - 2\varphi}\right] \tag{14}$$

$$M_{\theta=\pi-\varphi} = M_{\theta=\varphi} = \overline{M_1}X_1 + \overline{M_2}X_2 + M_P = PR.\left[sin\varphi - \frac{2cos\varphi}{\pi - 2\varphi}\right] \tag{15}$$

A comparison between A and B would make it that $M_{\theta=\frac{\pi}{2}} > M_{\theta=\varphi}$. Therefore, plastic hinges form at the ends of the arched plates and at the midspan. By substituting Equations (11) and (12), the yield load is obtained, which is equal to:

$$P_y = \frac{\xi b h^2 \sigma_y}{6R.\left(1 - \frac{2\cos\varphi}{\pi - 2\varphi}\right)} \tag{16}$$

Thus, the yield displacement is equal to $\Delta_y = \frac{P_y}{K_{steel}}$.

4. Numerical Modeling

In this section, we have employed Abaqus software (ABAQUS v.2022) to model the proposed braced member. To accurately represent the components, C3D8R elements were used to discretize the arched plate, loaded plates, and the dowel bar. To simplify the analysis, a nonlinear spring element—spring 2—was incorporated to replicate the behavior of the disc spring. The materials selected for the brace include Q235 B steel for the arched plates and Q345 steel for the remaining elements. The constitutive model for the steel adopts a bilinear kinematic hardening plasticity model. The modulus of elasticity after yielding is set at 5% of the value before yield, denoted as Ey = 0.05 E. For detailed information on the specific material properties, refer to Table 1.

Table 1. Mechanical parameters of the steel materials.

F_y (MPa)	F_u (MPa)	E (MPa)	v	Material
385	540	206,000	0.3	Q345
300	415	206,000	0.3	Q235

F_y: Yield stress (MPa). F_u: Ultimate stress (MPa). E: Young's modulus (MPa). v Poisson coefficient.

In order to analyze the hysteretic behavior of the proposed brace, a displacement-controlled loading scheme is utilized. The model is set up with one end fixed, while the other end undergoes a low-cycle repeated displacement. The loading displacement–time curve is depicted in Figure 4. Additionally, the dimensions of the proposed member are shown in Figure 5.

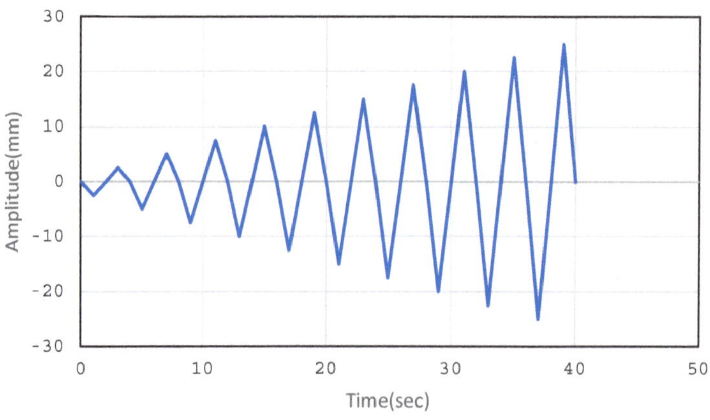

Figure 4. The loading history of the specimen.

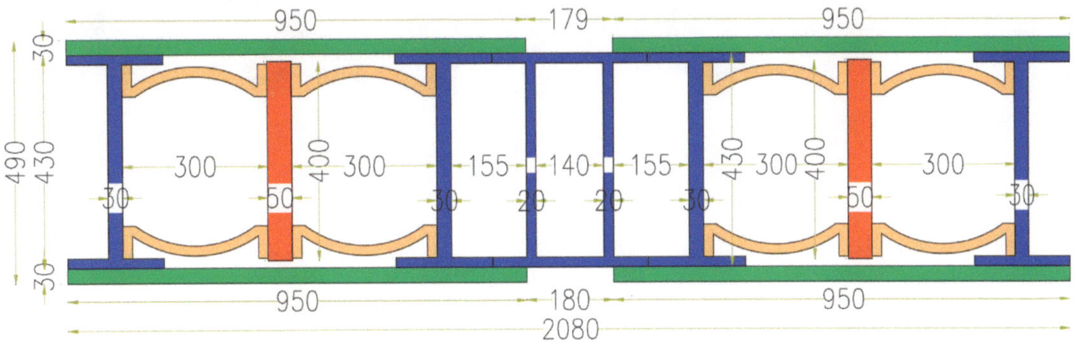

Figure 5. Dimension of proposed braced member.

5. Results of Analysis

In this section, the behavior of the proposed brace is analyzed using Abaqus finite element software (ABAQUS v.2022). The proposed member was modeled and subjected to cyclic loading to determine its behavior. The obtained results are illustrated in Figure 6. The analysis revealed that the proposed brace exhibits consistent flag-shaped hysteretic loops. The pinching effects in the member's hysteretic behavior are evident, caused by the pre-tensioning of the springs. The results highlight the significant influence of pinching effects and the occurrence of self-centering behavior when post-tensioning force is applied. Additionally, Figure 7 illustrates how the proposed brace operates. Figure 7a shows the condition of the brace without an applied load. In Figure 7b, it can be observed that when the brace is under tension, arched plate 1 is in tension, and arched plate 2 is in compression. Consequently, arched plate 3 is in compression, and arched plate 4 is in tension. When a compressive load is applied to the brace (Figure 7c), this process is reversed, with arched plates 1 and 4 becoming compressed and plates 2 and 3 becoming tensioned. This process prevents the formation of permanent deformation, thereby ensuring the centricity property in the proposed brace.

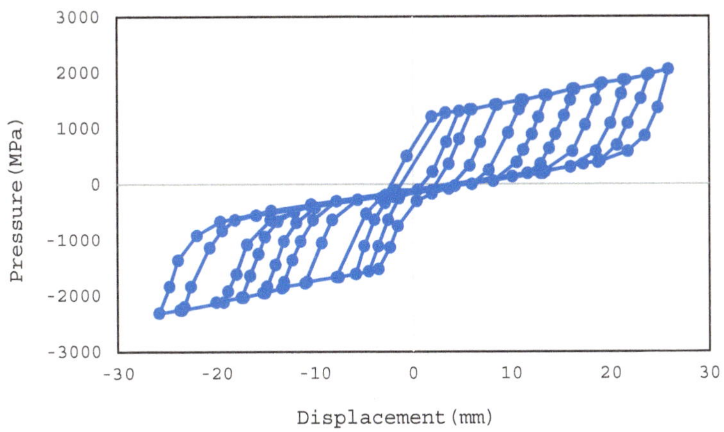

Figure 6. The proposed brace of cyclic behavior.

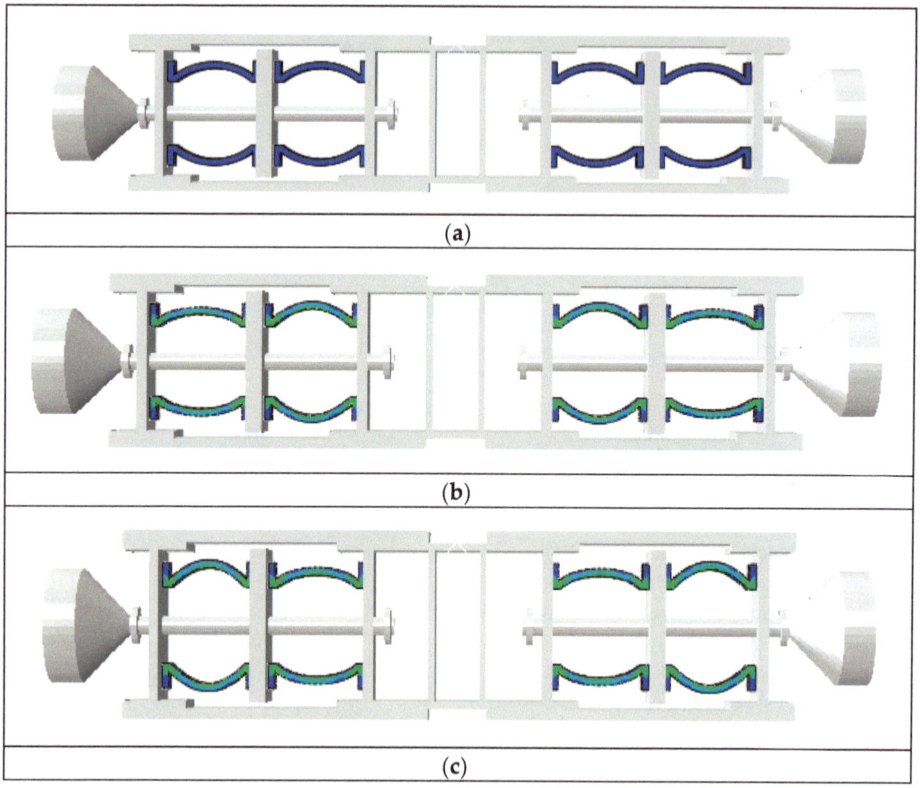

Figure 7. Function of the proposed member, (a) without stress; (b) Under tension; (c) Under compression.

Additionally, Figure 8 showcases the stress distribution within the proposed member. The figure indicates that the arched plates experience both tension and compression. Moreover, the self-centering of the proposed system is observed to arise from spring deflection and energy dissipation.

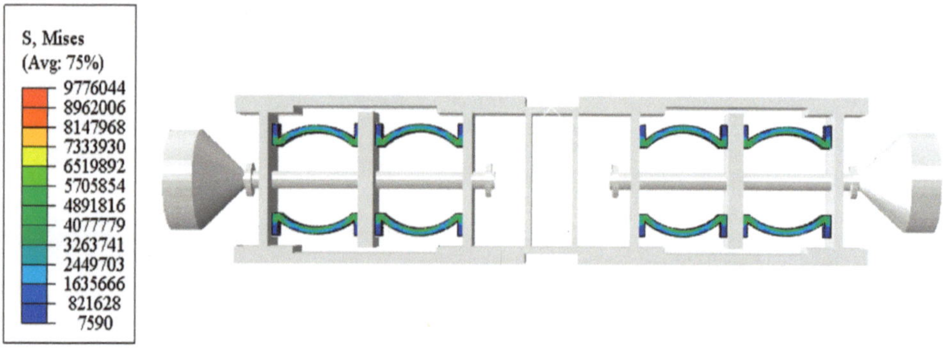

Figure 8. The stress status of proposed member.

6. Parametric Studies

This section examines the impact of various parameters on the proposed members. The parameters studied are as follows:

a. The arched plate.
b. The width of the plate.
c. The thickness of the plate.

To investigate these parameters, three groups of braced members were modeled. The first group consisted of various curves with similar width and thickness. The second group maintained a constant curve and thickness while varying the widths. The third group maintained a constant width and curve while varying the thicknesses. It should be noted that the length of the element was fixed at 500 mm. Table 2 provides an overview of the studied groups.

Table 2. Parameters of the proposed member.

Group	Label	Curvature	Width (mm)	Thickness (mm)	Length (mm)
1	BR-1	1/200	60	30	500
	BR-2	1/250	60	30	500
	BR-3	1/300	60	30	500
2	BR-4	1/300	60	30	500
	BR-5	1/300	80	30	500
3	BR-6	1/300	80	15	500
	BR-7	1/300	80	20	500
	BR-6	1/300	80	15	500

6.1. The Effects of Arched Plates

Within this section, we have examined the influence of arched plates on the proposed member. The models studied in this section are denoted as BR1, BR2, and BR3. The hysteretic curves of the proposed member are showcased in Figure 9. The results indicate that an increase in the number of arched plates leads to a higher ultimate strength. Specifically, the ultimate strengths of BR1, BR2, and BR3 are 398, 856, and 1105, respectively. Additionally, an increase in stiffness is observed, with BR1, BR2, and BR3 exhibiting stiffness values of 80, 200, and 245 kN/m, respectively. Furthermore, the member's energy dissipation increases, with values of 6.82, 14.07, and 27.91 kJ for BR1, BR2, and BR3, respectively. A comprehensive overview of the effects of arched plates can be found in Table 3.

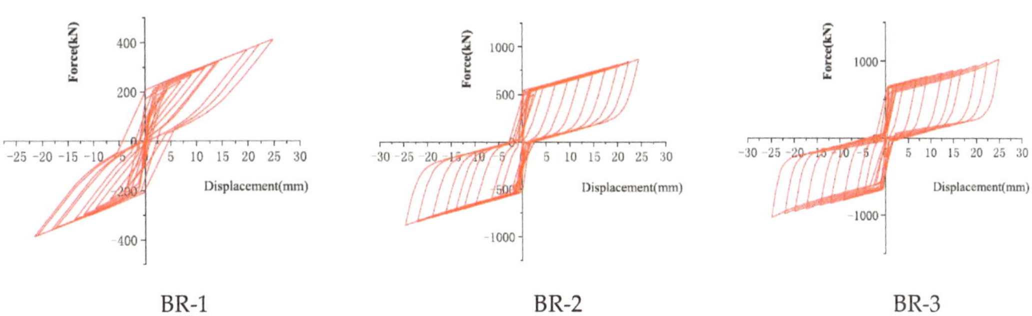

Figure 9. The effect of arc steel plate in the hysteric behavior of proposed brace.

Table 3. The effect of arched plates in the behavior of proposed brace.

Group	Name	Ultimate Strength (kN)	Stiffness (kN/mm)	Energy Dissipation (kJ)
1	BR-1	398	80	6.82
	BR-2	856	200	14.07
	BR-3	1105	245	27.91

6.2. The Effect of Width of Arch Plates

In this section, we have investigated the impact of arch width on the proposed member's behavior. The results demonstrate that an increase in arch width leads to an enhancement in both energy dissipation and initial stiffness of the proposed member, as depicted in Figure 10. It is crucial to select an appropriate width for the arch plates to ensure that the proposed member enters nonlinear behavior. Ideally, a width within the range of 50 to 80 mm is recommended for optimal performance.

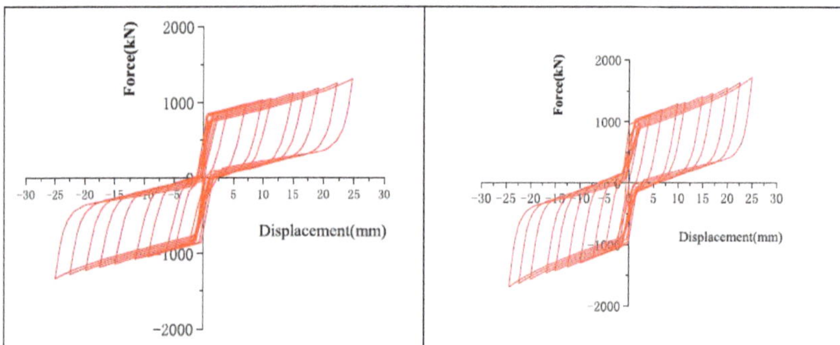

BR4 | BR5

Figure 10. The effect of arch width of proposed brace.

6.3. Effect of Arched Plates Thickness

In this section, we have examined the influence of arch plate thickness on the behavior of the proposed member. Two samples were analyzed, having thicknesses of 15 mm and 20 mm. Each sample was modeled, and the results were carefully studied. The findings indicate that the hysteretic behavior of the proposed member improves as the thickness of the plates increases. Consequently, it is evident that the proposed member enters a nonlinear behavior regime with the progressive increase in plate thickness. However, it is important to consider the economic aspect of the proposed member, as excessively increasing the thickness of arch plates may not be cost-effective. In this study, a thickness of 20 mm was selected for the member (Figure 11).

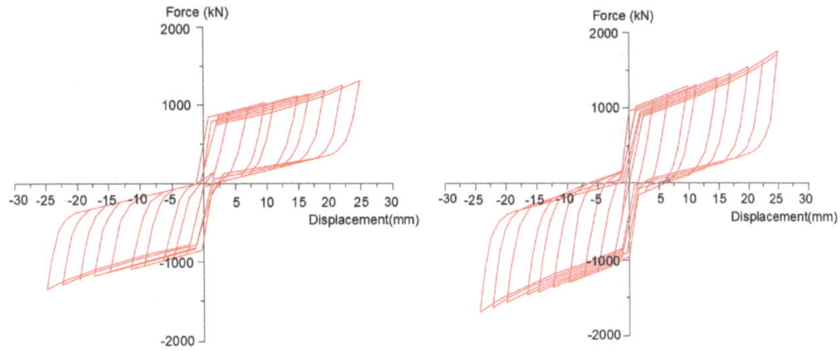

BR4 BR5

Figure 11. The effect of arch plate thickness in the behavior of proposed brace.

7. Experimental Specimen

In this section, the experimental investigation of the proposed bracing system is discussed. First, the components of the bracing system are described, followed by an explanation of the testing procedure. The experimental specimen consists of the following parts: springs, arched plates, loading plates, steel casing (including a central I-profile and four side I-profiles), dowel rods, and connections. The brace includes four groups of springs and eight groups of arched plates, constructed at a scale of 1:3. The assembly process of the bracing system is as follows: first, the dowel rod is inserted through the central hole of the profile and is secured by nut number 1 on the outer section of the end plates. The dowel rod also passes through the center of the spring. Once the spring is threaded onto the dowel rod, the loading plate is passed through the dowel rod, and a second spring is placed. The other end of the spring is connected to another profile and secured by nut number 2. Following this step, the springs are positioned on the other side, and the end plates are threaded onto the dowel rod and secured with nut number 1 again. Then, the arched plates, which provide the bracing system with its energy dissipation capability, are installed. It should be noted that these plates are connected to the loading and end plates, as well as the clamping plates. In the final stage, the two sections of the restraint system are connected by the central profile and installed on the supports at the ends. What sets this brace apart from other similar bracing systems is its simplicity in construction, assembly, and installation. The stiffness and concentricity of the system can be adjusted by the arrangement of the springs. The assembly stages are illustrated in Figure 12.

Springs of the brace system	Installation of the dowel bars

Figure 12. *Cont.*

Assembly of the clamping plates and connection of the profiles	Final condition of the proposed brace

Figure 12. Assembly of the brace element.

The experimental specimen was subjected to testing at the university, and the behavior of the restraint system was evaluated. Figure 13 depicts the experimental setup. As shown in Figure 13, the loading jack is placed on the top portion of the specimen, and the load is applied monotonically to the specimen. After loading, the final axial deformation of the restraint system is measured, and its behavior is determined. It is observed that under the applied load, the restraint system exhibits ductile behavior and does not undergo buckling.

Figure 13. Experimental setup.

8. Verification

This section presents the numerical modeling of the brace specimen in ABAQUS finite element software (ABAQUS v.2022). In the modeling process, as mentioned before, C3D8R solid elements were used to mesh the brace components. The mechanical properties of the materials employed are presented in Table 1. To simplify the numerical model, improve computational efficiency, and reduce convergence issues, the effects of the bolts were neglected in the modeling. Tie elements have been used between the restraining plates, profiles, and bolts. The applied loading in the numerical model follows a monotonic pattern. Comparing the numerical and experimental results shows an acceptable level of congruence. A more detailed comparison can be drawn from Table 4, wherein the ultimate strength, initial stiffness, and energy dissipation of the experimental specimen and the numerical model are tabulated. It can be observed that, when compared to the experimental results, the numerical model estimates the initial stiffness, ultimate capacity, and energy dissipation of the brace with error percentages of 21%, 23.4%, and 23.4%, respectively. These values indicate a relatively suitable concordance between the results obtained (Figure 14).

Table 4. Comparison between the overall results.

Parameter	Numerical	Experimental	Error
Initial stiffness (kN/mm)	324	266	21%
Ultimate capacity (kN)	1201	973	23.4%
Dissipated energy (kN.mm)	23,051	18,680	23.39%

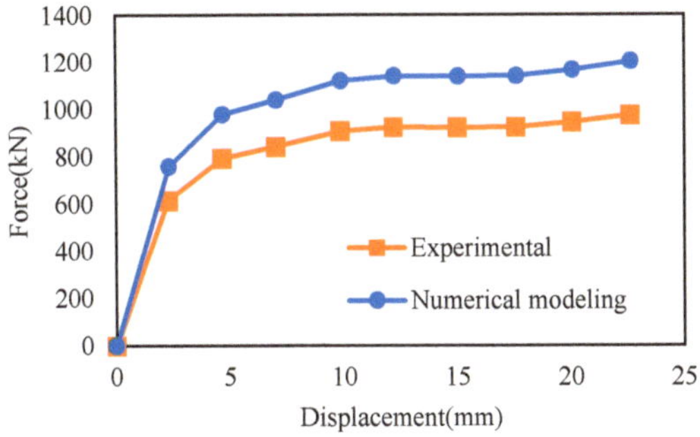

Figure 14. Comparison between the experimental and numerical results.

9. Conclusions

This article introduces a new concentric brace system with arch-shaped plates. The proposed brace element is modeled in the ABAQUS software, and its behavior under cyclic loading is investigated. Furthermore, the influence of various variables on the behavior of the proposed element is also studied. The parameters under investigation include the width, thickness, and curvature of the steel arched plates.

An experimental specimen at a scale of 1:3 was constructed and subjected to monotonic loading for examination. In the next stage, the experimental sample was numerically simulated, and the results of the modeling were compared with the results of the test. The results indicate that:

1. The proposed concentric brace element has high energy dissipation and stable concentric behavior, which can reduce the permanent deformations caused by seismic forces.
2. The modeling of the concentric brace element shows that the permanent deformation of the system depends on the pre-tensioning of the springs. This means that the permanent deformation of the system decreases with pre-tensioning.
3. In the proposed element, the arched plates play a primary role in energy dissipation through flexural yielding, and the springs provide the system with its concentricity.
4. The analysis results demonstrate that the proposed concentric bracing system exhibits flag-shaped hysteretic loops and maintains stable overall behavior.
5. The parametric study showed that increasing the width of the arched plates increases the energy dissipation capacity and decreases the load-carrying capacity of the element. Additionally, the analysis results indicate that increasing the thickness of the steel plate significantly enhances the energy dissipation capacity. Moreover, when the steel plate's thickness increases, the element's ability to undergo elastic deformation improves, resulting in a delayed entrance into the plastic phase.
6. The analysis results showed that by reducing the arched plates' curvature, the hysteretic response of the brace (characterized by fully formed hysteretic loops) exhibits completely formed (fat) loops and relatively smooth stiffness in each stage.

7. The experimental results indicate that the proposed bracing system does not experience a sudden buckling-induced drop in its load-carrying capacity, and the overall behavior of the system is suitably ductile. This feature is instrumental in preventing progressive collapse in structures.

Author Contributions: M.M.M.P.: methodology, software, validation, investigation, writing, visualization, formal analysis. S.M.N.: writing—review and editing, supervision, project administration, funding acquisition. A.S.M.: writing—review and editing, supervision, project administration, funding acquisition. All authors have read and agreed to the published version of the manuscript.

Funding: This research received no external funding.

Data Availability Statement: The scripts and models developed for this research article will be made available upon reasonable request.

Conflicts of Interest: The authors declare no conflict of interest.

References

1. McCormick, J.; Nagae, T.; Ikenaga, M.; Zhang, P.C.; Katsuo, M.; Nakashima, M. Investigation of the sliding behavior between steel and mortar for seismic applications in structures. *Earthq. Eng. Struct. Dyn.* **2009**, *38*, 1401–1419. [CrossRef]
2. Alam, M.S.; Nehdi, M.; Youssef, M.A. Seismic performance of concrete frame structures reinforced with superelastic shape memory alloys. *Smart Struct. Syst.* **2009**, *5*, 565–585. [CrossRef]
3. Chan, Y.W.S.; Zhou, Z.; Xiang, P. Experimental analysis of BFRP strengthened short steel tube with built-in FBG sensors under axial compression. *Thin-Walled Struct.* **2022**, *173*, 109047. [CrossRef]
4. Xu, L.; Yao, S.; Sun, Y. Development and validation tests of an assembly self-centering energy dissipation brace. *Soil Dyn. Earthq. Eng.* **2019**, *116*, 120–129. [CrossRef]
5. Dolce, M.; Cardone, D.; Marnetto, R. Implementation and testing of passive control devices based on shape memory alloys. *Earthq. Eng. Struct. Dyn.* **2000**, *29*, 945–968. [CrossRef]
6. Xu, L.; Fan, X.; Li, Z. Seismic assessment of buildings with prepressed spring self-centering energy dissipation braces. *J. Struct. Eng.* **2020**, *146*, 04019190. [CrossRef]
7. Zhong, C.; Christopoulos, C. Self-centering seismic-resistant structures: Historical overview and state-of-the-art. *Earthq. Spectra* **2022**, *38*, 1321–1356. [CrossRef]
8. Miller, D.J.; Fahnestock, L.A.; Eatherton, M.R. Development and experimental validation of a nickel–titanium shape memory alloy self-centering buckling-restrained brace. *Eng. Struct.* **2012**, *40*, 288–298. [CrossRef]
9. Xie, Q.; Zhou, Z.; Meng, S.-P. Behaviour of BFRP tendon systems under cyclic loading and its influence on the dual-tube SC-BRB hysteretic performance. *Constr. Build. Mater.* **2020**, *259*, 120388. [CrossRef]
10. Watanabe, A.; Hitomi, Y.; Saeki, E.; Wada, A.; Fujimoto, M. Properties of brace encased in buckling-restraining concrete and steel tube. In Proceedings of the Ninth World Conference on Earthquake Engineering, Tokyo-Kyoto, Japan, 2–9 August 1988; pp. 719–724.
11. Erochko, J.; Christopoulos, C.; Tremblay, R. Design, testing, and detailed component modeling of a high-capacity self-centering energy-dissipative brace. *J. Struct. Eng.* **2015**, *141*, 04014193. [CrossRef]
12. Eatherton, M.R.; Fahnestock, L.A.; Miller, D.J. Computational study of self-centering buckling-restrained braced frame seismic performance. *Earthq. Eng. Struct. Dyn.* **2014**, *43*, 1897–1914. [CrossRef]
13. Hu, J.W.; Lee, J.; Seo, J. Performance-based optimal design of self-centering friction damping brace systems between recentering capability and energy dissipation. *J. Mech. Sci. Technol.* **2014**, *28*, 3129–3136. [CrossRef]
14. Chou, C.-C.; Chung, P.-T. Development of cross-anchored dual-core self-centering braces for seismic resistance. *J. Constr. Steel Res.* **2014**, *101*, 19–32. [CrossRef]
15. Xu, L.-H.; Fan, X.-W.; Li, Z.-X. Development and experimental verification of a pre-pressed spring self-centering energy dissipation brace. *Eng. Struct.* **2016**, *127*, 49–61. [CrossRef]
16. Chou, C.-C.; Chung, P.-T.; Cheng, Y.-T. Experimental evaluation of large-scale dual-core self-centering braces and sandwiched buckling-restrained braces. *Eng. Struct.* **2016**, *116*, 12–25. [CrossRef]
17. Zhu, S.; Zhang, Y. Seismic behaviour of self-centring braced frame buildings with reusable hysteretic damping brace. *Earthq. Eng. Struct. Dyn.* **2007**, *36*, 1329–1346. [CrossRef]
18. Dong, H.; Du, X.; Han, Q.; Hao, H.; Bi, K.; Wang, X. Performance of an innovative self-centering buckling restrained brace for mitigating seismic responses of bridge structures with double-column piers. *Eng. Struct.* **2017**, *148*, 47–62. [CrossRef]

19. Zhou, Z.; Xie, Q.; Lei, X.; He, X.; Meng, S. Experimental investigation of the hysteretic performance of dual-tube self-centering buckling-restrained braces with composite tendons. *J. Compos. Constr.* **2015**, *19*, 04015011.
20. Tremblay, R.; Lacerte, M.; Christopoulos, C. Seismic response of multistory buildings with self-centering energy dissipative steel braces. *J. Struct. Eng.* **2008**, *134*, 108–120. [CrossRef]
21. Zhu, S.; Zhang, Y. Seismic analysis of concentrically braced frame systems with self-centering friction damping braces. *J. Struct. Eng.* **2008**, *134*, 121–131. [CrossRef]
22. Xu, L.H.; Fan, X.W.; Li, Z.X. Cyclic behavior and failure mechanism of self-centering energy dissipation braces with pre-pressed combination disc springs. *Earthq. Eng. Struct. Dyn.* **2017**, *46*, 1065–1080. [CrossRef]
23. Kim, H.-J.; Christopoulos, C. Friction damped posttensioned self-centering steel moment-resisting frames. *J. Struct. Eng.* **2008**, *134*, 1768–1779. [CrossRef]
24. Tapia, N.; Almazán, J.; Baquero, J. Development of a novel combined system of deformation amplification and added stiffness and damping: Analytical result and full scale pseudo-dynamic tests. *Eng. Struct.* **2016**, *119*, 61–80. [CrossRef]
25. Choi, E.; Jeon, J.-S.; Kim, W.J.; Kang, J.-W. Investigation of MRS and SMA dampers effects on bridge seismic resistance employing analytical models. *Int. J. Steel Struct.* **2018**, *18*, 1325–1335. [CrossRef]

buildings

MDPI

Article

Hysteretic Behavior Study on the RBS Connection of H-Shape Columns with Middle-Flanges or Wide-Flange H-Shape Beams

Saleem Mohammed Ali Ahmed Al-Saeedi, Linfeng Lu *, Osama Zaid Yahya Al-Ansi and Saddam Ali

School of Civil Engineering, Chang'an University, 75 Chang'an Middle Rd, Xi'an 710062, China; 2021028918@chd.edu.cn (S.M.A.A.A.-S.); 2022028916@chd.edu.cn (O.Z.Y.A.-A.); 2022028915@chd.edu.cn (S.A.)
* Correspondence: lulinfeng@chd.edu.cn

Abstract: Existing research on reduced beam section (RBS) connections in steel frames rarely addresses H-shaped beams with middle and wide flanges. Therefore, this study investigates the hysteretic behavior of RBS connections in H-shaped columns connected to H-shaped beams with middle and wide flanges. Using finite element analysis, the influence of key parameters (a, b, and c, where "a" represents the unweakened beam flange extension length, "b" represents the weakened beam flange length, and "c" represents the weakened beam flange depth, respectively) on structural performance was evaluated, focusing on rotational stiffness, load-carrying capacity, plastic rotation capacity, and ductility. The results indicate that increasing a enhances initial rotational stiffness and load capacity but reduces plastic rotation and ductility, making lower a values (near $0.5b_f$) optimal for ductile performance. Similarly, higher b values (up to $0.85b_f$) marginally reduce stiffness and load capacity, improving plastic rotation capacity, with a greater benefit in wide-flange beams. Meanwhile, a lower c value (around $0.20b_f$) offered balanced performance, with higher c values decreasing stiffness and load capacity but enhancing ductility. Overall, wider flanges improve plastic rotation and ductility but slightly decrease rotational stiffness, providing insights to guide RBS connection designs for seismic resilience.

Keywords: RBS connection; hysteretic behavior; finite element analysis; seismic performance; rotational stiffness; ductility; H-shaped beam

1. Introduction

Before 1994, rigid beam-to-column connections were widely regarded as having strong seismic performance and were commonly employed in earthquake-prone areas. This point of view was challenged after the Northridge earthquake in 1994. Many moment-resistance frames (MRFs) did not completely collapse during the quake, but unexpected brittle fractures occurred at or near the welds between the beam lower flange and the column flange [1,2]. These failures caused a wave of research on the seismic behavior of MRF connections [3–6]. Extensive research has been conducted to develop methods for improving the seismic performance of beam-to-column connections. Different researchers have introduced many new types of connections [7–9].

The reduced beam section (RBS) connection is designed by locally reducing the beam flange area a short distance from the column flange, effectively relocating the plastic hinge away from the beam-to-column joint. Previous laboratory experiments have demonstrated the effectiveness of RBS connections [4,10,11]. Recent studies have further optimized RBS designs using finite element analysis to enhance ductility and seismic performance [12], conducted parametric investigations on the behavior of RBS connections with jumbo beams and columns [13,14], and evaluated the cyclic performance of these connections in box-column applications [15]. Additionally, researchers have developed more accurate cyclic hysteretic models to predict the inelastic behavior of RBS connections [16], while others have explored the application of RBS designs for specific regional profiles using simulation tools [17]. Cimagala [12] pointed out that current standards offer general guidelines for RBS geometry but lack practical design tools and rules of thumb to simplify calculations and ensure optimal performance. Qi et al. [13] found that current design procedures may not provide sufficient ductility for RBS connections with jumbo sections, and while proposed improvements were identified, further experimental validation and practical design guidelines are needed. Bompa et al. [14] investigated RBS connections with member sizes exceeding the limits specified in current seismic provisions, finding that while they can perform well, very large sections increase the risk of weld fracture. This highlights the need for deeper RBS cuts to improve performance and reduce strain at the welds, suggesting a refinement of design requirements is necessary. Ghassemieh and Mirghaderi [15] evaluated the seismic performance of RBS moment connections in box columns subjected to different loading protocols, revealing key insights into connection behavior and strength degradation; however, further studies are needed to explore the long-term effects of these protocols on performance and stability. Horton et al. [16] developed a comprehensive database of modified Ibarra–Krawinkler (mIK) models to accurately capture the cyclic behavior of RBS connections under earthquake loading, revealing that conventional methods may lead to unreliable predictions. However, further research is needed to refine the mIK parameters for certain RBS geometries to enhance the accuracy of hysteresis behavior modeling. Although Indupriya and Anupriya [17] demonstrated that incorporating reduced beam sections (RBS) improves the performance of moment connections in Indian profiles under cyclic loading, their research fails to explore the impact of different RBS geometries and configurations on connection performance in various seismic scenarios.

Previous research on RBS connections, whether involving I-shaped beams with H-shaped or box columns, has predominantly focused on narrow-flange beams of normal or deep depth. No studies have yet explored middle-flange or wide-flange H-shaped beams. Gemechu and Lu [18] were the first to investigate the influence of widened beam-end flanges in mid-flange H-beams paired with box columns, revealing significant improvements in bending capacity, ductility, and seismic performance, along with optimal design parameters. However, further studies are necessary to examine wide-flange H-beams.

The unique characteristics of reduced beam-end flanges in middle-flange and wide-flange H-beams with H-columns deserve closer scrutiny, particularly regarding their effects on plastic hinge formation and stress distribution. While outward plastic hinge displacement can reduce weld stress, it introduces challenges such as increased deformation in the beam web and a higher risk of local buckling. This study addresses these gaps through comprehensive finite element simulations to evaluate the seismic performance of these connections.

The insights gained from this research will be crucial for refining connection designs, ultimately enhancing the stability and resilience of steel structures in contemporary construction practices. By understanding how these factors interact, we can develop more effective strategies to ensure the integrity and durability of structures in seismically active regions.

2. Design Standard for an RBS Connection

This study investigates an RBS connection joint that utilizes a hybrid connection of welds and bolts, illustrated in Figure 1.

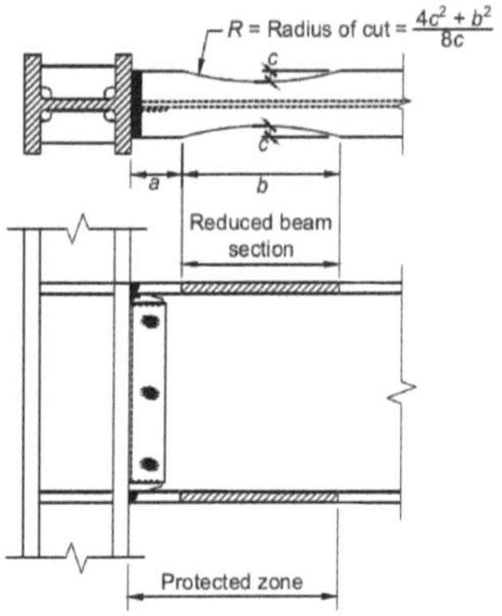

Figure 1. RBS connection details [19].

The requirements for several key geometric parameters shown in Figure 1 across different design standards are provided in Table 1. In Table 1, b_f is the width of the beam flange, and d_b is the depth of the beam.

Compared to the American standard, ANSI/AISC 358-16 [19], the Chinese standard, JGJ 99-2015 [20], uses a fixed, maximal cutting depth that reflects a safety-focused design philosophy, prioritizing reliable energy dissipation and structural performance at the expense of flexibility and optimization.

BS EN 1998-3 [21], with the fixation of parameters a and b, along with a relatively deep c, suggests a design philosophy focused on balancing structural strength and ductility, ensuring predictable plastic hinge formation while maintaining sufficient load-carrying capacity and connection integrity under seismic loads.

Table 1. Comparison of geometrical characteristics of reduced beam section.

Reference Standard	Geometrical Characteristics				
	a	b	c	e	R
ANSI/ AISC 358-16 [19]	$0.50b_f \leq a \leq 0.75b_f$	$0.65d_b \leq b \leq 0.85d_b$	$0.10b_f \leq c \leq 0.25b_f$	$a + b/2$	$(4c^2 + b^2)/8c$
BS EN 1998-3 [21]	$0.60b_f$	$0.75d_b$	$0.20b_f \leq c \leq 0.25b_f$	$a + b/2$	$(4c^2 + b^2)/8c$
JGJ99-2015 [20]	$0.50b_f \leq a \leq 0.75b_f$	$0.65d_b \leq b \leq 0.85d_b$	$0.25b_f$	$a + b/2$	$(4c^2 + b^2)/8c$

3. Numerical Analysis Methods

3.1. Numerical Modeling

The T-shaped joint of the external steel frame was selected, with the beam and column lengths being 3000 mm and 3600 mm, respectively, measured from the hinge center. The x-axis, y-axis, and z-axis represent the beam's length, depth, and width. The calculation diagram is shown in Figure 2a. Both the top and bottom of the column are designed with hinges, while the beam ends remain free.

As shown in Figure 2b, the top section of the column is coupled, with the coupling point constrained in the x and z directions for linear displacement. A rigid body is defined at the bottom of the column, with the rigid body reference point constrained in the x, y, and z directions for linear displacement. A concentrated force is applied at the top of the column, corresponding to an axial load ratio of 0.3. Displacement coupling is applied to all nodes within a 150 mm range at the beam-end loading point, and a displacement load is imposed on the coupled nodes in the y direction.

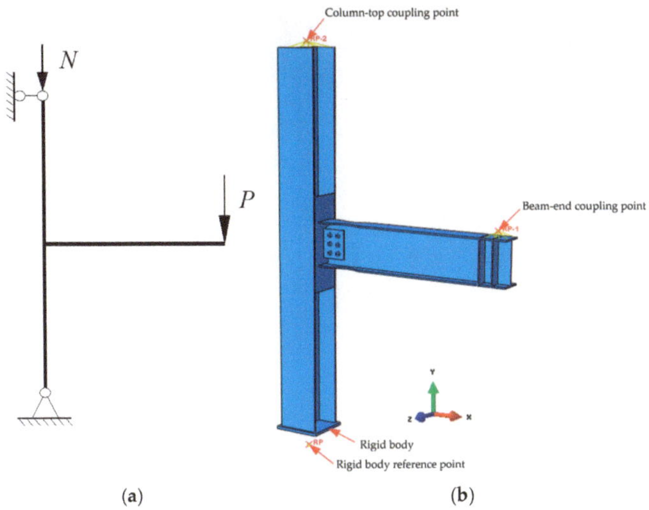

Figure 2. Calculation diagram and model boundary conditions. (**a**) Diagram; (**b**) boundary conditions.

ABAQUS (Version 6.1.4) finite element analysis technology is well established and reliable. The C3D8I element was selected due to its effectiveness for hysteretic behavior studies of RBS connections because it can capture complicated nonlinear stress–strain responses, handle cyclic loading easily, and properly mimic stress concentrations and plasticity in crucial locations. These characteristics make it excellent for studying the performance of H-shape columns with reinforced joint zones and varied beam flange configurations under seismic or other cyclic stress conditions. The mesh refinement criteria in hysteretic behavior investigations of RBS connections are designed to capture stress concentrations, plastic hinge formation, geometric discontinuities, and nonlinear behaviors. Mesh sizes are determined using geometric, material, loading factors, and convergence studies. The goal is to strike a compromise between accuracy and computing efficiency, with crucial regions sufficiently tuned to produce credible findings. The mesh size for the hot-rolled H-shapes is 25 mm, the steel plates are 10 mm, and the high-strength bolt is 2 mm. This same is true of plastic hinges and contact areas to enhance accuracy, whereas a coarser mesh is applied to non-critical regions to optimize computational efficiency, as illustrated in Figure 3.

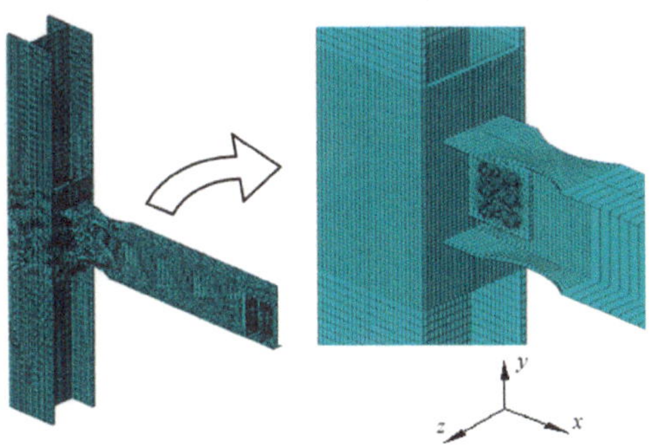

Figure 3. Mesh of a specimen model (3000 mm × 3600 mm).

Welding connections are represented using the "TIE" command in ABAQUS. The TIE command is widely used for simulating weld behavior in FEA because it simplifies the modeling process while accurately capturing the load transfer and displacement continuity between welded parts. It provides a computationally efficient and reliable approximation of weld behavior for studies focusing on global structural behavior, such as hysteretic performance in RBS connections. The pre-loading value is set according to the JGJ82-2010 [22] standard, and the "bolt load" function in ABAQUS is used to simulate it. When pre-tension is applied to high-strength bolts, multiple contact relationships are established. Key contact interfaces include plate-to-beam web, bolt-head-to-web, and bolt-shank-to-wall, and all are simulated using ABAQUS's "surface-to-surface" contact method. The stiffer surface is assigned as the master, while the less stiff surface acts as the slave. Normal contact is "hard" contact, meaning the surfaces remain in contact under compression and separate when tension occurs. Tangential contact is modeled with Coulomb friction, using the actual friction factor of 0.44 [23] for the bolt connection surface.

3.2. Material Property

As illustrated in Figure 4, the steel and bolt materials are modeled using a three-line isotropic hardening approach, defined by a stress–strain curve with σ_y, σ_u, and σ_{st} representing the yield, ultimate, and failure stresses, respectively, and ε_y, ε_u, and ε_{st} denoting the corresponding strains. The mechanical properties of Q235 steel are averaged from the stress–strain data reported by Nie et al. [24], while those of grade 8.8 M20 bolts are derived from the findings of Kontolati et al. [25], as summarized in Table 2.

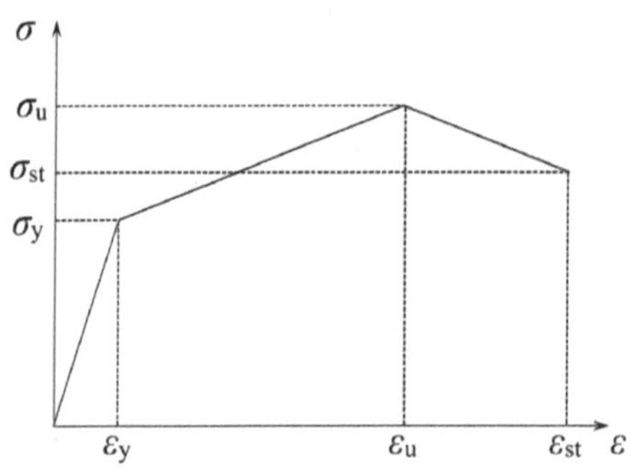

Figure 4. Stress–strain model.

Table 2. Material properties of Q235 Steel and grade 8.8 bolt.

Material Samples	E/MPa	σ_y/MPa	ε_y	σ_u/MPa	ε_u	σ_{st}/MPa	ε_{st}
Q235 steel	197,000	320	0.001748	451.667	0.14361	360.643	0.19653
8.8 grade M20 bolt	161,000	645.0	0.004	793.5	0.025	636.9	0.121

3.3. Validation

3.3.1. Validation Example 1

The SP5 specimen, designed by Song et al. [26], has geometrical dimensions, with the ABAQUS model shown in Figure 5a,b. The column and beam sections are H250 × 250 × 10 × 16 and H374 × 180 × 8 × 12, respectively, and made of Q235 steel. The loading protocol for the specimen follows ANSI/AISC 341-05 [27], which includes six cycles at inter-story drift angles of 0.375% rad, 0.5% rad, and 0.75% rad, along with four cycles at 1% rad. For drift angles of 1.5% rad, 2% rad, 3% rad, and 4% rad, two cycles are applied for each angle, with two additional cycles for every subsequent 1% rad increment. Loading is stopped when the bearing capacity drops to 85% of the maximum load. Displacement-controlled cyclic loading is applied at the beam end according to the test protocol [26]. During the test, a column top force of 500 kN is applied, with a bolt pre-tension of 190 kN and a friction coefficient of 0.35 between the connection plate, web, and bolts.

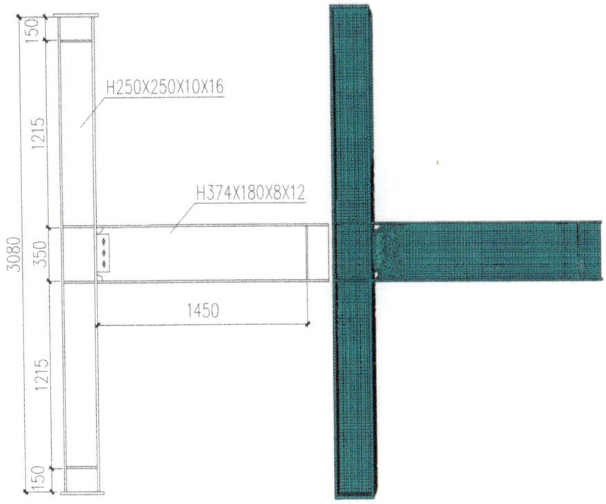

(a) (b)

Figure 5. SP5 specimen. (**a**) Schematic diagram of dimensions; (**b**) FE model.

The P–Δ hysteresis curve from the SP5 specimen experiment is shown in Figure 6a, while Figure 6b presents the P–Δ curves from the finite element analysis of the SP5 specimen. Despite some differences between the experimental and finite element hysteresis curves, the shapes are similar, showing a spindle form with near symmetry and minimal pinching. The maximum load from the finite element model was 171.41 kN compared to the experimental maximum load of 169.89 kN, resulting in an error of just 0.9%. The maximum plastic rotation of the SP5 test was 0.26 rad, while the numerical analysis predicted 0.274 rad, with a 5.4% error. This indicates that the numerical analysis method used in this study is highly reliable.

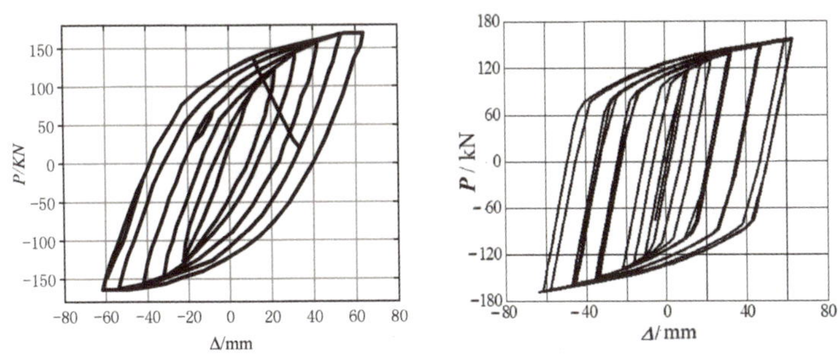

(a) (b)

Figure 6. P–Δ curves of SP5 specimen. (**a**) Test; (**b**) ABAQUS.

3.3.2. Validation Example 2

Gao's SPS-2 specimen [28] was examined using the dimensions shown in Figure 7. The beam and column were fabricated from Q235 steel, with hot-rolled steel sections of HN300 × 150 × 6.5 × 9 for the beam and HW250 × 250 × 9 × 14 for the column. The material properties of Q235 steel include a yield strength of 299.2 MPa and a tensile strength of 420.6 MPa. Based on ANSI/AISC 341-05 [27], the loading protocol involved displacement-controlled cyclic loading applied at the beam end, with testing halting once the load reduced to 85% of the maximum. Key parameters for the experiment included a column top force of 850 kN (compression ratio = 0.4), a bolt pre-loading of 155 kN, and a friction coefficient of 0.35.

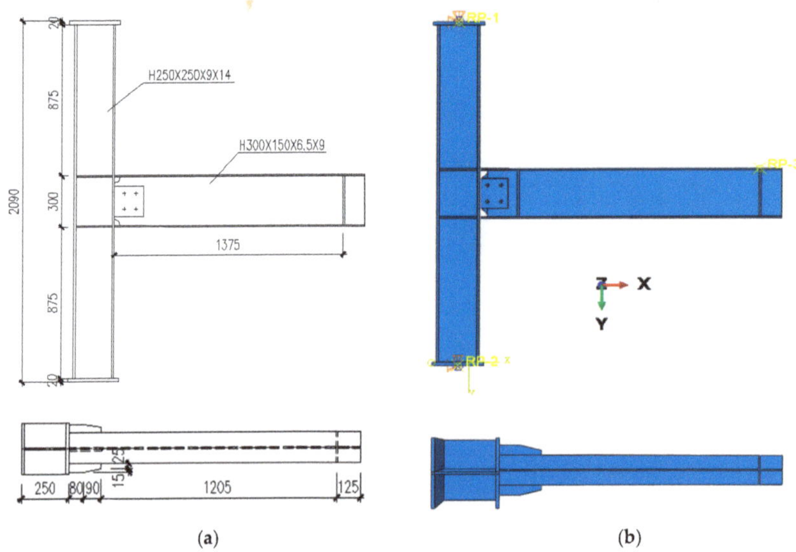

Figure 7. SPS-2 specimen. (**a**) Schematic diagram of dimensions; (**b**) FE model.

Figure 8a,b present a comparison of the experimental and finite element P–Δ hysteresis curves, while Figure 9a,b illustrate the consistent failure modes observed in both cases. The hysteresis curves exhibited a spindle-like symmetry with minimal pinching. The finite element model predicted a maximum load of 164.23 kN, which closely aligned with the experimental value of 153.73 kN (an error of 6.8%). The test reached a maximum plastic rotation of 0.392 rad, while the numerical analysis estimated 0.375 rad, resulting in an error of −4.43%, confirming the reliability of the analysis method.

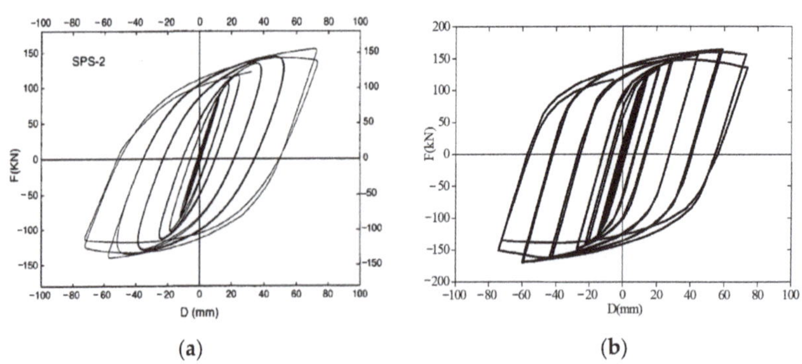

Figure 8. P–Δ curves of SPS-2 specimen. (**a**) Test; (**b**) ABAQUS.

(a) (b)

Figure 9. Failure modes of SPS-2 specimen. (**a**) Test; (**b**) ABAQUS.

4. Design of Analysis Specimens with Middle-Flange or Wide-Flange H-Shape Beams

4.1. Specimens with Middle-Flange H-Shape Beams

The column's cross-sectional dimensions are HW502 × 470 × 20 × 25, with a height of 3600 mm, while the beam has dimensions of HM340 × 250 × 9 × 14 and a span of 3000 mm. The analysis focused on three key design parameters for the widened flange connection, a, b, and c, as illustrated in Figure 1. The study considered three series: MRA, MRB, and MRC. For each series, three joint configurations were modeled for finite element analysis, as detailed in Table 3. The steel used was Q235, with material properties provided in Table 2. The design parameters adhere to the standards of ANSI/AISC 358-16 [19] and JGJ99-2015 [20].

Table 3. Properties of specimens.

Specimen No.	*a* (mm)		*b* (mm)		*c* (mm)	
MRA-1	$0.5b_f$	125	$0.75d_b$	255	$0.20b_f$	50
MRA-2	$0.65b_f$	162.5	$0.75d_b$	255	$0.20b_f$	50
MRA-3	$0.75b_f$	187.5	$0.75d_b$	255	$0.20b_f$	50
MRB-1	$0.65b_f$	162.5	$0.65d_b$	221	$0.20b_f$	50
MRB-2	$0.65b_f$	162.5	$0.75d_b$	255	$0.20b_f$	50
MRB-3	$0.65b_f$	162.5	$0.85d_b$	289	$0.20b_f$	50
MRC-1	$0.65b_f$	162.5	$0.75d_b$	255	$0.20b_f$	50
MRC-2	$0.65b_f$	162.5	$0.75d_b$	255	$0.225b_f$	56.25
MRC-3	$0.65b_f$	162.5	$0.75d_b$	255	$0.25b_f$	62.5
WRA-1	$0.5b_f$	175	$0.75d_b$	262.5	$0.20b_f$	70
WRA-2	$0.65b_f$	227.5	$0.75d_b$	262.5	$0.20b_f$	70
WRA-3	$0.75b_f$	262.5	$0.75d_b$	262.5	$0.20b_f$	70
WRB-1	$0.65b_f$	227.5	$0.65d_b$	227.5	$0.20b_f$	70
WRB-2	$0.65b_f$	227.5	$0.75d_b$	262.5	$0.20b_f$	70
WRB-3	$0.65b_f$	227.5	$0.85d_b$	297.5	$0.20b_f$	70
WRC-1	$0.65b_f$	227.5	$0.75d_b$	262.5	$0.20b_f$	70
WRC-2	$0.65b_f$	227.5	$0.75d_b$	262.5	$0.225b_f$	78.75
WRC-3	$0.65b_f$	227.5	$0.75d_b$	262.5	$0.25b_f$	87.5

In Table 3, the naming convention for the specimens is as follows: in MRA, MRB, and MRC, "M" represents H-shaped steel specimens with middle flanges; "R" indicates that the connection is of the RBS type; and "A", "B", and "C" correspond to the varying parameters a, b, and c, respectively. The numbers "−1", "−2", and "−3" denote the three parallel specimens within a group.

4.2. Specimens with Wide-Flange H-Shape Beams

Specimens with wide-flange H-shape beams were defined as the W series of specimens, with HW502 × 470 × 20 × 25 for columns (height 3600 mm) and HM350 × 350 × 12 × 19 for beams (span 3000 mm). The steel was Q235, with material properties provided in Table 2. Three series were analyzed: WRA, WRB, and WRC. Three joint configurations were modeled for each series for finite element analysis (see Table 3).

In Table 3, the naming convention for the wide-flange H-shaped steel series specimens, WRA, WRB, and WRC, is the same as for the middle-flange H-shaped specimens described in Section 4.1. The only difference is that the letter "W" replaces "M" to indicate wide-flange H-shaped steel specimens.

5. Analysis Results and Discussion

5.1. A Series Joints

We defined all specimens with the varying parameter "*a*" as Series A specimens, including three middle-flange and three wide-flange beam specimens. Their failure modes, hysteresis curves, and skeleton curves are shown in Figures 10 and 11, respectively. The load-carrying capacities of the specimens, determined from the skeleton curves, are listed in Table 4. R_{ki} represents the initial rotational stiffness. M_y, M_{max}, and M_u denote the yield, peak, and ultimate moments, respectively, with corresponding rotational displacements θ_y, θ_{max}, and θ_u. According to FEMA 350 [29], θ_p is defined as the rotation capacity. The ductility factor, μ, is calculated as $\mu = \theta_u/\theta_y$.

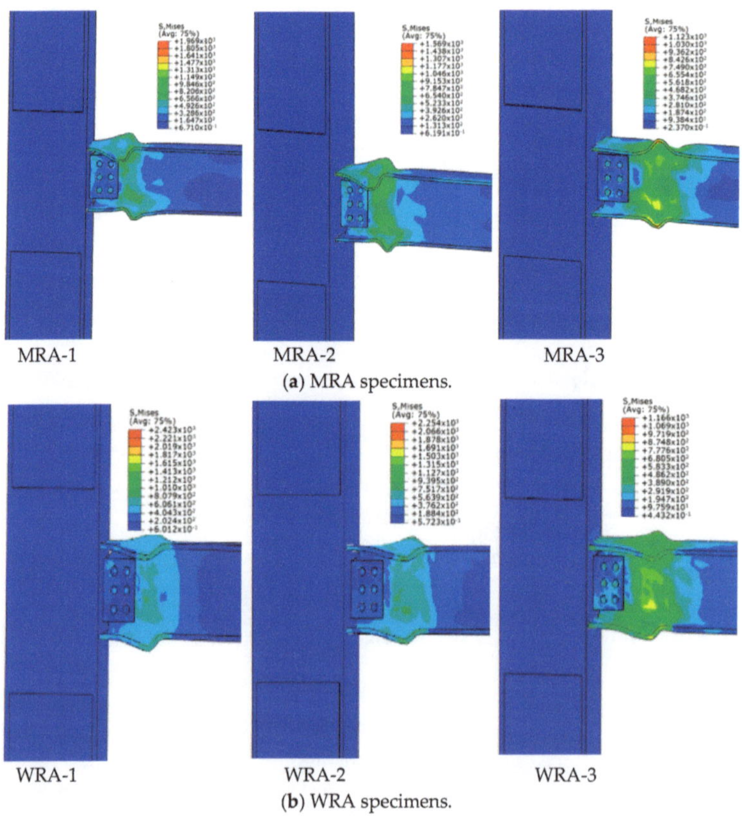

Figure 10. Stress contour plot of A series joints at failure time.

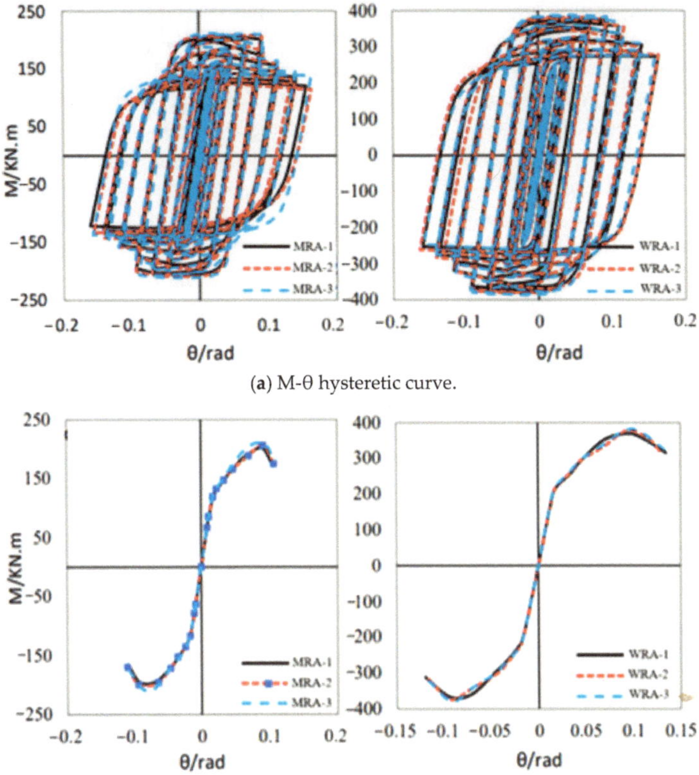

(a) M-θ hysteretic curve.

(b) M-θ skeleton curve.

Figure 11. M-θ hysteretic curve and skeleton curve of A series specimens.

Table 4. Mechanical performance indexes of A series joints.

Joint Number	R_{ki} /kN·m·rad^{-1}	M_y /kN·m	θ_y /rad	M_{max} /kN·m	θ_{max} /rad	M_u /kN·m	θ_u /rad	θ_p /rad	μ
MRA-1	7537.78	168	0.048	201.48	0.0904	171.30	0.1058	0.0999	2.20
MRA-2	7554.80	172	0.052	206.71	0.0910	175.70	0.1069	0.0981	2.06
MRA-3	7566.02	178	0.050	209.53	0.0938	178.10	0.1082	0.0995	2.16
WRA-1	12,815.30	315	0.054	369.61	0.0936	314.17	0.1353	0.1264	2.51
WRA-2	12,845.50	310	0.055	376.89	0.0937	320.36	0.1283	0.1195	2.33
WRA-3	12,866.20	319	0.056	380.29	0.0939	323.25	0.1335	0.1246	2.38

Observing Figure 10 reveals that plastic hinges formed at the reduced sections in all specimens. However, the plastic hinge regions varied depending on the value of "*a*". When "*a*" was at the lower limit ($0.5b_f$), severe buckling deformation occurred in the beam-end flange plate, whereas deformation was smaller when "*a*" was at the upper limit ($0.75b_f$).

As shown in the hysteresis and skeleton curves in Figure 11, as the "a" parameter increases, the bearing capacity of the specimen slightly improves. Further analysis of Table 4 shows that as the value of "*a*" increases, the joints' rotational stiffness and load-carrying capacities increase. The initial rotational stiffness increased very slightly and negligibly: for the MRA specimens, the middle and upper values showed increases of 0.23% and 0.37%, respectively, compared to the lower limit ($0.5b_f$); for the WRA specimens, these increases were 0.24% and 0.40%, respectively. The increase in maximum load-carrying capacity (peak moment) was smaller for both groups, with the MRA specimens showing increases of 2.6% and 4.0%, and the WRA specimens showing increases of 2% and 2.9%. Although the joints' plastic rotation capacities and ductility factors initially decreased (with lower mid values) and then increased as "*a*" increased, both θ_p and μ at the upper limit of "*a*" were smaller than those at the lower limit. Comparing the MRA and WRA specimens, the WRA specimens showed greater reductions at the mid-value of "*a*" (with decreases of 5.5% in θ_p and 7.2% in μ) than the MRA specimens (with decreases of 1.8% in θ_p and 6.4% in μ). Similarly, at the upper limit, the WRA specimens had larger reductions (1.4% in θ_p and 5.2% in μ) than the MRA specimens (0.4% in θ_p and 1.8% in μ). This indicates that widening the beam flange led to degradation in the plastic rotation capacity and ductility of the RBS connections.

Based on the observations and data analysis, in conclusion, an "*a*" value near the lower limit ($0.5b_f$) is recommended for achieving better ductility and plastic rotation capacity, especially in wide-flange beams (WRA specimens). While a higher "*a*" improves stiffness and load-carrying capacity, it reduces ductility and rotation capacity, which is crucial for RBS connections. Designs prioritizing ductility should thus use a lower "*a*" value, whereas stiffness-focused designs can consider a moderate increase, with caution against excessive flange widening. Generally, increasing the value of "*a*" reduces the plastic rotation capacity and ductility of RBS connections due to changes in plastic hinge location, moment gradient, and strain localization. These effects have significant practical design implications, particularly for seismic performance, where sufficient ductility and energy dissipation are critical.

5.2. B Series Joints

We defined all specimens with the varying parameter "*b*" as Series B specimens, which included three middle-flange beam specimens and three wide-flange beam specimens. Their failure modes, hysteresis curves, and skeleton curves are shown in Figures 9 and 10, respectively. The load-carrying capacities of the specimens, determined from the skeleton curves, are listed in Table 5.

Table 5. Mechanical performance indexes of B series joints.

Joint Number	R_{ki} /kN·m·rad^{-1}	M_y /kN·m	θ_y /rad	M_{max} /kN·m	θ_{max} /rad	M_u /kN·m	θ_u /rad	θ_p /rad	μ
MRB-1	7594.83	173	0.048	207.30	0.0940	176.21	0.1010	0.0960	2.10
MRB-2	7554.80	172	0.052	206.70	0.0910	175.70	0.1110	0.0981	2.13
MRB-3	7520.65	170	0.052	205.11	0.0937	174.06	0.1120	0.1017	2.15
WRB-1	12,890.95	318	0.053	375.92	0.0938	319.50	0.1223	0.1035	2.31
WRB-2	12,845.49	317	0.055	374.70	0.0937	317.40	0.1296	0.1123	2.36
WRB-3	12,792.21	317	0.056	374.11	0.0931	316.70	0.1334	0.1186	2.38

Observing Figure 12 reveals that plastic hinges formed at the reduced sections in all specimens. However, the plastic hinge regions varied depending on the value of "*b*". When "*b*" was at the lower limit ($0.65d_b$), severe buckling deformation occurred in the beam-end flange plate, with less buckling in the web. At the upper limit (0.85 d_b), flange and web buckling deformations were more pronounced, resulting in a larger plastic hinge region.

As shown in the hysteresis and skeleton curves in Figure 13, as the "*b*" parameter increases, the bearing capacity of the specimen slightly improves. Further analysis of Table 5 shows that as "*b*" increases, the joints' rotational stiffness and load-carrying capacities slightly decrease, but the reduction is minimal—less than 1% for both the MRB and WRB specimens, making this difference negligible. These results confirm the effectiveness of these configurations in obtaining ductile behavior while maintaining strength and stiffness, providing them with acceptable options for practical structural designs, especially in seismic applications. The joints' plastic rotation capacity and ductility factor increase with an increase in "b". Comparing the MRB and WRB specimens, the increase at the mid-value of "b" for the MRB specimens (2.2% in θ_p and 1.4% in μ) was smaller than that for the WRB specimens (8.5% in θ_p and 2.2% in μ). Similarly, at the upper limit, the increase in the MRB specimens (5.9% in θ_p and 2.4% in μ) was smaller than in the WRB specimens (14.6% in θ_p and 3.0% in μ). This indicates that widening the beam flange improves the plastic rotation capacity and ductility of the RBS connections. However, further increases in WRB specimens require additional design refinements to ensure consistent and reliable behavior.

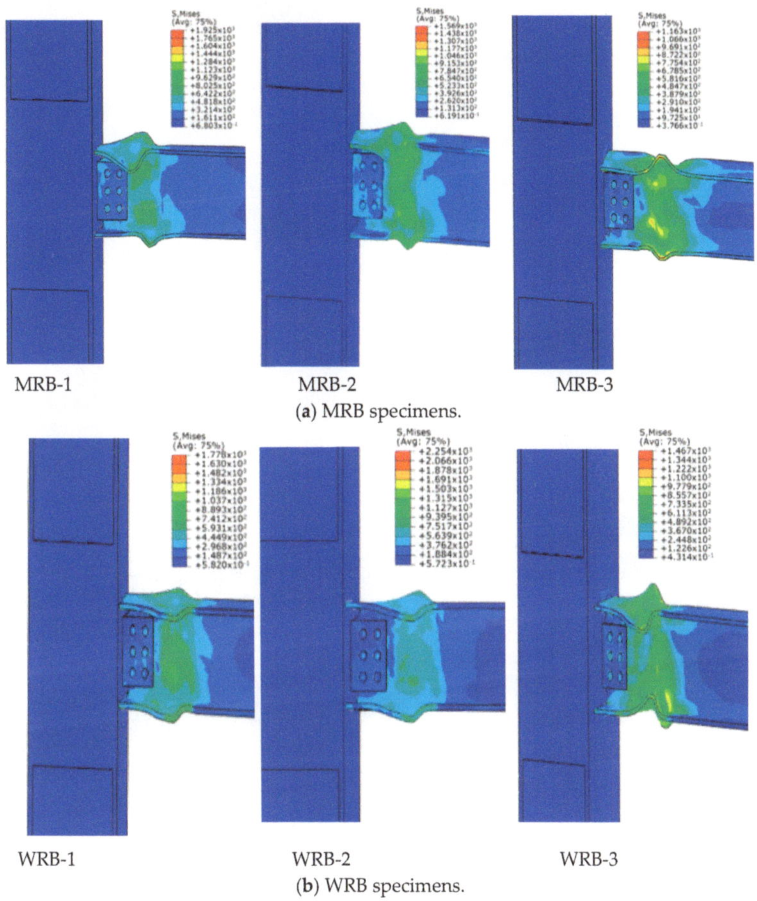

Figure 12. Stress contour plot of B series joints at failure time.

In conclusion, a higher "b" value close to the upper limit ($0.85d_b$) is recommended, as it enhances the plastic rotation capacity and ductility of the RBS connections with minimal impact on rotational stiffness and load-carrying capacity. Widening the beam flange further benefits plastic hinge formation, particularly in WRB specimens, by reducing buckling effects and improving overall ductility.

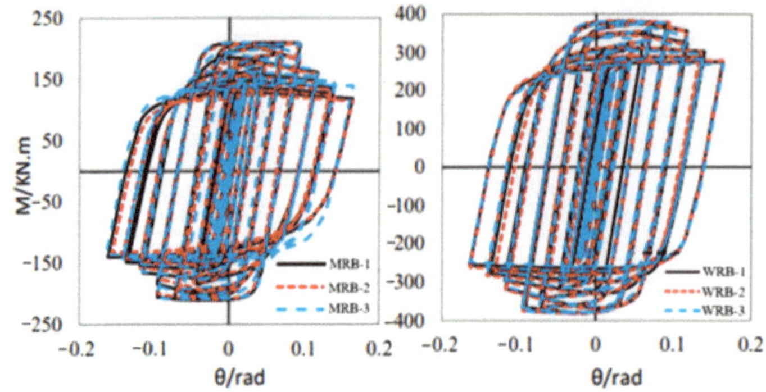

(**a**) M-θ hysteretic curve.

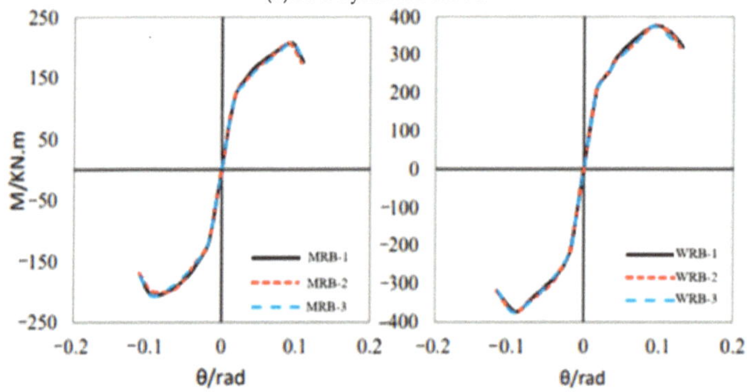

(**b**) M-θ skeleton curve.

Figure 13. M-θ hysteretic curve and skeleton curve of B series specimens.

5.3. C Series Joints

Six specimens with varying parameters "*c*" were defined as Series C specimens. The failure modes, hysteresis curves, and skeleton curves for these specimens are presented in Figures 14 and 15, respectively. The load-carrying capacities, as determined from the skeleton curves, are detailed in Table 6.

Table 6. Mechanical performance indexes of C series joints.

Joint Number	R_{ki} /kN·m·rad^{-1}	M_y /kN·m	θ_y /rad	M_{max} /kN·m	θ_{max} /rad	M_u /kN·m	θ_u /rad	θ_p /rad	μ
MRC-1	7554.80	172	0.052	206.71	0.0939	175.70	0.1070	0.0981	2.06
MRC-2	7439.26	174	0.046	195.75	0.0933	166.4	0.1110	0.1016	2.41
MRC-3	7310.81	145	0.037	183.92	0.0931	156.33	0.113	0.1042	3.05
WRC-1	12,845.49	310	0.055	376.89	0.0937	320.36	0.1283	0.1195	2.33
WRC-2	12,610.51	305	0.051	354.39	0.0920	301.23	0.1330	0.1224	2.61
WRC-3	12,341.15	275	0.047	327.82	0.0913	278.65	0.1388	0.1245	2.95

Observing Figure 11 reveals that plastic hinges formed at the reduced sections in all specimens. However, the regions of the plastic hinges varied depending on the value of "*c*". When "*c*" was set at $0.20b_f$, severe buckling deformation occurred in the beam-end flange plate, while the web experienced minimal buckling. At the upper limit ($0.25b_f$), the deformation of the beam-end flange plate decreased (with the deformation of the wide flange WRC specimens being smaller than that of the MRC specimens), while the buckling deformation of the web increased, resulting in a larger plastic hinge region.

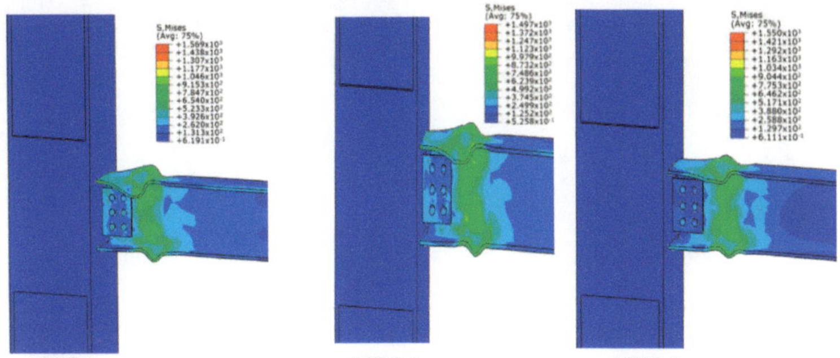

(**a**) MRC specimens.

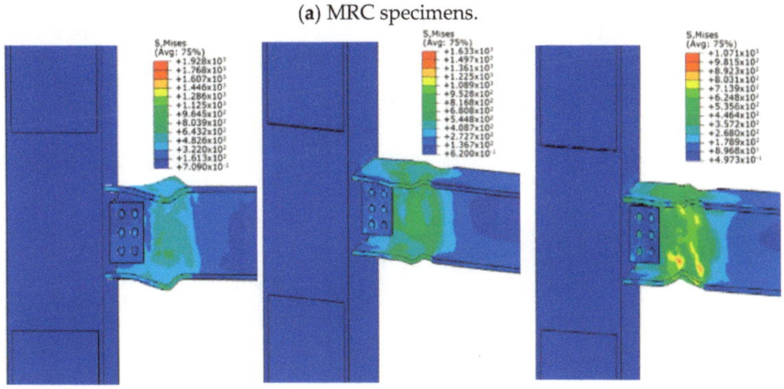

(**b**) MRC specimens.

Figure 14. Stress contour plot of C series joints at failure time.

As shown in the hysteresis and skeleton curves in Figure 15, as the "c" parameter increases, the bearing capacity of the specimen greatly decreases. Further analysis of Table 6 shows that as the value of "c" increases, the joints' rotational stiffness and load-carrying capacities decrease, with a reduction from −1.4% to −3.9% in initial rotational stiffness and a reduction in load-carrying capacity ranging from −5.3% to −13.0%. The joints' plastic rotation capacity and ductility factor increase as "c" increases. Comparing the MRC and WRC specimens, the increases at the mid-value of "c" for the MRC specimens (3.6% in θ_p and 17.0% in μ) were greater than those for the WRC specimens (2.4% in θ_p and 12.0% in μ). Similarly, at the upper limit, the increases for the MRC specimens (6.2% in θ_p and 48.0% in μ) were larger than those for the WRC specimens (4.2% in θ_p and 26.6% in μ). This indicates that widening the beam flange results in a smaller increase in the plastic rotation capacity and ductility of the RBS connections.

Based on this analysis, a "c" value between the lower limit ($0.20b_f$) and the mid-limit ($0.225b_f$) is recommended to balance rotational stiffness, load-carrying capacity, and ductility in RBS connections. While increasing "c" enhances plastic rotation capacity and ductility, it leads to great decreases in initial rotational stiffness (less than −2%) and load-carrying capacity (−5.3% to −6.0%). Furthermore, widening the beam flange lessens the positive impact of a higher "c" value on plastic rotation capacity and ductility. Therefore, a lower "c" value offers a more optimal performance balance.

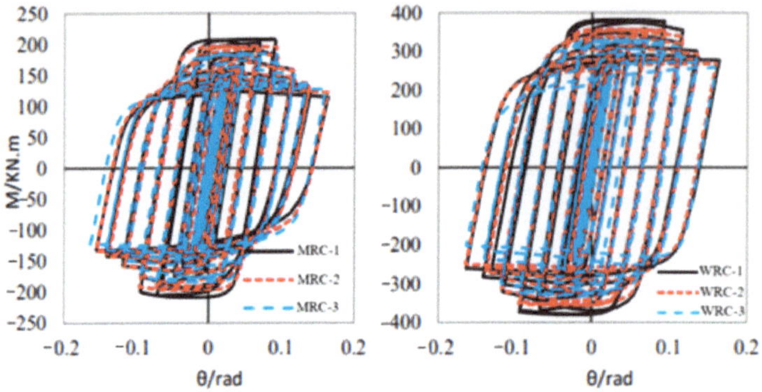

(a) M-θ hysteretic curve.

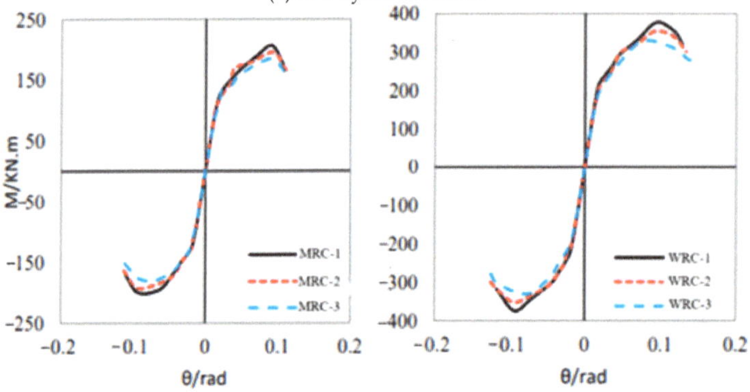

(b) M-θ skeleton curve.

Figure 15. M-θ hysteretic curve and skeleton curve of C series specimens.

5.4. Discussion

RBS (reduced beam section) connections are designed to improve seismic performance by concentrating plastic deformations in the beam. The parameters *a*, *b*, and *c* affect the connection's stiffness, load-bearing capacity, plastic rotation capacity, and ductility in the following ways:

1. Parameter *a* (Unweakened Beam Flange Extension Length):

A longer "*a*" increases stiffness and load-bearing capacity but reduces plastic rotation and ductility. This is because the unweakened section delays the formation of a plastic hinge, limiting flexibility and energy absorption. The results of this study on RBS connections for mid- and wide-flange H-shaped steel beams, along with previous findings on RBS connections for narrow-flange H-shaped steel beams [9], provide strong support for this conclusion. An "*a*" value near the lower limit ($0.50b_f$) is recommended, aligning with the ranges suggested in ANSI/AISC 358-16 [19] and JGJ99-2015 [20]. However, the value specified in BS EN 1998 [21] ($0.60b_f$) may be overly large, prioritizing load-bearing capacity at the expense of connection ductility.

2. Parameter *b* (Weakened Beam Flange Length):

A larger "*b*" reduces stiffness but improves plastic rotation and ductility by allowing for more deformation. This enhances energy dissipation, but excessive weakening can decrease load-bearing capacity. Unlike RBS connections for narrow-flange H-shaped steel beams [9], the "*b*" values for mid- and wide-flange H-shaped steel beam RBS connections in this study are better suited to the upper limit ($0.85d_b$) of the range recommended by

ANSI/AISC 358-16 [19] and JGJ99-2015 [20]. In contrast, the value specified in BS EN 1998 [21] ($0.75d_b$) is more appropriate for narrow-flange beams falling near the median of the ranges in the other two standards.

3. Parameter *c* (Weakened Beam Flange Depth):

A smaller "*c*" improves load-bearing capacity but reduces plastic rotation and ductility. Larger "*c*" values allow for more deformation, improving ductility and energy absorption. This aligns with findings from studies on narrow-flange H-shaped steel beam RBS connections [9]. The flange width is significantly larger for mid- and wide-flange H-shaped steel beams than for narrow-flange beams. When the flange reduction ratio adopts the mid-to-lower-limit values ($0.20b_f$~$0.225b_f$) specified in BS EN 1998 [21], it effectively balances load-bearing capacity and ductility in RBS connections. In contrast, the values in JGJ99-2015 [20] result in excessive section weakening, overly compromising load-bearing capacity, while the lower limit design in ANSI/AISC 358-16 [19] overly sacrifices ductility.

Optimizing *a*, *b*, and *c* is essential for balancing stiffness, load-bearing capacity, and ductility in RBS connections, which improves seismic performance.

6. Conclusions

This study examined the hysteretic behavior of reduced beam section (RBS) connections in H-shaped columns with middle-flange and wide-flange H-shaped beams. Key parameters (*a*, *b*, and *c*) were varied to investigate their impact on structural performance, including rotational stiffness, load-carrying capacity, plastic rotation capacity, and ductility. The following conclusions can be drawn:

1. Increasing "a" slightly enhanced the initial rotational stiffness and load-carrying capacity but significantly decreased plastic rotation capacity and ductility. Therefore, lower "*a*" values near the minimum ($0.5b_f$) are recommended for enhanced ductility, especially for wide-flange beams (WRA specimens).
2. Higher values of "*b*" led to slightly reduced rotational stiffness and load capacity but with minimal impact (less than 1%). Conversely, plastic rotation capacity and ductility improved with an increasing "*b*" value. Thus, a "*b*" value close to the upper limit ($0.85b_f$) is recommended to achieve optimal plastic behavior, particularly in WRA specimens.
3. As "*c*" increased, rotational stiffness and load-carrying capacity decreased while plastic rotation capacity and ductility improved. However, increasing "*c*" was less significant in wide-flange beams. A lower "*c*" value ($0.20b_f$) is preferable for balanced structural performance.
4. Widening the beam flange generally enhanced plastic rotation capacity and ductility but decreased rotational stiffness. This study's insights can inform design decisions to optimize RBS connections for seismic resilience, achieving a balance between stiffness, load capacity, and ductility in mid- and wide-flange H-shaped beam connections.

Author Contributions: Conceptualization, L.L.; investigation. S.M.A.A.A.-S.; validation, S.M.A.A.A.-S., O.Z.Y.A.-A. and S.A.; resources, L.L.; writing—original draft preparation, S.M.A.A.A.-S. and L.L.; writing—review and editing, L.L.; project administration, L.L.; funding acquisition, L.L. All authors have read and agreed to the published version of the manuscript.

Funding: This research was funded by the Nature Science Foundation of China (NSFC), grant number 51278061.

Data Availability Statement: The original contributions presented in the study are included in the article; further inquiries can be directed to the corresponding author.

Conflicts of Interest: The authors declare that they have no conflicts of interest.

References

1. Miller, D.K. Lessons learned from the Northridge earthquake. *Eng. Struct.* **1998**, *20*, 249–260. [CrossRef]
2. Mahin, S.A. Lessons from damage to steel buildings during the Northridge earthquake. *Eng. Struct.* **1998**, *20*, 261–270. [CrossRef]
3. Tsai, K.C.; Wu, S.; Popov, E.P. Cyclic performance of steel beam-column moment joints. *Eng. Struct.* **1995**, *17*, 596–602. [CrossRef]
4. Chen, S.J.; Yeh, C.H.; Chu, J.M. Ductile steel beam-to-column connections for seismic resistance. *J. Struct. Eng.* **1996**, *122*, 1292–1299. [CrossRef]
5. Popov, E.P.; Yang, T.S.; Chang, S.P. Design of steel MRF connections before and after 1994 Northridge earthquake. *Eng. Struct.* **1998**, *20*, 1030–1038. [CrossRef]
6. Engelhardt, M.D.; Sabol, T.A. Reinforcing of steel moment connections with cover plates: Benefits and limitations. *Eng. Struct.* **1998**, *20*, 510–520. [CrossRef]
7. Rezaifar, O.; Younesi, A. Finite element study the seismic behavior of connection to replace the continuity plates in (NFT/CFT) steel columns. *Steel Compos. Struct.* **2016**, *21*, 73–91. [CrossRef]
8. Zahrai, S.M.; Mirghaderi, S.R.; Saleh, A. Increasing plastic hinge length using two pipes in a proposed web reduced beam section, an experimental and numerical study. *Steel Compos. Struct.* **2017**, *23*, 421–433. [CrossRef]
9. Lu, L.; Xu, Y.; Liu, J.; Lim, J. Cyclic performance and design recommendations of a novel weak-axis reduced beam section connection. *Steel Compos. Struct.* **2018**, *27*, 337–353.
10. Lignos, D.G.; Kolios, D.; Miranda, E. Fragility assessment of reduced beam section moment connections. *J. Struct. Eng.* **2010**, *136*, 1140–1150. [CrossRef]
11. Crisan, A.; Dubina, D. Bending–shear interaction in short coupling steel beams with reduced beam section. *J. Constr. Steel Res.* **2016**, *122*, 190–197. [CrossRef]
12. Cimagala, J.P. Optimization of Reduced Beam Sections (RBS) for Ductile Detailing of Seismic Joint Connections Using Finite Element Analysis (FEA). *Comput. Eng. Phys. Model.* **2021**, *4*, 43–54.
13. Qi, L.; Leon, R.T.; Eatherton, M.R.; Paquette, J.L. Parametric investigation on the design of RBS moment connections with jumbo beams and columns. *J. Constr. Steel Res.* **2021**, *177*, 106436. [CrossRef]
14. Bompa, D.V.; Elghazouli, A.Y.; Bogdan, T.; Eatherthon, M.R.; Leon, R.T. Inelastic cyclic response of RBS connections with jumbo sections. *Eng. Struct.* **2023**, *281*, 115758. [CrossRef]
15. Ghassemieh, M.; Mirghaderi, S.R. Cyclic dependency assessment of RBS moment connection in box-column. *J. Constr. Steel Res.* **2021**, *177*, 106472. [CrossRef]
16. Horton, T.A.; Hajirasouliha, I.; Davison, B.; Ozdemir, Z.; Abuzayed, I. Development of more accurate cyclic hysteretic models to represent RBS connections. *Eng. Struct.* **2021**, *245*, 112899. [CrossRef]
17. Indupriya, G.; Anupriya, B. Simulation of RBS moment connections using the Finite Element Method for Indian profiles. *Mater. Today Proc.* **2022**, *64*, 1023–1028. [CrossRef]
18. Gemechu, T.D.; Lu, L. Analysis of Factors Affecting the Seismic Performance of Widened Flange Connections in Mid-Flange H-Beams and Box Columns. *Buildings* **2024**, *14*, 3170. [CrossRef]
19. *ANSI/AISC 358-16*; Prequalified Connections for Special and Intermediate Steel Moment Frames for Seismic Applications, including Supplements No. 1 and No. 2 Seismic Provisions for Structural Steel Buildings. American Institute of Steel Construction: Chicago, IL, USA, 2016.
20. *JGJ99-2015*; Technical Specification for Steel Structures of Tall Buildings. China Architecture & Building Press: Beijing, China, 2015.
21. *BS EN 1998-3*; Eurocode 8: Design of Structures for Earthquake Resistance Part 3: Assessment and Retrofitting of Buildings. British Standards Institution: London, UK, 11 January 2006.
22. *JGJ 82-2011*; Technical Specification for High Strength Bolt Connections of Steel Structures. China Architecture & Building Press: Beijing, China, 2011.
23. Lu, L.; Zhu, P.; Ding, S.; Ma, Z.; Li, R.; Nie, S.; Wang, W.; Raftery, G.M. Impact of anti-corrosion coatings and maintenance on high-strength bolt friction connections in C4 marine environment. *Structures* **2024**, *68*, 107098. [CrossRef]
24. Nie, W.; Lu, L.; Li, R.; Hao, H.; Luo, T. Global buckling investigation of T-shaped built-up columns composed of two hot-rolled H-shapes. *Structures* **2023**, *58*, 105485. [CrossRef]
25. Katiana, K.; Apostolos, K.; Olympia, P. Numerical investigation of weak axis I profile connections. In Proceedings of the 9th Hellenic National Conference on Steel Structures, Larissa, Thessaly, Greece, 5–7 October 2017.
26. Song, Z.; Gu, Q.; Guo, B. Cyclic-loading Performance of Steel Beam-to-column Moment Connections. *J. Build. Struct.* **2001**, *22*, 53–57.
27. *ANSI/AISC 341-05*; Seismic Provisions for Structural Steel Buildings. American Institute of Steel Construction: Chicago, IL, USA, 2005.

28. Gao, P. Experimental Evaluation on the Behavor of Side-Plate Reinforced Section and Widen Flange Section of Steel Frame. Master's Thesis, Qingdao University of Technology, Qingdao, China, 2009.
29. *FEMA-350*; Recommended Seismic Design Criteria for New Steel Moment-Frame Buildings. FEMA: Washington, DC, USA, 2000.

buildings

MDPI

Article

Effects of Cover-Plate Geometry on the Mechanical Behavior of Steel Frame Joints with Middle-Flange and Wide-Flange H-Beams

Osama Zaid Yahya Al-Ansi, Linfeng Lu *, Saleem Mohammed Ali Ahmed AL-Saeedi and Bingyou Liu

School of Civil Engineering, Chang'an University, 75 Chang'an Middle Rd., Xi'an 710062, China; 2022028916@chd.edu.cn (O.Z.Y.A.-A.); 2021028918@chd.edu.cn (S.M.A.A.A.-S.); 2023228019@chd.edu.cn (B.L.)
* Correspondence: lulinfeng@chd.edu.cn

Abstract: This study investigates the mechanical behavior of cover-plate reinforced connections in steel frames with I-section columns and middle- or wide-flange H-beams, addressing gaps in current design standards. Finite element analyses validated by experimental data were employed to explore the effects of cover-plate geometry—shape, length, and thickness—on seismic performance. Results demonstrate that cover plates improve load-bearing capacity and ductility by relocating plastic hinges outward from joint regions. Specifically, cover-plate connections increased ductility by 25%, yield moment by 15%, and initial rotational stiffness by 7% compared to non-reinforced connections. The shape of the top cover plate had minimal impact on mechanical behavior. The cover-plate length and thickness significantly influenced seismic ductility and load-bearing capacity. The cover-plate thickness should be at least 0.3 times the beam flange thickness (not less than 6 mm) while ensuring the combined thickness of the cover plate and beam flange does not exceed the column flange thickness. These recommendations address the conservatism of existing standards, balancing material efficiency and seismic performance. Optimal cover-plate lengths of 0.7 to 0.9 times the beam depth were also identified. These findings provide practical guidelines for designing resilient steel frame connections in seismic regions.

Keywords: cover plate; connections; cyclic loading; H-shape beam; plastic hinge

1. Introduction

The Northridge and Kobe earthquakes occurred 40 years ago, leading to significant damage and failure of beam–column joints [1–3]. Both earthquakes revealed weaknesses in traditional welded and bolted connections in steel frames. They had a major impact on connection design standards and academic research [4]. Beam-to-column connections are crucial for steel building or rack frames [5]. Special attention is required for performance analysis under seismic, fire, and other disasters.

After the Northridge earthquake, the Federal Emergency Management Agency (FEMA) introduced new guidelines, such as FEMA-350 [6], recommending enhanced seismic design criteria for steel moment-resisting frames. This guideline emphasized increased connection toughness and reliability to withstand severe earthquake loading [6]. Similarly, in Japan, the Kobe earthquake prompted amendments to the Building Standard Law, focusing on strengthening steel structure design for improved energy dissipation and plastic deformation capacity [7]. The academic community responded by conducting experimental and analytical studies on various connection types, including reinforced connections with welded cover plates, to understand failure mechanisms and improve seismic detailing [8,9]. Some design standards [10–12] have adopted the cover-plate connection as a post-earthquake retrofitting method for steel frame connections. Additionally, it has been incorporated as a connection method for new steel frames in FEMA-350 [6] and in the Chinese standard JGJ 99-2015 [13].

Research on the traditional form of cover-plate reinforced connections has been ongoing. I-section columns are commonly used in steel frames. However, due to inevitable defects in weak joint regions, replacing the column web or welding reinforcing plates in a certain range at the joint area is necessary [14], which complicates construction. Lu et al. proposed new joint region strengthening connections [15,16] to address this and investigated and reported on the design methods for weak-axis cover-plate connections [17]. Shi's team extended the application of cover-plate connection joints from conventional low-carbon and low-alloy steel frames to high-strength steel frames [18,19]. Their research demonstrated that cover-plate connection joints exhibit excellent seismic performance and high ductility when using high-strength steel.

Building on the traditional cover-plate connection design, some scholars have introduced improvements. For instance, Casita et al. [20] combined the traditional cover-plate connection with the reduced beam section (RBS) connection. Numerical investigations revealed that the new connection outperformed traditional cover-plate and RBS connections regarding plastic hinge location, while the cover-plate connection exhibited the best energy dissipation capacity. Meng et al. [21] introduced a new connection type featuring double-leg energy dissipation cover plates (DEDCPs) and conducted experimental and numerical studies. Their findings indicated that the hysteretic curves of the DEDCPs were well-formed and demonstrated strong energy dissipation capabilities. Ribeiro et al. [22] used the Solid Isotropic Material with Penalization (SIMP) method to study the bolt cover-plate connection used in traditional H-shaped steel beam splicing, extending the application of topology optimization techniques and exploring advanced computer-aided design technologies. Nassiraei [23] conducted a different technical exploration, where he developed a comprehensive probabilistic framework to evaluate tubular X-joint resistance strengthened with collar plates under tensile and fire conditions, integrating extensive FE analyses and advanced statistical modeling. This work is a key reference for researchers studying tubular joint behavior, offering rigorous FE analyses and advanced statistical modeling for probabilistic structural assessment under extreme conditions. Other studies have used bolt cover-plate technology to address the splicing issues of modular steel buildings. For example, Cao's team developed a module-to-module bolted connection [24–26]. Deng et al. developed a novel liftable connection [27,28]. The extensive experimental and numerical analysis results of these two new cover-plate connections are highly beneficial for promoting the adoption and use of prefabricated steel buildings.

Despite the emergence of various new cover-plate connection technologies, traditional cover-plate connections continue to dominate steel frame structures in China. The data of cover-plate connections used in the standards [12,13,29] are based on research conducted on connections of narrow-flanged H-beams ($b_f/d_b \leq 1/2$, where b_f is the beam flange width and d_b is the beam depth, as shown in Table 1). With the extensive promotion of hot-rolled H-shapes in China, middle-flange and wide-flange hot-rolled H-shapes are also used in beam components [30]. Therefore, the applicability of design standards to cover-plate connections for H-shaped beams with middle and wide flanges needs further investigation, and existing research on cover-plate connections reveals certain gaps. Existing research [31,32] has proven that reliable numerical analysis methods were used to investigate the seismic performance of joints or connections. This study uses finite element numerical methods to analyze the seismic performance of cover-plate connections between H-shaped steel columns and H-shaped steel beams with medium and wide flanges. The results are compared with existing design criteria for cover-plate connections based on narrow-flange H-shaped beams, and new design recommendations are provided.

Table 1. List of notations and abbreviations.

Notation	Definition	Notation	Definition
b_f	beam flange width	CP	specimens with cover plates
d_b	beam depth	WCP	specimens without cover plates
x-axis	length direction of the beam	CPS	specimens with varying shapes of top cover plate
y-axis	height direction of the column	CPL	specimens with varying lengths of rectangular cover plates
z-axis	width direction of the beam	CPT	specimens with varying thicknesses of rectangular cover plates
σ_y	yield stress	b_{cp}	width of the cover plate
σ_u	maximum stress	t_{cp}	thickness of the cover plate
σ_{st}	failure stress	t_{bf}	thickness of the beam flange
ε_y	yield strain	l_{cp}	length of the cover plate
ε_u	maximum strain	b_{cp1}	width of the top cover plate
ε_{st}	failure strain	b_{cp2}	the width of the bottom cover
MPa	Mega pascal	θ_y	yield rotational displacements
R_k	initial rotational stiffness	θ_{max}	peak rotational displacements
M_y	yield moment	θ_u	terminated rotational displacements
M_{max}	peak moment	θ_p	plastic rotation capacity
M_u	terminated moment		

2. Numerical Analysis Methods and Materials

2.1. Material Property

As shown in Figure 1, the steel and bolt materials adhere to a three-line isotropic hardening model corresponding to a simplified engineering stress–strain curve. In this context, σ_y, σ_u, and σ_{st} denote the yield, maximum, and failure stresses, respectively, while ε_y, ε_u, and ε_{st} represent the corresponding strains. The properties of Q235 steel and grade 8.8 M20 bolts are from Refs. [33,34]. Table 2 displays the engineering stresses and strains of Q235 steel and an 8.8-grade bolt. The Poisson's ratio is taken as 0.3. In ABAQUS (6.1.4) analysis, it is necessary to convert engineering stress and strain into true stress and strain. The analysis applies the Mises yield criterion and mixed hardening rule reflecting the Bauschinger effect and cyclic hardening of steel.

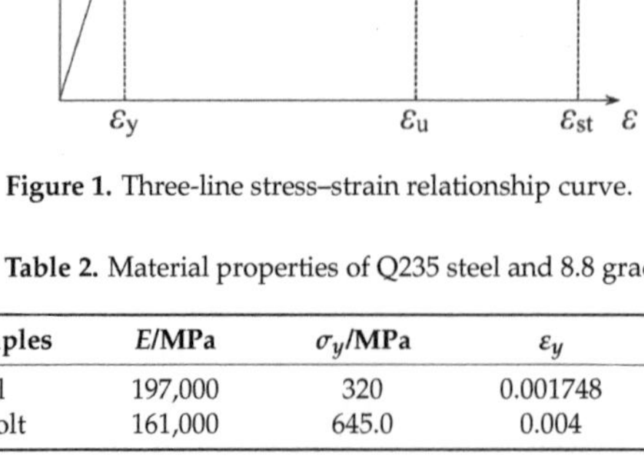

Figure 1. Three-line stress–strain relationship curve.

Table 2. Material properties of Q235 steel and 8.8 grade bolt.

Material Samples	E/MPa	σ_y/MPa	ε_y	σ_u/MPa	ε_u	σ_{st}/MPa	ε_{st}
Q235 steel	197,000	320	0.001748	451.667	0.14361	360.643	0.19653
8.8-grade bolt	161,000	645.0	0.004	793.5	0.025	636.9	0.121

2.2. Numerical Modeling

The T-shaped joint of the external steel frame was selected, with the beam and column lengths being 3000 mm and 3600 mm, respectively, measured from the hinge center. The skin steel plates on both sides of the reinforced joint region were 8 mm thick, with a width equal to the column depth and a height extending 150 mm above and below the beam depth, welded onto the column flanges. The calculation diagram is shown in Figure 2a.

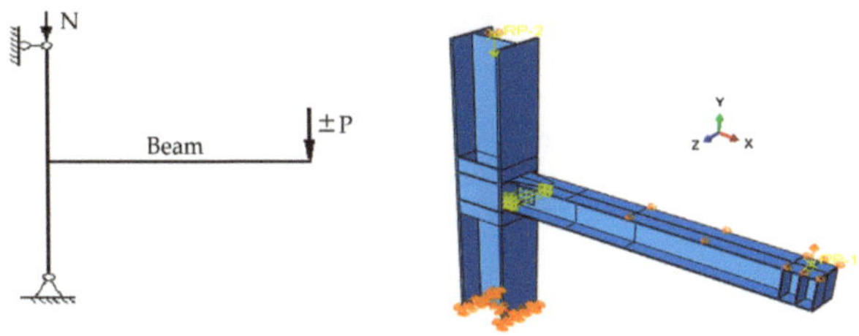

(**a**) (**b**)

Figure 2. Schematic diagram of boundary condition and node loading. (**a**) Calculation diagram; (**b**) Abaqus.

The length direction of the beam, the height direction of the column, and the width direction of the beam are defined as the x-axis, y-axis, and z-axis, respectively. As shown in Figure 2b, the top section of the column is coupled, with the coupling point restricted in the x and z directions for linear displacement. At the base of the column, linear displacements in all three directions are constrained while rotations are allowed. At the top of the column, only linear displacements in the x and z directions are restricted. A concentrated force was applied at the top of the column, corresponding to an axial load ratio of 0.3 to conform to the majority of middle- and top-floor joints in multi-story and high-rise steel frame buildings. Displacement coupling was applied to all nodes within a 150 mm range at the beam-end loading point, and cyclic displacement loading was used to simulate seismic effects to the coupled nodes in the y direction. The loading protocol for the specimen followed ANSI/AISC 341-05 [35], which includes six cycles at inter-story drift angles of 0.375% rad, 0.5% rad, and 0.75% rad, and four cycles at 1% rad. For drift angles of 1.5% rad, 2% rad, 3% rad, and 4% rad, two cycles were applied for each angle, with two additional cycles for every subsequent 1% rad increment. Loading was terminated when the bearing capacity decreased to 85% of the maximum load.

ABAQUS (Version 6.1.4) finite element analysis technology is well-established and reliable. The C3D8I elements, which are 8-node linear hexahedral elements with integrated hourglass control, were utilized for the structural analysis. These elements were chosen for their ability to accurately represent the geometry of the beam–column joints and the cover-plate areas, and they are well-suited for modeling solid structures. Using these elements ensures that the structural behavior is captured with high fidelity.

A mesh convergence analysis was conducted to verify the adequacy of the mesh size and ensure the accuracy of the result. The process involved testing the finite element model with progressively smaller mesh sizes and comparing key output parameters, including the load–displacement curves and stress distributions at critical locations. Convergence was achieved when further mesh refinement produced negligible changes in the results, confirming that the chosen mesh size reliably captured the structural behavior. The analysis demonstrated that the selected mesh sizes struck an optimal balance between precision and computational efficiency. During the meshing of the finite element model, finer meshes were applied at the beam–column joint, expected plastic hinge locations on the beam, and contact areas such as bolt holes and bolts. For the model, the mesh size was 25 mm for hot-rolled H-shapes, a 10 mm mesh size was applied to the 600 mm region at the beam-to-column connection to better capture the plastic hinge zone, 5 mm for the cover and web connection plate, and 2 mm for high-strength bolts. Figure 3 illustrates the 3D finite element local model, highlighting the meshed structure, particularly the critical areas such as the beam–column joint and cover plates. Refined mesh regions were clearly marked to enhance the visualization of the meshing strategy, as shown in Figure 3c.

(**a**) (**b**) (**c**)

Figure 3. Three-dimensional finite element local model. (**a**) CP finite element model; (**b**) WCP finite element mode; (**c**) local mesh refinement.

Welded connections were modeled in ABAQUS using the "TIE" command, a common FEA approach for simulating weld behavior. This method simplifies the modeling process, ensuring accurate load transfer and displacement continuity between welded components. It offers a computationally efficient and reliable way to approximate weld performance, particularly for analyzing global structural behavior such as the hysteretic response of RBS connections. Preloading of high-strength bolts was applied based on the JGJ82-2010 [36], using the "bolt load" function in ABAQUS. When preloading is applied, multiple contact interactions are established, including plate-to-beam web, bolt head-to-web, and bolt shank-to-wall interfaces. These interactions were modeled using the "surface-to-surface" contact method, assigning the stiffer surface as the master and the less stiff surface as the slave. Normal contact was "hard", maintaining contact under compression and allowing separation under tension. Tangential contact is governed by Coulomb friction, with a friction factor of 0.44 [37] applied to the bolt connection surfaces.

2.3. ABAQUS Analysis Validation

The CPS-1 Specimen, as shown in Figure 4, originally tested by Wang [38], was chosen for validation purposes. This specimen was fabricated using Q235 steel and incorporated a hybrid bolt–weld connection featuring M20 high-strength bolts. The setup involved a typical simply supported configuration with pinned connections at both beam and column ends, and an axial force of 850 kN was applied to the top of the column. Rectangular cover plates were used for both the top and bottom. The column section was HW 250 × 250 × 9 × 14, with a total length of 2050 mm, while the beam had a length of 1750 mm and a section of HN 300 × 150 × 6.5 × 9. Under displacement-controlled cyclic loading, as described by Wang [38] and conducted following ANSI/AISC 341-05 [35], forces were applied at the beam end. Loading ceased when the bearing capacity dropped to 85% of its maximum value.

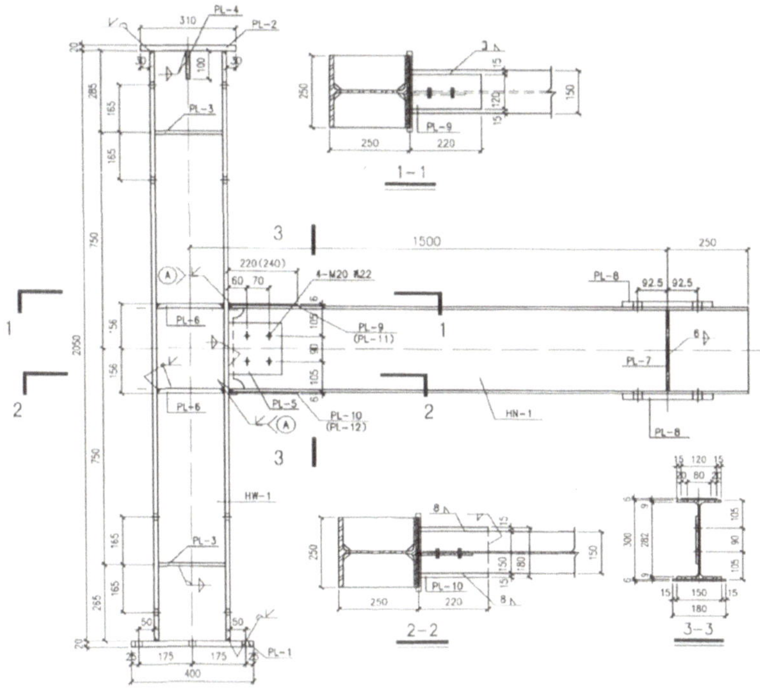

(a)

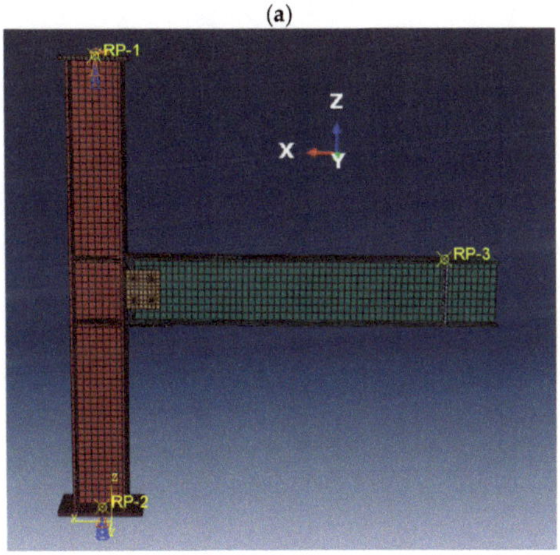

(b)

Figure 4. CPS-1 Specimen. (**a**) Schematic diagram of dimensions (Unit: mm); (**b**) ABAQUS model.

The P-Δ hysteresis curve obtained from the CPS-1 experiment is displayed in Figure 5a, while Figure 5b illustrates the corresponding P-Δ curves derived from ABAQUS simulations. Although slight discrepancies exist between the experimental and simulated hysteresis curves, their shapes exhibit high similarity. The peak load recorded during the test was 156.83 kN, whereas the ABAQUS simulation produced a maximum load of 159.5 kN, yielding an error of just 1.7%. The experimental peak load occurred at a beam-end displacement of approximately 56 mm, whereas the simulation reached its maximum load at 58 mm, demonstrating close agreement between the two. The maximum plastic rotation angle measured in the test was 4.47% rad, compared to 4.75% rad predicted by the simulation, with a difference of 6.1%. The high consistency between the ABAQUS numerical analysis and experimental results indicates that the finite element modeling and analysis in this study are reliable. Additionally, Figure 6 demonstrates that the failure mode observed in the numerical analysis is generally consistent with the experimental results. This methodology can be further employed to investigate the seismic performance of cover-plate connections.

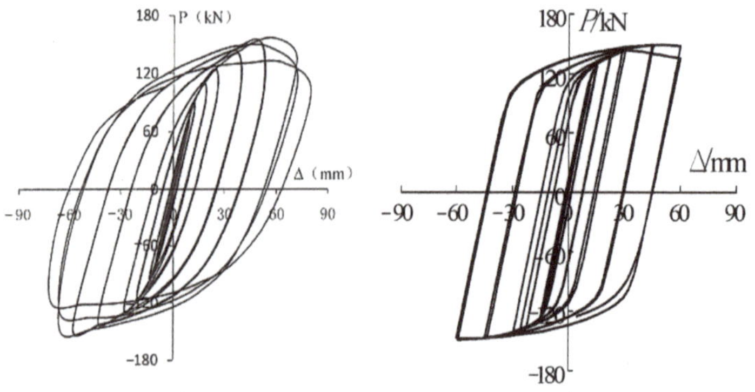

Figure 5. P-Δ curves of CPS-1 Specimen. (**a**) Test; (**b**) ABAQUS.

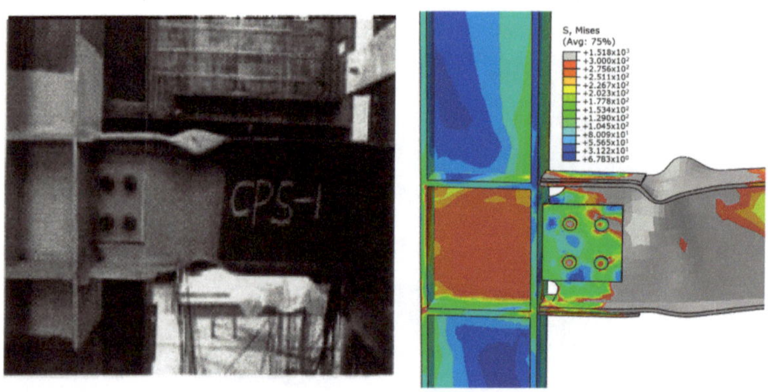

Figure 6. Failure modes of CPS-1 Specimen. (**a**) Test; (**b**) ABAQUS.

3. Specimen Design

The columns selected hot-rolled wide-flange H-shape HW 502 × 470 × 20 × 25, and the chosen beams hot-rolled wide-flange H-shape HW 350 × 350 × 12 × 19, hot-rolled middle-flange H-shape HM 390 × 300 × 10 × 16 and HM 340 × 250 × 9 × 14 to build analysis specimens. The research designed different specimens according to the research objectives, as shown in Tables 3 and 4. The research employed three different shapes of top cover plates (trapezoid, wedge, and rectangular) for beams with three different cross-sections, as shown in Figure 7. The researcher uniformly designed the bottom cover plates as rectangular; see the specific data for each specimen as shown in Tables 3 and 4. Explanation of specimen naming: "CP" denotes specimens with cover plates, "WCP" denotes specimens without cover plates, the "CPS" series represents specimens comparing variations in top cover-plate shapes, the "CPL" series represents specimens with varying lengths of rectangular cover plates, and the "CPT" series represents specimens with varying thicknesses of rectangular cover plates.

Table 3. Details of CP and WCP specimens.

Group No.	Spe.	Beam Size	Beam Length (mm)	Column Size	Column Height (mm)	Cover Plate		
						Size of the Cover Plate		t_{cp}/t_{bf}
						Top	Bottom	
1st group	CP1	HW 350 × 350 × 12 × 19	3000	HW 502 × 470 × 20 × 25	3600	Figure 7(aii)	365 × 370 × 6	0.32
	WCP1	HW 350 × 350 × 12 × 19		HW 502 × 470 × 20 × 25		-	-	-
2nd group	CP2	HM 390 × 300 × 10 × 16	3000	HW 502 × 470 × 20 × 25	3600	Figure 7(bii)	315 × 320 × 6	0.38
	WCP2	HM 390 × 300 × 10 × 16		HW 502 × 470 × 20 × 25		-	-	-
3rd group	CP3	HM 340 × 250 × 9 × 14	3000	HW 502 × 470 × 20 × 25	3600	Figure 7(cii)	265 × 270 × 6	0.24
	WCP3	HM 340 × 250 × 9 × 14		HW 502 × 470 × 20 × 25		-	-	-

Table 4. Full details of CPS, CPL, and CPT series specimens.

Member	Section of Column	Section of Beam	Bottom Cover Plate	Top Cover Plate	Beam Flange Thickness (mm)	Cover-Plate Thickness/Beam Flange Thickness
CPS-1-1			365 × 370 × 6			
CPS-1-2		HW 350 × 350 × 12 × 19	365 × 370 × 6	Figure 7a	-	-
CPS1-3			365 × 370 × 6			
CPS-2-1			315 × 320 × 6			
CPS-2-2	HW 502 × 470 × 20 × 25	HM 390 × 300 × 10 × 16	315 × 320 × 6	Figure 7b	-	-
CPS-2-3			315 × 320 × 6			
CPS-3-1			265 × 270 × 6			
CPS-3-2		HM 340 × 250 × 9 × 14	265 × 270 × 6	Figure 7c	-	-
CPS-3-3			265 × 270 × 6			
CPL-1-1			335 × 370 × 6	335 × 330 × 6		
CPL-1-2		HW 350 × 350 × 12 × 19	365 × 370 × 6	365 × 330 × 6	-	-
CPL1-3			395 × 370 × 6	395 × 330 × 6		
CPL-2-1			285 × 320 × 6	285 × 280 × 6		
CPL-2-2	HW 502 × 470 × 20 × 25	HM 390 × 300 × 10 × 16	315 × 320 × 6	315 × 280 × 6	-	-
CPL-2-3			345 × 320 × 6	345 × 280 × 6		
CPL-3-1			235 × 270 × 6	235 × 230 × 6		
CPL-3-2		HM 340 × 250 × 9 × 14	265 × 270 × 6	265 × 230 × 6	-	-
CPL-3-3			295 × 270 × 6	295 × 230 × 6		
CPT-1-1			365 × 370 × 4	365 × 330 × 4	19	0.21
CPT-1-2		HW 350 × 350 × 12 × 19	365 × 370 × 6	365 × 330 × 6	19	0.32
CPT1-3			365 × 370 × 8	365 × 330 × 8	19	0.42
CPT-2-1			315 × 320 × 4	315 × 280 × 4	16	0.25
CPT-2-2	HW 502 × 470 × 20 × 25	HM 390 × 300 × 10 × 16	315 × 320 × 6	315 × 280 × 6	16	0.38
CPT-2-3			315 × 320 × 8	315 × 280 × 8	16	0.50
CPT-3-1			265 × 270 × 4	265 × 230 × 4	14	0.29
CPT-3-2		HM 340 × 250 × 9 × 14	265 × 270 × 6	265 × 230 × 6	14	0.43
CPT-3-3			265 × 270 × 8	265 × 230 × 8	14	0.57

To facilitate the analysis and description that follows, several geometric parameters are defined as follows: b_{cp} is the width of the cover plate, t_{cp} is the thickness of the cover plate, b_f is the width of the beam flange, t_{bf} is the thickness of the beam flange, l_{cp} is the length of the cover plate, d_b is the beam depth, b_{cp1} is the width of the top cover plate, and b_{cp2} is the width of the bottom cover plate. Several mechanical properties of the joints are defined as follows: R_k represents the initial rotational stiffness. M_y, M_{max}, and M_u denote the yield, peak, and terminated moments, respectively, with corresponding rotational displacements θ_y, θ_{max}, and θ_u. According to FEMA 350 [6], θ_p is defined as the plastic rotation capacity.

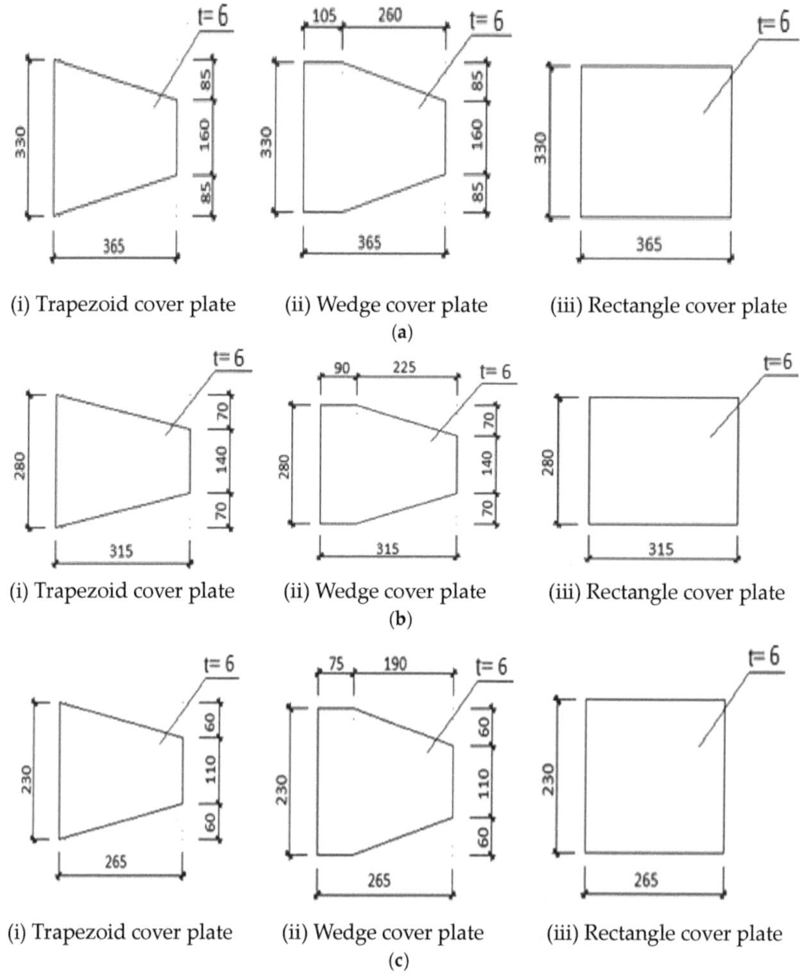

Figure 7. Various cover-plate dimensions (Unit: mm). (**a**) Cover plates for HW 350 × 350 × 12 × 19; (**b**) cover plates for HM 390 × 300 × 10 × 16; (**c**) cover plates for HM 340 × 250 × 9 × 14.

4. Analysis of the Reinforcement Effect of Cover Plates

The failure modes of the three groups of CP and WCP specimens are shown in Figure 8. The hysteresis curves and skeleton curves of the joints are shown in Figures 9–11, respectively. The mechanical properties calculated from the skeleton curves are listed in Table 5.

Table 5. Mechanical properties of CP and WCP specimens.

Group No.	Member	R_k (kN.m/rad)	M_y (kN.m)	θ_y (rad)	M_{max} (kN.m)	θ_{max} (rad)	Mu (kN.m)	θ_u (rad)	θ_p
1st group	CP-1	66,736.88	790	0.021	984.79	0.043	837.08	0.072	0.068
	WCP-1	62,311.39	704	0.020	872.29	0.044	741.45	0.059	0.055
2nd group	CP-2	64,133.85	627	0.014	774.97	0.032	658.72	0.062	0.058
	WCP-2	59,981.84	568	0.015	699.45	0.030	594.53	0.053	0.049
3rd group	CP-3	38,794.91	440	0.023	546.91	0.044	464.88	0.054	0.065
	WCP-3	36,308.14	355	0.016	443.29	0.032	376.79	0.054	0.049

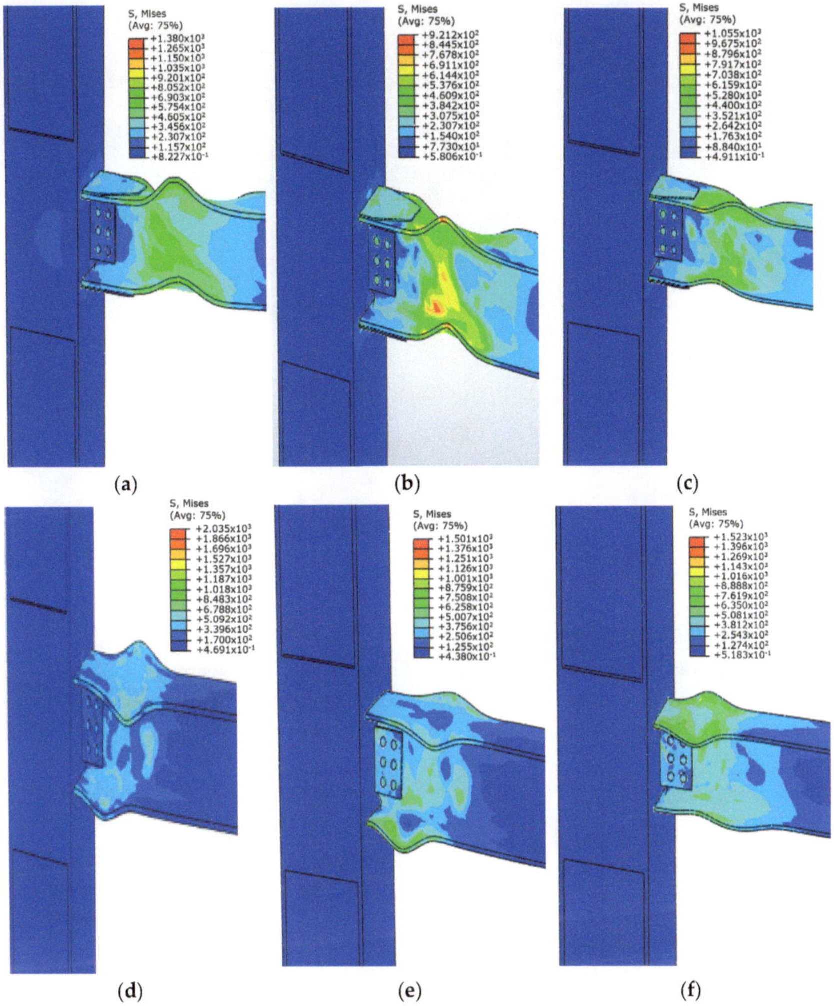

Figure 8. Failure modes of CP and WCP specimens. (**a**) CP-1; (**b**) CP-2; (**c**) CP-3; (**d**) WCP-1; (**e**) WCP-2; (**f**) WCP-3.

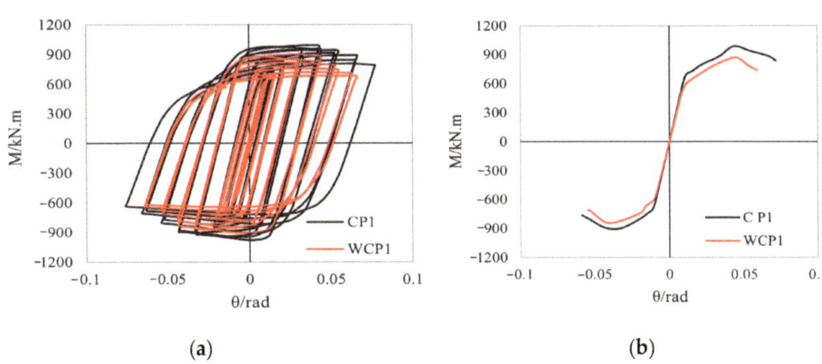

Figure 9. M-θ curves for the 1st group. (**a**) Hysteresis curve; (**b**) skeleton curve.

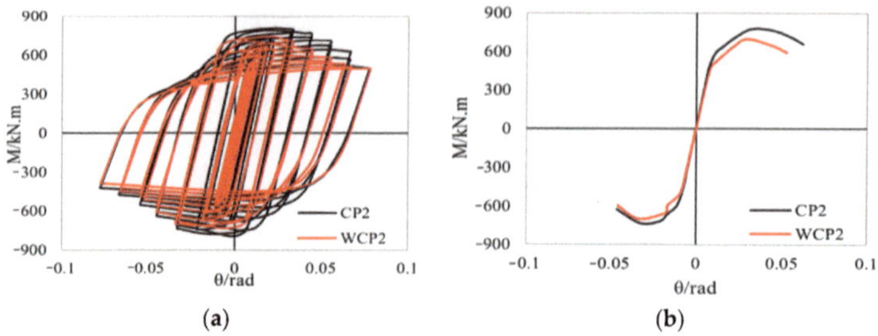

Figure 10. M-θ curves for the 2nd group. (**a**) Hysteresis curve; (**b**) skeleton curve.

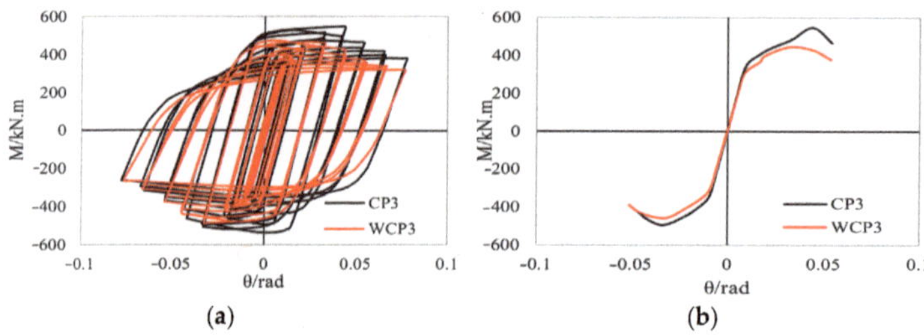

Figure 11. M-θ curves for the 3rd group. (**a**) Hysteresis curve; (**b**) skeleton curve.

As shown in Figure 8, compared with the plastic hinge formation locations in the corresponding WCP specimens, all CP joints enable the beam to form plastic hinges at the outer ends of the cover plates. This indicates that, despite the widened beam flanges, the appropriate configuration of cover plates successfully achieved the intended relocation of plastic hinges. Notably, in the CP-3 specimen, the cover-plate thickness was less than one-fourth of the beam flange thickness, yet it still effectively shifted the plastic hinge outward.

Figures 9–11 illustrate the hysteresis and skeleton curves for CP and WCP specimens. Through the comparison of the hysteresis curves, it can be observed that the curves of the CP specimens are more pronounced than those of the corresponding WCP specimens, indicating that the CP specimens have greater energy dissipation capacity. The comparison of the skeleton curves shows that the elastic stiffness of the CP specimens is similar to that of the corresponding WCP specimens, and the yield rotational deformation is also close. However, the three load indicators (yield moment, peak moment, and terminated moment) of the CP specimens are all higher than those of the corresponding WCP specimens, and the terminated rotational deformation is larger. These results indicate that the CP specimens have better load-bearing capacity and ductility than the WCP specimens, demonstrating that the cover-plate reinforced connection, as one of the post-Northridge connections, has excellent seismic performance.

M_y, M_{max}, and M_u can be calculated through the skeleton curve. Calculations of θ_y, θ_{max}, and θ_u are similar to calculations of the displacements described by Shim et al. [39]. The calculation method for θ_p follows the procedure from the author's previous research [16]. The calculated data for each parameter of the joints are listed in Table 5. Data in Table 5 show that the initial rotational stiffness of the three CP specimen groups increased by approximately 7% on average compared to the WCP specimens. The yield moment My, a key design control parameter, showed an average increase of about 15%. However, in terms of plastic rotational capacity θ_p, which measures the seismic ductility of the joints, the CP specimens achieved an average increase of 25% over the WCP specimens. These results reveal that the cover plates and the strengthened joint regions in this study's middle- and wide-flange hot-rolled H-beam joints effectively enhanced load-bearing capacity and improved seismic ductility, demonstrating high ductility and load-bearing capacity characteristics [40]. The improvement in the load-bearing capacity and ductility of the cover-plate reinforced joints compared to those without cover plates primarily results from the cover plate transferring the plastic hinge outside the beam–column connection, reducing stress concentration and enhancing both load-bearing capacity and ductility. Additionally, the cover plate increases the stiffness and stability of the joint, reducing local buckling and instability while improving the overall bending capacity and delaying yielding and failure. As a result, the cover plate significantly enhances the joint's load-bearing capacity and ductility, improving the joints' seismic performance.

5. Effect Parameters Analysis of Cover Plates and Discussion

5.1. The Influences of Top Cover-Plate Shape

To facilitate welding, the top cover plate is designed to be narrower than the top flange, while the bottom cover plate is generally wider than the bottom flange. There are three types of top cover plates—trapezoidal, wedge, and rectangular—whereas the bottom cover plate is typically rectangular. To examine the effects of top cover-plate shape on the cyclic response of cover-plate connections, three group CPS series of joints were utilized to analyze the effects (shown in Table 4). The geometric dimensions of the three top cover-plate types are provided in Figure 7. The lower cover plates are all rectangular, with dimensions listed in Table 4.

The failure modes of CPS series joints are illustrated in Figure 12, and the analysis results are summarized in Table 6. As shown in Figure 12 and Table 6, the variation in the top cover-plate shape had a minimal impact on initial rotational stiffness (the average fluctuation amplitude is less than ±1%), load-bearing capacity (the average fluctuation amplitude is less than ±5%), and ductility (the average fluctuation amplitude of θ_p is about ±5%). Additionally, all three group joints' maximum plastic rotation angles exceeded 0.03 rad, meeting FEMA 267 [29]. Therefore, altering the shape of the top cover plate had minimal impact on the mechanical behavior of cover-plate joints, whether the joints used I-section columns with middle-flange or wide-flange beams. However, trapezoidal and wedge-shaped cover plates can reduce the total length of filet welds and lessen the welding heat effect on the beam flange steel. Conversely, rectangular cover plates offer greater ease of fabrication.

Table 6. Numerical analysis results of all CPS, CPL, and CPT series specimens.

Member	R_k (kN.m/rad)	M_y (kN.m)	θ_y (rad)	M_{max} (kN.m)	θ_{max} (rad)	M_u (kN.m)	θ_u (rad)	θ_p (rad)
CPS-1-1	66,583.86	805	0.0216	971.05	0.0423	825.39	0.073	0.069
CPS-1-2	66,736.88	790	0.0212	984.79	0.0434	837.08	0.072	0.068
CPS1-3	66,995.35	728	0.0152	922.89	0.0432	784.46	0.069	0.065
CPS-2-1	64,001.49	623	0.0142	769.54	0.0445	654.11	0.065	0.061
CPS-2-2	64,133.85	627	0.0142	774.97	0.0319	658.72	0.062	0.058
CPS-2-3	64,358.96	632	0.0143	779.03	0.0425	662.18	0.064	0.060
CPS-3-1	38,721.33	455	0.022	539.72	0.0444	458.76	0.055	0.051
CPS-3-2	38,794.91	440	0.02285	546.91	0.0440	464.88	0.054	0.050
CPS-3-3	38,937.06	430	0.01885	523.5	0.0446	444.97	0.051	0.047
CPL-1-1	66,677.97	720	0.0152	921.81	0.0445	783.54	0.067	0.063
CPL-1-2	66,995.35	728	0.015	922.89	0.0432	784.46	0.069	0.065
CPL1-3	67,306.95	820	0.0197	976.31	0.0445	829.86	0.060	0.056
CPL-2-1	64,011.88	628	0.0145	768.2	0.0308	652.97	0.055	0.050
CPL-2-2	64,358.96	635	0.015	779.03	0.0425	662.18	0.064	0.060
CPL-2-3	64,700.05	640	0.0144	788.19	0.0437	669.97	0.068	0.064
CPL-3-1	38,677.23	405	0.0177	508.08	0.0445	431.87	0.050	0.046
CPL-3-2	38,937.06	430	0.0189	523.5	0.0446	444.97	0.051	0.047
CPL-3-3	39,194.10	395	0.0143	497.41	0.0438	422.8	0.055	0.051

Table 6. *Cont.*

Member	R_k (kN.m/rad)	M_y (kN.m)	θ_y (rad)	M_{max} (kN.m)	θ_{max} (rad)	M_u (kN.m)	θ_u (rad)	θ_p (rad)
CPT-1-1	65,600.07	712	0.014	889.75	0.0438	756.29	0.061	0.056
CPT-1-2	66,263.70	724	0.0148	919.34	0.0445	781.44	0.064	0.050
CPT-1-3	66,860.34	725	0.0148	921.46	0.0427	783.24	0.073	0.048
CPT-2-1	63,091.55	610	0.0132	755.72	0.0446	642.36	0.064	0.059
CPT-2-2	63,695.87	608	0.0129	756.95	0.0332	643.41	0.05	0.056
CPT-2-3	64,228.23	612	0.0131	763.67	0.0333	649.44	0.056	0.051
CPT-3-1	38,198.82	415	0.0191	519.99	0.0446	441.99	0.056	0.052
CPT-3-2	38,548.34	388	0.0146	482.89	0.044	410.46	0.053	0.048
CPT-3-3	38,850.70	436	0.0186	514.58	0.0436	437.40	0.052	0.048

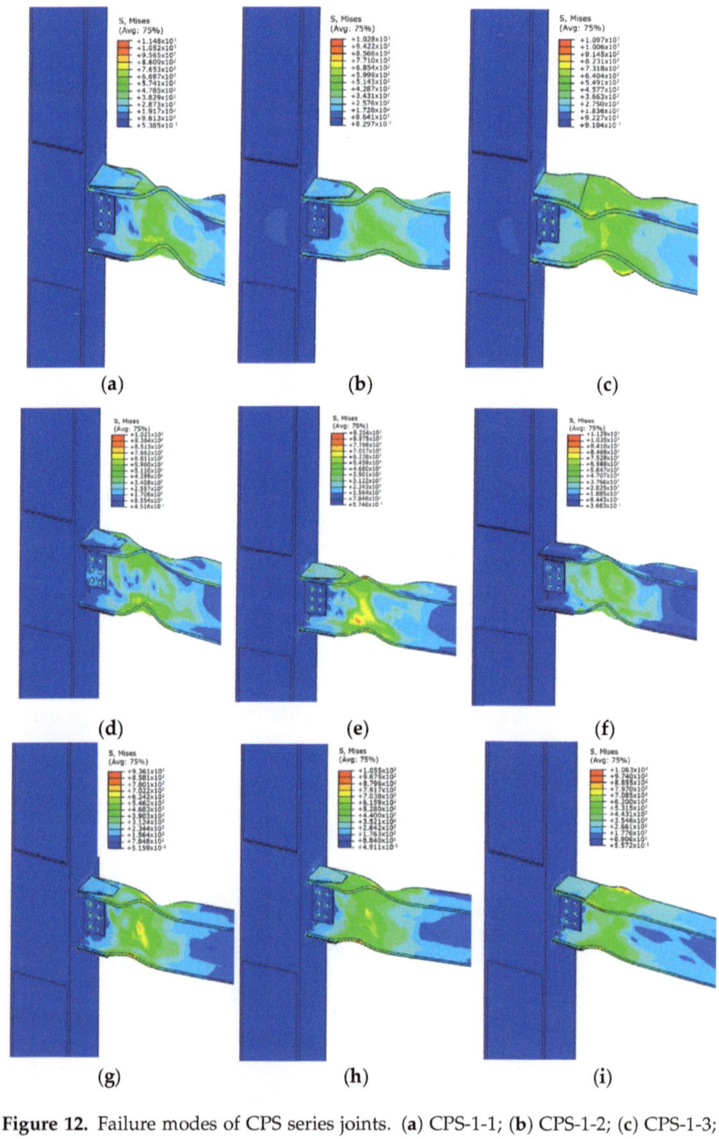

Figure 12. Failure modes of CPS series joints. (**a**) CPS-1-1; (**b**) CPS-1-2; (**c**) CPS-1-3; (**d**) CPS-2-1; (**e**) CPS-2-2; (**f**) CPS-2-3; (**g**) CPS-3-1; (**h**) CPS-3-2; (**i**) CPS-3-3.

5.2. The Influences of Cover-Plate Length

FEMA 267 [29] recommends the minimum cover-plate length be at least half of the beam depth for cover-plate moment connections, and it is the same as BS EN 1998-3 [12]. JGJ99-2015 [13] recommends a range of 0.5 to 0.75 times the beam depth. To better understand the mechanical behavior of cover-plate connections with varying cover-plate lengths, series joints were designed, as shown in Table 4, as follows: CPL-1: l_{cp} = (0.96~1.12) d_b, CPL-2: l_{cp} = (0.73~0.88) d_b, and CPL-3: l_{cp} = (0.69~0.87) d_b. Following the conclusions in Section 5.1, the rectangular cover plates were selected for all specimens for convenience, with specific details provided in Table 4.

The failure modes of the CPL series joints are illustrated in Figure 13, with numerical analysis results summarized in Table 6. As shown in Figure 13, plastic hinges were successfully formed in the CPL-2 and CPL-3 series specimens outside the cover plate. In contrast, for the CPL-1 series specimens, except for CPL-1-1, the cover plates of CPL-1-2 and CPL-1-3 were located within the plastic hinge region and exhibited significant deformation. A cover plate that is too long may extend into the plastic hinge region of the beam, causing considerable deformation. This is mainly due to the excessive bending moment in the plastic hinge region, which leads to bending and local yielding. Additionally, stress concentration and deformation amplification effects are intensified. Furthermore, an overly long cover plate may lead to local buckling and uneven plastic deformation, further increasing the deformation. Therefore, for cover-plate connections using hot-rolled H-shaped beams with middle or wide flanges, the observations indicate that the cover-plate length should not exceed 0.9 times the beam depth. Additionally, based on the analysis data for the CP-2 and CP-3 series specimens in Table 6, it was found that variations in cover-plate length (within 0.90 times the beam depth) had a minor effect on the initial rotational stiffness and bearing capacity of the joints, with fluctuations within ±3%. However, as the length increased, the plastic rotational capacity θ_p increased significantly, with an improvement exceeding 10%. This indicates that changes in cover-plate length have a notable impact on the seismic ductility of the joints.

Based on the observations and data analysis, a reasonable cover-plate length, 0.70~0.90 times the beam depth, is recommended for better ductility and plastic rotation capacity.

5.3. The Influences of Cover-Plate Thickness

For cover-plate connections, BS EN 1998-3 [12] requires the cover-plate thickness to be 1.2 times the beam flange thickness, while JGJ99-2015 [13] specifies that it should be no less than the beam flange thickness. FEMA 267 [29] requires that the combined thickness of the cover plate and beam flange must exceed twice the beam flange thickness but remain less than the column flange thickness. However, from the perspective of ensuring a strong-column–weak-beam mechanism, the FEMA 267 [29] requirement that the combined thickness of the cover plate and beam flange must not exceed the thickness of the column flange is more critical. This implies that the BS EN 1998-3 [12] and JGJ99-2015 [13] are not comprehensive enough, especially when the cover-plate thickness must be smaller than the beam flange thickness. In fact, China's national architectural design standard atlas 16G519 [41] only specifies that the cover-plate thickness should be greater than or equal to 6mm without following the provisions of the JGJ99-2015 [13]. Therefore, all specimens designed in this study adhered to the principle of FEMA 267 [29]. Notably, the CPS and CPL series specimens analyzed earlier demonstrated an ideal plastic hinge failure mode in the beams. This means that using a cover plate much thinner than the beam flange thickness can still ensure the characteristics of the reinforced connection. The beam plastic hinge is easily formed and located far from the column flange.

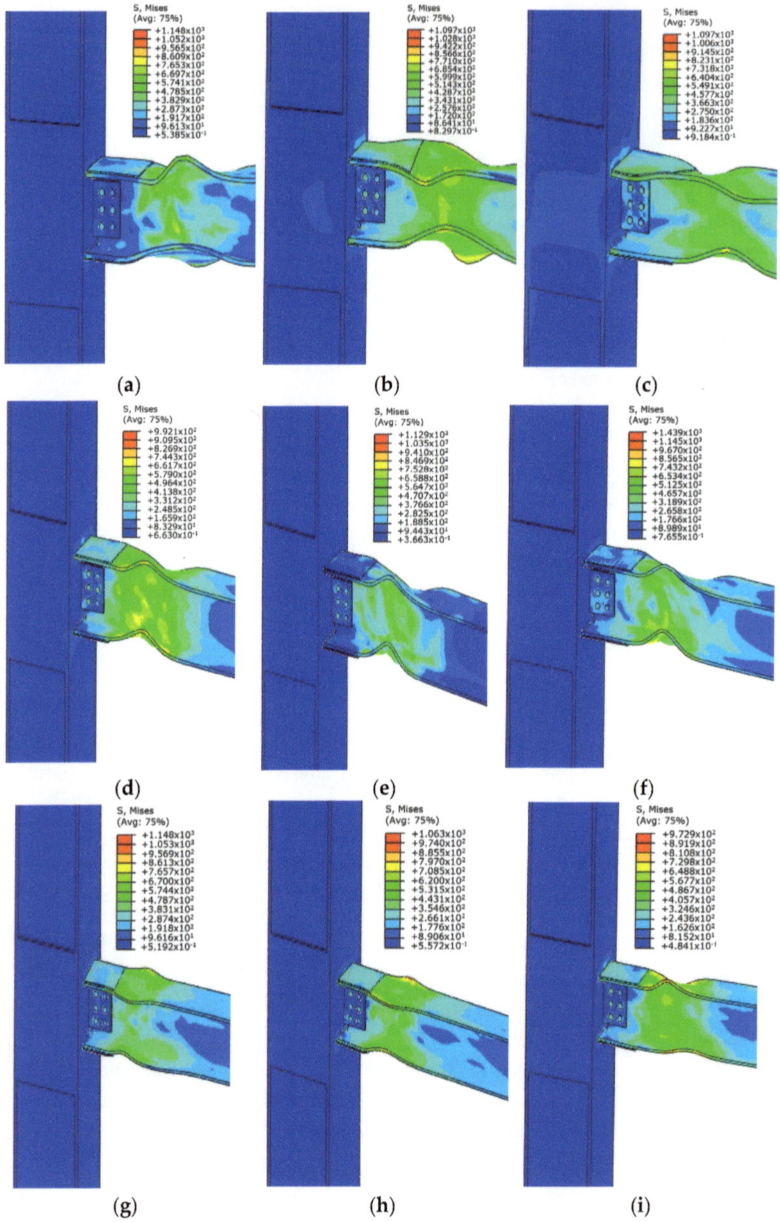

Figure 13. Failure modes of CPL series joints. (**a**) CPL-1-1; (**b**) CPL-1-2; (**c**) CPL-1-3; (**d**) CPL-2-1; (**e**) CPL-2-2; (**f**) CPL-2-3; (**g**) CPL-3-1; (**h**) CPL-3-2; (**i**) CPL-3-3.

From another perspective, this suggests that the thickness requirements for cover plates in BS EN 1998-3 [12] and JGJ99-2015 [13] may be overly conservative. In practice, excessively thick plates necessitate very thick groove welds to connect the beam flange ends to the column flanges, increasing the risk of weld defects and potential fracture failure during earthquakes. Therefore, the CPT series specimens were designed to explore the minimum feasible thickness for cover plates, as detailed in Table 4.

Increasing the cover-plate thickness enhances the connection's initial rotational stiffness and bearing capacity. However, its effects on joint plastic rotation capacity require quantification through numerical analysis. As shown in Figure 14, the plastic hinge of specimen CPT-1-1 extended into the cover plate, while that of CPT-2-1 slightly reached the edge. The plastic hinges, for others, were formed outside the cover plate. By examining the ratio of cover-plate thickness to beam flange thickness (Table 4), it was found that when the cover-plate thickness is greater than or equal to 0.3 times the beam flange thickness, the combined thickness of the cover plate and beam flange is less than the column flange thickness, and cover-plate connections can achieve the performance of reinforced connections for hot-rolled H-shaped beams with middle or wide flanges. This significantly enhances the ability to extend plastic hinge formation outward. Additionally, the data in Table 6 show that for the CPT specimens in this study, the increase in cover-plate thickness improved the initial rotational stiffness and bearing capacity by less than 5%, which can be considered negligible. The joint's plastic rotational capacity reductions are more than 5%, warranting consideration of its impact. The mechanism behind this phenomenon can be explained as follows: increasing the cover-plate thickness increases the local stiffness of the beam end, thereby enhancing the connection's load-bearing capacity. However, as the stiffness increases, the plastic deformation capacity of both the cover plate and the beam end is restricted. Although the plastic hinge is pushed out to the cover plate, this also reduces plastic rotation capacity. In contrast, a thinner cover plate offers less deformation constraint due to its lower stiffness, causing the cover-plate region to undergo plastic deformation prematurely. This prevents the plastic hinge from moving outward while reducing the load-bearing capacity.

However, the plastic rotational capacity of all specimens exceeded 0.03 rad, which is the minimum rotational capacity requirement specified by FEMA 267 [29]. Based on the observations and data analysis, and considering the practical experience from the Chinese standard drawing set 16SG519 [41], it is recommended that the minimum thickness of the cover plate be 0.3 times the flange thickness of the beam, not less than 6mm. Additionally, the total thickness of the cover plate and beam flange should not exceed the thickness of the column flange, considering the balance between the connection's stiffness and ductility.

5.4. Discussion

This study presents a comprehensive analysis of cover-plate connections for hot-rolled H-shaped beams with middle or wide flanges, comparing the results with existing design guidelines, including FEMA 267 [29], BS EN 1998-3 [12], JGJ99-2015 [13], and China's national standard 16SG519 [41]. This research provides valuable insights into the behavior of cover-plate connections, particularly regarding the influences of cover-plate length and thickness.

1. Cover-Plate Length:

The study's findings suggest that cover-plate length significantly impacts the connection's seismic ductility. Specifically, the research recommends a cover-plate length of up to 0.90 times the beam depth to achieve optimal ductility and bearing capacity, aligning with the observations that cover plates longer than this can induce excessive deformation and stress concentration. This is consistent with the guidelines of FEMA 267 [29], BS EN 1998-3 [12], and JGJ99-2015 [13], which suggest a minimum cover-plate length of at least half the beam depth. However, the recommended cover-plate length of 0.70~0.90 times the beam depth in this study exceeds the upper limit of 0.75 times the beam depth specified by BS EN 1998-3 [12] and JGJ99-2015 [13]. This breakthrough is particularly suitable for hot-rolled H-shaped beams with middle or wide flanges, as it allows for better plastic rotation capacity and seismic ductility, which are critical for enhancing the overall performance of the connection.

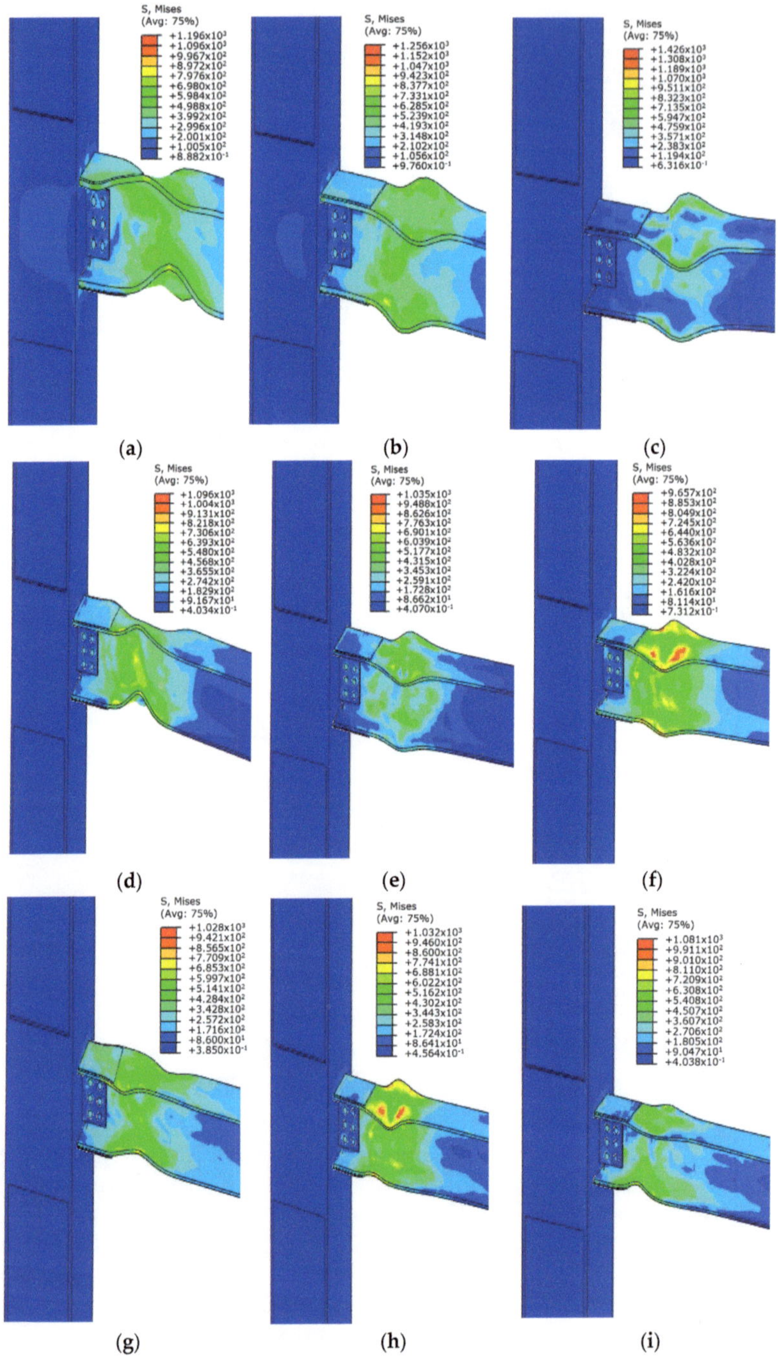

Figure 14. Failure modes of CPT series joints. (**a**) CPT-1-1; (**b**) CPT-1-2; (**c**) CPT-1-3; (**d**) CPT-2-1; (**e**) CPT-2-2; (**f**) CPT-2-3; (**g**) CPT-3-1; (**h**) CPT-3-2; (**i**) CPT-3-3.

2. Cover-Plate Thickness:

The research challenges the conservatism of current design standards regarding cover-plate thickness. While BS EN 1998-3 [12] and JGJ99-2015 [13] specify that the cover-plate thickness should be 1.2 times or 1.0 times the beam flange thickness, the study shows that a thinner cover plate, as long as it is at least 0.3 times the beam flange thickness and no less than 6mm, can still effectively reinforce the connection without compromising its performance. This suggests that the prescribed thickness in current standards may be overly conservative, leading to unnecessary material use and potential fabrication challenges.

The recommendation for cover-plate thickness in this study, although deviating from the provisions of BS EN 1998-3 [12] and JGJ99-2015 [13], aligns with FEMA 267 [29] and 16SG519 [41] and may be more suitable for the connections of middle-flange and wide-flange hot-rolled H-shaped steel beams studied in this research.

6. Conclusions

This research presents a comprehensive investigation into the impact of cover-plate geometry on the mechanical performance of steel frame connections, with a specific focus on joints using middle-flange and wide-flange H-shapes. The study contributes novel insights into the design of steel frame connections, particularly in improving seismic resilience and joint ductility. Key findings include the following:

1. Adopting cover-plate reinforced connections can effectively shift plastic hinge formation outward in frame connections using middle-flange or wide-flange H-shaped steel beams. This strategy significantly enhances the seismic performance and ductility of the joint. Compared to WCP specimens, CP specimens increased ductility by approximately 25% and improved initial rotational stiffness and yield moment by 7% and 15%, respectively.
2. Adjusting the length of the cover plate impacts seismic ductility. A reasonable cover-plate length, 0.70~0.90 times the beam depth, is recommended for better ductility and plastic rotation capacity, although it exceeds the upper limit of 0.75 times the beam depth specified by BS EN 1998-3 and JGJ99-2015.
3. The cover-plate thickness plays a crucial role in seismic ductility. This study recommends that the cover-plate thickness be at least 0.3 times the beam flange thickness (not less than 6mm) and that the combined thickness of the cover plate and beam flange not exceed the column flange thickness to maintain a balance between connection stiffness and ductility, even though this does not comply with the provisions of BS EN 1998-3 and JGJ99-2015.

For further investigating the application of cover-plate connections in modern steel frames, future research could focus on the performance of cover-plate reinforced connections specifically for middle- and wide-flange H-shaped steel beams, considering the effects of slab interaction under various loading conditions, to further refine design guidelines and enhance their applicability.

Author Contributions: Conceptualization, L.L.; software, O.Z.Y.A.-a.; validation, O.Z.Y.A.-a., S.M.A.A.A.-S. and B.L.; resources, L.L.; writing—original draft preparation, O.Z.Y.A.-a. and L.L.; writing—review and editing, L.L., O.Z.Y.A.-a., S.M.A.A.A.-S. and B.L.; project administration, L.L.; funding acquisition, L.L. All authors have read and agreed to the published version of the manuscript.

Funding: This research was funded by the Nature Science Foundation of China (NSFC), grant number 51278061.

Data Availability Statement: The original contributions presented in the study are included in the article, further inquiries can be directed to the corresponding author.

Conflicts of Interest: The authors declare that they have no conflict of interests.

References

1. Kinki, A.I.J. *Reconnaissance Report on Damage to Steel Building Structures Observed from the 1995 Hyogoken-Nanbu Earthquake;* Steel Committee of Kinki Branch, Architectural Institute of Japan: Osaka, Japan, 1995.
2. Mahin, S.A. Lessons from damage to steel buildings during the Northridge earthquake. *Eng. Struct.* **1998**, *20*, 261–270. [CrossRef]
3. Muguruma, H.; Nishiyama, M.; Watanabe, F. Lessons learned from the Kobe earthquake—A Japanese perspective. *Pci J.* **1995**, *40*, 28–42. [CrossRef]
4. Malley, J.O.; Carter, C.J.; Saunders, C.M. Seismic design guidelines and Provisions for steel-framed buildings: FEMA 267/267A and 1997 AISC Seismic Provisions. *Earthq. Spectra* **2000**, *16*, 179–203. [CrossRef]
5. Simoncelli, M.; Zucca, M.; Stochino, F. Fire Resistance of Steel Rack Frames: Assessment, Reinforcement and Collapse Mitigation Strategies. *Fire Technol.* **2024**, 1–23. [CrossRef]
6. *FEMA-350;* Recommended Seismic Design Criteria for New Steel Moment-Frame Buildings. FEMA: Washington, DC, USA, 2000.
7. Nakashima, M.; Roeder, C.W.; Maruoka, Y. Steel moment frames for earthquakes in United States and Japan. *J. Struct. Eng.* **2000**, *126*, 861–868. [CrossRef]
8. Engelhardt, M.D.; Sabol, T.A. Reinforcing of steel moment connections with cover plates: Benefits and limitations. *Eng. Struct.* **1998**, *20*, 510–520. [CrossRef]
9. Kim, T.; Whittaker, A.S.; Gilani, A.S.J.; Bertero, V.V.; Takhirov, S.M. Experimental evaluation of plate-reinforced steel moment-resisting connections. *J. Struct. Eng.* **2002**, *128*, 483–491. [CrossRef]
10. *FEMA-351;* Recommended Seismic Evaluation and Upgrade Criteria for Existing Welded Steel Moment-Frame Buildings. FEMA: Washington, DC, USA, 2000.
11. *FEMA-355D;* State of the Art Report on Connection Performance. FEMA: Sacramento, CA, USA, 2000.
12. *BS EN 1998-3;* Eurocode 8: Design of Structures for Earthquake Resistance Part 3: Assessment and Retrofitting of Buildings. British Standards Institution: London, UK, 2006.
13. *JGJ99-2015;* Technical Specification for Steel Structures of Tall Buildings. China Architecture & Building Press: Beijing, China, 2015.
14. *GB 50011-2010;* Code for Seismic Design of Buildings. China Architecture & Building Press: Beijing, China, 2010.
15. Lu, L.; Xu, Y.; Zheng, H. Investigation of composite action on seismic performance of weak-axis column bending connections. *J. Constr. Steel Res.* **2017**, *129*, 286–300. [CrossRef]
16. Lu, L.; Xu, Y.; Lim, J.B.P. Mechanical performance of a new I-section weak-axis column bending connection. *Steel Compos. Struct.* **2018**, *26*, 31–44.
17. Lu, L.; Xu, Y.; Zheng, H.; Lim, J.B.P. Cyclic response and design procedure of a weak-axis cover-plate moment connection. *Steel Compos. Struct.* **2018**, *26*, 329–345.
18. Chen, X.; Shi, G. Experimental study on seismic behaviour of cover-plate joints in high strength steel frames. *Eng. Struct.* **2019**, *191*, 292–310. [CrossRef]
19. Jiang, S.; Shi, G.; Zhang, N.; Zhao, H.; Sun, T. Experimental study on seismic behavior of cover-plate connections between steel beams and high strength steel box columns. *Thin-Walled Struct.* **2023**, *193*, 111208. [CrossRef]
20. Casita, C.B.; Suswanto, B.; Iranata, D.; Matsumura, M. Numerical Investigation to Assess the Cyclic Performance of RBS Connection using Cover Plates. *E3S Web Conf.* **2023**, *434*, 02017. [CrossRef]
21. Meng, B.; Du, Q.; Li, F.; Duan, S.; Li, L. Seismic performance of steel connections with double-leg energy dissipation cover plates. *Structures* **2022**, *43*, 1945–1961. [CrossRef]
22. Ribeiro, T.; Bernardo, L.; Carrazedo, R.; De Domenico, D. Eurocode-compliant topology optimisation and analysis of a steel cover-plate in a splice moment connection. *Mater. Today Proc.* **2022**, *65*, 1056–1063. [CrossRef]
23. Nassiraei, H. Probabilistic analysis of strength in retrofitted X-Joints under tensile loading and fire conditions. *Buildings* **2024**, *14*, 2105. [CrossRef]
24. Zhai, S.Y.; Lyu, Y.F.; Cao, K.; Li, G.Q.; Wang, W.Y.; Chen, C. Experimental study on bolted-cover plate corner connections for column-supported modular steel buildings. *J. Constr. Steel. Res.* **2022**, *189*, 107060. [CrossRef]
25. Zhai, S.Y.; Lyu, Y.F.; Cao, K.; Li, G.Q.; Wang, W.Y.; Chen, C. Experimental-numerical investigation and design of bolted-cover plate connections under bending for modular steel buildings. *J. Build. Eng.* **2023**, *75*, 107057. [CrossRef]
26. Cao, K.; Zhai, S.Y.; Lyu, Y.F.; Li, G.Q.; Wang, W.Y.; Chen, C. Working mechanism evaluations of full-scale joints with bolted-cover plate connection for modular steel buildings. *Thin-Walled Struct.* **2024**, *199*, 111772. [CrossRef]
27. Deng, E.F.; Lian, J.Y.; Zhang, Z.; Qian, H.; Zhang, G.C.; Zhang, P.; Sheikh, S.A. Axial mechanical behavior of an innovative liftable connection for modular steel construction. *Thin-Walled Struct.* **2023**, *182*, 110256. [CrossRef]
28. Deng, E.F.; Wang, Y.H.; Zong, L.; Zhang, Z.; Zhang, J.F. Seismic behavior of a novel liftable connection for modular steel buildings: Experimental and numerical studies. *Thin-Walled Struct.* **2024**, *197*, 111563. [CrossRef]

29. *FEMA-267;* Interim Guidelines: Evaluation, Repair, Modification and Design of Steel Moment Frames, Report No. SAC-95-02. FEMA: Washington, DC, USA, 1995.
30. Gemechu, T.D.; Lu, L. Analysis of Factors Affecting the Seismic Performance of Widened Flange Connections in Mid-Flange H-Beams and Box Columns. *Buildings* **2024**, *14*, 3170. [CrossRef]
31. Valente, M. Numerical investigations of steel beam-to-column connections with reinforcing plates. *Appl. Mech. Mater.* **2012**, *234*, 78–83. [CrossRef]
32. Valente, M.; Castiglioni, C.A.; Kanyilmaz, A. Dissipative devices for earthquake resistant composite steel structures: Bolted versus welded solution. *Bull. Earthq. Eng.* **2016**, *14*, 3613–3639. [CrossRef]
33. Nie, W.; Lu, L.; Li, R.; Hao, H.; Luo, T. Global buckling investigation of T-shaped built-up columns composed of two hot-rolled H-shapes. *Structures* **2023**, *58*, 105485. [CrossRef]
34. Katiana, K.; Apostolos, K.; Olympia, P. Numerical investigation of weak axis I profile connections. In Proceedings of the 9th Hellenic National Conference on Steel Structures, Larissa, Thessaly, Greece, 5–7 October 2017.
35. *ANSI/AISC 341-05;* Seismic Provisions for Structural Steel Buildings. American Institute of Steel Construction: Chicago, IL, USA, 2005.
36. *JGJ 82-2011;* Technical Specification for High Strength Bolt Connections of Steel Structures. China Architecture & Building Press: Beijing, China, 2011.
37. Lu, L.; Zhu, P.; Ding, S.; Ma, Z.; Li, R.; Nie, S.; Wang, W.; Raftery, G.M. Impact of anti-corrosion coatings and maintenance on high-strength bolt friction connections in C4 marine environment. *Structures* **2024**, *68*, 107098. [CrossRef]
38. Wang, P. Experimental Evaluation of Flange-Plate Reinforced and Cover-Plate Reinforced Steel Moment-Resistance Connections. Master's Thesis, Qingdao University of Technology, Qingdao, China, 2009.
39. Shim, H.J.; Lee, E.T.; Kim, S.B.; Kim, S.S. Development and performance evaluation of weak-axis column bending connections for advanced constructability. *Int. J. Steel Struct.* **2014**, *14*, 369–380. [CrossRef]
40. Lu, L.; Zhang, J.; Zhang, G.; Peng, H.; Liu, B.; Hao, H. The Influence of Box-Strengthened Panel Zone on Steel Frame Seismic Performance. *Buildings* **2023**, *13*, 3042. [CrossRef]
41. *16G519;* Detail Drawings of Steel Structure Joints for Multi-Story and High-Rise Civil Buildings. China Planning Press: Beijing, China, 2016.

buildings

MDPI

Article

Numerical Study of the Effect of Out-of-Plane Distance in the Lateral Direction at the Mid-Span of a Steel Beam on the Sectional Moment Capacity

Ahmed Mohamed Sayed [1,*], Nageh M. Ali [1,2], Mishal H. Aljarbou [1], Abdullah Alzlfawi [1], Salman Aldhobaib [1], Hani Alanazi [1] and Abdulmalik H. Altuwayjiri [1]

1 Department of Civil and Environmental Engineering, College of Engineering, Majmaah University, Al Majmaah 11952, Saudi Arabia; nageh82@mu.edu.sa (N.M.A.); m.aljarbou@mu.edu.sa (M.H.A.); a.alzlfawi@mu.edu.sa (A.A.); s.aldhobaib@mu.edu.sa (S.A.); hm.alanazi@mu.edu.sa (H.A.); a.altuwayjiri@mu.edu.sa (A.H.A.)

2 Department of Civil Engineering, College of Engineering, Assiut University, Assiut 71511, Egypt

* Correspondence: a.sayed@mu.edu.sa

Abstract: Steel I-beams may be subject to deviation from their normal path towards the lateral direction due to obstacles along their axis line. This deviation in the lateral direction, i.e., the out-of-plane distance, affects the behavior of the steel beams and may reduce their ultimate capacity. To obtain this effect, finite element modeling (FEM) was used to model these beams with and without an out-of-plane distance at the mid-span beam length with several different variables. These variables were the out-of-plane distance, cross-section dimensions, beam length, and steel yield stress. The reliability of using FEM simulation was confirmed by comparing the experimental test results of 25 available steel beams in previous studies. The results indicate the high accuracy of the simulation of this beam in terms of ultimate capacity, structural behavior, and deformation patterns. After verifying the results, 116 broad-flange I-beam (BFIB) steel beams with different out-of-plane distances were modeled. The results showed that using an out-of-plane distance equal to the flange width of the BFIB-300 cross-section caused a 60% decrease in the ultimate capacity. The reduction ratios in the ultimate moment capacity in out-of-plane steel beams were directly proportional to the out-of-plane distance, cross-sectional dimensions, and steel yield stress, while the beam length had no effect. Failure in beams containing an out-of-plane distance occurs as a result of a global buckling in the upper flange, which contains tensile stresses at the outer edge and compressive stresses at the inner edge, with stress concentration occurring at the point of contact of the out-of-plane part with the main beam. The prediction results of the design codes were compared with the results of experimental tests and the FEM analysis of the beams with and without out-of-plane distances. For all the beams with out-of-plane distances, all the design codes were unable to predict this ultimate capacity.

Keywords: BFIB steel beam; out-of-plane distance; sectional moment capacity; current design codes; finite element modeling

1. Introduction

Steel beams are the essential element when implementing covered areas or multi-story steel buildings. Normally, and in most cases, the axis of the steel beam in the longitudinal direction should be straight and in exactly one line. In some cases, such as the presence of irregular edges of the area to be covered or the presence of an obstacle in the longitudinal direction of the beam, a deviation occurs in the direction of its axis towards its lateral direction, which is called an out-of-plane distance. This deviation can have a significant effect on the efficiency of steel beams and their ultimate bending moment capacity. Many previous studies have taken many variables into account when analyzing and monitoring the effectiveness of steel beams [1–4]. These steel beams have been studied under the influence of different types of loads, such as flexural loads [5–8] and shear loads [9–12], lateral torsional buckling [13–17], cyclic loads [18–21], and axial compression loads [22], or under the influence of temperature [23–25]. Saliba et al. [26] analyzed and compared the results of 34 steel beams collected from experimental tests and those available in previous studies and the predictions of different versions of Eurocode 3 [27,28]. The comparison results showed that EN 1993-1-4 [27] was conservative in predicting the steel beam capacities, while EN 1993-1-5 [28] was more accurate. The average ratios between the test results and the code predictions were 1.22 and 1.13 for EN 1993-1-4 [27] and EN 1993-1-5 [28], respectively, with a coefficient of variation, COV, equal to 14% and 11%, respectively. In general, all the prediction values from the codes were less than the actual capacity of the steel beams, meaning they all fell on the safe side of the design; this is consistent with the same result in [29,30]. Graciano [31] conducted experimental tests on steel beams using a longitudinal web stabilizer in the compression zone. The results showed that placing a longitudinal stabilizer at a distance equal to 0.3 from the height of the web of the beam increased the ultimate strength of the beam; however, if the distance of the longitudinal stabilizer exceeded 40 times the thickness of the web, this did not affect the efficiency of the steel beam. Others have studied the presence of damage, cuts, or openings in the cross-section of the steel beam, whether in the web or the flange [19,32–35]. Sayed [19] conducted a numerical study on steel beams with damage to the upper and lower flanges of the cross-section. The results indicated that the presence of any percentage of damage to the upper flange directly affected the efficiency and ultimate capacity of these steel beams. The presence of damage at percentages of the flange width equal to 10% and 40% on one side of the upper flange resulted in reductions in the ultimate bending moment capacity of the beam of 3.1% and 27.1%, respectively. When these ratios occurred from both sides of the upper flange, it caused reductions in capacity equal to 4.9% and 42.5%, respectively. As for the damage in the lower flange, no effect appeared until the damage percentage was equal to 30% of the width of the flange on both sides. Hou et al. [36] conducted experimental tests on seven steel beam specimens with a complete cut at the lower edge of the cross-section at the mid-length of the beam—a control beam—and used damage to the web at rates equal to 15% and 28% of the web height at the cut location at the lower edge. It was found that the damage in the steel beam web with flange cuts in the tension zone directly affected the efficiency and ultimate capacity. Meanwhile, the damage rates of 15% and 28% caused reductions in the ultimate load capacity by rates of 75.76% and 57.54%, respectively.

Recently, the use of finite element modeling (FEM) simulation has become essential in modeling structural elements, as it saves time, effort, and costs when implementing these elements compared to experimental tests. It is also possible to implement some structural elements that cannot be implemented through experimental tests due to the large size of the models, the presence of danger during some measurements, or the time required to implement these tests. For these reasons, researchers have tended to use FEM simulation and have measured the accuracy and reliability of the modeling of various structural elements under many factors, variables, and different loads [37–40]. FEM analysis was used in modeling steel beams with and without a corrugated web [41–44] and cold-formed sections [45–48]. Perera and Mahendran [49] carried out FEM analysis on six steel beams that had already been studied experimentally. The comparison results showed excellent agreement in the deformed shapes, moment–deflection relationship, and ultimate capacity, with an average ratio of 1.03 with a COV of 2.0%. Tao et al. [50] performed FEM analysis on 340 specimens with different geometric dimensions and cross-sectional variations, where there were 44 rectangular, 142 circular, and 154 square sections of steel tubes. The results showed high accuracy with an average capacity ratio of 1.019 with a standard deviation of 0.072. This result is considered one of the most important conclusions due to the large number of specimens used in the verification. From all of this evidence, it is concluded that FEM simulation can model structural elements accurately and reliably to obtain different results in terms of behavior and ultimate capacity.

Due to the abnormal conditions in the implementation of steel I-beams, where there are deviations in the edges of the coverage area or there are obstacles in the longitudinal direction of the center line of the beam, there is a high probability that these beams will be implemented with an out-of-plane distance in the lateral direction of the beam axis. By reviewing the previous literature, it became clear that most of the factors and variables that can affect the behavior and efficiency of steel beams have been studied extensively, except for this type of steel beam, where there is an out-of-plane distance in the lateral direction of the beam length, whether by experimental study or numerical analysis. Perhaps the shortcoming in studying these steel beams with an out-of-plane distance was due to the lack of exposure of researchers or the engineering community to the implementation of this type of steel beam at the construction site, or it may have been due to uncertainty or a lack of knowledge of the extent of its impact on the efficiency of this type of structural elements. With the increasing need for change in the structural systems, whether already existing or new, especially regarding bridges in which the columns are made of reinforced concrete and the steel beams are of the simple support type, it is possible to be exposed to this type of steel beam with a deviation in the beam axis, which causes the presence of an out-of-plane distance during implementation. Here, the following question can be asked: Why conduct this study on steel beams with an out-of-plane distance? This study establishes the first basic principles for choosing between two methods: first, implementing such steel beams with an out-of-plane distance, taking into account the results that are obtained through this study; second, avoiding it and the consequences resulting from its use and using other solutions to treat this type of steel beam. This will only become clear after studying these steel beams with an out-of-plane distance.

Therefore, in this research, an FEM study is conducted on broad-flange I-beam (BFIB) steel beams with an out-of-plane distance in the lateral direction at the mid-span length under many variables that may affect the beam efficiency and behavior. These variables include out-of-plane distance, cross-section dimensions, beam length, and steel yield stress. To achieve the purpose of this study, it was divided into four main stages. The first stage is to verify the accuracy and reliability of using FEM modeling in the simulation of such types of structural elements by comparing the results of experimental tests of some available beams in previous studies with the results of FEM analysis in terms of ultimate capacity, structural behavior, and deformed shape. The second stage, after ensuring the reliability of the FEM analysis, is to conduct a simulation of BFIB steel beams with all the proposed variables to obtain the extent of their effect on the steel beam's efficiency and behavior. Third, these beam capacities are compared with the prediction results from current design codes [51,52] to determine the accuracy of their use in calculating the ultimate capacity of this type of steel beam. Fourth, an attempt is made to propose a new model that can take different variables into account to calculate the ultimate capacity of steel beams with out-of-plane distance.

2. Current Design Models

Many previous studies have presented new models or modified existing models to calculate the steel beam moment capacity to achieve the highest possible accuracy. These proposed new models are mostly based on existing design codes. By reviewing the previous studies, it became clear that most of the studies were concerned with ANSI/AISC 360-22 [51] and Eurocode-3 [52] in designing or comparing the results of experimental tests or FEM analysis. ANSI/AISC 360-22 provides models that can predict the ultimate bending moment capacity of steel beams with a value that is usually less than or equal to the actual ultimate capacity of the steel beam, i.e., the values are on the safe side of the design. In conducting the forecasting process, ANSI/AISC 360-22 relies on classifying the cross-sections of steel beams into three main categories: compact, noncompact, and slender. These classifications depend directly on the dimensions of the flange for the cross-section in terms of width and thickness, the dimensions of the beam web in terms of height and thickness, and the type of cross-section used to determine the appropriate limits for it. The ultimate capacity of the bending moment for steel beams can be calculated based on the critical classification of the cross-section categories in terms of the flange or web, as shown in the following equations:

For the beam web:

$$(\lambda = \frac{d_w}{t_w}) \leq (\lambda_P = 3.76\sqrt{\frac{E}{f_y}}) \quad \text{for the compact category} \tag{1}$$

$$3.76\sqrt{E/f_y} < \frac{d_w}{t_w} \leq 5.70\sqrt{E/f_y} \quad \text{for the noncompact category} \tag{2}$$

$$\frac{d_w}{t_w} > (\lambda_r = 5.70\sqrt{\frac{E}{f_y}}) \quad \text{for the slender category} \tag{3}$$

For the beam flange:

$$(\lambda = \frac{b}{t_f}) \leq (\lambda_P = 0.38\sqrt{\frac{E}{f_y}}) \quad \text{for the compact category} \tag{4}$$

$$0.38\sqrt{E/f_y} < b/t_f \leq 1.0\sqrt{E/f_y} \quad \text{for the noncompact category} \tag{5}$$

$$b/t_f > (\lambda_r = 1.0\sqrt{\frac{E}{f_y}}) \quad \text{for the slender category} \tag{6}$$

where d_w is beam web height, t_w is beam web thickness, t_f is flange thickness, $b = b_f/2$, b_f is total flange width, f_y is the steel yield stress, and E is the steel modulus of elasticity. After determining the category of the beam cross-section, the appropriate equation (Equations (7) to (9)) can be used to calculate the ultimate capacity of the bending moments.

$$M_P = f_y \times Z \quad \text{for the compact category} \tag{7}$$

$$M_{el} = M_P - (M_P - f_y \times S)\left(\frac{\lambda - \lambda_P}{\lambda_r - \lambda_P}\right) \leq M_P \quad \text{for the noncompact category} \tag{8}$$

$$M_{ef} = f_y \times S_{ef} \quad \text{for the slender category} \tag{9}$$

where M_p is the plastic moment capacity, M_{el} is the elastic moment capacity, M_{ef} is the effective moment capacity, Z is the plastic section modulus, S is the elastic section modulus, and S_{ef} is the effective section modulus.

As for Eurocode-3 [52], there are four design categories for the cross-section of the steel beam. These categories depend on the dimensions of the flange and the web of the cross-section in terms of the ratio between the width and thickness, the cross-section type, and the type of stresses affecting the section that determine the limits for each design category.

For the beam web:

$$Class\ 1: \ \frac{d_w}{t_w} \leq 72\ \sqrt{235/f_y} \tag{10}$$

$$Class\ 2: \ 72\ \sqrt{235/f_y} < \frac{b}{t_f} \leq \ 83\ \sqrt{235/f_y} \tag{11}$$

$$Class\ 3: \ 83\ \sqrt{235/f_y} < \frac{d_w}{t_w} \leq \ 124\ \sqrt{235/f_y} \tag{12}$$

$$Class\ 4: \ \frac{d_w}{t_w} > 124\ \sqrt{235/f_y} \tag{13}$$

For the beam flange:

$$Class\ 1: \ \frac{b}{t_f} \leq 9\ \sqrt{235/f_y} \tag{14}$$

$$Class\ 2: \ 9\ \sqrt{235/f_y} < \frac{b}{t_f} \leq \ 10\ \sqrt{235/f_y} \tag{15}$$

$$Class\ 3: \ 10\ \sqrt{235/f_y} < \frac{b}{t_f} \leq \ 14\ \sqrt{235/f_y} \tag{16}$$

$$Class\ 4: \ \frac{b}{t_f} > 14\ \sqrt{235/f_y} \tag{17}$$

Class-1 and Class-2 are similar to Equation (7), which uses a plastic section to calculate the ultimate beam moment capacity according to ANSI/AISC 360-22. In Class-4, the beam bending moment is calculated as in the previous Equation (9), while the elastic section in Equation (18) for Class-3 is used to calculate the ultimate beam moment capacity.

$$M_{el} = f_y \times S \tag{18}$$

3. Finite Element Model Study

3.1. Modeling and Characteristics of Steel Materials

To perform the FEM simulation of steel beams, ANSYS software (version 22) was used. To model the steel beams, the SOLID186 element [53] was used, as this element can simulate the different properties of the steel used in terms of large deflection, plasticity, higher strain capabilities, hyperelasticity, and creep. This element has 3-DOFs in the three directions with twenty points distributed at the vertices and midpoint of the element's perimeter. This element is suitable for asymmetric meshes and can model the deformations of elastic–plastic materials. To perform the simulation of these properties of the steel used, the relationship between stress and strain was obtained from the experimental results of axial tensile tests; in addition, the modulus of elasticity was determined, as reported in previous studies [54–57], and Poisson's ratio of 0.3 for steel was used. The stress–strain relationships and modulus of elasticity used in the simulation of the steel beams in the FEM modeling process are listed in Table 1.

Table 1. The relationship between stress and strain for steel applied to the FEM modeling.

Sayed and Alanazi [54]		Shi et al. [55]		Shokouhian and Shi [56]		Yang et al. [57]			
f_y = 282 MPa		f_y = 557 MPa		f_y = 408.2 MPa		f_y = 769 MPa		f_y = 754 MPa	
Stress (MPa)	**Strain**	**Stress (MPa)**	**Strain**	**Stress (MPa)**	**Strain**	**Stress (MPa)**	**Strain**	**Stress (MPa)**	**Strain**
282	0.0013	557.00	0.0027	408.20	0.0021	769.00	0.0040	754.00	0.0039
291	0.0143	563.16	0.0063	414.13	0.0090	782.84	0.0128	767.86	0.0082
301	0.0279	570.51	0.0096	442.08	0.0180	787.86	0.0256	776.42	0.0126
317	0.0439	609.69	0.0199	474.90	0.0297	799.14	0.0352	780.99	0.0165
347	0.0825	640.44	0.0312	502.04	0.0430	811.14	0.0579	792.94	0.0279
369	0.1262	673.73	0.0491	526.75	0.0648	815.18	0.0823	800.38	0.0395
376	0.1543	693.70	0.0731	529.50	0.0934	816.00	0.0987	811.00	0.0718
377	0.1795	695.58	0.1079	527.00	0.1400	815.75	0.1125	799.24	0.1000
339	0.1946	687.32	0.1266	515.50	0.2200	807.91	0.1266	775.00	0.1168
299	0.2057	672.82	0.1428	481.00	0.2475	772.48	0.1487	747.88	0.1312
E = 211 GPa		E = 206 GPa		E = 197.6 GPa		E = 190.8 GPa		E = 194.9 GPa	

3.2. Model Studies of Structure

Numerical analysis modeling was used to simulate 141 specimens of solid and flat-flange steel beams. These steel beams have many variables that directly affect their efficiency and ultimate load capacity. Two main groups were used for these samples, as follows:

The first group consists of 25 experimental test specimens of steel beams available in previous studies [56–60], as listed in Table 2. These beam specimens contain many variables, and each variable has a wide range of values applied in the experimental tests. Among these variables are the beam flange width, b_f, ranging from 159.15 to 271.00 mm; the flange thickness, t_f, ranging from 6.35 to 16.38 mm; the web beam height, d_w, ranging from 220.70 to 661.00 mm; the web thickness, t_w, ranging from 6.17 to 15.00 mm; and the steel yield stress used, f_y, varying from 408.2 to 886.00 MPa, in addition to the length of the steel beam, L, ranging from 1360 to 6000 mm. Some of these steel beam specimens were tested experimentally under the influence of three-point loading, as in [57,58,60], while others were tested under the influence of four-point loading, as in [55,56,59]. The loading was applied to the FME models by force control. All these specimens are collected from experimental tests and contain many variables and are primarily used to study the reliability and accuracy of using FEM simulation.

Through Equations (1) to (12), the design class classification of the steel beam sections collected from previous studies can be determined, as listed in Table 3. Through the geometric dimensions shown in Table 2, the sections were classified based on the critical condition in the design category, whether for the beam flange or the beam web. It is noted that most critical classifications follow the values resulting from the dimensions of the beam flange, whether for ANSI/AISC 360-22 [51] or Eurocode-3 [52] because the permissible limits for determining the section category for the flange have lower values than their counterparts for the beam web. Specimens were carefully selected to include all design categories based on the current design code classifications. For ANSI/AISC 360-22, there are specimens of the compact, noncompact, and slender categories, and for Eurocode-3 [52], there are the categories Class-1, Class-2, Class-3, and Class-4. This diversity of design categories provides a high degree of validation and reliability when comparing experimental test results to existing design code prediction results or when comparing them to FEM simulation results.

Table 2. Details of steel I-beams from previous studies used to verify FEM simulation.

FE Model Based on	Beam Specimens	Steel Beam Dimensions						f_{yf} (MPa)	f_{yw} (MPa)
		b_f (mm)	t_f (mm)	d_w (mm)	t_w (mm)	L (mm)	a (mm)		
Shi et al. [55]	I-460-1	167.8	13.87	520.9	11.96	3400	1000	557.0	510.5
	I-460-2	260.1	13.94	610.8	12.08	3400	1000	557.0	510.5
	I-890-3	166.5	6.35	661.0	6.17	2000	600	886.0	886.0
Saliba and Gardner [58]	I-600 × 200 × 12 × 10-1	200.1	12.4	599.2	10.2	1360	600	484.0	433.0
	I-600 × 200 × 12 × 8-2	199.8	12.3	600.0	8.4	2560	1200	484.0	431.0
	I-600 × 200 × 12 × 10-2	200.4	12.6	600.1	10.6	2560	1200	484.0	433.0
	I-600 × 200 × 12 × 15-2	200.1	15.3	599.0	15.0	2560	1200	484.0	564.0
Shokouhian and Shi [56]	C1	169.0	11.91	357.4	7.87	3399	1000	408.2	442.8
	C2	263.6	11.85	358.48	7.82	3401	1000	408.2	442.8
Yang et al. [57]	Y1-3	178.98	13.59	373.26	9.57	3500	1500	769.0	781.0
	Y2-3	197.47	13.63	372.44	9.63	3500	1500	769.0	781.0
	Y6-3	201.46	13.50	323.94	9.65	3500	1500	769.0	781.0
	Y8-3	159.95	16.10	319.15	9.64	3500	1500	754.0	781.0
	Y9-3	159.15	16.26	370.11	9.90	3500	1500	754.0	781.0
	Y10-3	177.81	16.03	369.95	9.81	3500	1500	754.0	781.0
	Y12-4	177.40	16.38	320.19	9.50	4000	1200	754.0	781.0
	Y13-4	159.34	16.35	368.8	9.68	4000	1200	754.0	781.0
	Y14-4	182.36	16.11	368.32	9.51	4000	1200	754.0	781.0
Beg and Hladnik [59]	B1	271.00	12.40	222.70	10.40	5000	1800	797.0	775
	C1	251.00	12.60	221.30	10.40	5000	1800	776.0	775
	D1	220.80	12.40	220.90	10.40	5000	1800	873.0	830
	E1	198.80	12.60	220.70	10.40	5000	1800	797.0	830
Xiong et al. [60]	C1	180.17	10.48	249.26	9.01	5000	2500	525.0	541.0
	C2	179.19	10.49	247.74	9.00	6000	3000	525.0	541.0
	C3	179.15	10.47	431.03	8.79	5000	2500	525.0	541.0

The second group consists of 116 broad-flange I-beam (BFIB) specimens containing several variables with out-of-plane distances in the mid-span lateral direction. These variables are the out-of-plane distance, the size of the cross-section, the length of the steel beam, and the steel yield stress. These beams were simulated in 19 basic models that differed based on the BFIB cross-section size, beam length, and steel yield stress, as listed in Table 4. Each model contained a standard specimen with no out-of-plane distance, and seven specimens with out-of-plane distances of 300, 400, 500, 600, 800, 1000, and 2000 mm in the lateral direction of the middle beam span, as shown in Figure 1. All the BFIB steel beams were tested under three-point loading to failure load using hinged point support, as shown in Figure 1. Ten 10 mm thick stiffener plates were used at the supports, quarter span, and load points. To implement the part of the out-of-plane distance, the same type of cross-section was used for implementing the main beam. The installation was conducted by connecting the upper and lower flanges with the corresponding flanges of the main beam at its beginning and end. The entire height of the web out-of-plane part was also connected with the web of the main beam, as shown in Figure 1. This interconnection was performed by perfectly bonding all the available points in the mesh division. The value of the out-of-plane distance from the middle of the main beam flange of the two separate parts was also determined. To obtain the true effect of all the variables on these beams with an out-of-plane distance, these geometric dimensions were chosen to resemble those actually used in reality.

Table 3. Compactness classification of steel beam cross-section from previous studies.

FE Model Based on	Beam Specimens	ANSI/AISC 360-22 [51]			Eurocode-3 [52]		
		Flange	Web	Section	Flange	Web	Section
Shi et al. [55]	I-460-1	C	C	C	Class-1	Class-1	Class-1
	I-460-2	NC	C	NC	Class-3	Class-2	Class-3
	I-890-3	NC	S	S	Class-4	Class-4	Class-4
Saliba and Gardner [58]	I-600 × 200 × 12 × 10-1	NC	C	NC	Class-3	Class-2	Class-3
	I-600 × 200 × 12 × 8-2	NC	C	NC	Class-3	Class-3	Class-3
	I-600 × 200 × 12 × 10-2	NC	C	NC	Class-3	Class-2	Class-3
	I-600 × 200 × 12 × 15-2	C	C	C	Class-2	Class-1	Class-2
Shokouhian and Shi [56]	C1	C	C	C	Class-2	Class-1	Class-2
	C2	NC	C	NC	Class-4	Class-1	Class-4
Yang et al. [57]	Y1-3	NC	C	NC	Class-3	Class-1	Class-3
	Y2-3	NC	C	NC	Class-3	Class-1	Class-3
	Y6-3	NC	C	NC	Class-3	Class-1	Class-3
	Y8-3	C	C	C	Class-1	Class-1	Class-1
	Y9-3	C	C	C	Class-1	Class-1	Class-1
	Y10-3	C	C	C	Class-2	Class-1	Class-2
	Y12-4	C	C	C	Class-2	Class-1	Class-2
	Y13-4	C	C	C	Class-1	Class-1	Class-1
	Y14-4	C	C	C	Class-2	Class-1	Class-2
Beg and Hladnik [59]	B1	NC	C	NC	Class-4	Class-1	Class-4
	C1	NC	C	NC	Class-4	Class-1	Class-4
	D1	NC	C	NC	Class-4	Class-1	Class-4
	E1	NC	C	NC	Class-4	Class-1	Class-4
Xiong et al. [60]	C1	NC	C	NC	Class-3	Class-1	Class-3
	C2	NC	C	NC	Class-3	Class-1	Class-3
	C3	NC	C	NC	Class-3	Class-2	Class-3

Note: C = compact; NC = noncompact; S = slender.

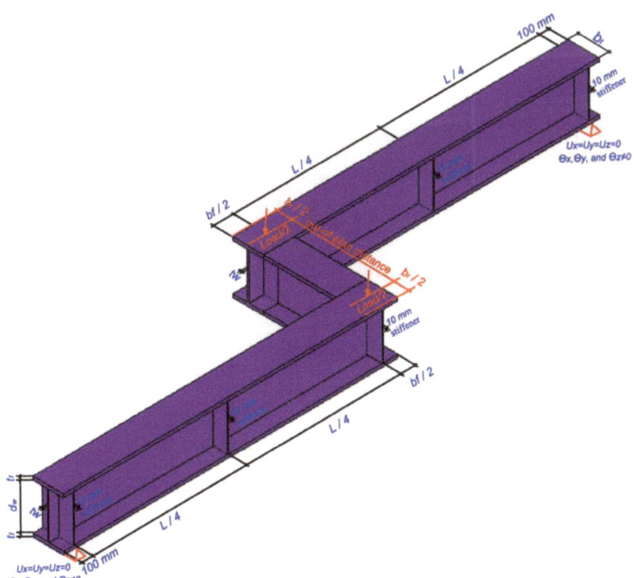

Figure 1. Boundary conditions and geometric details of the steel I-beams and out-of-plane distance used in FEM simulation.

Table 4. Details of the steel I-beams created in this study by FEM simulation.

Beam Specimens	Steel Beam Cross-Section Dimensions (mm)						f_y (MPa)	Cases of Out-of-Plane Distance, OOP, (mm)						
	b_f	t_f	d_w	t_w	L	a		1	2	3	4	5	6	7
BFIB500-L10-Fy282	300	30	440	16	10,000	5000	282	300	400	500	600	800	1000	2000
BFIB500-L9-Fy282	300	30	440	16	9000	4500	282	300	400	500	600	800	1000	2000
BFIB500-L8-Fy282	300	30	440	16	8000	4000	282	300	400	500	600	800	1000	2000
BFIB500-L6-Fy282	300	30	440	16	6000	3000	282	300	400	500	600	800	1000	2000
BFIB300-L10-Fy282	300	20	276	12	10,000	5000	282	300	400	500	600	800	1000	2000
BFIB300-L9-Fy282	300	20	276	12	9000	4500	282	300	400	500	---	---	---	---
BFIB300-L8-Fy282	300	20	276	12	8000	4000	282	300	400	500	---	---	---	---
BFIB300-L6-Fy282	300	20	276	12	6000	3000	282	300	400	500	---	---	---	---
BFIB400-L10-Fy282	300	26	348	14	10,000	5000	282	300	400	500	600	800	1000	2000
BFIB400-L9-Fy282	300	26	348	14	9000	4500	282	300	400	500	---	---	---	---
BFIB400-L8-Fy282	300	26	348	14	8000	4000	282	300	400	500	---	---	---	---
BFIB400-L6-Fy282	300	26	348	14	6000	3000	282	300	400	500	---	---	---	---
BFIB600-L10-Fy282	300	32	536	17	10,000	5000	282	300	400	500	600	800	1000	2000
BFIB600-L9-Fy282	300	32	536	17	9000	4500	282	300	400	500	---	---	---	---
BFIB600-L8-Fy282	300	32	536	17	8000	4000	282	300	400	500	---	---	---	---
BFIB600-L6-Fy282	300	32	536	17	6000	3000	282	300	400	500	---	---	---	---
BFIB500-L10-fy408	300	30	440	16	10,000	5000	408	300	400	500	600	800	1000	2000
BFIB500-L10-fy557	300	30	440	16	10,000	5000	557	300	400	500	600	800	1000	2000
BFIB500-L10-fy769	300	30	440	16	10,000	5000	769	300	400	500	600	800	1000	2000

3.3. Selection of the Optimal Mesh Size for Element Convergence

The sensitivity and convergence of the element mesh size dimensions play an effective and important role in the accuracy and results of the FEM analysis, as does the time required to perform the numerical analysis process. Therefore, the accuracy and convergence of the optimal mesh were tested by conducting five experiments on the largest specimen with the longest length equal to 6000 mm, where C2 was available from previous studies, including that by Xiong et al. [60]. Different trials were made by changing the dimensions of the mesh element size in terms of height, width, and length, respectively, where trial 1 had dimensions of 10 × 10 × 50 mm, trial 2 had dimensions of 10 × 10 × 20 mm, trial 3 had dimensions of 10 × 10 × 10 mm, trial 4 had dimensions of 5 × 5 × 20 mm, and trial 5 had dimensions of 5 × 5 × 10 mm. In each trial, the ratio between the maximum bending moment resulting from the FEM analysis process and its corresponding value from the experimental test ($M_{u.FEM}/M_{u.Exp}$) was found. The relationship between each experiment and its corresponding measured ratio was plotted to determine the optimal size of the mesh elements, as shown in Figure 2. From the figure, it is clear that the optimal size of the mesh dimensions was 5 × 5 × 20 mm in height, width, and length, respectively, as it gave the best possible accuracy for the FEM results and gave the best time required to conduct the FEM analysis process. Therefore, these dimensions were adopted in conducting the remaining FEM analysis processes for all the specimens, whether from previous studies or those proposed in this study.

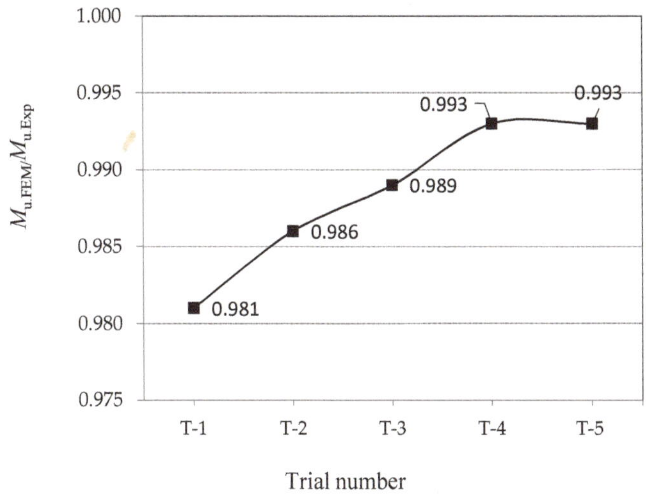

Figure 2. The relationship between the size of the mesh dimensions and the accuracy ratio in the ultimate bending moments.

4. Verifying Results from FEM Modeling and Current Design Codes

4.1. Verifying Results from FEM

The values of the steel beam moment capacities resulting from the experimental tests included in the previous studies [55–60] and the results of the FEM simulation of the same beams were compared. The process of comparing these results is the basic element for relying on the results and reliability of the FEM simulation in modeling these types of structural elements, as shown in Table 5 and Figure 3. Since these specimens contain many variables with a wide range of data and different design categories according to the classification of the current design codes, the accuracy and reliability of the modeling are demonstrated by the FEM analysis of these steel beams. By comparing the ratios between the steel beam capacity resulting from the FEM modeling and the experimental test results, the average of the ratios $M_{u.FEM}/M_{u.Exp}$ was found to be equal to 1.007, with a coefficient of correlation, r, equal to 0.999 and a coefficient of variation, COV, equal to 1.573%. Through these values, the use of the FEM modeling in simulating these types of steel beams is characterized by high accuracy and strong reliability in predicting the steel beam capacity. This result is consistent with many previous studies [30,49,50] when comparing the results of experimental tests with the results of FEM simulations.

An important element that demonstrates the accuracy of FEM simulations is the relationship between the applied moments and the corresponding mid-span deflection. Therefore, the beam bending moment and deflection relationship was obtained for six specimens with different variables included in previous studies, in which there were three steel beam specimens from Shi et al. [55] and three specimens from Saliba and Gardner [58]. The relationship resulting from the FEM simulation of the same steel beams mentioned above was determined, as shown in Figure 4. Figure 4 shows the accuracy and reliability of FEM simulation in modeling these steel beams, where the consistency between the relationships is excellent, and this result is consistent with many previous studies [26,29,49].

Table 5. Comparison of existing design codes and FEM results for steel I-beams with experimental test results.

FE Model Based on	Beam Specimens	$M_{u\cdot Exp}$ (kN.m)	FEM Results		Eurocode-3		ANSI/AISC 360-22	
			$M_{u\cdot FEM}$ (kN.m)	$M_{u\cdot FEM}/M_{u\cdot Exp}$	M_{EC3} (kN.m)	$M_{EC3}/M_{u\cdot Exp}$	M_{AISC} (kN.m)	$M_{AISC}/M_{u\cdot Exp}$
Shi et al. [55]	I-460-1	1225.40	1220.71	0.996	1107.42	0.904	1107.42	0.904
	I-460-2	1999.70	2029.06	1.015	1634.49	0.817	1802.57	0.901
	I-890-3	797.10	804.00	1.009	778.91	0.977	778.91	0.977
Saliba and Gardner [58]	I-600 × 200 × 12 × 10-1	1102.80	1126.05	1.021	1003.66	0.910	1054.88	0.957
	I-600 × 200 × 12 × 8-2	1172.00	1184.33	1.011	948.38	0.809	1053.68	0.899
	I-600 × 200 × 12 × 10-2	1395.00	1380.12	0.989	1029.32	0.738	1163.15	0.834
	I-600 × 200 × 12 × 15-2	2162.00	2139.76	0.990	1669.12	0.772	1669.12	0.772
Shokouhian and Shi [56]	C1	447.91	455.66	1.017	414.72	0.926	414.72	0.926
	C2	616.70	613.08	0.994	502.49	0.815	572.10	0.928
Yang et al. [57]	Y1-3	1012.23	1037.91	1.025	858.61	0.848	976.23	0.964
	Y2-3	1117.22	1149.06	1.028	931.68	0.834	1043.37	0.934
	Y6-3	962.23	957.55	0.995	798.76	0.830	887.64	0.922
	Y8-3	955.01	964.44	1.010	842.67	0.882	842.67	0.882
	Y9-3	1158.33	1191.08	1.028	1018.66	0.879	1018.66	0.879
	Y10-3	1230.55	1251.79	1.017	1091.67	0.887	1091.67	0.887
	Y12-4	1065.00	1034.81	0.972	927.58	0.871	927.58	0.871
	Y13-4	1184.36	1179.61	0.996	1013.63	0.856	1013.63	0.856
	Y14-4	1274.40	1290.05	1.012	1103.45	0.866	1103.45	0.866
Beg and Hladnik [59]	B1	703.80	724.61	1.030	515.43	0.732	670.31	0.952
	C1	646.20	640.70	0.991	506.92	0.784	614.36	0.951
	D1	644.40	638.66	0.991	518.41	0.804	620.38	0.963
	E1	559.80	567.00	1.013	502.72	0.898	515.72	0.921
Xiong et al. [60]	C1	335.3	344.88	1.029	292.81	0.873	313.81	0.936
	C2	326.39	324.12	0.993	289.58	0.887	301.58	0.924
	C5	634.68	631.40	0.995	561.05	0.884	574.05	0.904

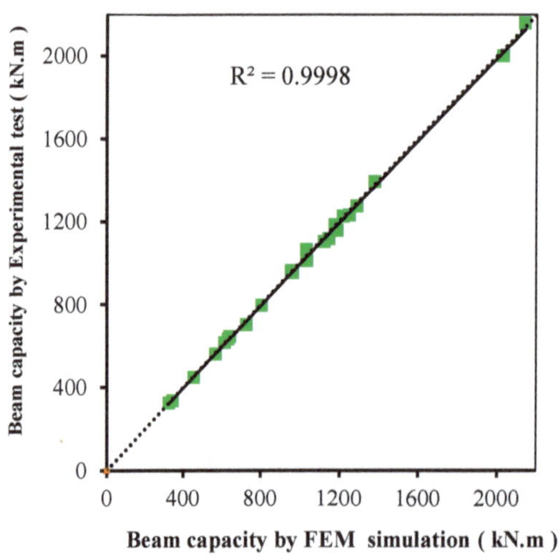

Figure 3. Comparison of FEM results with experimental test results from previous studies.

The same steel beams from the previous studies that were used to compare the moment-to-deflection relationship [55,58] were used to compare the deformation shapes and failure locations with the FEM simulations, as shown in Figure 5. This comparison is important and complementary to the process of studying the reliability of simulation using FEM analysis. From the deformation shapes and their locations, it became clear that the cause of the failure was a local buckling in the upper flange of the cross-section of the steel beam. This buckling occurs at the distance between the loading points when using the four-point loading system or near the applied loading point in the case when the three-point loading system is used; it was measured in both the experimental test and the FEM simulation results. The results of this comparison show the accuracy and reliability of the FEM simulation in modeling these steel beams, as there is an excellent match. This result is consistent with previous studies, such as those of Sayed and Al-Jarbou [29] and Perera and Mahendran [49].

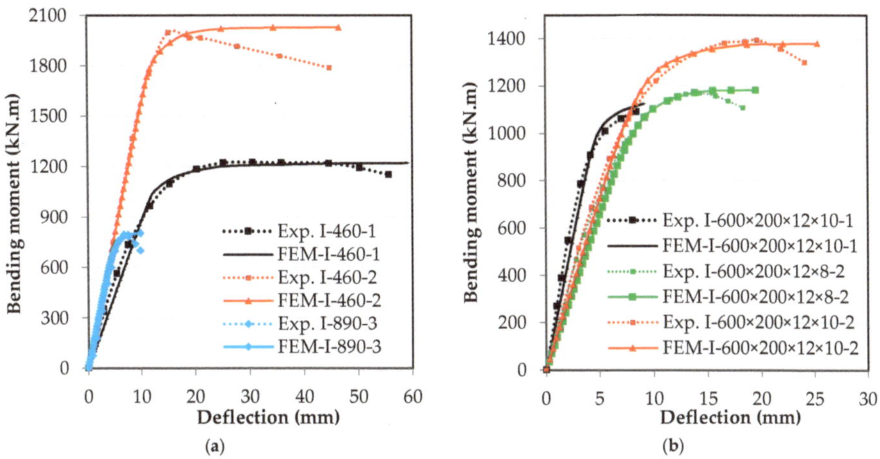

Figure 4. The beam bending moment and corresponding deflection relationships were obtained from FEM results and experimental tests. (**a**) Steel I-beam from Shi et al. [55]. (**b**) Steel I-beam from Saliba and Gardner [58].

4.2. Verifying Results from Current Design Codes

By reviewing the previous studies, it is found that most of them use the current codes Eurocode-3 and ANSI/AISC 360-22 in the comparison process. This is because these codes are the basic element for building the rest of the proposed design models to predict the beam capacities. Through the available Equations (1) to (10) and the classification of the obtained design categories, as shown in Table 3, the predicted values of the steel beam moment capacity were calculated, as listed in Table 5. Through the table, the relationship between the ultimate beam capacity resulting from the prediction of the design codes and the ultimate capacity of the same beams resulting from the experimental tests was determined, as shown in Figure 6. From Table 5 and Figure 6, it is clear that the predictions of the current design codes [51,52] apply completely when the design classification of the steel beam cross-section is a compact category for ANSI/AISC 360-22 and with the Class-1 and Class-2 classifications of Eurocode 3. The difference in prediction values occurs when the classification is a noncompact category for ANSI/AISC 360-22 and a Class-3 classification for Eurocode 3, while the conformity returns if the classification is a slender category for ANSI/AISC 360-22 and a Class-4 classification for Eurocode 3. From Figure 6, all the values predicted from the current design codes [51,52] are less than the actual ultimate moment capacity resulting from the experimental tests of steel beams. This result is consistent with Sayed and Aljarbou [29] and Perera and Mahendran [30]. According to ANSI/AISC 360-22, the beam moment capacity can be predicted with an average $M_{AISC}/M_{u.Exp}$ ratio of 0.908, with an r of 0.998 and a COV of 5.12%. As for Eurocode-3, it can be predicted with an average $M_{EC3}/M_{u.Exp}$ ratio of 0.851, with an r of 0.997 and a COV of 6.84%. All these results are for steel beams without any out-of-plane distances.

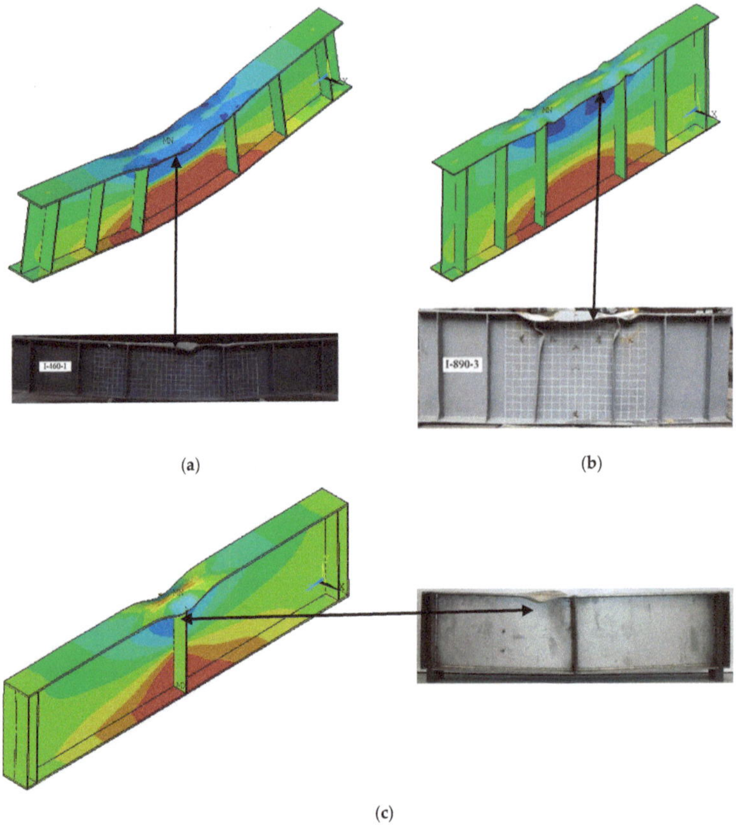

Figure 5. The deformations and buckling failure locations were obtained by FEM analysis and experimental tests. (**a**) I-460-1 Shi et al. [55]. (**b**) I-890-3 Shi et al. [55]. (**c**) I-600 × 200 × 12 × 10-2 Saliba and Gardner [58].

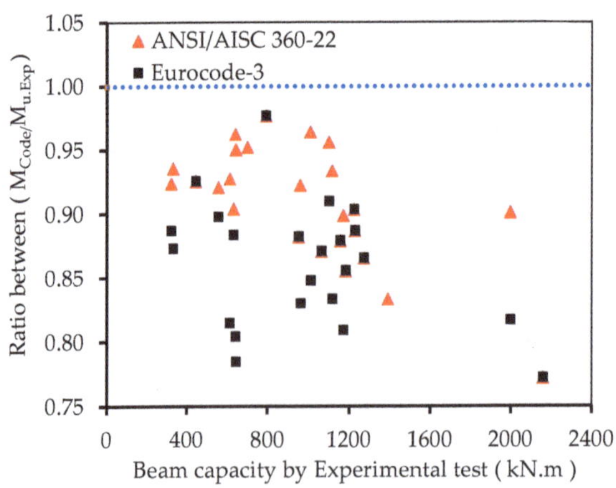

Figure 6. The ratio of bending moment resulting from both existing design codes and experimental tests.

5. Results and Discussion

To find out the effect of the presence of out-of-plane distance in the lateral direction of the length of the steel beam on its efficiency, a second group containing 116 specimens was proposed and analyzed using FEM analysis. This group included several variables related to cross-section size, steel beam length, and yield stress. All of these variables were used at different out-of-plane distances. The effect of all these variables on the behavior of the steel beam, stress distribution in the cross-section, and BFIB steel beam capacity was obtained.

5.1. Bending Moment and Deflection Relationship

The relationship between bending moment and the corresponding mid-span deflection can directly explain the behavior of steel beams. By analyzing the results of the specimens, it became clear that when there was any out-of-plane distance in the middle of the beam span, a significant decrease occurred in the steel beam capacity, at different rates depending on the value of the out-of-plane distance and the steel beam cross-section dimensions. For example, when using an out-of-plane distance of 300 mm in steel beams with BFIB-600 and BFIB-300 sections, where this distance represents the value of the beam flange width, there was a decrease in the beam capacity, which reached, respectively, 50% and 60% of the beam capacity without an out-of-plane distance. It was also observed that at this out-of-plane value, the deflection value increased when compared to the beam without an out-of-plane distance, as shown in Figure 7. With the increase in the out-of-plane distance to 2000 mm for the same beams, the reduction in the ultimate bending moment capacity reached 6% and 11%, respectively. It was also observed that the behavior of the steel beam changed from the first level of load application to the steel beam.

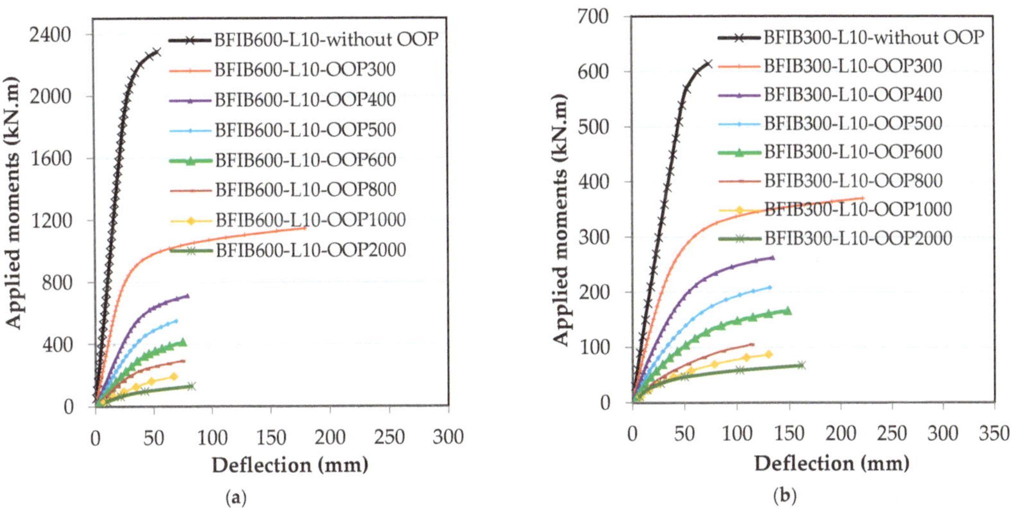

Figure 7. Bending moment and the corresponding deflection relationships obtained from FEM simulation with different out-of-plane distances. (**a**) BFIB-600-L10-fy282 MPa. (**b**) BFIB-300-L10-fy282 MPa.

5.2. Shape of Deformations and Stress Distribution

By FEM analysis of BFIB steel beams with or without an out-of-plane distance, the deformations and stress distributions of each specimen were obtained. It was observed that steel beams without out-of-plane distance failed due to local buckling at the top flange of the beam cross-section. When there is an out-of-plane distance, stresses are concentrated in the upper flange. These stresses are the result of global buckling along the entire beam length of the upper flange, starting from the support point to the middle beam length. These stresses are in the form of compressive stresses at the inner edge of the upper flange, and they are met by tensile stresses in the same flange, but on the outer edge. The location of the concentration of these stresses is at the point of contact of the upper flange of the out-of-plane length with the upper flange of the main beam, as shown in Figure 8. Stress concentration, whether compressive or tensile, continues with the increase in the out-of-plane distance and is always at values greater than the yield stress of the steel used. The compressive and tensile stresses of the BFIB-500 section steel beam with an out-of-plane distance of 300 mm were 410 MPa, and 311 MPa, respectively, while the yield stress used was 282 MPa. These stresses appeared despite the clear reduction in the beam capacity to 55% of the beam capacity without an out-of-plane distance. With the out-of-plane distance increasing to 1000 mm, these stresses reached 306 MPa, and 352 MPa, respectively, which was also higher than the yield stress of the steel used despite the sharp decrease in the ultimate bending moment capacity, which reached 10.0%.

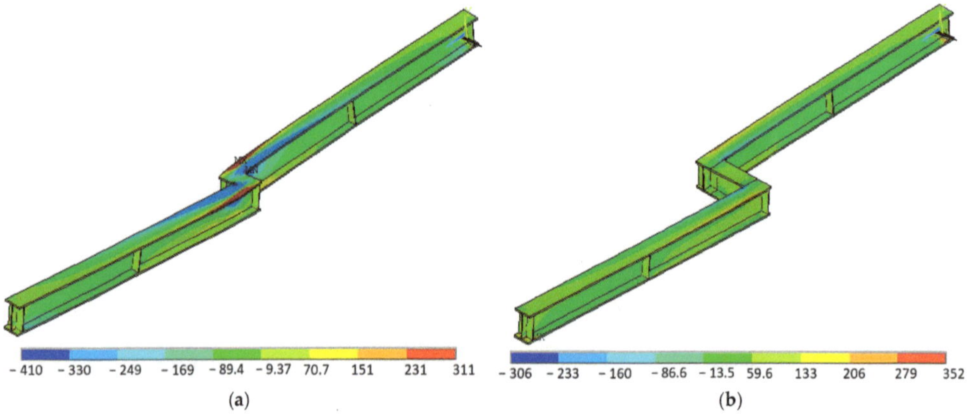

Figure 8. Deformations and stress distributions resulting from FEM simulations of the BFIB500-L10-fy282 specimen. (**a**) BFIB500-L10-fy282-OOP300. (**b**) BFIB500-L10-fy282-OOP1000.

Here, the stress distribution shape through the cross-section of the steel beam with an out-of-plane distance is accurately analyzed. The stress distribution shape through the cross-section of the longitudinal part of the steel beam was measured at the stress concentration point where the out-of-plane part contacts the longitudinal axis of the beam. The stress distribution through the cross-section was measured at different ratios of 3, 15, 25, 50, 75, and 100% of the applied load value, as shown in Figure 9. From Figure 9, it is clear that the cross-section is exposed to an irregular stress distribution on both the upper and lower flange, and this appears from the first application of the load. These stresses on a single flange, whether upper or lower, are exposed to two types of stresses, tensile stresses at the outer edge of the beam axis and compressive stresses at the inner edge, while this distribution is reflected on the lower flange, where the tensile stresses are on the inner edge and the compressive stresses are on the outer edge of the beam axis. This is evidence that the effect of the out-of-plane distance on the structural system of the steel beam causes a global buckling from the first application of the load, as shown in Figure 9a. Through this distribution of stress, it can be said that the effect of buckling on steel beams with an out-of-plane distance begins from the first moment of applying the load, which is the main factor in causing the failure of this structural system. It accelerates the increase in stress values at the stress concentration point to values exceeding the yield stress of the steel used.

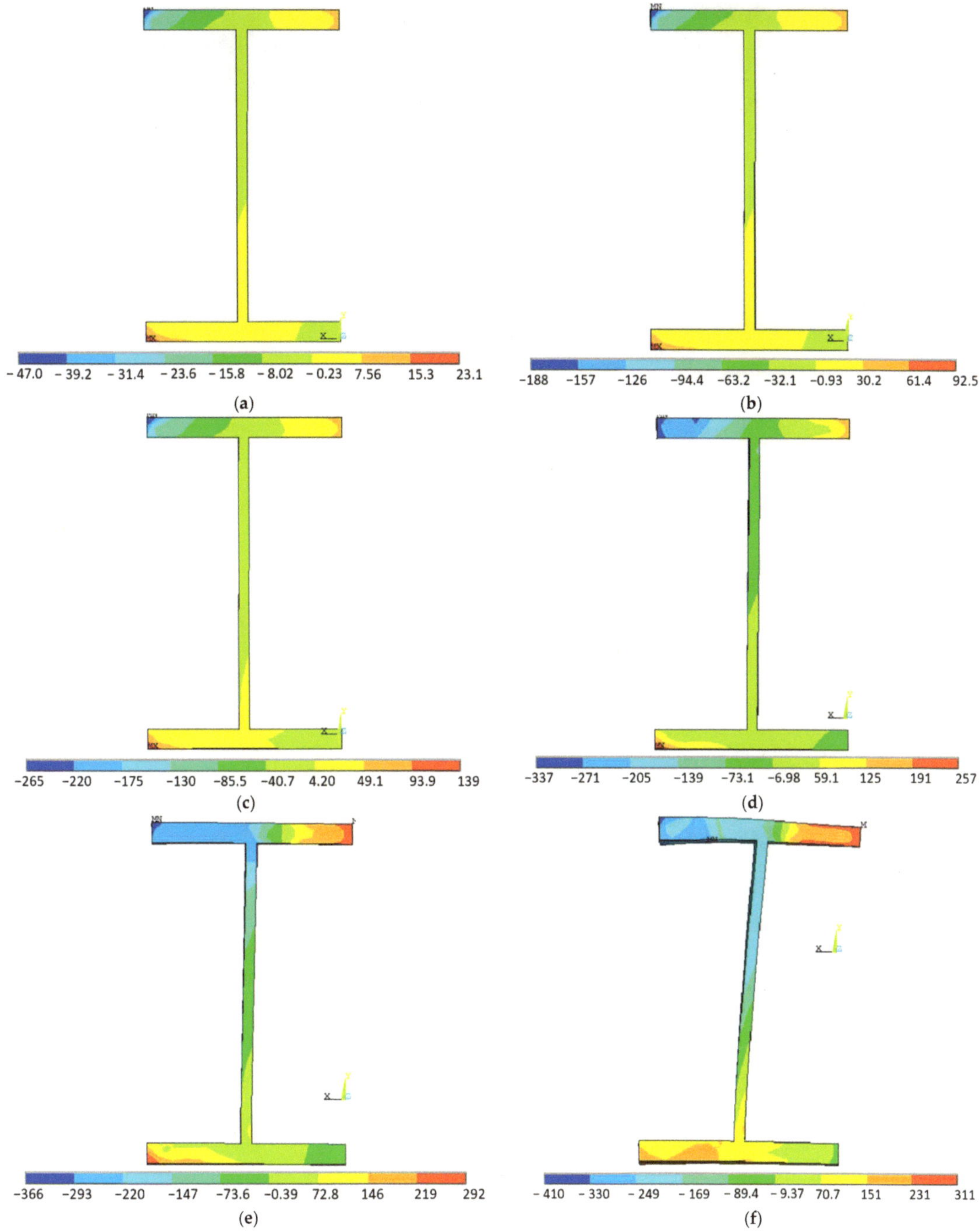

Figure 9. Stress distributions through the cross-section at the critical location of stress concentration at different applied load ratios for the BFIB500-L10-fy282-OOP300 specimen. (**a**) At 3% of the applied load. (**b**) At 15% of the applied load. (**c**) At 25% of the applied load. (**d**) At 50% of the applied load. (**e**) At 75% of the applied load. (**f**) At 100% of the applied load.

5.3. Effect of Variables and Out-of-Plane Distance on Beam Efficiency

To obtain the effect of different variables such as cross-section size, steel beam length, yield stress, and the out-of-plane distance on the efficiency of the BFIB steel beams, Figures 10–15, were constructed; these figures represent the relationship between the steel beam capacity with out-of-plane distance and the beam without out-of-plane distance for each variable separately.

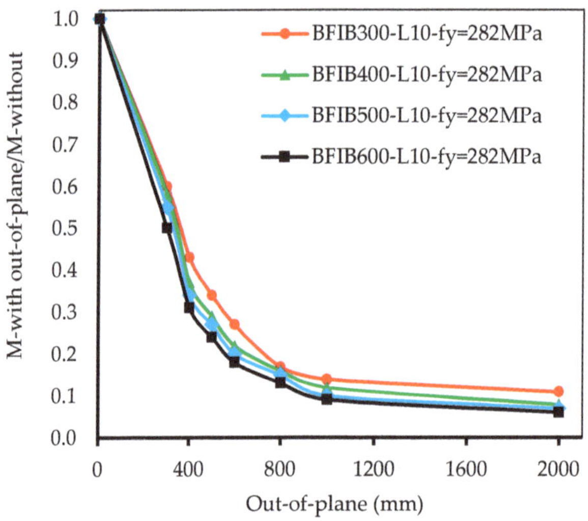

Figure 10. Relationship between moment capacity ratio and out-of-plane value for beam cross-section effect.

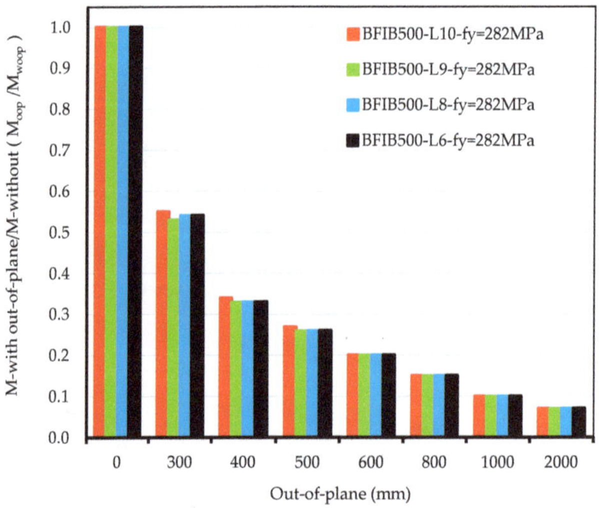

Figure 11. Relationship between moment capacity ratio and out-of-plane value for beam length effect with BFIB-500 section.

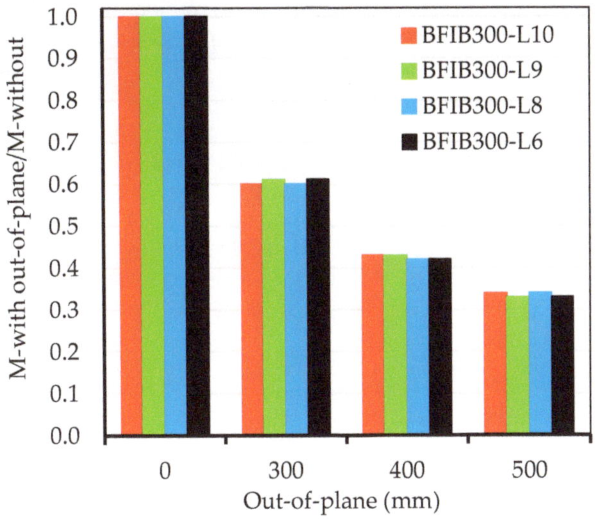

Figure 12. Relationship between moment capacity ratio and out-of-plane value for beam length effect with BFIB-300 section.

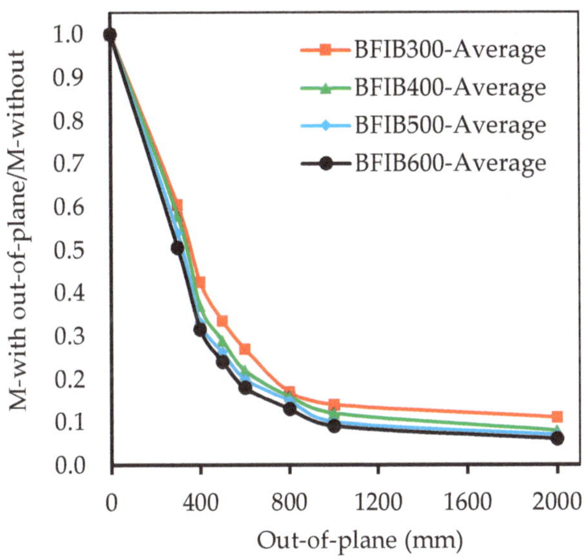

Figure 13. Relationship between average moment capacity ratios and out-of-plane value for beam length and cross-section effects.

5.3.1. Effect of Beam Cross-Sectional Dimensions

One of the most important elements that directly affect the efficiency of the steel element is the dimensions of the cross-section, as it is the basic element used in determining the ultimate capacity of bending moments, as well as in choosing the type of design category for the cross-section in terms of the compact, noncompact, and slender categories. Figure 10 shows the relationship between the effect of cross-sectional dimensions on the percentage reduction in beam capacity due to the presence of an out-of-plane distance. Different cross-sections were used to study this variable, namely BFIB-300, BFIB-400, BFIB-500, and BFIB-600, with the same out-of-plane distances. Figure 10 shows that the effect of cross-sectional dimensions is directly proportional to the percentage reduction in beam capacity. By fixing the length of the steel beam as well as the yield stress of the steel used and using an out-of-plane distance of 300 mm with steel beams with the cross-sections BFIB-300, BFIB-400, BFIB-500 and BFIB-600, the ratio of the beam capacity with out-of-plane distance to those without it (M_{OOP}/M_{woop}) decreased by ratios equal to 0.60, 0.58, 0.55, and 0.50, respectively. With the change in the out-of-plane distance to 1000 mm, the decrease ratios of M_{OOP}/M_{woop} reached 0.14, 0.12, 0.10, and 0.09, respectively. But after the out-of-plane distance reached 1000 mm, the decrease ratio was small compared to the beginning; when the out-of-plane distance was increased to 2000 mm, the decrease ratios of M_{OOP}/M_{woop} reached 0.11, 0.08, 0.07, and 0.06, respectively. Through this, as the cross-sectional dimensions of the BFIB steel beam increase, the reduction ratio in the steel beam capacity with out-of-plane distance increases.

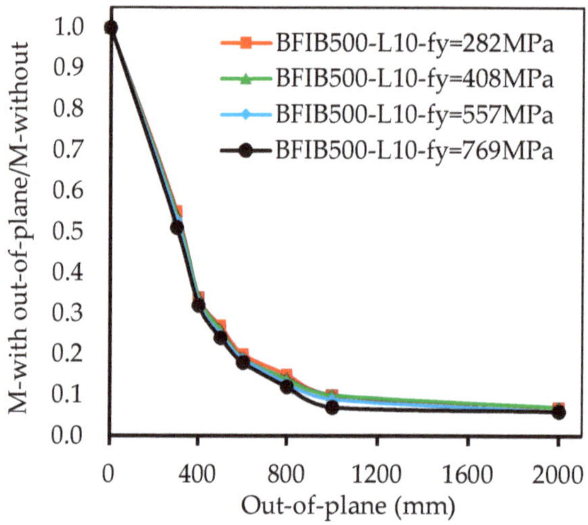

Figure 14. The relationship between average moment capacity ratios and out-of-plane value for steel beam yield stress effect.

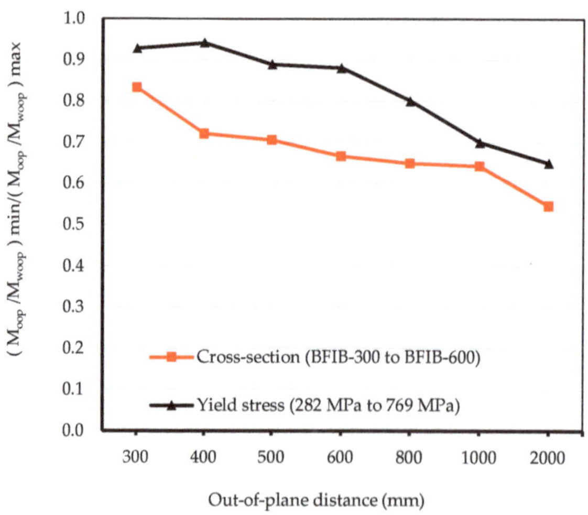

Figure 15. Minimum-to-maximum reduction ratio and out-of-plane distance relationship.

5.3.2. Effect of Beam Length

Since the length of the steel beam is the variable that causes the presence of an out-of-plane distance, it was necessary to use it as a variable to know the extent of its effect on the efficiency of the steel beams with the presence of out-of-plane distances. To verify this, steel beams of variable lengths of 6000, 8000, 9000, and 10,000 mm were used with the same section of the BFIB-500 beam and the same yield stress as the steel used. The relationship between the percentages of decrease in the ultimate capacity of bending moments with the effect of out-of-plane distances was determined, as shown in Figure 11. From Figure 11, it is clear that this variable does not affect the efficiency of the steel beam or the value of the decrease in the ultimate capacity of the bending moments with the presence of out-of-plane distance. When using an out-of-plane distance of 300 mm for steel beams of lengths 6000, 8000, 9000, and 10,000 mm, the reduction ratio of M_{OOP}/M_{woop} was 0.55, 0.53, 0.54, and 0.54, respectively. When the out-of-plane distance was increased to 2000 mm, the reduction ratio for all the lengths was 0.07.

To verify this result, the same specified lengths were used with the other cross-sections, BFIB-300, BFIB-400, and BFIB-600, and the same result was obtained, as shown in Figure 12 for cross-section BFIB-300. When using an out-of-plane distance of 300 mm for steel beams of lengths 6000, 8000, 9000, and 10000 mm, the reduction ratios of M_{OOP}/M_{woop} were 0.60, 0.61, 0.60, and 0.61, respectively. This was also measured at all out-of-plane distances used with this variable. Therefore, the two variables were combined, where the cross-sectional dimensions (which have a direct proportional effect on the percentages of decrease in the ultimate capacity) with the length of the steel beam (which has no effect) and the average of the percentages of these two variables were taken, as shown in Figure 13. It is clear from Figure 13 that it is identical to Figure 12, as the main variable for these two figures is only the dimensions of the cross-section, and there is no effect of the length of the steel beam.

5.3.3. Steel Yield Stress Effect

The yield stress of the steel used is one of the most important variables that directly affect the efficiency of steel elements as well as their ultimate capacity. Therefore, different values of this variable were used, such as 282, 408, 557, and 769 MPa, for the same beam cross-section and length, and the extent of its effect on the percentage of decrease in the ultimate capacity of bending moments with an out-of-plane distance was determined, as shown in Figure 14. As can be seen in the figure, the effect of changing the yield stress on the ultimate capacity was measured by changing the out-of-plane distance. By using an out-of-plane distance of 300 mm for steel beams with yield stresses 282, 408, 557, and 769 MPa, the decrease ratios of M_{OOP}/M_{woop} reached 0.55, 0.54, 0.53, and 0.51, respectively. The maximum value of the decrease with the variable yield stress was reached at an out-of-plane distance of 2000 mm, and the decrease value was 0.07 with yield stresses of 282 and 408 MPa, while the decrease ratio was 0.06 with yield stresses of 557 and 769 MPa.

5.3.4. Effect of Out-of-Plane Distance on Range and Rate of Change for All Variables

The previous results show the extent of the effect of an out-of-plane distance on the variables affecting the ultimate capacity, such as the dimensions of the cross-section and the yield stress of the steel used. For this purpose, the ratio between the highest rate of reduction in the M_{OOP}/M_{woop} to the lowest rate of reduction for the same variable is obtained. When using the BFIB-300 beam cross-section dimension, the reduction rates were higher than when using the BFIB-600 with all the out-of-plane distances used. Therefore, the ratio of the M_{OOP}/M_{woop} for BFIB-300 to BFIB-600 is used to obtain the relationship between the effects of the out-of-plane distances on the cross-section dimensions. Also, the decreasing rate ratio was found for steel yield stress from 282 MPa to 769 MPa, as shown in Figure 15. From Figure 15, the rates of change due to the effect of the cross-section dimension and steel yield stress are 16.7%, and 7.3%, respectively, for out-of-plane distances of 300 mm, while the rates of change are 45.5%, and 35.0%, respectively, for out-of-plane distances of 2000 mm. From these values, the effect of the out-of-plane distances is higher for the dimensions of the cross-section than the effect of the yield stress of the steel used.

5.4. The Discussion

When steel beams with an out-of-plane distance are used, the relationship between the bending moments and the corresponding deflection increases significantly. The ultimate bending moment capacity decreases with increasing out-of-plane distance, although the corresponding deflection values increase. This is due to the loss of this type of out-of-plane steel beams to the arch system in resisting deflection, as shown in Figure 7. With the presence of an out-of-plane distance, stresses are concentrated at the outer edges of the upper flange. These stresses are of different types: one type is tensile stress at the outer edge of the beam, while the other type is compressive stress at the inner edge of the beam. These stresses result from the occurrence of a global buckling along the entire length of the steel beam in the upper flange. This dent causes an acceleration in the failure process of the beam and its inability to resist external loads. It is the main reason for the failure of this system, as it causes a large increase in stress values exceeding the yield stresses of the steel used under the influence of small load values, as shown in Figures 8 and 9.

The effect of the cross-sectional dimensions on steel beams with an out-of-plane distance is a direct effect; as the cross-sectional dimensions increase, the percentage of reduction in the ultimate capacity of the bending moments increases. This increase in the reduction is due to the approach of the cross-sectional dimensions, especially the web, to another design classification according to the design codes. The web category classification ratios approach the end of the compact category and the beginning of the noncompact category. In addition, with the increase in the web beam height, the possibility of the buckling of the upper flange increases, and this is enhanced by the presence of the concentration of two different types of stresses at the upper flange edges. As for the effect of the length of the beam, no effect was observed on the percentage of decrease in the ultimate capacity of bending moments when using an out-of-plane distance. This is because the buckling occurring at the upper flange occurs along the entire length of the beam, in addition to the fact that the critical location in which stresses are concentrated is precisely defined, where part of an out-of-plane distance connects with the longitudinal beam. The effect of the yield stresses of the steel used on the percentage of reduction in the ultimate capacity of the bending moments of steel beams with out-of-plane distance is directly proportional. As the yield stress increases, the percentage of reduction in the ultimate capacity increases. This is because, with the increase in yield stress values, the design category of the beam cross-sections changes according to the design codes. Through Equations (1) to (17), it was found that the yield stress was inversely proportional to the change in the limits of the design categories of the cross-sections. Thus, with the increase in the yield stress values, they approach the end of the compact category and the beginning of the noncompact category. This result is consistent with previous studies, such as those of Sayed and Al-Jarbou [29].

6. Prediction of Beam Capacity with Out-of-Plane Distance

The second group was used in the FEM analysis process; it contained 116 specimens of steel beams, divided into 19 specimens without an out-of-plane distance and 97 specimens with an out-of-plane distance. Different variables were used, including four types of different cross-sections,, BFIB-300, BFIB-400, BFIB-500, and BFIB-600; the lengths of the steel beams changed from 6000 mm to 10,000 mm, and different types of yield stresses were also used for the steel used, where it changes from 282 MPa to 769 MPa, as listed in Table 6. In addition, the values calculated from the current design codes, ANSI/AISC 360-22 [51] and European-3 [52], are also listed in Table 6.

Table 6. Summary of FEM results for steel I-beams with and without out-of-plane distance.

FEM Specimens	OOP (mm)	Shear Span (m)	FEM Results		ANSI/AISC 360-22 [51] Eurocode-3 [52]	
			M_{FEM} (kN.m)	M_{OOP}/M_{woop}	M_{Code} (kN.m)	M_{Code}/M_{FEM}
BFIB500-L10-fy282-without OOP	0	5.0	1648.57	1.00	1442.40	0.87
BFIB500-L10-fy282-OOP300	300	5.0	905.30	0.55	---	1.59
BFIB500-L10-fy282-OOP400	400	5.0	561.24	0.34	---	2.57
BFIB500-L10-fy282-OOP500	500	5.0	437.09	0.27	---	3.30
BFIB500-L10-fy282-OOP600	600	5.0	337.79	0.20	---	4.27
BFIB500-L10-fy282-OOP800	800	5.0	247.05	0.15	---	5.84
BFIB500-L10-fy282-OOP1000	1000	5.0	169.49	0.10	---	8.51
BFIB500-L10-fy282-OOP2000	2000	5.0	117.27	0.07	---	12.30
BFIB500-L9-fy282-without OOP	0	4.5	1700.46	1.00	1442.40	0.85
BFIB500-L9-fy282-OOP300	300	4.5	902.65	0.53	---	1.60
BFIB500-L9-fy282-OOP400	400	4.5	563.22	0.33	---	2.56
BFIB500-L9-fy282-OOP500	500	4.5	435.76	0.26	---	3.31
BFIB500-L9-fy282-OOP600	600	4.5	336.89	0.20	---	4.28
BFIB500-L9-fy282-OOP800	800	4.5	249.01	0.15	---	5.79
BFIB500-L9-fy282-OOP1000	1000	4.5	169.70	0.10	---	8.50
BFIB500-L9-fy282-OOP2000	2000	4.5	117.18	0.07	---	12.31
BFIB500-L8-fy282-without OOP	0	4.0	1695.00	1.00	1442.40	0.85
BFIB500-L8-fy282-OOP300	300	4.0	914.66	0.54	---	1.58
BFIB500-L8-fy282-OOP400	400	4.0	558.53	0.33	---	2.58
BFIB500-L8-fy282-OOP500	500	4.0	439.43	0.26	---	3.28
BFIB500-L8-fy282-OOP600	600	4.0	335.27	0.20	---	4.30
BFIB500-L8-fy282-OOP800	800	4.0	247.38	0.15	---	5.83
BFIB500-L8-fy282-OOP1000	1000	4.0	168.90	0.10	---	8.54
BFIB500-L8-fy282-OOP2000	2000	4.0	117.18	0.07	---	12.31
BFIB500-L6-fy282-without OOP	0	3.0	1691.78	1.00	1442.40	0.85
BFIB500-L6-fy282-OOP300	300	3.0	918.72	0.54	---	1.57
BFIB500-L6-fy282-OOP400	400	3.0	564.39	0.33	---	2.56
BFIB500-L6-fy282-OOP500	500	3.0	439.43	0.26	---	3.28
BFIB500-L6-fy282-OOP600	600	3.0	336.28	0.20	---	4.29
BFIB500-L6-fy282-OOP800	800	3.0	248.69	0.15	---	5.80
BFIB500-L6-fy282-OOP1000	1000	3.0	169.10	0.10	---	8.53
BFIB500-L6-fy282-OOP2000	2000	3.0	117.18	0.07	---	12.31
BFIB300-L10-fy282-without OOP	0	5.0	614.00	1.00	541.23	0.88
BFIB300-L10-fy282-OOP300	300	5.0	369.82	0.60	---	1.46
BFIB300-L10-fy282-OOP400	400	5.0	262.61	0.43	---	2.06
BFIB300-L10-fy282-OOP500	500	5.0	207.80	0.34	---	2.60
BFIB300-L10-fy282-OOP600	600	5.0	166.38	0.27	---	3.25
BFIB300-L10-fy282-OOP800	800	5.0	104.63	0.17	---	5.17
BFIB300-L10-fy282-OOP1000	1000	5.0	87.19	0.14	---	6.21
BFIB300-L10-fy282-OOP2000	2000	5.0	66.84	0.11	---	8.10
BFIB300-L9-fy282-without OOP	0	4.5	620.81	1.00	541.23	0.87
BFIB300-L9-fy282-OOP300	300	4.5	379.92	0.61	---	1.42
BFIB300-L9-fy282-OOP400	400	4.5	264.64	0.43	---	2.05
BFIB300-L9-fy282-OOP500	500	4.5	207.94	0.33	---	2.60
BFIB300-L8-fy282-without OOP	0	4.0	623.80	1.00	541.23	0.87
BFIB300-L8-fy282-OOP300	300	4.0	375.85	0.60	---	1.44
BFIB300-L8-fy282-OOP400	400	4.0	263.73	0.42	---	2.05
BFIB300-L8-fy282-OOP500	500	4.0	209.25	0.34	---	2.59

Table 6. *Cont.*

FEM Specimens	OOP (mm)	Shear Span (m)	FEM Results		ANSI/AISC 360-22 [51] Eurocode-3 [52]	
			M_{FEM} (kN.m)	M_{OOP}/M_{woop}	M_{Code} (kN.m)	M_{Code}/M_{FEM}
BFIB300-L6-fy282-without OOP	0	3.0	629.72	1.00	541.23	0.86
BFIB300-L6-fy282-OOP300	300	3.0	381.45	0.61	---	1.42
BFIB300-L6-fy282-OOP400	400	3.0	264.85	0.42	---	2.04
BFIB300-L6-fy282-OOP500	500	3.0	209.25	0.33	---	2.59
BFIB400-L10-fy282-without OOP	0	5.0	1109.65	1.00	960.99	0.87
BFIB400-L10-fy282-OOP300	300	5.0	640.83	0.58	---	1.50
BFIB400-L10-fy282-OOP400	400	5.0	410.68	0.37	---	2.34
BFIB400-L10-fy282-OOP500	500	5.0	317.75	0.29	---	3.02
BFIB400-L10-fy282-OOP600	600	5.0	246.06	0.22	---	3.91
BFIB400-L10-fy282-OOP800	800	5.0	178.25	0.16	---	5.39
BFIB400-L10-fy282-OOP1000	1000	5.0	129.51	0.12	---	7.42
BFIB400-L10-fy282-OOP2000	2000	5.0	93.00	0.08	---	10.33
BFIB400-L9-fy282-without OOP	0	4.5	1093.16	1.00	960.99	0.88
BFIB400-L9-fy282-OOP300	300	4.5	638.65	0.58	---	1.50
BFIB400-L9-fy282-OOP400	400	4.5	411.94	0.38	---	2.33
BFIB400-L9-fy282-OOP500	500	4.5	320.85	0.29	---	3.00
BFIB400-L8-fy282-without OOP	0	4.0	1097.59	1.00	960.99	0.88
BFIB400-L8-fy282-OOP300	300	4.0	641.31	0.58	---	1.50
BFIB400-L8-fy282-OOP400	400	4.0	407.20	0.37	---	2.36
BFIB400-L8-fy282-OOP500	500	4.0	322.40	0.29	---	2.98
BFIB400-L6-fy282-without OOP	0	3.0	1120.15	1.00	960.99	0.86
BFIB400-L6-fy282-OOP300	300	3.0	645.91	0.58	---	1.49
BFIB400-L6-fy282-OOP400	400	3.0	408.59	0.36	---	2.35
BFIB400-L6-fy282-OOP500	500	3.0	320.85	0.29	---	3.00
BFIB600-L10-fy282-without OOP	0	5.0	2283.04	1.00	1926.68	0.84
BFIB600-L10-fy282-OOP300	300	5.0	1147.39	0.50	---	1.68
BFIB600-L10-fy282-OOP400	400	5.0	714.58	0.31	---	2.70
BFIB600-L10-fy282-OOP500	500	5.0	549.28	0.24	---	3.51
BFIB600-L10-fy282-OOP600	600	5.0	415.01	0.18	---	4.64
BFIB600-L10-fy282-OOP800	800	5.0	292.95	0.13	---	6.58
BFIB600-L10-fy282-OOP1000	1000	5.0	195.30	0.09	---	9.87
BFIB600-L10-fy282-OOP2000	2000	5.0	130.20	0.06	---	14.80
BFIB600-L9-fy282-without OOP	0	4.5	2277.41	1.00	1926.68	0.85
BFIB600-L9-fy282-OOP300	300	4.5	1157.15	0.51	---	1.67
BFIB600-L9-fy282-OOP400	400	4.5	714.59	0.31	---	2.70
BFIB600-L9-fy282-OOP500	500	4.5	549.28	0.24	---	3.51
BFIB600-L8-fy282-without OOP	0	4.0	2276.29	1.00	1926.68	0.85
BFIB600-L8-fy282-OOP300	300	4.0	1146.23	0.50	---	1.68
BFIB600-L8-fy282-OOP400	400	4.0	718.89	0.32	---	2.68
BFIB600-L8-fy282-OOP500	500	4.0	546.84	0.24	---	3.52
BFIB600-L6-fy282-without OOP	0	3.0	2253.80	1.00	1926.68	0.85
BFIB600-L6-fy282-OOP300	300	3.0	1159.71	0.51	---	1.66
BFIB600-L6-fy282-OOP400	400	3.0	713.41	0.32	---	2.70
BFIB600-L6-fy282-OOP500	500	3.0	546.84	0.24	---	3.52
BFIB500-L10-fy408-without OOP	0	5.0	2341.41	1.00	2087.90	0.89
BFIB500-L10-fy408-OOP300	300	5.0	1267.99	0.54	---	1.65
BFIB500-L10-fy408-OOP400	400	5.0	792.44	0.34	---	2.63
BFIB500-L10-fy408-OOP500	500	5.0	616.74	0.26	---	3.39
BFIB500-L10-fy408-OOP600	600	5.0	453.87	0.19	---	4.60
BFIB500-L10-fy408-OOP800	800	5.0	327.79	0.14	---	6.37
BFIB500-L10-fy408-OOP1000	1000	5.0	222.47	0.10	---	9.39
BFIB500-L10-fy408-OOP2000	2000	5.0	156.88	0.07	---	13.31

Table 6. *Cont.*

FEM Specimens	OOP (mm)	Shear Span (m)	FEM Results		ANSI/AISC 360-22 [51] Eurocode-3 [52]	
			M_{FEM} (kN.m)	M_{OOP}/M_{woop}	M_{Code} (kN.m)	M_{Code}/M_{FEM}
BFIB500-L10-fy557-without OOP	0	5.0	3160.82	1.00	2848.99	0.90
BFIB500-L10-fy557-OOP300	300	5.0	1660.14	0.53	---	1.72
BFIB500-L10-fy557-OOP400	400	5.0	1046.59	0.33	---	2.72
BFIB500-L10-fy557-OOP500	500	5.0	790.21	0.25	---	3.61
BFIB500-L10-fy557-OOP600	600	5.0	588.72	0.19	---	4.84
BFIB500-L10-fy557-OOP800	800	5.0	417.23	0.13	---	6.83
BFIB500-L10-fy557-OOP1000	1000	5.0	268.67	0.09	---	10.60
BFIB500-L10-fy557-OOP2000	2000	5.0	192.71	0.06	---	14.78
BFIB500-L10-fy769-without OOP	0	5.0	4368.50	1.00	3933.35	0.90
BFIB500-L10-fy769-OOP300	300	5.0	2227.32	0.51	---	1.77
BFIB500-L10-fy769-OOP400	400	5.0	1398.09	0.32	---	2.81
BFIB500-L10-fy769-OOP500	500	5.0	1062.51	0.24	---	3.70
BFIB500-L10-fy769-OOP600	600	5.0	790.61	0.18	---	4.98
BFIB500-L10-fy769-OOP800	800	5.0	524.22	0.12	---	7.50
BFIB500-L10-fy769-OOP1000	1000	5.0	323.26	0.07	---	12.17
BFIB500-L10-fy769-OOP2000	2000	5.0	266.46	0.06	---	14.76

First stage: The simulation results of the BFIB steel beams without and with out-of-plane spacing were used to verify the accuracy of the existing codes, Eurocode-3 and ANSI/AISC 360-22. First, the FEM analysis results of the BFIB cross-section steel beams without out-of-plane distance were compared with the prediction results obtained from existing design code calculations [49,50], as shown in Figure 16. By determining the dimensions of the cross-sections and calculating the design category for each section according to the design codes, it became clear that all the specimens from the type BFIB fell into the compact category according to ANSI/AISC 360-22 and the Class-1 category according to Eurocode-3. From Figure 16, it is clear that the values of the ultimate beam moment capacity without out-of-plane distance resulting from the design codes are always less than the ultimate beam capacity resulting from the FEM simulation. This means that it falls on the safe side of the design. The average M_{Code}/M_{FEM} ratio was 0.867 with a correlation coefficient value of 0.999 and a COV of 2.072%, and this was the case for all the design codes of ANSI/AISC 360-22 and European-3. This result is in complete agreement with the conclusions available from previous studies and with Sayed and Aljarbou [29] and Pereira and Mahendran [30].

Secondly, the predicted results from the current design codes were compared with the beam capacity with out-of-plane distance, as shown in Figures 17–20. The ratio between M_{wCode}/M_{oopFEM} and the out-of-plane distance was compared for the variable cross-section dimensions, as shown in Figure 17. It is clear from Figure 17 that all the specimens with out-of-plane distance fall outside the design safe zone, where the predicted values generated by the current design codes are always larger than the actual values generated by FEM simulation. It was found that when using an out-of-plane distance of 300 mm with the cross-sections of dimensions BFIB-300, BFIB-400, BFIB-500, and BFIB-600, the ratios between the values predicted from the design codes and the values of the FEM results of M_{wCode}/M_{oopFEM} were equal to 1.46, 1.50, 1.59, and 1.68, respectively. This difference increased with the increase in the out-of-plane distance, and when this distance reached 2000 mm for the same cross-section of steel beams, the ratios between the M_{wCode}/M_{oopFEM} values were equal to 8.10, 10.33, 12.30, and 14.80, respectively. From this result, it is clear that the effect of the cross-section dimensions is directly proportional to the increase in the ratio between the prediction of the design codes and the results of the numerical analysis.

This is explained by the fact that with the increase in the dimensions of the cross-section of the beam, the height of the web increases; thus, the ratio between the height of the web and the thickness increases, which brings these ratios closer to the maximum limit of the design category according to the existing codes. Thus, the cross-sectional resistance changes as the design category of the beam approaches the end of the compact category and the beginning of the noncompact category; this causes the clear difference shown in Figure 17.

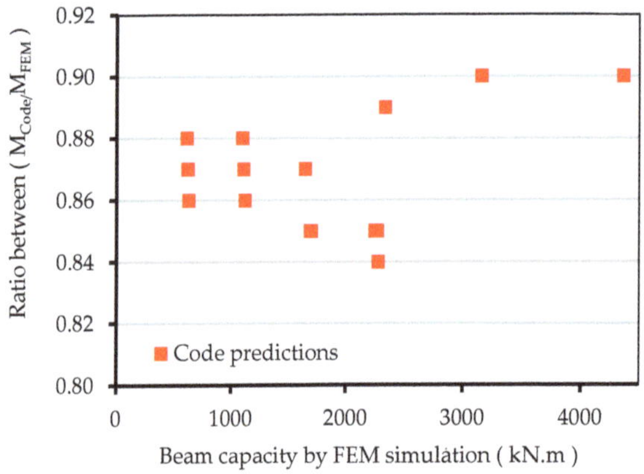

Figure 16. Bending moment ratio between current design codes [51,52] and FEM results for the beam without out-of-plane distance.

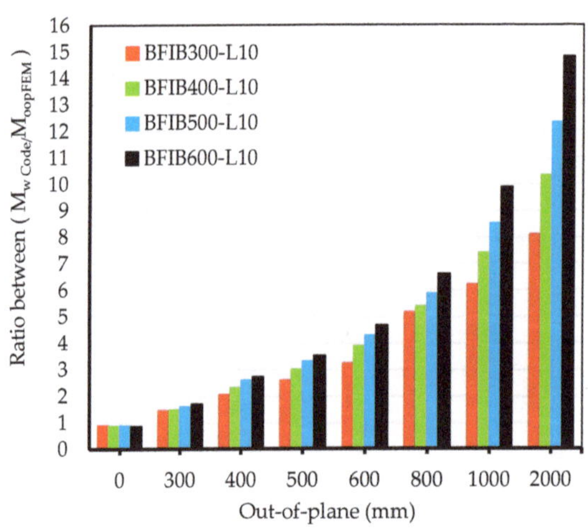

Figure 17. Comparison of existing design codes [51,52] and out-of-plane value for beam cross-section effect.

Regarding the effect of the length of the steel beam on the prediction values of the ultimate bending moment capacity for the current design codes, the relationship between the prediction ratio from the design codes and the ultimate bending moment capacity for the same steel beams with out-of-plane distances was established, as shown in Figure 18. From Figure 18 it is clear that the effect of the length of the steel beam does not affect the prediction values resulting from the current design codes. When using an out-of-plane distance of 300 mm for steel beams of lengths 6000, 8000, 9000, and 10,000 mm, the reduction ratios M_{wCode}/M_{oopFEM} were 1.59, 1.60, 1.58, and 1.57, respectively. When the out-of-plane distance was increased to 2000 mm, the reduction ratio for all the lengths was 12.31 for all the beam lengths. This result is similar to the same result obtained previously when studying the effect of steel beam length on the reduction value of the ultimate bending moment capacity of BFIB steel beams with out-of-plane distances.

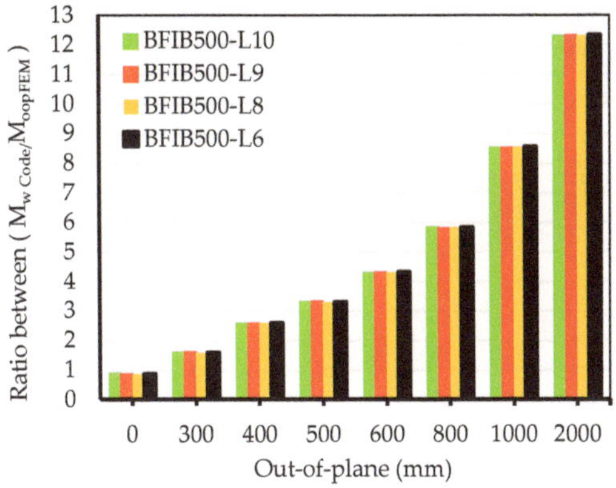

Figure 18. Comparison of existing design codes [51,52] and out-of-plane value for beam length effect with BFIB 500 section.

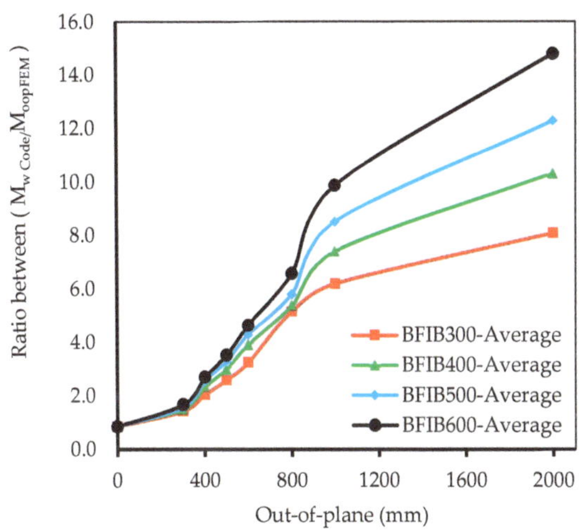

Figure 19. Comparison of average moment capacity ratios from current design codes [51,52] and out-of-plane value for beam length and cross-section effects.

Therefore, the two variables were combined, where the cross-sectional dimensions (which have a direct proportional effect on the percentages of decrease in the ultimate capacity) with the length of the steel beam (which has no effect) and the average of the percentages of these two variables were taken, as shown in Figure 13. Therefore, the average M_{wCode}/M_{oopFEM} ratio for BFIB-300 to BFIB-600 is used to obtain the relationship between the effects of the out-of-plane distances on the cross-section dimensions. From Figure 19, the range and rate of the change in the variable with out-of-plane distances of 300 mm and 2000 mm are 15.97% and 82.72%, respectively.

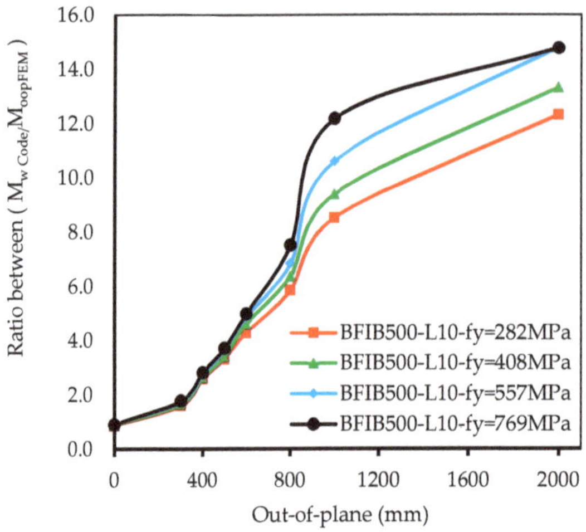

Figure 20. Comparison of average moment capacity ratios from current design codes [51,52] and out-of-plane value for steel beam yield stress effect.

The effect of changing the yield stress of the steel used on the prediction values of the existing design codes was also studied by changing the out-of-plane distance, as shown in Figure 20. By using an out-of-plane distance of 300 mm with steel beams with yield stresses of 282, 408, 557, and 769 MPa, the ratio of M_{wCode}/M_{oopFEM} reached 1.59, 1.65, 1.72, and 1.77, respectively. The maximum value of the ratio with variable yield stress was reached at an out-of-plane distance of 2000 mm, and the ratios of M_{wCode}/M_{oopFEM} reached 12.30, 13.31, 14.78, and 14.76, respectively. The ratio of M_{wCode}/M_{oopFEM} for the steel yield stresses of 282 MPa and 769 MPa was used to obtain the relationship between the effects of out-of-plane distances and predicted values from the current design codes. From Figure 20, it is clear that the prediction values from the existing design codes are directly proportional to the increase in yield stress values. This is due to the fact that yield stresses are inversely proportional to the maximum limits of the design categories (Equations (1) to (17)), which changes the resistance of the cross-sections due to their transition from one design category to another or from the increase in the ratios to the end of the current category and the beginning of a new design category. Through these proportions, the range and rate of the change in the variable with out-of-plane distances of 300 mm and 2000 mm are 11.32% and 20.00%, respectively.

Second stage: Using the above results, a new model can be obtained to predict the reduction ratio of the ultimate BFIB beam moment capacity with off-plan distance. From Table 6 and Figures 10–15, the variables have a direct relationship with the reduction ratio in the BFIB beam capacity when used as an out-of-plane value, in terms of out-of-plane distance, cross-sectional dimensions, and steel beam yield stress. As a result, factor K takes the effect of the variables, as was proposed in Equation (19). The value of the out-of-plane distance was added as a direct value to the relationship, as well as the value of the yield stress of the steel used. Since all the sections fell within the compact category according to the design codes, the cross-section dimensions were replaced by the plastic section modulus of the cross-section. The relationship between the M_{OOP}/M_{woop} ratios and the K factor was plotted, as shown in Figure 21.

$$K = OOP \times S_{Pl} \times f_y \quad (19)$$

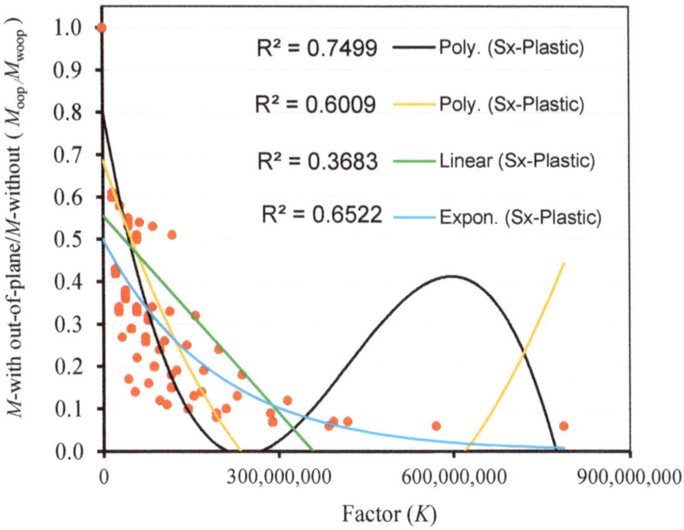

Figure 21. The relationship between moment capacity reduction ratio and factor K was obtained from FEM analysis.

Four different types of functions were used for all of these values; the functions were linear, exponential, second-degree polynomial, and third-degree polynomial. This was to obtain the best possible function that could be relied upon when proposing a model that could predict the percentage of reduction in the ultimate capacity of bending moments for steel beams with an out-of-plane distance. It is clear from Figure 21 that the linear effect of the variables does not give sufficient accuracy in predicting the ultimate BFIB steel beam capacity with the out-of-plane distance. This is evident from a coefficient of determination R^2 which comes from the third-degree polynomial relationship, in which the largest possible value is 0.7499. In order to obtain higher accuracy in the results, it was necessary to change the effect of the variables from linear to nonlinear, through trial and error. The power of the variables was changed, and the coefficient of determination was measured each time to reach the best possible value. From this result, a modification of the K_m factor was obtained, and it became that used in Equation (20). After applying the factor K_m, the calculated values were very large numbers, as the variables were placed in the equation in millimeters; so, the number 10,000 was used to achieve mathematically accepted values.

$$K_m = OOP \times {S_{Pl}}^{0.18} \times {f_y}^{0.10} / 10000 \quad (20)$$

The relationship between the M_{oop}/M_{woop} ratio and the new factor K_m was plotted, as shown in Figure 22. Through the third-degree polynomial function, which has the highest coefficient of determination value, Equation (21) was obtained with a coefficient of determination equal to 0.9907. The new models can predict the ultimate BFIB steel beam capacity with out-of-plane distance as a ratio of the ultimate capacity of steel beams without out-of-plane distance with high accuracy and a coefficient of correlation of 0.995. From Figure 22 and the third-order polynomial function used to propose the new model, it is clear that increasing the value of the factor K_m above 6.0 gives incorrect prediction values for the reduction ratios of the steel beams with an out-of-plane distance. Therefore, a limit of this factor K_m was set which should not exceed 6.0 when applying Equation (21) to predict the reduction ratios of these steel beams.

$$\frac{M_{\mathrm{oop}}}{M_{woop}} = -0.0168{K_{\mathrm{m}}}^{3} + 0.2009{K_{\mathrm{m}}}^{2} - 0.7592K_{\mathrm{m}} + 1.0048 \quad where\ K_m \leq 6.0 \tag{21}$$

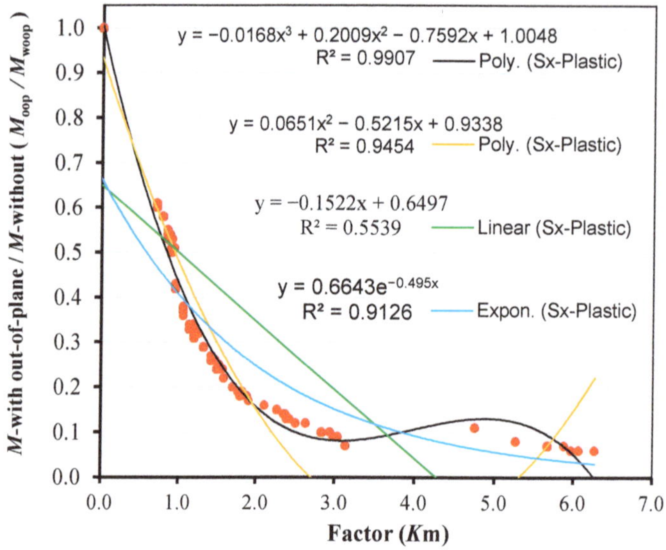

Figure 22. Relationship between moment capacity reduction ratio and the modification factor K_{m}.

7. Conclusions

The possibility of an out-of-plane distance for steel beams in the lateral direction along their length can occur as a result of deviations in the edges of the area to be covered or the presence of obstacles in the natural direction along the length of the beam. To study this research point, firstly, the FEM simulation results were verified by comparing them with the beams available in previous studies without out-of-plane distance in their length. The results were accurate with an average beam capacity ratio of 1.007, a coefficient of correlation of 0.999, and a COV of 1.573%. Secondly, FEM modeling was performed on BFIB steel beams with and without out-of-plane distance to take into account different variables, such as out-of-plane distance, cross-section dimensions, length of steel beams, and yield stress of steel beams. Through this study, the following conclusions were reached:

- The existing design codes were used to predict the BFIB beam capacity without out-of-plane distance in the compact category; both the FEM simulation results and the experimental test results were less than the steel beam capacity.
- The existing design codes were used to predict the beam capacity, and any application of the out-of-plane distance was more than the steel beam capacity, i.e., they were on the unsafe side of the design.
- The reduction ratios of the ultimate moment capacity in out-of-plane BFIB steel beams were directly proportional to the out-of-plane distance, cross-sectional dimensions, and yield stress of the steel beams, while the length of the steel beams had no effect.
- The effect of cross-sectional dimensions and steel yield stress on the reduction ratios was nonlinear, except for the out-of-plane distance which was linear.
- When an out-of-plane distance of 300 mm was applied to the BFIB-300 steel beam, this distance was only equal to the width of the cross-section flange of the steel beam; the reduction ratio in the ultimate moment capacity was 0.60. As the cross-sectional dimensions increased, the reduction ratio increased, reaching 0.50 with the BFIB-600 steel beam.

- In steel beams that contain an out-of-plane distance, failure is always due to the concentration of compressive stresses at the inner edge; there is a concentration of tensile stresses at the outer edge of the upper flange at the point of contact with the out-of-plane distance that results from global buckling along the entire length of the beam.

This study resulted in several research points that can be used in future studies. Conducting experimental tests on these steel beams with an out-of-plane distance included the use of several variables that were proven to affect the efficiency of these structural elements. Studying these steel beams with an out-of-plane distance using different support ends, such as their connection to steel columns, caused a difference in the structural system and thus a difference in the results. Various other locations along the beam axis where there is an out-of-plane distance need to be considered, such as one-quarter and one-third of the beam length. Other cross-sections used in the manufacture of steel beams, whether hot-formed or built-up sections, can also be studied.

Author Contributions: Conceptualization, A.M.S., N.M.A. and H.A.; methodology, A.M.S., M.H.A., A.A. and N.M.A.; software, A.M.S., H.A. and A.A.; validation, A.M.S., N.M.A., H.A., A.H.A. and S.A.; formal analysis, A.M.S., N.M.A., A.H.A. and H.A.; investigation, A.M.S., A.A., H.A., A.H.A. and S.A.; resources, A.M.S. and A.H.A.; data curation, A.M.S., A.H.A. and N.M.A.; writing—original draft preparation, A.M.S.; writing—review and editing, N.M.A., M.H.A., A.A., H.A., A.H.A. and S.A.; visualization, A.M.S., M.H.A., H.A. and A.H.A.; supervision, A.M.S.; project administration, A.M.S., N.M.A., M.H.A., A.A., A.H.A., H.A. and S.A. All authors have read and agreed to the published version of the manuscript.

Funding: This research received no external funding.

Data Availability Statement: Data will be made available on request.

Acknowledgments: The author extends the appreciation to the Deanship of Postgraduate Studies and Scientific Research at Majmaah University for funding this research work through the project number (R-2025-1531).

Conflicts of Interest: The authors declare no conflicts of interest.

References

1. Lee, C.H.; Han, K.H.; Uang, C.M.; Kim, D.-K.; Park, C.-H.; Kim, J.-H. Flexural strength and rotation capacity of I-shaped beams fabricated from 800 MPa steel. *J. Struct. Eng.* **2013**, *139*, 1043–1058. [CrossRef]
2. Thomas, S.J.; Earls, C.J. Cross-sectional compactness and bracing requirements for HPS483W girders. *J. Struct. Eng.* **2003**, *129*, 1569–1583. [CrossRef]
3. Han, Y.; Li, N.; Zhang, G. Study on flexural behavior of corroded i-shaped steel beams strengthened with hybrid CFRP/GFRP sheets. *Materials* **2023**, *16*, 5080. [CrossRef]
4. Earls, C.J.; Shah, B.J. High performance steel bridge girder compactness. *J. Constr. Steel Res.* **2002**, *58*, 859–880. [CrossRef]
5. Green, P.S.; Sause, R.; Ricles, J.M. Strength and ductility of HPS flexural members. *J. Constr. Steel Res.* **2002**, *58*, 907–941. [CrossRef]
6. Earls, C.J. Uniform moment behavior of high performance steel I-shaped beams. *J. Constr. Steel Res.* **2001**, *57*, 711–728. [CrossRef]
7. Xue, J.; Ma, S.; Chen, X.; Wu, Q.; Wang, Y.; Wang, Y.; Akbar, M.; Yang, N. Experimental investigation and design of novel hollow flange beams under bending. *Buildings* **2024**, *14*, 1413. [CrossRef]
8. El-Mahdy, G.M.; El-Saadaway, M.M. Ultimate strength of singly symmetric I-section steel beams with variable flange ratio. *Thin-Walled Struct.* **2015**, *87*, 149–157. [CrossRef]
9. Hamoda, A.; Yehia, S.A.; Ahmed, M.; Sennah, K.; Abadel, A.A.; Shahin, R.I. Experimental and numerical investigation of shear strengthening of simply supported deep beams incorporating stainless steel plates. *Buildings* **2024**, *14*, 3680. [CrossRef]
10. Elamary, A.S.; Mohamed, M.A.; Sharaky, I.A.; Mohamed, A.K.; Alharthi, Y.M.; Ali, M.A.M. Utilizing artificial intelligence approaches to determine the shear strength of steel beams with flat webs. *Metals* **2023**, *13*, 232. [CrossRef]
11. Nikoomanesh, M.R.; Goudarzi, M.A. Experimental and numerical evaluation of shear load capacity for sinusoidal corrugated web girders. *Thin-Walled Struct.* **2020**, *153*, 106798. [CrossRef]

12. Jáger, B.; Kövesdi, B.; Dunai, L. Numerical investigations on bending and shear buckling interaction of I-Girders with slender WEB. *Thin-Walled Struct.* **2019**, *143*, 106199. [CrossRef]
13. Lim, J.; Kang, Y.-J.; Lee, J.; Kim, S.; Lee, K. Experimental and analytical study of horizontally curved i-girders subjected to equal end moments. *Metals* **2021**, *11*, 1132. [CrossRef]
14. Andrade, A.; Camotim, D. Lateral-torsional buckling of singly symmetric web-tapered beams: Theory and application. *J. Eng. Mech. ASCE* **2005**, *131*, 586–597. [CrossRef]
15. Andrade, A.; Camotim, D.; Dinis, P.B. Lateral-torsional buckling of singly symmetric web-tapered thin-walled I-beams: 1D model vs. shell FEA. *Comput. Struct.* **2007**, *85*, 1343–1359. [CrossRef]
16. Mascolo, I.; Modano, M.; Fiorillo, A.; Fulgione, M.; Pasquino, V.; Fraternali, F. Experimental and numerical study on the lateral-torsional buckling of steel c-beams with variable cross-section. *Metals* **2018**, *8*, 941. [CrossRef]
17. Asgarian, B.; Soltani, M.; Mohri, F. Lateral-torsional buckling of tapered thin-walled beams with arbitrary cross-sections. *Thin-Walled Struct.* **2013**, *62*, 96–108. [CrossRef]
18. Tan, L.; Zhou, Y.; Kong, H.; Yue, Z.; Wang, Q.; Zhou, L. Dynamic monitoring of steel beam stress based on PMN-PT sensor. *Buildings* **2024**, *14*, 2831. [CrossRef]
19. Sayed, A.M. Numerical study of the effects of flange damage on the steel beam capacity under static and fatigue flexural loads. *Results Eng.* **2023**, *17*, 100835. [CrossRef]
20. Nasery, M.M. Investigation on behaviours along weak axes of steel beam under low velocity impact loading: Experimental and numerical. *Buildings* **2023**, *13*, 2331. [CrossRef]
21. Narendra, P.V.R.; Singh, K.D. Elliptical hollow section steel cantilever beams under extremely low cycle fatigue flexural load–a finite element study. *Thin-Walled Struct.* **2017**, *119*, 126–150. [CrossRef]
22. Yuan, H.X.; Wang, Y.Q.; Gardner, L.; Shi, Y.J. Local-overall interactive buckling of welded stainless steel box section compression members. *Eng. Struct.* **2014**, *67*, 62–76. [CrossRef]
23. Cai, M.; Liu, S.; Wang, F. Temperature effect of composite girders with corrugated steel webs considering local longitudinal stiffness of webs. *Buildings* **2024**, *14*, 1939. [CrossRef]
24. Abed, F.; Abdul-Latif, A.; Yehia, A. Experimental study on the mechanical behavior of en08 steel at different temperatures and strain rates. *Metals* **2018**, *8*, 736. [CrossRef]
25. Hu, W.; Zhang, Z.; Shi, J.; Chen, Y.; Li, Y.; Feng, Q. Refined analysis of spatial three-curved steel box girder bridge and temperature stress prediction based on WOA-BPNN. *Buildings* **2024**, *14*, 415. [CrossRef]
26. Saliba, N.; Real, E.; Gardner, L. Shear design recommendations for stainless steel plate girders. *Eng. Struct.* **2014**, *59*, 220–228. [CrossRef]
27. *EN 1993-1-4*; Eurocode 3: Design of Steel Structures–Part 1.4: General Rules Supplementary Rules for Stainless Steel. CEN: Brussels, Belgium, 2006.
28. *EN 1993-1-5*; Eurocode 3: Design of Steel Structures–Part 1.5: Plated Structural Elements. CEN: Brussels, Belgium, 2006.
29. Sayed, A.M.; Aljarbou, M.H. Numerical study of the effect of circular openings in the upper surface of rectangular hollow flange steel beams on the sectional moment capacity. *Results Eng.* **2024**, *24*, 103620. [CrossRef]
30. Perera, N.; Mahendran, M. Section moment capacity tests of hollow flange steel plate girders. *J. Constr. Steel Res.* **2018**, *148*, 97–111. [CrossRef]
31. Graciano, C. Strength of longitudinally stiffened webs subjected to concentrated loading. *J. Struct. Eng.* **2005**, *131*, 268–278. [CrossRef]
32. Silwal, P.K.; Parvin, A.; Alhusban, M. Numerical investigation on strengthening of steel beams for corrosion damage or web openings using carbon fiber reinforced polymer sheets. *Buildings* **2024**, *14*, 1069. [CrossRef]
33. Zhang, X.; Yu, S.; Feng, S.; Fan, D.; Zhang, F.; Cao, H. Study on the bending performance of prefabricated h-shaped steel beams with different bolt hole types. *Buildings* **2024**, *14*, 2988. [CrossRef]
34. Yu, N.T.; Kim, B.; Yuan, W.B.; Li, L.Y.; Yu, F. An analytical solution of distortional buckling resistance of cold-formed steel channel-section beams with web openings. *Thin-Walled Struct.* **2019**, *135*, 446–452. [CrossRef]
35. El-Ariss, B.; Elkholy, S. Feasibility of reusing damaged steel beams in temporary structures. *Infrastructures* **2021**, *6*, 69. [CrossRef]
36. Hou, W.; Wang, L.; Shi, D. Flexural behaviour of strengthened damaged steel beams using carbon fibre-reinforced polymer sheets. *Sci. Rep.* **2022**, *12*, 10134. [CrossRef]
37. Kumar, S.; Gupta, P.K.; Iqbal, M.A. Experimental and numerical study on self-compacting alkali-activated slag concrete-filled steel tubes. *J. Constr. Steel Res.* **2024**, *214*, 108453. [CrossRef]
38. Wang, Y.-B.; Li, G.-Q.; Chen, S.-W.; Sun, F.-F. Experimental and numerical study on the behavior of axially compressed high strength steel box-columns. *Eng. Struct.* **2014**, *58*, 79–91. [CrossRef]
39. Rabi, M.; Ferreira, F.P.V.; Abarkan, I.; Limbachiya, V.; Shamass, R. Prediction of the cross-sectional capacity of cold-formed CHS using numerical modelling and machine learning. *Results Eng.* **2023**, *17*, 100902. [CrossRef]
40. Al-Kaseasbeh, Q. Analysis of hydrocarbon fire-exposed cold-formed steel columns. *Results Eng.* **2023**, *20*, 101400. [CrossRef]

41. Chan, C.L.; Khalid, Y.A.; Sahari, B.B.; Hamouda, A.M.S. Finite element analysis of corrugated web beams under bending. *J. Constr. Steel Res.* **2002**, *58*, 1391–1406. [CrossRef]
42. Zhang, Y.C.; Chen, Y. Experimental study and finite element analysis of square stub columns with straight ribs of concrete-filled thin-walled steel tube. *J. Build. Struct.* **2006**, *27*, 16–22.
43. Shao, Y.B.; Mohamed, H.S.; Wang, L.; Wu, C.S. Experimental and numerical investigation on stiffened rectangular hollow flange beam. *Int. J. Steel Struct.* **2020**, *20*, 1564–1581. [CrossRef]
44. Erdal, F.; Tunca, O.; Tas, S.; Ozcelik, R. Experimental and finite element study of optimal designed steel corrugated web beams. *Adv. Struct. Eng.* **2021**, *24*, 1814–1827. [CrossRef]
45. Isleem, H.F.; Chukka, N.D.K.R.; Bahrami, A.; Oyebisi, S.; Kumar, R.; Qiong, T. Nonlinear finite element and analytical modelling of reinforced concrete filled steel tube columns under axial compression loading. *Results Eng.* **2023**, *19*, 101341. [CrossRef]
46. Uzzaman, A.; Lim, J.B.P.; Nash, D.; Rhodes, J.; Young, B. Cold-formed steel sections with web openings subjected to web crippling under two-flange loading conditions—Part I: Tests and finite element analysis. *Thin-Walled Struct.* **2012**, *56*, 38–48. [CrossRef]
47. Xue, J.; Ma, S.; Chen, X.; Wu, Q.; Akbar, M. Finite element modeling of assembling rivet-fastened rectangular hollow flange beams in bending. *J. Constr. Steel Res.* **2023**, *211*, 108177. [CrossRef]
48. Senapati, S.; Sangle, K.K. Nonlinear static analysis of cold-formed steel frame with rigid connections. *Results Eng.* **2022**, *15*, 100503. [CrossRef]
49. Perera, N.; Mahendran, M. Finite element analysis and design for section moment capacities of hollow flange steel plate girders. *Thin-Walled Struct.* **2019**, *135*, 356–375. [CrossRef]
50. Tao, Z.; Wang, Z.-B.; Yu, Q. Finite element modelling of concrete-filled steel stub columns under axial compression. *J. Construct. Steel Res.* **2013**, *89*, 121–131. [CrossRef]
51. *ANSI/AISC 360-22*; Specification for Structural Steel Buildings. American Institute of Steel Construction: Chicago, IL, USA, 2022.
52. *EN 1993-1-1: 2005*; Eurocode 3: Design of Steel Structures-Part 1-1: General Rules and Rules for Buildings. European Committee for Standardization: Brussels, Belgium, 2005.
53. ANSYS. *ANSYS Help User's Manual, Version 22*; Swanson Analysis Systems, Inc.: Technology Drive Canonsburg, PA, USA, 2022.
54. Sayed, A.M.; Alanazi, H. Performance of steel metal prepared using different welding cooling methods. *Case Stud. Constr. Mater.* **2022**, *16*, e00953. [CrossRef]
55. Shi, Y.; Xu, K.; Shi, G.; Li, Y. Local buckling behavior of high strength steel welded I-section flexural members under uniform moment. *Adv. Struct. Eng.* **2018**, *21*, 93–108. [CrossRef]
56. Shokouhian, M.; Shi, Y.J. Flexural strength of hybrid steel I-beams based on slenderness. *Eng. Struct.* **2015**, *93*, 114–128. [CrossRef]
57. Yang, B.; Dong, M.; Han, Q.; Elchalakani, M.; Xiong, G. Flexural behavior and rotation capacity of welded i-beams made from 690-mpa high-strength steel. *J. Struct. Eng.* **2021**, *147*, 04020320. [CrossRef]
58. Saliba, N.; Gardner, L. Experimental study of the shear response of lean duplex stainless steel plate girders. *Eng. Struct.* **2013**, *46*, 375–391. [CrossRef]
59. Beg, D.; Hladnik, L. Slenderness limit of class 3 I cross-sections made of high strength steel. *J. Constr. Steel Res.* **1996**, *38*, 201–217. [CrossRef]
60. Xiong, G.; Kang, S.-B.; Yang, B.; Wang, S.; Bai, J.; Nie, S.; Hu, Y.; Dai, G. Experimental and numerical studies on lateral torsional buckling of welded Q460GJ structural steel beams. *Eng. Struct.* **2016**, *126*, 1–14. [CrossRef]

buildings

MDPI

Correction

Correction: Mata, R.; Nuñez, E. Cyclic Behavior of Concrete-Filled Tube Columns with Bidirectional Moment Connections Considering the Local Slenderness Effect. *Buildings* 2024, *14*, 2240

Ramón Mata and Eduardo Nuñez *

Department of Civil Engineering, Universidad Católica de la Santísima Concepción, Concepción 4061735, Chile; ramon.mata@uss.cl
* Correspondence: enunez@ucsc.cl

Affiliation Modification

In the published publication [1], there was an error regarding the affiliation for "Ramón Mata". In addition to affiliation 1, the updated affiliation should NOT include the following:

"2 Facultad de Ingeniería, Arquitectura y Diseño, Universidad San Sebastián, Lientur 1457, Concepción 4080871, Chile"

The authors state that the scientific conclusions are unaffected. This correction was approved by the Academic Editor. The original publication has also been updated.

Reference

1. Mata, R.; Nuñez, E. Cyclic Behavior of Concrete-Filled Tube Columns with Bidirectional Moment Connections Considering the Local Slenderness Effect. *Buildings* **2024**, *14*, 2240. [CrossRef]

buildings

MDPI

Article

Lateral-Torsional Buckling Behavior of Low-Carbon H-Shape Bolted Composite Beams

Kojo Kumah Darkwah, Linfeng Lu *, Bingyou Liu, Zhengzhong Huang and Hanlin Hao

School of Civil Engineering, Chang'an University, 75 Chang'an Middle Rd, Xi'an 710061, China; 2020028900@chd.edu.cn (K.K.D.); 2023228019@chd.edu.cn (B.L.); 2024128046@chd.edu.cn (Z.H.); 2021128031@chd.edu.cn (H.H.)
* Correspondence: lulinfeng@chd.edu.cn

Abstract: The construction industry faces increasing pressure to reduce energy consumption and carbon emissions. Promoting sustainable practices, such as using hot-rolled H-shapes and minimizing welding, can address these challenges. This study introduces low-carbon H-shape bolted composite beams (HBCBs) as a sustainable alternative. Experimental and numerical investigations into the lateral-torsional buckling behavior of HBCBs demonstrated that high-strength bolts significantly enhance structural stability and capacity. A reliable simulation method was developed and validated based on experimental results. The study reveals that bolt spacing exceeding 1/15 of the beam span has negligible effects on load-bearing capacity, and HBCB performance reaches approximately 95% of plug-welded counterparts. Despite this, longitudinal slip reduction remains challenging, warranting further research to optimize performance.

Keywords: H-shape beam; lateral-torsional buckling; longitudinal slippage; low-carbon; high-strength bolt

1. Introduction

The traditional construction sector faces unprecedented challenges in energy saving and carbon emission reduction (CEs). The construction industry significantly impacts the environment, accounting for about 30% of global CEs, 40% of world energy consumption [1,2], and 51.3% of total CEs in China [3,4]. Consequently, it must play a crucial role in reducing emissions. Most embodied CEs occur before the operation phase [5], making material selection, efficient building processes, design for disassembly, and optimized logistics essential [6].

Choosing low-carbon materials can notably reduce a building's initial carbon footprint. Low-carbon materials refer to materials that minimize carbon emissions throughout their entire life cycle, including raw material extraction, production, transportation, construction, usage, and disposal. The carbon footprint of a material is typically assessed based on its embodied carbon (kgC/kg). According to the data from Hammond and Jones [7], the embodied carbon of steel, concrete, and timber are 0.482, 0.035, and 0.125 kgC/kg, respectively. Timber generally has a lower carbon footprint, and some types of timber can absorb CO_2 during their growth. Therefore, timber generally has lower CEs than concrete and steel [8], but poor design can result in higher emissions [9]. Prefabricated steel structures, on the other hand, often conserve energy and reduce CEs more effectively than concrete [10], making them a promising long-term solution for the construction industry. During construction, carbon emissions for a steel frame are 10.1 tCO_2, whereas for reinforced concrete (RC), they are 45.5 tCO_2 [2], further highlighting the advantages of steel structures in reducing environmental impact, especially by replacing the traditional RC shear wall structure with a new steel plate shear wall system [11,12] and other new technologies [13,14]. Therefore, in this study, the concept of low carbon specifically refers to reducing carbon emissions during construction.

Studies have shown that the component production phase is the primary source of carbon emissions (CEs) in steel structures, accounting for 60–73% of total lifecycle emissions [3]. Within this phase, welding-related CEs are particularly significant, comprising 57.2% of emissions in the prefabrication process [15]. Therefore, reducing welding can significantly lower overall carbon emissions. Hence, steel components should utilize hot-rolled shapes (HRSs) or cold-formed shapes (CFSs) to minimize welding. Although CFSs with fasteners have the lowest carbon emissions [16], their application is limited. Conversely, the utilization rate of HRSs remains low. For example, in China, the world's largest steel producer, the engineering utilization rate of HRSs is only 15–20%, with no significant changes in recent years [6]. Developing high-performance HRS built-up components is crucial. Lu et al. [17,18] introduced innovative designs such as flanged cruciform H-shape columns (FCHCs) and T-shaped built-up columns (NTBCs). Also, bolts should replace welding where possible, exemplified by developing weld-free columns [19].

Beams, which use more steel than columns in steel buildings, face limitations due to insufficient manufacturing and market standards for hot-rolled H-shapes. To address this, the authors propose H-shape bolted composite beams (HBCBs), similar to using high-strength bolts to connect steel plate girders' two sub-beams to solve issues of shipping and field lifting [20]. HHLBs can be used for new construction and can also strengthen existing H-beams. Zhu et al. [20] investigated the bearing capacity of composite and I-section plate beams of the same height and flange plates through experimental comparison. Although all specimens exhibited lateral-torsional buckling with slight web yielding near loading points, the composite plate beam's bearing capacity was 6.1% higher than the I-section one. However, they did not thoroughly address composite plate beams' lateral-torsional buckling or overall stability.

Xu et al. [21–23] used a similar strategy connecting two cold-formed channels in modular light steel buildings. Their findings revealed that laminated equal channel beams had greater loading capacity and bending stiffness than independent bending double beams [21]. Increasing the number of bolt connections constrained slipping behavior, suggesting that an appropriate number of bolts should be used in engineering practice for optimal efficiency [22]. The results [23] showed that the influence of design factors on EIe (bending stiffness) and Pu (ultimate capacity) were ranked as follows: $n > \beta > f > b > p$ for EIe and $p > b > \beta > f > n$ for Pu. Here, n represents the number of bolts, f interfacial friction, β layer height ratio, p load type, and b boundary condition. However, Xu et al.'s [21–23] study did not focus on hot-rolled H-shapes.

In summary, several research gaps remain in studying low-carbon H-shape bolted composite beams (HBCBs). Systematic experimental and numerical investigations into their lateral-torsional buckling behavior are lacking, and the main influencing factors are still unclear. Therefore, this paper explores the lateral-torsional buckling behavior of HBCBs and the effects of different factors through experimental and numerical studies. The main contributions are as follows: first, lateral-torsional buckling tests were conducted on an HBCB specimen and a bolted-free comparison specimen; second, a complete ABAQUS analysis methodology was proposed and validated through experiments; and finally, numerical investigations were carried out on the influence of high-strength bolt pre-loadings and the influence of high-strength bolt spacings. These studies lay the foundation for future research on the mechanical performance analysis and design methods of HBCBs.

2. Experimental Investigations

2.1. Specimens Design and Original Data

The cross-section and geometric symbols of an HBCB specimen are shown in Figure 1. Two extension lengths were arranged outside of two supports to fix the beam and the supports, making the specimen length L, where $L = L_e + 2 \times 150$ mm. Thus, the geometric length L of all test beams was 3000 mm, with an effective length Le of 2700 mm.

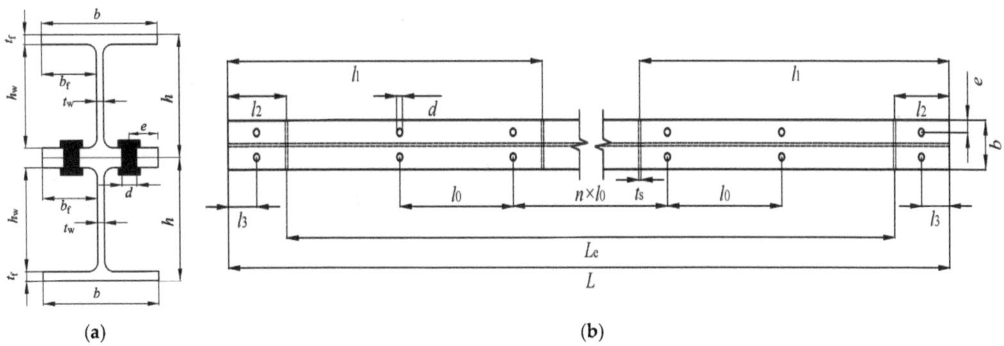

Figure 1. Cross-sectional configurations and dimensional symbols of the HBCB specimen. (**a**) Cross-section; (**b**) top view.

In this study, an HBCB used two Q235B HW100 × 100 × 6 × 8 H-shapes combined with 8.8-grade M12 high-strength bolts. A corresponding no-connection laminated beam (NCLB) specimen was also created to study the effect of bolts. According to GB50017-2017 [24], beams met the S1 cross-section requirements, which allows for the entire section's plasticity. A pair of 6 mm thick transverse stiffeners were placed at the supports, quarter points, and mid-span of the beam's effective length.

The bolts were spaced at 300 mm spacings equal to $1/9\ L_e$. At the supports, two bolts were positioned 75 mm apart. According to GB50017-2017 [24], a pre-tensioned force of 38.5 kN was given to a high-strength bolt of the HHLB, and the bolts' center spacing l_0, end spacing l_3, and edge margin spacing e complied with the requirements.

The actual geometric dimensions of the specimens were measured before and after assembly, as detailed in Tables 1 and 2. Figure 2 presents the simplified engineering stress–strain curve, and the data from the coupon tensile test are listed in Table 3.

Table 1. Dimensions of cross-sections (Unit: mm).

Specimen	t_f	t_w	b	b_f	h	h_w
The upper beam of HBCB	7.9	6.2	100.2	47.2	99.6	83.8
The lower beam of HBCB	8.1	6.2	100.4	47.0	100.0	83.6
The upper beam of NCLB	8.1	6.2	100.1	47.1	100.1	83.8
The lower beam of NCLB	8.1	6.0	100.3	47.0	100.2	83.9

Table 2. Dimensions of specimens (Unit: mm).

Specimen	L	L_e	H	l_0	l_1	l_2	l_3	d	e	t_s
HBCB	2999.74	2699.44	199.58	299.6	824.4	149.4	74.4	13.9	25.2	5.9
NCLB	3000.02	2699.96	200.30	299.9	824.9	149.9	74.9	13.9	25.1	5.8

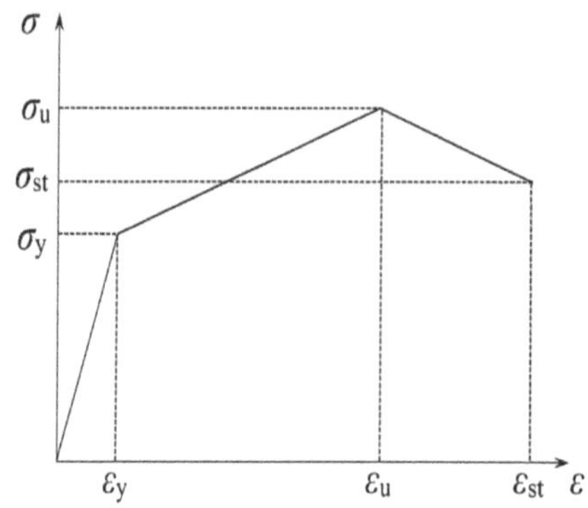

Figure 2. Three-line stress–strain relationship curve.

Table 3. Mechanical properties of steel and bolts.

Samples	E/MPa	σ_y/MPa	ε_y	σ_u/MPa	ε_u	σ_{st}/MPa	ε_{st}
Web 100-1	203,000	345.7	0.00191	464.1	0.17724	368.7	0.24158
Web 100-2	202,000	359.3	0.00218	469.2	0.18205	368.9	0.26537
Web 100-3	208,000	362.8	0.00240	471.0	0.17979	369.2	0.25779
Mean	204,333	355.9	0.00216	468.1	0.17969	368.9	0.25491
Flange 100-1	202,000	320.2	0.00163	440.8	0.25002	345.2	0.35135
Flange 100-2	210,000	308.2	0.00163	445.6	0.38066	343.8	0.38066
Flange 100-3	204,000	291.0	0.00390	441.9	0.34566	345.6	0.34566
Mean	205,333	306.5	0.00239	442.8	0.32545	344.9	0.35922
Bolt	161,000	645.0	0.004	793.5	0.025	636.9	0.121

2.2. Test Setup

Figure 3a shows the loading system and on-site photos of the specimen. An actuator, a rigid extension rod, a load distribution beam, and two loading point solid round steel bars were used in the loading system to apply two-point concentrated loads at the four quarter points of the upper flange of the specimen, causing the beam to form a pure bending between the two loading points. Two particular beam end supports, including a simple-supported support and splint support at each beam end, were designed, as shown in Figure 3b. At the supporting end, the beam cannot freely warp and can rotate around the major and minor axes without lateral displacement and rotate around its longitudinal axis.

Before testing loading, all specimens were pre-loaded with 5 mm and held for 10 min to check the loading equipment and data collection system and eliminate the gaps between the specimen parts and test setup. After pre-loading, they were unloaded to zero. A displacement loading controlled the formal testing loading; the speed was 1 mm/min. The displacement increment of each step was 1 mm, held for 2 min to check and judge whether there were any abnormalities in the specimen. After the overall instability of a specimen, we continued loading until the load dropped to 80% of the peak load or the maximum elongation capacity of the actuator was reached, and at this time, we finished the test.

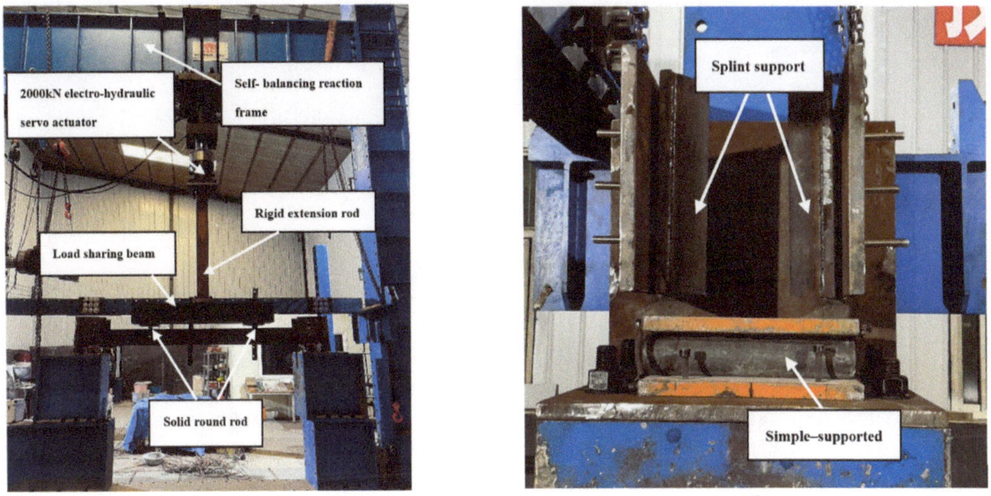

(a) (b)

Figure 3. Photo of the test setup on site. (**a**) Test setup; (**b**) Beam end support.

2.3. Arrangement of Displacement Meters and Strain Gauges

Figure 4a shows the specimen's displacement meters arrangement. Because the beam moved downward during the test, a smooth glass piece was pasted outside the mid-span stiffener of the specimen to ensure that the displacement meters D1, D2, and D3 measuring the lateral displacement of the beam could slip up and down freely without changing its horizontal direction. Among them, D2 was located at the superimposed flange of the upper and lower beams. The displacement meter D4 measured the vertical displacement of the beam at mid-span. Displacement meters D5 and D6 measured the longitudinal relative slip of the upper and lower steel beams. Displacement meters D7 and D8 measured the vertical displacement at the support. With the displacement meter D4, the actual deflection at the mid-span of the beam was calculated. Because the pure bending section of the beam should be the first to undergo overall instability failure, strain gauges were only arranged on the mid-span area of the specimen, as shown in Figure 4b. A data acquisition system hooked up to the loading system automatically recorded loads, displacements, and strains.

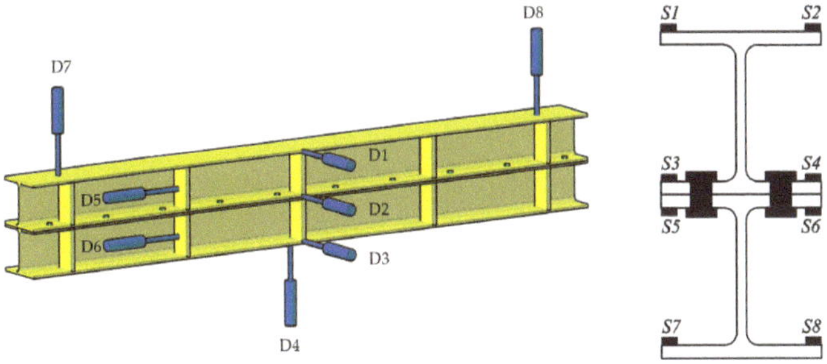

(a) (b)

Figure 4. Displacement meters and strain gauges of specimens. (**a**) Arrangement of displacement meters; (**b**) arrangement of strain gauges.

3. Test Results and Discussion

3.1. Test Observations

3.1.1. HBCB Specimen

In the initial loading stage, the HBCB remained in the elastic phase. A clear and loud sound was heard when the loading displacement reached 14 mm (approximately 92 kN), indicating a longitudinal relative slip between the upper and lower flanges connected by bolts. Real-time data from displacement meters D1, D2, and D3 showed that the mid-span section of the specimen tilted. After that, the load continued to increase, the mid-span deflection grew significantly, and the load increment rate began to slow.

At the loading displacement of 22 mm (approximately 104 kN), another sudden sound occurred, and the load dropped to about 93 kN. Real-time monitoring with displacement meters D5 and D6 revealed longitudinal relative slip along the beam's length. As loading progressed, the load data continued to rise, and at a displacement of 124 mm (approximately 127 kN), the actuator reached its maximum range due to equipment limitations, resulting in a mid-span deflection of 117 mm, at which point the test was terminated.

The HBCB specimen did not exhibit significant mid-span section torsion or lateral shifting. Minor displacements were noted at the stiffeners of the upper and lower beams, and the connecting bolts of the composite beam were tilted, indicating longitudinal relative slip. At the end of the test, the mid-span failure mode of the HBCB specimen is shown in Figure 5a, while the overall lateral deformation and deflection after unloading are presented in Figure 5b,c.

Figure 5. The failure modes of the HBCB specimen. (**a**) Lateral-torsional buckling and longitudinal slippage; (**b**) residual plastic deformation out of the plane; (**c**) residual plastic deformation in the plane.

The results suggest that lateral-torsional buckling of the specimens occurred, in which the upper and lower beams worked together (see Figure 5). No local buckling was found in the test.

3.1.2. NCLB Specimen

In the initial loading phase, the specimen remained in the elastic stage. As loading continued, when the displacement reached 20 mm (load of 84 kN), the rate of increase in bearing capacity began to slow. Real-time data from displacement meters D1, D2, and D3 showed that the mid-span section of the specimen tilted. When the displacement increased to 25 mm (load of 89 kN), a clear and loud sound was heard, and real-time data monitoring from displacement meters D5 and D6 revealed a longitudinal relative slip at the stiffeners of the upper and lower beams of the specimen. At this point, the specimen's bearing capacity suddenly dropped to 86 kN. When the displacement loading was up to 31 mm (load of 93 kN), a second distinct sound occurred, and real-time data from displacement meters D1, D2, and D3 showed that the mid-span section of the specimen tilted strongly, and the bearing capacity dropped to 90 kN. The bearing capacity data continued to rise as loading progressed. The specimen reached its maximum bearing capacity when the displacement reached 47 mm (load of 95 kN). During this phase, the specimen emitted continuous sounds, and data from the displacement meters revealed that with each sound, lateral shifting and longitudinal relative slip occurred at both the upper and lower parts of the composite beam. The displacement was further increased to 110 mm, at which point the bearing capacity was 82 kN, approximately 80% of the peak load. Due to actuator range limitations, the test was terminated, and the mid-span deflection was 106 mm.

During the test, it was observed that a lack of bolt constraints exhibited poorer overall performance. Consequently, significant torsion and lateral shifting occurred at the mid-span section of the upper and lower beams, and substantial longitudinal relative slip was also observed. The final failure mode was independent lateral-torsional buckling for upper and lower beams. The failure mode and local phenomena are shown in Figure 6a, and the overall lateral deformation and deflection after unloading are shown in Figure 6b,c.

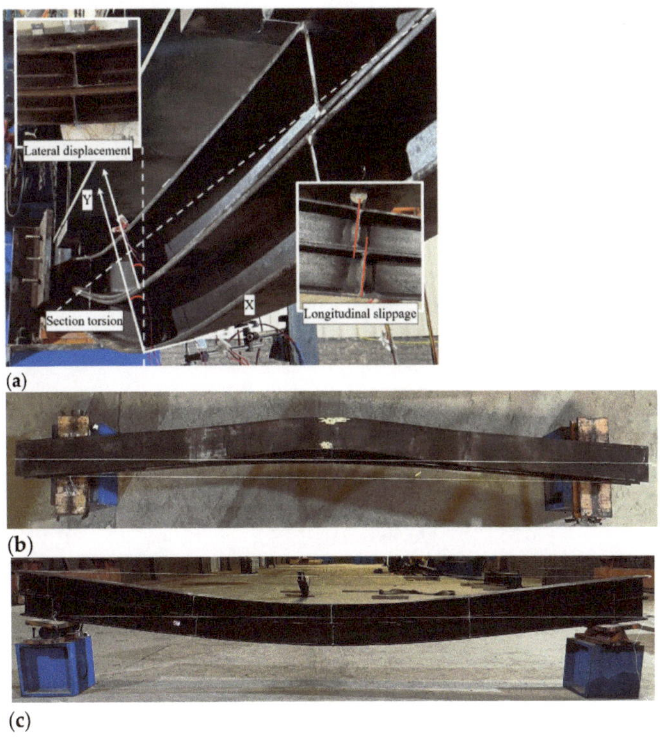

Figure 6. The failure modes of the NCLB specimen. (**a**) Lateral-torsional buckling and longitudinal slippage; (**b**) residual plastic deformation out of the plane; (**c**) residual plastic deformation in the plane.

In addition to longitudinal relative slip, the two beams (upper and lower beams) underwent a relative transverse slip, as shown in Figure 6, which revealed that the lateral movement and torsion of the upper and lower beams are not synchronized. It implied that the lateral-torsional buckling of the upper and lower beams does not develop simultaneously. No local buckling was also found.

3.2. Specimens' Load and Displacement Curves

3.2.1. Load vs. Mid-Span Deflection

Consider the self-weight of the specimen, and the load vs. mid-span deflection is given based on the test data, where the displacement is the average of the differences between the measurement data of displacement meters D4, D7, and D8. Due to insufficient estimation of specimen deflection, the HBCB specimen failed to be loaded to the predetermined end load because the loading actuator reached its limit. Therefore, it has no declining section after the peak value. Figure 7a shows that the HBCB specimen experienced noticeable elastic and elastoplastic stages, and the NCLB specimen underwent three stages from the start of loading to the end of the test: elastic, elastoplastic, and failure.

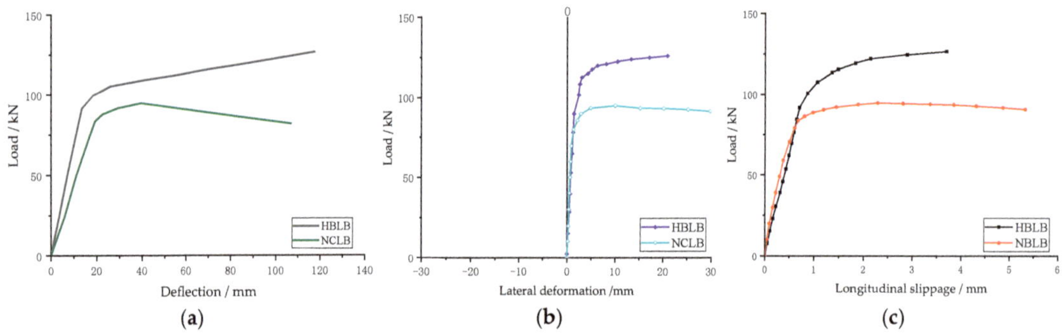

Figure 7. Specimens' load and displacement curves. (**a**) Curves of load–deflection; (**b**) curves of load–lateral deformation; (**c**) Curves of load–longitudinal slippage.

The critical load (P_{cr}), maximum load (P_{max}), and final load (P_{end}) corresponding to the three stages of the specimen are shown in Table 4. The data indicate that with the addition of bolt connections, the critical load and maximum load of the HBCB specimen increased by 9.31% and 33.38%, respectively, compared with the NCLB specimen. The bolt assembly effect is significant.

Table 4. The bearing capacity of the test and ABAQUS (Unit: kN).

Specimen	Test			ABAQUS		
	P_{cr}	P_{max}	P_{end}	$P_{cr, FE}$	$P_{max, FE}$	$P_{end, FE}$
HBCB	91.66	126.59	126.59	99.59	126.05	102.10
Error	-	-	-	+8.65%	−0.43%	-
NCLB	83.85	94.91	82.52	91.27	96.88	80.47
Error	-	-	-	+8.85%	+2.08%	−2.48%

3.2.2. Load vs. Mid-Span Lateral Deformation

The calculation formula for the mid-span lateral deformation at the centroid of the HBCB is Equation (1).

$$u = \frac{1}{2}(\Delta_1 - \Delta_3) \tag{1}$$

Δ_1 and Δ_3 are the measured values of the mid-span displacement meters D1 and D3 in Figure 4a.

Because there was relative lateral slippage between the upper and lower beams of the NCLB specimen and because the real-time data of the relative lateral slippage of the upper and lower beams could not be measured during the test, the accurate lateral displacement of the specimen cannot be calculated. Because the lateral deformation value is an auxiliary index, this study still uses Equation (1) for approximate calculation.

Figure 7b shows the load vs. mid-span lateral deformation of specimens. At lower loads, both specimens exhibited almost no lateral displacement at the mid-span cross-section. However, once the load increased to the critical load (P_{cr}), lateral displacement at the mid-span cross-section began to occur and gradually increased until lateral-torsional buckling occurred. After buckling, the lateral displacement of the mid-span cross-section grew rapidly while the load-carrying capacity decreased sharply, consistent with the typical characteristics of global instability in flexural members. Due to the limitations of the loading device, the HBCB specimen could not be loaded to its ultimate load capacity, resulting in the absence of a descending curve for lateral displacement at the mid-span. However, this does not affect the identification of its lateral-torsional buckling behavior.

3.2.3. Load vs. Relative Longitudinal Slippage

The calculation formula for the relative longitudinal slippage is Equation (2).

$$s = |\Delta_5 - \Delta_6| \tag{2}$$

Δ_5 and Δ_6 are the measured values of the displacement gauges D5 and D6 in Figure 4a.

Figure 7c shows that the longitudinal relative slip between the upper and lower beams of the specimen gradually increased with the load. When the load-carrying capacity reached the critical load of the HBCB specimen, the longitudinal relative slip between the upper and lower beams accelerated until the specimen eventually failed due to instability.

In addition, although the longitudinal relative slip of the HBCB specimen was smaller than that of the NCLB specimen, the collaborative performance between the upper and lower beams of the structure was still far from perfect. This indicates that the number of high-strength bolts used in the specimen might be insufficient to effectively restrict the development of longitudinal relative slip between the upper and lower beams, necessitating further research in the future.

3.3. Specimens' Load and Strain Curve

Figure 8 illustrates the load–strain curves for eight strain gauge measurement points located at the mid-span cross-section of the two specimens, with the strain gauge arrangement shown in Figure 4a. The figure shows that at lower load levels, the measured strain data from strain gauges S1, S2, S7, and S8 are symmetric about the neutral axis, and their curves exhibit a linear growth trend. This indicates that the specimens primarily experienced in-plane bending during the initial loading phase.

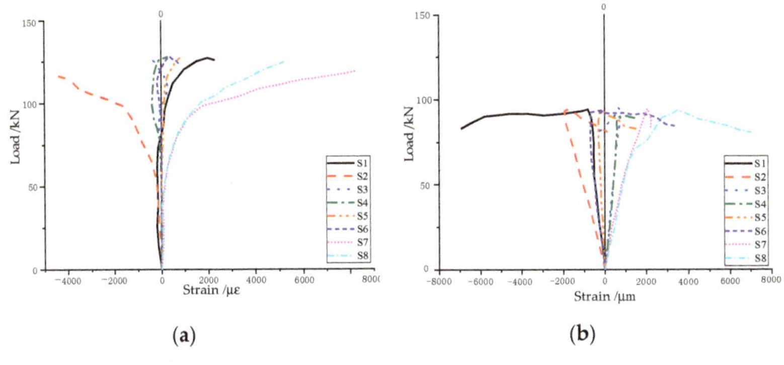

(**a**) (**b**)

Figure 8. Specimens' load and strain curves. (**a**) HBCB specimen; (**b**) NCLB specimen.

As the load increased, the curves displayed clear inflection points and deviations, with their symmetry disrupted, indicating that lateral deformation and torsion occurred in the mid-span cross-section alongside bending, both of which intensified with further loading. Furthermore, Figure 8 shows that the compressive strain in the flanges of the specimens exceeded the average yield strain of the steel flanges obtained from material testing (see Table 3), confirming that the specimens had elastoplastic deformation consistent with observed experimental phenomena.

Comparing the load–strain curves at the mid-span for the specimens, it was found that strain gauges S5 and S6 of the HBCB specimen remained in a state of low strain (close to the neutral axis) until the final load was reached, whereas S5 and S6 of the NCLB specimen were consistently under compression. This phenomenon demonstrates that the high-strength bolted HBCB specimen behaved as an integrated beam, working as a single unit under load, whereas the non-bolted laminated NCLB specimen functioned with the upper and lower beams independently bearing the load.

4. ABAQUS Analysis Methodology and Validation

The two-point loading tests on the HBCB and NCLB specimens effectively demonstrated the beams' lateral-torsional buckling failure mode, indicating the tests' success. The high-strength bolt-connected hot-rolled H-shape composite beams can be considered as a unified beam working together under load, while the upper and lower beams independently load the unbolted pure laminated beam specimens. However, the HBCB specimen still exhibited longitudinal relative slip between the upper and lower beams, indicating that the number of high-strength bolts used in the test was insufficient. Numerical investigations are essential to investigate bolts' influence on the lateral-torsional buckling, load-bearing capacity, deformation, and relative slip of the HBCB. Therefore, this study proposes a reliable numerical analysis method that can be utilized for future investigations involving various parameter variations. The detailed ABAQUS 6.1.4 numerical analysis methodology for model development and verification is outlined below.

4.1. Finite Element Model

4.1.1. Model Component Dimensions

In ABAQUS, the dimensions of the modeled specimens are identical to the actual measured dimensions of the two composite beam specimens used in the experiment. The measured geometric data are shown in Tables 1 and 2. The specimen uses high-strength bolts with diameters corresponding to their nominal diameters (e.g., M12 uses 12 mm). The bolt holes in the experimental specimen are based on actual measured values (Table 2), while the bolt hole diameters in the numerical model are set as the nominal bolt diameter plus 2 mm.

4.1.2. True Stress–Strain Relationship of Steel and Bolts

The ABAQUS analysis accounts for material nonlinearity. According to the stress–strain model of steel and bolts shown in Figure 2, the engineering stress and strain listed in Table 3 must be converted into true stress and strain before inputting the material constitutive model into ABAQUS. The conversion method can be found in Reference [25].

4.1.3. Contact Settings

In ABAQUS, contact settings typically involve defining contact surfaces, the type of contact, and the contact properties.

The contact relationships in the HBCB specimen primarily exist between the lower surface of the upper flange of the upper beam and the upper surface of the lower flange of the lower beam, the bolt cap and the flange surface, the bolt shank and the flange hole wall, and the lower surface of the lower flange of the lower beam and the upper surface of the support base plate. The complex contact relationships can lead to difficulties in model convergence and excessively long computation times during analysis. Therefore, to reduce the convergence difficulty and time in finite element analysis, the high-strength bolt model is simplified by replacing the hexagonal bolt cap with a circular bolt cap, and the influence of the bolt threads is ignored. "Surface-to-Surface Contact" is applied between the flange overlap faces, between the bolt cap and the flange surface, and between the lower surface of the lower flange of the lower beam and the upper surface of the support base plate, with interaction properties defined.

For normal behavior, "Hard Contact" is selected in the pressure-overclosure option, with the "Allow Separation" option being enabled to prevent penetration between model components. For tangential behavior, Coulomb friction is used to describe the sliding behavior between contact surfaces, with the friction coefficient being set to 0.3 according to GB 50017-2017 [24] for "Q235 steel with wire-brushed surfaces to remove rust or untreated clean rolled surfaces". The stiffeners are connected to the flanges and webs of the hot-rolled H-shaped steel beams using fillet welds, with "Tie" constraints being applied in the interaction settings.

The bolt contact settings were removed for the specimen NCLB, while other contact settings remained consistent with those of the HBCB specimen.

4.1.4. Boundary Conditions and Load Application

Appropriate boundary conditions and load applications were defined in the numerical model as follows to simulate the experimental setup accurately.

The experiment utilized simple and splint supports (Figure 3), with line-point contact between the angle knife edge and the side of the beam. At the support locations, the specimen ends were allowed to undergo minor rotations around the X-axis (strong axis) and Y-axis (weak axis), as well as longitudinal sliding along the Z-axis (specimen's longitudinal axis) and free warping. However, lateral and vertical displacements along the X-axis and Y-axis were restricted.

In the numerical analysis, these boundary conditions needed to be implemented in the finite element model by applying displacement and rotational degree-of-freedom constraints and support boundary conditions to ensure consistency with the actual experimental setup. The simple support at one end is the hinge, and the boundary condition of the left end reference point is set as U1 = U2 = U3 = UR2 = UR3 = 0; the simple support at the other end needs to release the degree of freedom in the Z-axis, and the boundary condition of the right end reference point is set as U1 = U2 = UR2 = UR3 = 0; we set the boundary conditions at the stiffeners, flanges, and support pads at both ends of the model as U1 = 0.

When applying concentrated loads, reference points RP-1 and RP-2 should be created at the loading points. A "Coupling" constraint should be established within a 3 mm range on either side of the loading points, and the concentrated load should be applied along the Y-axis direction.

During the BUCKLE analysis phase, it is necessary to determine the most unfavorable model of the component, i.e., the first-order buckling mode. Therefore, a unit concentrated force is applied, and the resulting eigenvalue represents the buckling load of the component.

In the nonlinear analysis phase, loading is performed by applying displacement to the component.

An example of an HBCB specimen is shown in Figure 9a.

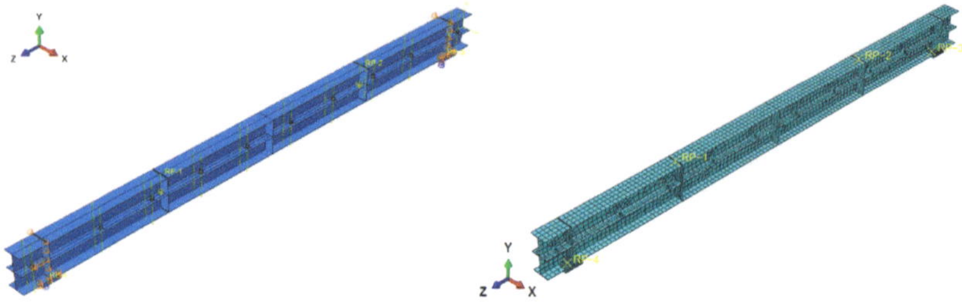

(a) (b)

Figure 9. Finite element model of the HBCB specimen. (**a**) Boundary condition; (**b**) element and mesh.

4.1.5. Element and Mesh

For nonlinear buckling analysis, shell elements provide better analysis accuracy compared with solid elements. Therefore, except for the bolts, which are modeled using solid element C3D8R, all other components are modeled using shell element S4R in the ABAQUS model.

Through extensive modeling and analysis, the final mesh sizes selected were 25 mm for the H-beam model, 10 mm for the stiffener model, and 2 mm for the high-strength bolt model. Additionally, sections with bolt holes were subdivided, and mesh refinement was applied to these areas to achieve optimal mesh quality. The overall model mesh division is shown in Figure 9b.

4.1.6. Bolt Pre-Tension Force

The bolt pre-load is simulated by defining the bolt load before applying the displacement load. Figure 10 shows the location and direction of the applied bolt pre-load. In the finite element simulation, the middle face of the bolt shank was cut, and the bolt load was applied to this middle face in the axial direction of the bolt shank.

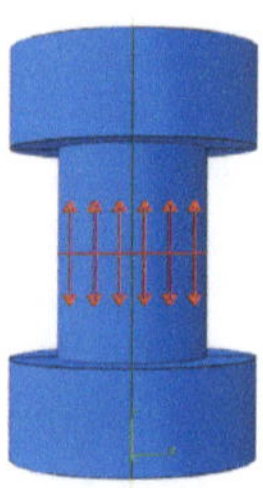

Figure 10. Addition of bolt pre-load in finite element modeling.

4.1.7. Initial Geometric Imperfections

The initial geometric imperfections (initial bending) were introduced using the "consistent defect mode method". This method utilizes the first-order buckling mode obtained from the BUCKLE analysis as the most critical mode. The node coordinates were adjusted according to this mode and scaled by a specified defect coefficient. According to GB 50017-2017 [24], the initial bending was set to one-thousandth of the component's span. In ABAQUS, these imperfections were incorporated using the "*IMPERFECTION" command.

4.1.8. Residual Stress

The residual stress distribution for hot-rolled H-shapes was given by ECCS [26], as shown in Figure 11a. Here, σ_{rc} represents residual compressive stress, and σ_{rt} represents residual tensile stress, with $\sigma_{rc} = -\sigma_{rt} = 0.3f_y$; f_y is the steel's yield strength, using σ_y in

Table 3. It was assumed that the local bolt holes do not alter the distribution of residual stresses. These residual stresses were combined to obtain the residual stress distribution for the HBCB, shown in Figure 11b. In ABAQUS, a predefined field was created in the load module, selecting "stress" and the field area according to the residual stress distribution and introducing the residual stress as the initial stress in the model.

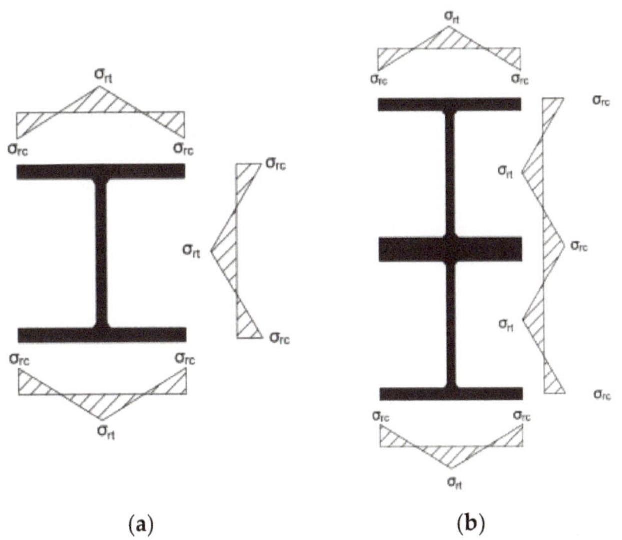

Figure 11. Residual stress distribution. (**a**) Residual stress distribution of an H-beam; (**b**) residual stress distribution of the HBCB and NCLB.

4.2. Finite Element Verification

4.2.1. Comparison of Failure Modes

The modeling and analysis of the HBCB and NCLB specimens were performed following the method in Section 4.1. The finite element simulation failure modes obtained are shown in Figures 12a and 13a. When compared with the experimental failure modes shown in Figures 12b and 13b, it is observed that the FE failure modes are highly similar to the experiments.

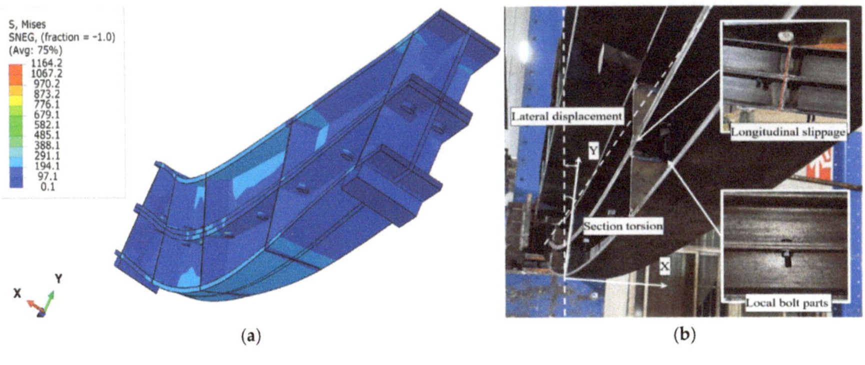

Figure 12. Comparison of failure modes of the HBCB specimen. (**a**) ABAQUS analysis; (**b**) experiment.

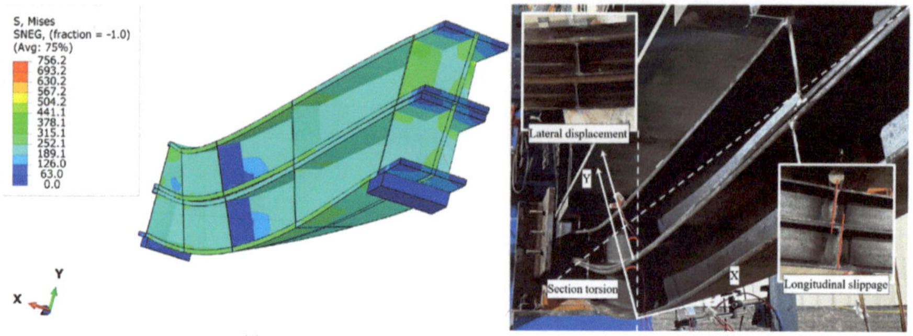

(a) (b)

Figure 13. Comparison of failure modes of the NCLB specimen. (**a**) ABAQUS analysis; (**b**) experiment.

4.2.2. Comparison of Load–Deformation Curves

Figures 14 and 15 show the load–deformation curves, load–mid-span deflection curves, load–mid-span lateral displacement curves, and load–longitudinal slip curves for the HBCB and NCLB specimens.

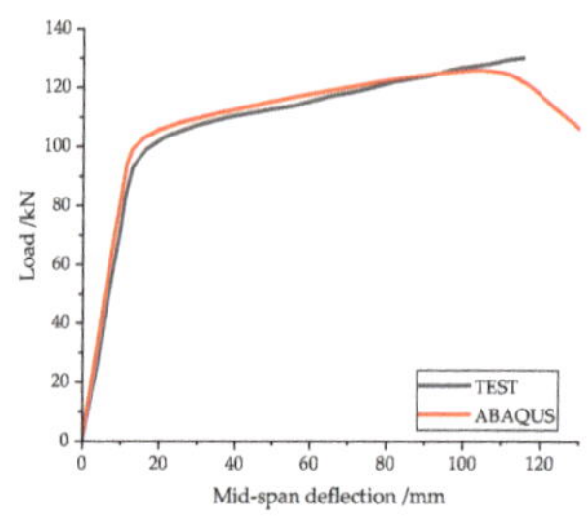

(a)

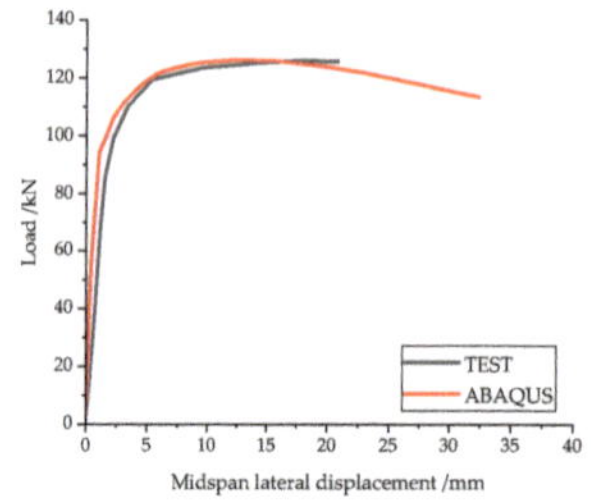

(b)

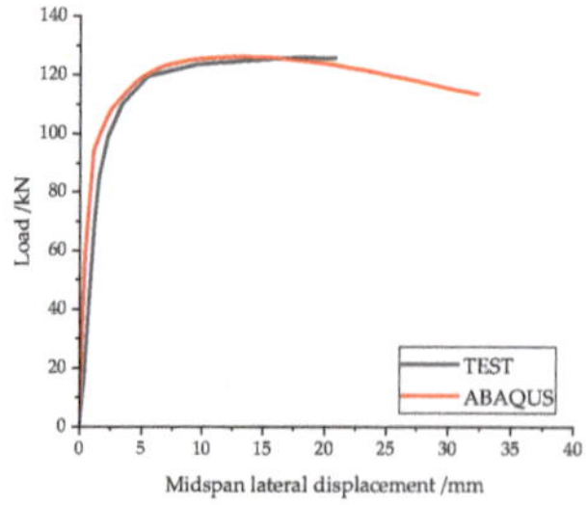

(c)

Figure 14. Comparison of load–deformation curves of the HBCB specimen. (**a**) Load–Mid-span deflection; (**b**) load–mid-span lateral displacement; (**c**) load–longitudinal slip.

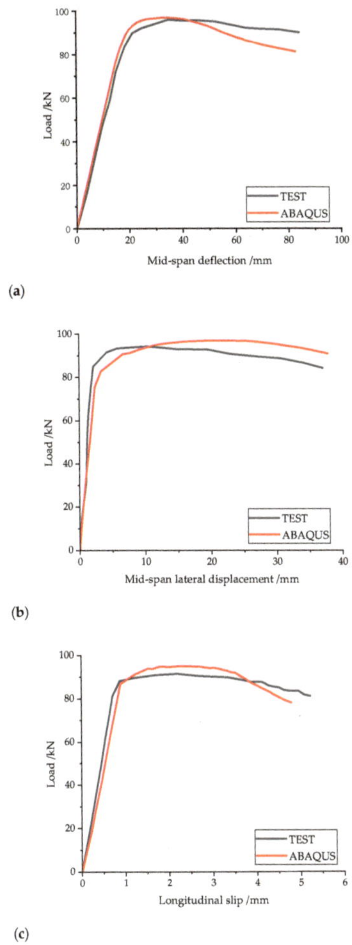

Figure 15. Comparison of load–deformation curves of the NCLB specimen. (**a**) Load–Mid-span deflection; (**b**) load–mid-span lateral displacement; (**c**) load–longitudinal slip.

By comparing Figures 14 and 15, it can be observed that the ABAQUS simulation analysis performs well for the bolted HBCB specimens, with the curve trends closely matching the experimental results. The simulation aligns well with the experiments before lateral-torsional buckling occurs for the boltless NCLB specimens but shows larger deviations in the post-buckling (elastic-plastic) stage. The specimens' critical load (corresponding to elastic lateral-torsional buckling), peak load, and ultimate load (corresponding to the test end) are extracted and listed in Table 4. The error between the critical loads obtained from the ABAQUS analysis and the experimental values is less than 9%, with the error for the peak loads being less than 3%. For the relatively complex composite beams with high-strength bolts discussed in this paper, focusing on the lateral-torsional buckling of the beam, both experimental and numerical simulation approaches are highly challenging. However, the current error magnitude is within an acceptable range. The FE model cannot fully capture material nonlinearity, contact behavior, and bolt slip (NCLB). Test specimens have manufacturing errors, residual stresses, and boundary condition differences. Discrepancies mainly appear in the post-buckling (elastic-plastic) stage, but the overall error is within 9%, ensuring reliability. The current differences mostly reflect the simulation of

NCLB, while the accuracy for HBCB is good. Therefore, the gap in NCLB simulation does not affect future research.

These results demonstrate the reliability and validity of the ABAQUS simulation method used in this study, making it a suitable tool for further research on HBCB specimens.

5. Analysis of the Influence of High-Strength Bolts

The influence of bolts on the structural performance of specimens is a critical aspect of this study. For bolted HBCB specimens, the bolts significantly enhance the stability and overall behavior under load, as evidenced by the closer agreement between ABAQUS simulation results and experimental data. The presence of bolts delays lateral-torsional buckling, resulting in higher critical and peak loads compared with boltless NCLB specimens.

All specimens are modeled based on the experimental HBCB specimen as the prototype, with the same Q235 steel and M12 8.8-grade high-strength bolt specifications remaining unchanged, while only pre-tension forces and bolt spacings are varied.

5.1. The Influence of High-Strength Bolt Pre-Loadings

Due to potential improper operations during the actual assembly of composite beams, the pre-tension force applied to high-strength bolts may deviate from the values specified in the standards. To investigate the effect of bolt pre-tension on the load-bearing capacity of H-shaped steel composite beams, the degree of pre-tension force is described as $c_d = Nr/Ns$, representing the deviation between the actual pre-tension force and the standard requirement. Nr is the actual applied pre-loading (kN), and Ns is the required pre-loading (kN) according to GB50017-2017 [24].

High-strength bolts of M12 8.8 grade were selected, with a bolt spacing of L/10. A finite element model was established with designed pre-loading application degrees of 0.6, 1.0, and 1.4, corresponding to cases of insufficient, standard, and excessive pre-tension, respectively, in practical engineering. Figure 16a shows the load–mid-span deflection curves of the composite beams, while Figure 16b illustrates the load–longitudinal relative slip curves.

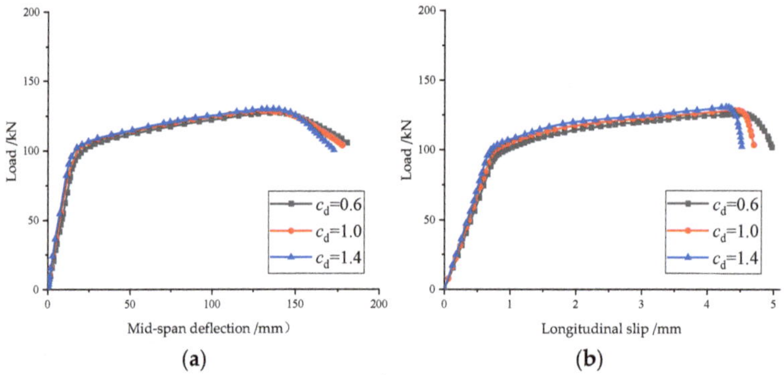

Figure 16. Effect of bolt pre-loadings vs. deformation. (**a**) Mid-span deflection curves; (**b**) Longitudinal slip curves.

Figure 16 shows that the slope of the curves in the elastic stage and the elastic-plastic ultimate load increase with the degree of bolt pre-tension force. However, during the failure stage, the overall deformation of the composite beam decreases as the bolt pre-tension force increases. This indicates that higher bolt pre-tension force improves the composite beam's overall stiffness and load-bearing capacity. In the deflection curve (Figure 16a), higher pre-tension increases stiffness and ultimate load while reducing final deformation. In the slip curve (Figure 16b), higher pre-tension reduces longitudinal slip, enhancing overall stability, but the effect is limited.

According to Table 5, when the actual pre-tension force of M12 8.8-grade high-strength bolts decreases from 38.5 kN to 23.1 kN, the maximum load-bearing capacity of the composite beam reduces by 1.50%, and the maximum longitudinal slip (u_{dmax}) increases by 5.96%. Conversely, when the pre-tension force increases from 38.5 kN to 53.9 kN, the maximum load-bearing capacity rises by 3.20%, and the maximum longitudinal relative slip (u_{dmax}) decreases by 4.04%.

Table 5. The ultimate bearing capacity and the maximum longitudinal slip value of composite beams under different degrees of pre-tension application (Unit: kN).

Bolt Type	Pre-Load Level	N_r (kN)	N_s (kN)	P_{max} (kN)	u_{dmax} (mm)
M12 8.8	0.6	23.1	38.5	124.16	4.98
	1.0	38.5	38.5	126.05	4.70
	1.4	53.9	38.5	130.10	4.51

The comparison indicates that the degree of pre-tension force can enhance the composite beam's ultimate load-bearing capacity and longitudinal slip resistance, though the influence is not significant.

5.2. The Influence of High-Strength Bolt Spacings

In order to further investigate the influence of high-strength bolt spacings, this paper analyzes the different bolt spacings carried out to establish eight bolt spacings of L, L/2, L/4, L/5, L/8, L/10, L/15, L/16, and T, where L represents the length of the member and T represents the connections to the spliced upper and lower beams with plug welds at the bolt holes to replace the high-strength bolts. All specimens are modeled based on the experimental HBCB specimen as the prototype, with the same Q235 steel and M12 8.8-grade high-strength bolt specifications remaining unchanged, while only the bolt spacing arrangement is varied.

Figure 17 shows that as the bolt spacing decreases (equivalent to increasing the number of bolts), the load-bearing capacity of the specimen improves, and longitudinal relative slip is suppressed.

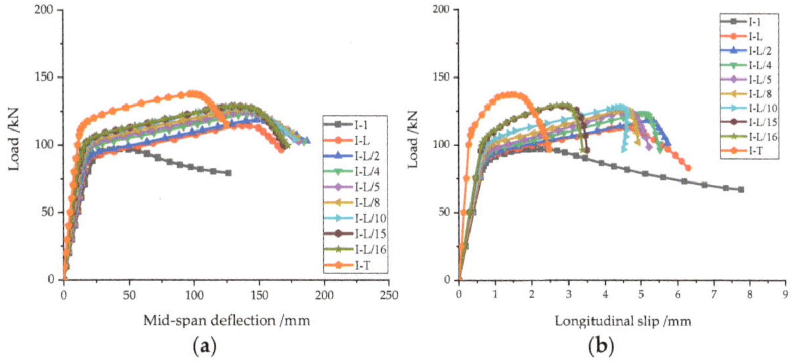

Figure 17. Effect of different bolt spacing on load vs. deformation. (**a**) Mid-span deflection curves; (**b**) longitudinal slip curves.

The data in Table 6 show that when the bolt spacing is reduced from L/10 to L/15, the maximum load changes by only +0.37%. Similarly, reducing the spacing from L/10 to L/16 results in an increase of just +0.76% in maximum load. Moreover, the maximum load for a spacing of L/10 is 93% of the corresponding welded connection, and for L/16, it is 93.7%. This indicates that when the bolt spacing is reduced to 1/10 of the span, a further reduction in spacing has a negligible effect on improving the load-bearing capacity. However, the data in Table 6 also show that when the bolt spacing is reduced from L/10 to L/15, the maximum longitudinal slip decreases by 21.5%. When the spacing is further reduced from L/15 to L/16, the reduction in maximum longitudinal slip is only 3.4%. Balancing load-bearing capacity and reducing longitudinal slip, the current data suggest that the optimal bolt spacing is L/15. While this spacing reduces the relative longitudinal slip between the upper and lower beams, it does not achieve the same performance as welding. Further research is needed to explore ways to minimize longitudinal slip and enhance the load-bearing capacity of HBCBs.

Table 6. The ultimate load-bearing capacity of high-strength bolt connected composite beams under varying bolt spacings.

Specimen	Bolt Spacing l_0/mm	l_0/L	Bolt Counts	P_{max}/kN	D_{max}/mm	u_{dmax}/mm
HW-HW	—	—	0	97.11	126.51	7.76
HW-HW-L	3000	1	4	114.43	167.29	6.30
HW-HW-L/2	1500	1/2	6	118.64	187.09	5.73
HW-HW-L/4	750	1/4	10	123.82	184.53	5.52
HW-HW-L/5	600	1/5	12	124.13	180.63	5.22
HW-HW-L/8	375	1/8	16	125.35	179.87	4.93
HW-HW-L/10	300	1/10	20	128.19	177.77	4.46
HW-HW-L/15	200	1/15	32	128.67	170.08	3.50
HW-HW-L/16	187.5	1/16	34	129.16	172.41	3.38
HW-HW-T	—	—	—	137.90	125.12	2.47

6. Conclusions

This study aims to reduce CO_2 emissions during the manufacturing phase of steel components and proposes a novel "low-carbon H-shape bolted composite beam (HBCB)". Experimental investigations were conducted on its lateral-torsional buckling behavior, and a comprehensive numerical analysis method was developed. After validating the feasibility of the numerical method, further investigations were carried out on the influence of high-strength bolts. Based on the findings from experimental and numerical analyses, the following conclusions can be drawn:

1. The proposed low-carbon H-shape bolted composite beams (HBCBs) demonstrated reliable lateral-torsional buckling behavior, validated through experimental and numerical investigations.
2. High-strength bolt connections significantly enhanced the stability and load-bearing capacity of the beams compared with boltless laminated counterparts.
3. Bolt spacing critically influences performance when the bolt spacing exceeds 1/15 of the beam span. At a spacing of L/15, the load-bearing capacity of HBCBs reached approximately 93.3% of that of beams with plug-weld connections.
4. High-strength bolts reduced longitudinal slip effectively, but they did not fully replicate the performance of welded connections. Further research is necessary to optimize bolt arrangements and reduce longitudinal slips to enhance the overall mechanical performance of HBCBs.

7. Limitations, Advantages, and Future Perspectives

7.1. Limitations

Despite the significant contributions of this study, several limitations should be acknowledged:

1. Simplifications in finite element modeling: The numerical model simplifies the bolt geometry by replacing the hexagonal bolt cap with a circular cap and ignoring bolt threads to improve convergence and computational efficiency. While this simplification does not affect overall trends, minor discrepancies may exist compared with actual behavior.
2. Longitudinal slip effects: The study found that high-strength bolted composite beams (HBCBs) exhibited longitudinal relative slip between the upper and lower beams. Although increasing bolt spacing beyond L/15 had negligible effects on load-bearing capacity, slip control remains a challenge and requires further optimization.
3. Limited parameter investigation: The current study primarily focuses on bolt pre-loadings and bolt spacing, while other influential factors, such as material properties, flange–web interaction, and residual stresses, were not extensively investigated.
4. Lack of cyclic loading tests: This research focuses on monotonic loading conditions, whereas real-world structures often experience cyclic loading. The performance of HBCBs under repeated loads needs further exploration.

7.2. Advantages

1. Low-carbon strategy: This study emphasizes reducing welding-related CO_2 emissions, which constitute 57.2% of the prefabrication process emissions. By using hot-rolled H-shapes and bolted connections, this study presents an environmentally friendly alternative to traditional welded structures.
2. Enhanced structural performance: Experimental and numerical results demonstrate that HBCBs achieve approximately 93.3% of the load-bearing capacity of plug-welded connections while significantly reducing construction emissions.
3. Reliable numerical modeling: The proposed ABAQUS finite element model was validated against experimental data, with errors in critical and peak loads remaining within 9% and 3%, respectively, ensuring its applicability for further parametric studies.
4. Engineering applicability: This study provides practical insights into optimizing bolt spacing (L/15 as the optimal threshold) and bolt pre-loading (increased pre-tension enhances stiffness and reduces slip), guiding real-world design and construction.

7.3. Future Perspectives

1. Further optimization of bolt arrangements: While L/15 was identified as an effective bolt spacing, additional investigations into alternative patterns (e.g., staggered arrangements) could further enhance performance and slip resistance.
2. Cyclic performance studies: Future research should explore the long-term durability of HBCBs under seismic loading to ensure their reliability in extreme conditions.
3. Incorporation of residual stresses: More detailed modeling of residual stresses, particularly around bolt holes, could refine the accuracy of numerical predictions.
4. Hybrid connection approaches: Investigating the combination of bolted and minimal welding techniques may help balance load-bearing capacity, slip control, and sustainability goals.

This study lays a solid foundation for developing low-carbon, high-strength bolted composite beams, and future research can build upon these findings to further enhance their mechanical performance and structural efficiency.

Author Contributions: Conceptualization, L.L.; investigation. K.K.D., B.L. and H.H.; validation, B.L., Z.H. and H.H.; resources, L.L.; writing—original draft preparation, K.K.D., L.L., B.L., Z.H. and H.H.; writing—review and editing, L.L., B.L. and Z.H.; project administration, L.L.; funding acquisition, L.L. All authors have read and agreed to the published version of the manuscript.

Funding: This research received no external funding.

Data Availability Statement: The original contributions presented in the study are included in the article; further inquiries can be directed to the corresponding author.

Conflicts of Interest: The authors declare that they have no conflicts of interest.

References

1. Gan, V.; Wong, C.; Tse, K.; Chen, J.; Lo, I.; Chan, C. Parametric modelling and evolutionary optimization for cost-optimal and low-carbon design of high-rise reinforced concrete buildings. *Adv. Eng. Inform.* **2019**, *42*, 100962. [CrossRef]
2. Arogundade, S.; Dulaimi, M.; Ajayi, S. Holistic Review of Construction Process Carbon-Reduction Measures: A Systematic Literature Review Approach. *Buildings* **2023**, *13*, 1780. [CrossRef]
3. You, J.; Xu, X.; Wang, Y.; Xiang, X.; Luo, Y. Life cycle carbon emission assessment of large-span steel structures: A case study. *Structures* **2023**, *52*, 842–853. [CrossRef]
4. Zhang, J.; Zhang, Y.; Chen, Y.; Wang, J.; Zhao, L.; Chen, M. Evaluation of Carbon Emission Efficiency in the Construction Industry Based on the Super-Efficient Slacks-Based Measure Model: A Case Study at the Provincial Level in China. *Buildings* **2023**, *13*, 2207. [CrossRef]
5. *BS EN 15978:2011*; Sustainability of Construction Works-Assessment of Environmental Performance of Buildings-Calculation Method. British Standards Institute: London, UK, 2011.
6. Li, Q.; Yue, Q.; Feng, P.; Xie, N.; Liu, Y. Development Status and Prospect of Steel Structure Industry Based on Carbon Peak and Carbon Neutrality Target. *Prog. Steel Build. Struct.* **2022**, *24*, 1–6.
7. Hammond, G.; Jones, C. Embodied energy and carbon in construction materials. *Proc. Inst. Civ. Eng. Energy* **2008**, *161*, 87–98. [CrossRef]
8. Dani, A.; Roy, K.; Masood, R.; Fang, Z.; Lim, J. A comparative study on the life cycle assessment of New Zealand residential buildings. *Buildings* **2022**, *12*, 50. [CrossRef]
9. Hart, J.; D'Amico, B.; Pomponi, F. Whole-life embodied carbon in multistory buildings: Steel, concrete and timber structures. *J. Ind. Ecol.* **2021**, *25*, 403–418. [CrossRef]
10. Mostafavi, F.; Tahsildoost, M.; Zomorodian, Z.S. Energy efficiency and carbon emission in high-rise buildings: A review (2005–2020). *Build. Environ.* **2021**, *206*, 108329. [CrossRef]
11. Rousta, A.M.; Azandariani, M.G.; Ardakani, M.A.S.; Shoja, S. Cyclic behavior of an energy dissipation system with the vertical steel panel flexural-yielding dampers. *Structures* **2022**, *45*, 629–644. [CrossRef]
12. Sadeghi, M.; Shoja, S.; Ardakani, M.A.S.; Jahangiri, M. Innovative improvement towards steel plate shear walls employing the grid stiffeners along with the metallic yielding dampers. *J. Constr. Steel Res.* **2024**, *223*, 109080. [CrossRef]
13. Hu, S.; Liu, S.; Zeng, S.; Zhang, B.; Xu, Z. Investigating seismic performance of a novel self-centering shear link in EBF utilizing experimental and numerical simulation. *J. Constr. Steel Res.* **2025**, *224*, 109129. [CrossRef]
14. Liu, S.; Chen, S.; Zeng, S.; Zhang, B.; Hu, S. Seismic performance analysis of K-shaped EBF with an innovative crack-resistant composition beam. *J. Build. Eng.* **2025**, *101*, 111818. [CrossRef]
15. Huang, Z.; Zhou, H.; Tang, H.; Zhao, Y.; Lin, B. Carbon emissions of prefabricated steel structure components: A case study in China. *J. Clean. Prod.* **2023**, *406*, 137047. [CrossRef]
16. Roy, K.; Dani, A.A.; Ichhpuni, H.; Fang, Z.; Lim, J.B.P. Improving sustainability of steel roofs: Life cycle assessment of a case study roof. *Appl. Sci.* **2022**, *12*, 5943. [CrossRef]
17. Lu, L.; Wang, D.; Dai, Z.; Luo, T.; Ding, S.; Hao, H. Global Buckling Investigation of the Flanged Cruciform H-shapes Columns (FCHCs). *Appl. Sci.* **2021**, *11*, 11458. [CrossRef]
18. Nie, W.; Lu, L.; Li, R.; Hao, H.; Luo, T. Global buckling investigation of T-shaped built-up columns composed of two hot-rolled H-shapes. *Structures* **2023**, *58*, 105485. [CrossRef]
19. Lin, X.; Okazaki, T.; Chung, Y.L.; Nakashima, M. Flexural performance of bolted built-up columns constructed of H-SA700 steel. *J. Constr. Steel Res.* **2013**, *82*, 48–58. [CrossRef]
20. Zhu, Y.; Zhao, J. Experimental and analytical studies on bearing capacity of bolt connected composite plate girders. *Int. J. Steel Struct.* **2017**, *17*, 1317–1329. [CrossRef]
21. Xu, B.; Xia, J.; Chang, H.; Ma, R.; Zhang, L. Flexural behaviour of pairs of laminated unequal channel beams with different interfacial connections in corner-supported modular steel buildings. *Thin-Walled Struct.* **2020**, *154*, 106792. [CrossRef]

22. Xu, B.; Xia, J.; Ma, R.; Chang, H.; Yang, C.; Zhang, L. Investigation on interfacial slipping response of laminated channel beams with bolt connections in modular steel buildings. *J. Build. Eng.* **2023**, *63*, 105441. [CrossRef]
23. Xu, B.; Xia, J.; Chang, H.; Chen, X.; Yang, C.; Zhang, L. Numerical-analytical investigation on bending performance of laminated beams in modular steel buildings. *J. Constr. Steel Res.* **2024**, *217*, 108630. [CrossRef]
24. *GB 50017-2017*; Standard for Design of Steel Structures. China Architecture & Building Press: Beijing, China, 2017.
25. Xin, H.; Veljkovic, M. Evaluation of high strength steels fracture based on uniaxial stress-strain curves. *Eng. Fail. Anal.* **2021**, *120*, 105025. [CrossRef]
26. ECCS. *Ultimate Limit State Calculation of Sway Framed with Rigid Joints*; Technical Committee 8 (TC 8) of European Convention for Constructional Steelwork (ECCS): Brussels, Belgium, 1984.

buildings MDPI

Article

Experimental Research and Numerical Simulation on the Hysteresis Performance of Strengthened Beam–Column Joints

Xiuying Yang [1,*], Yanling Li [2], Shixin Duan [3], Guodong Zhao [4], Jie Song [4] and Denghu Jing [5]

[1] School of Architecture and Engineering, Liaocheng University, Liaocheng 252000, China
[2] Dongying Hekou Construction Project Quality and Safety Evaluation Center, Dongying 257000, China
[3] Shandong Quality Inspection and Testing Center of Construction Engineering Co., Ltd., Jinan 250108, China
[4] Shandong Academy of Building Research Co., Ltd., Jinan 250108, China
[5] School of Civil Engineering, Southeast University, Nanjing 211189, China
* Correspondence: yangxiuying@lcu.edu.cn

Abstract: Beam–column joints in reinforced concrete frames are subjected to complex forces and are prone to damage under seismic actions. This paper proposes a method to strengthen beam–column joints using angle steel and split bolts. The hysteretic performance of the strengthened components is investigated through test and finite element numerical simulation. The influencing parameters, including concrete strength grade, axial compression ratio, stirrup characteristic value, angle steel leg length, and angle steel leg thickness, are analyzed. The results show that angle steel can simultaneously enhance the strength and stiffness of the strengthened joints. With an increase in concrete strength grade, the load-carrying capacity of the strengthened components continuously increases. However, when the axial compression ratio exceeds 0.6, the load-carrying capacity of the strengthened components significantly decreases. The size of the stirrup characteristic value has little influence on the shear resistance of the strengthened joints. The leg length and leg thickness of the angle steel have certain effects on the strengthening effectiveness. The method of outward movement of plastic hinges can effectively improve the seismic performance of bi-directionally loaded spatial joints.

Keywords: strengthened beam–column joints; angle steel; hysteretic performance; influencing parameters; numerical analysis

1. Introduction

In structures, the beam–column connection joint is a crucial part that plays a significant role in transmitting internal forces between beams and columns and coordinating deformations. They are subjected to complex loading conditions, including axial forces, bending moments, and shear forces from columns, as well as bending moments, shear forces, and torsional effects from beams [1–3]. Additionally, factors such as shrinkage, creep, temperature variations, and foundation settlement can also influence the behavior of joints [4–6]. Under seismic actions, joints are particularly vulnerable and prone to damage, and failure of joints can lead to the loss of load-carrying capacity of the entire frame and structural collapse [7–9]. With the increase in the service life of buildings, changes in their functions, and the threat of various natural disasters, the beam–column connection joints of existing buildings often face various damages, weakening the load-bearing capacity and seismic performance of the joints. Therefore, strengthening and renovation work is urgently needed.

At present, there are many strengthening methods for beam–column connection joints, such as the enlarged cross-section method, the method of bonding fiber-reinforced polymer, and the method of bonding steel plate [10–17]. The enlarged cross-section method can improve the load-bearing capacity of the joints by increasing the area of concrete and steel bars, but it will increase the self-weight of the structure and affect the use of building space. The method of bonding FRP is easy to construct, lightweight and high-strength, and can effectively improve the seismic performance of the joints, but there are problems with the bonding durability with the base material. The method of bonding steel plates can significantly improve the shear and bending resistance of the joints, but additional treatments are required for fire and corrosion protection. These strengthening methods have their own advantages and disadvantages, and in practical engineering applications, a comprehensive selection needs to be made according to factors such as the failure mode of the joints, structural characteristics, and use requirements [18–20].

Although existing research has achieved certain results in the strengthening of beam–column connection joints, they still need to be further improved. In view of this, conducting research on the strengthening of beam–column connection joints has important theoretical significance and engineering application value [21,22].

Taking the internal deep beam–column joints as the research objects, the study on the hysteretic performance of joints being strengthened with angle steel and bolts were conducted.

Considering the critical role of joints in frame structures and the need to ensure their satisfactory seismic performance, a new strengthening scheme for beam–column joints is proposed in this study. Angle steel is added at the beam–column joints, and the angle steel on both sides is fixed and clamped by bolts, so as to improve the mechanical performance of the joints, and also make the plastic hinge at the joints move outward, so as to protect the joints.

This approach can dissipate seismic energy and ensure the integrity of the bond between the beam reinforcement and concrete, which provides better confinement to the core region of the joint, enhancing its stiffness and shear strength, and thereby improving the seismic performance of the frame structure.

2. Testing Process

To verify the effectiveness of angle steel strengthened frame joints, relevant experimental studies were conducted.

2.1. Test Specimen

This experiment focuses on the intermediate joints of a general multi-story and multi-span frame under horizontal forces. The beam–column combination between the beam–column reverse bending points is selected as the specimen, and the specimen is shown in Figure 1.

Angle steel is set at the beam–column joint and fixed by tension bolts. This method causes the plastic hinge at the end of the beam to move outward, improves the stress state of the joint, enhances its ductility, and improves the bending strength of the beam–column section, thereby reinforcing the beam–column joint.

Two specimens were made, one of which was strengthened with angle steel L200 $\times$ 20 at the beam–column node, numbered RJ-1. The other one is not strengthened as a comparison, numbered J-1. The concrete strength grade used is C30 with a compressive strength of 29.7 MPa, the longitudinal reinforcement of beams and columns is HRB400 ribbed steel bars with a yield strength of 437.5 MPa, and the stirrups of beams, columns, and nodes are HPB235 plain round steel bars with a yield strength of 302.5 MPa. The diameter of stirrup is 8 mm, and the stirrup spacing is 150 mm. The dimensions and reinforcement of the two specimens are identical, and the cross-section is shown in Figure 2. The experimental strengthened node is shown in Figure 3.

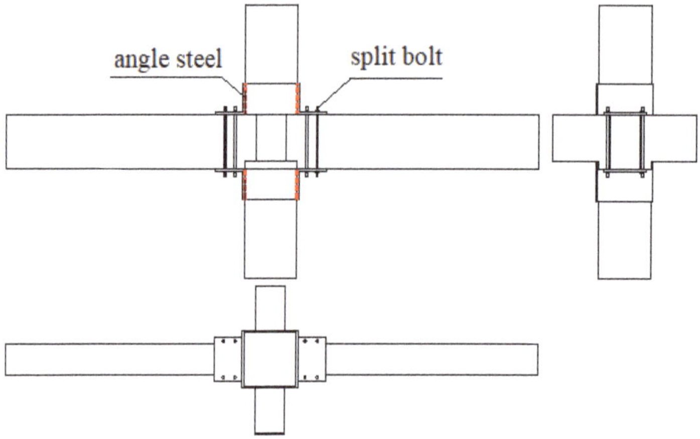

Figure 1. Beam–column combination joint.

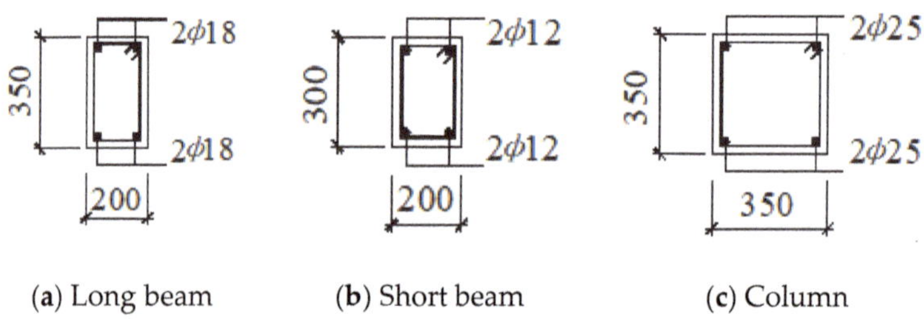

Figure 2. Cross-section of beams and columns.

Figure 3. Experimental strengthened joint.

2.2. Test Device

The loading device of the test is shown in Figure 4. The 320 t hydraulic jack is installed at the upper end of the column to apply axial pressure to the node. Two sets of jacks with the same specifications are arranged at the outer ends of the left and right beams and controlled by an oil pump to achieve synchronous low cycle repeated loading. Steel plate cushion blocks are pre-set at the upper and lower ends of the column and on both sides, and four steel blocks welded with circular steel bars are used to clamp the column head on both sides of the column end, simulating the hinge support of the column end.

Figure 4. Schematic diagram of loading device.

The loading method adopts a mixed control of load and displacement. Firstly, the hydraulic jack at the upper column end applies axial force to the specimen to a predetermined axial compression ratio. Subsequently, a low-cycle repeated load is applied by using two jacks at the outer ends of the beams, one upwards and the other downwards, in a symmetrical manner.

At the initial stage of loading, the load is controlled by a graded cyclic loading. When the beam end reaches yield, the loading is controlled by displacement. The multiple of the displacement at the yield of the beam end is taken to load step by step, and the cycle is repeated three times for each displacement value.

3. Analysis of Test Results

The load-displacement curves at the beam end of two specimens J-1 and RJ-1 obtained from the test results are shown in Figure 5.

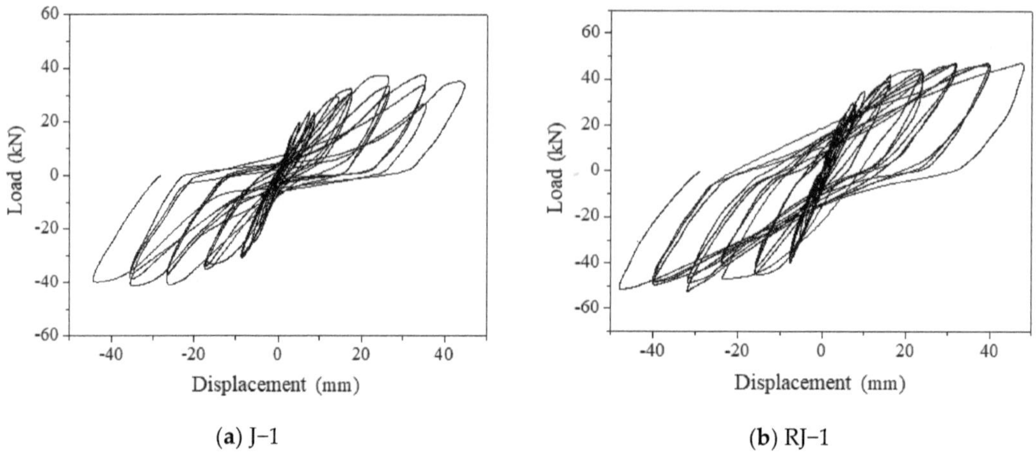

Figure 5. Hysteresis curve at beam end.

The comparison specimen J-1 maintains good linear elasticity at the beginning of loading, without residual deformation, and has very low energy dissipation capacity. When loaded to 9 mm (30 kN), the hysteresis curve forms a full hysteresis loop. However, at 18 mm (33 kN), the hysteresis curve began to experience "necking" phenomenon, and this phenomenon became more pronounced as the displacement of the beam end increased. When the displacement increases to 36 mm (38 kN), the hysteresis curve shows a horizontal segment during loading, indicating almost zero stiffness. Finally, the hysteresis curve shows an inverted S-shape and gradually reverses towards the displacement axis, resulting in a decreasing initial stiffness under loading.

At the beginning of loading, the specimen RJ-1 maintained good linear elasticity, with almost no residual deformation and minimal energy dissipation capacity. A full hysteresis loop is formed on the 8 mm (35 kN), and as the displacement increases, the hysteresis loop becomes fuller until the end of loading, and there is no necking phenomenon on the curve.

Through comparative tests of angle steel strengthened beam–column nodes, it can be concluded that the beam reinforcement in the strengthened specimen nodes is effectively protected, the anchoring length increases, and the bonding performance of the beam reinforcement is strengthened. At the same time, strengthening causes the plastic hinge to move outward, avoiding cracks at the beam–column interface, achieving continuous constraints on the nodes by the beam, improving the seismic performance of the node combination, and exhibiting a full hysteresis loop in the hysteresis curve. Illustrating that the angle steel strengthening scheme proposed in this paper can significantly improve the seismic performance of beam–column joints.

4. Finite Element Modeling

The finite element model is established by ABAQUS based on the experimental conditions conducted by the authors. The concrete has a strength grade of C30, while the longitudinal reinforcement for beams and columns consists of HRB400 steel bars. The stirrups for beams, columns, and joints are made of HPB235 steel bars. The angle steel used for joint strengthening is L200 $\times$ 20. The material properties are set according to the previous experiment. Axial compression is applied at the upper end of the column, while axial tension or compression is applied at the outer end of the beam, allowing for synchronized low-cycle cyclic loading [23,24].

4.1. Mesh of the FEA Model

The concrete elements have a side length of 50 mm, with a local refinement of 25 mm in the joint region, resulting in a total of 6196 elements. The reinforcement elements have a side length of 25 mm, and a total of 2992 elements are used. The side length of the strengthened angle steel is determined based on the thickness of the angle steel legs, and the mesh is approximately divided into hexahedral elements.

4.2. Element Selection

The finite element model of the strengthened concrete joint adopts a segregated approach, including three types of elements: concrete, reinforcement, and angle steel [25]. The concrete and angle steel are represented using three-dimensional solid elements (C3D8R), while the reinforcement is modeled using three-node truss elements (T3D2) in three-dimensional space [26].

Using the truss element approach, the connection between concrete and reinforcement is established by creating separate models for concrete and reinforcement (longitudinal and stirrup reinforcement) within their respective parts [27]. The sections and properties are assigned in the property section, and the interaction section uses the "embed" feature to embed the reinforcement skeleton into the solid concrete model, thus achieving the connection between concrete and reinforcement [28]. The truss elements do not need to have the same mesh as the solid elements, their nodes are automatically constrained to the nodes of the solid elements. ABAQUS assumes that the behavior of concrete is independent of the surrounding bonded reinforcement. The interaction effects, such as bond-slip, between concrete and reinforcement are approximated by introducing "tension stiffening" in the concrete model [29].

In the simulation of the strengthened concrete joint, rigid elements are added at the top and bottom surfaces of the column and the upper and lower surfaces of the beam ends, forming rigid faces to simulate the test loading pad. The computational model is illustrated in Figure 6.

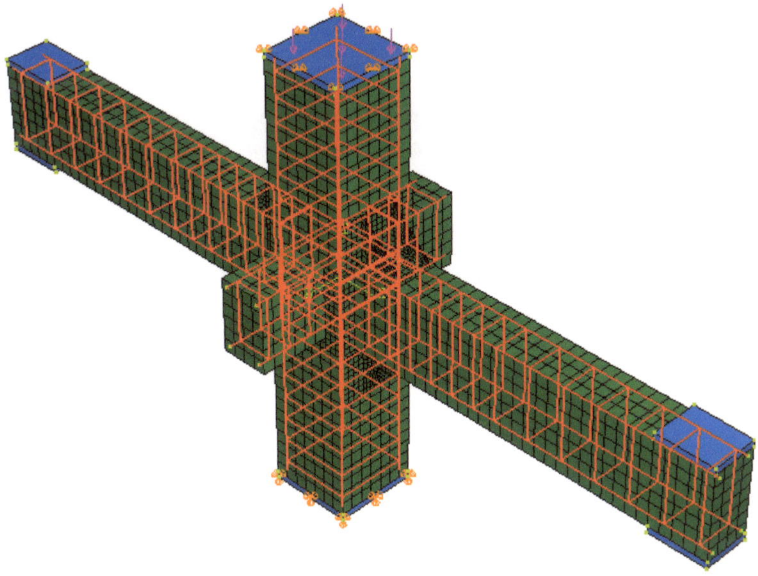

Figure 6. Finite element analysis model.

4.3. Material Model

In the finite element simulation, the damaged plasticity model is adopted for a constitutive model of concrete. The damage index is introduced to simulate the characteristic that the unloading stiffness of concrete decreases with the increase in damage by reducing the elastic stiffness matrix of concrete. Based on the damaged plasticity model of concrete, the formation of the yield surface is mainly controlled by equivalent plastic strain.

The equivalent plastic strain of the model is calculated, which is the cumulative result of the plastic strain in the whole deformation process of the model. The equivalent plastic strain greater than 0 indicates that the material has yielded. In the finite element analysis model, concrete cracking is considered based on the fracture energy cracking criterion, which is defined by the calculation of cracking strain.

The plastic deformation of steel is described by Mises yield surface and the associated flow rule.

4.4. Loading Method

To accurately simulate the loading process of the test, the following boundary conditions can be applied to the model based on the test setup: horizontal displacement constraints in two directions at the top of the column, and horizontal and vertical displacement constraints in three directions at the bottom of the column [30].

The loading is performed using displacement control. Initially, a vertical axial force is applied at the top of the column, followed by displacement loading applied at the left and right ends of the beam. The displacement increment for each step is 5 mm, and multiple sub-steps are used to achieve complete displacement.

4.5. Verification of Model Effectiveness

The finite element analysis data are compared with the test results conducted by the authors to validate the effectiveness of the finite element model and analysis method. Displacement loading is employed, and to facilitate comparison with the test results, the displacement is controlled to be equal to the experimental yield displacement. The loading process is divided into 12 steps to complete the entire process.

The comparison between the simulated specimen's hysteresis curve and the experimental specimen's hysteresis curve is shown in Figure 7, indicating a good agreement between the two. The equivalent plastic strain (PEEQ) of the simulated specimen, which represents the cumulative plastic strain throughout the deformation process [31], is shown in Figure 8. A PEEQ value greater than 0 indicates material yielding, which corresponds well with the actual failure mode of the experimental specimen.

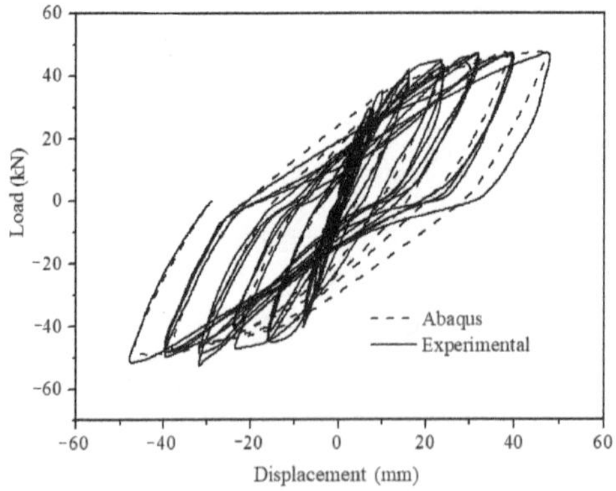

Figure 7. Comparison between simulated and experimental values.

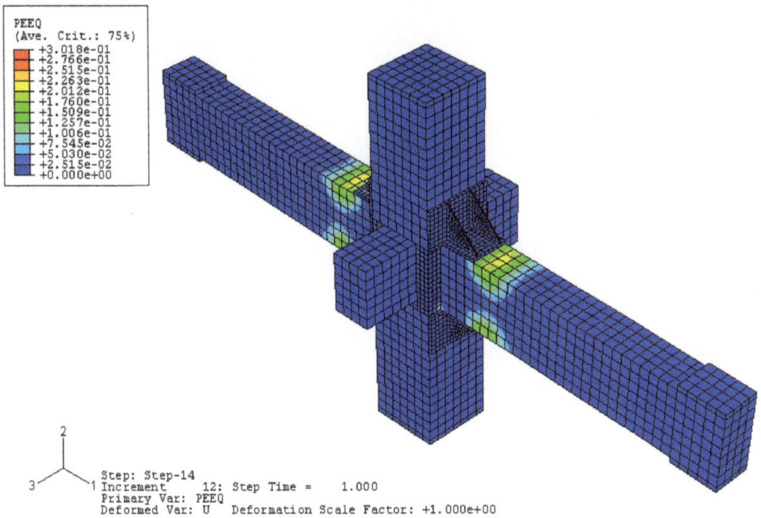

Figure 8. Equivalent plastic strain for simulation analysis.

Therefore, the analytical model used in this study can effectively simulate the ultimate bearing capacity and failure process of strengthened concrete nodes with angle steel strengthening. It is also observed that the plastic hinges of the beams are effectively shifted outward with the angle steel strengthening, leading to an improvement in the seismic performance of the nodes.

5. Numerical Analysis of Influencing Parameters

5.1. Influencing Parameters

To comprehensively study the improvement in the structural performance of existing strengthened concrete nodes with angle steel strengthening, the previously established computational model is used to analyze the parameters that affect the strengthening effect [32]. These parameters include concrete strength grade, axial compression ratio, stirrup characteristic values, angle steel leg length, angle steel leg thickness, as well as the analysis of a group of spatial joints. Except for the concrete strength grade, the form, dimensions, and materials of the nodes in the analysis are consistent with the experiments described in this paper. The mechanical properties of the concrete are presented in Table 1. The loading regime of the specimens adopts displacement control with Δ = 5 mm, and it is completed in 12 substeps. The strengthened angle steel without ribs is analyzed, and the detailed parameters of the calculated specimens are shown in Table 2, totaling 39 components.

Table 1. Mechanical performance of concrete.

Component	f_{cu} (MPa)	f_t (MPa)	E_c ($\times 10^4$ MPa)
JC-1, RJC-1	13	1.45	2.2
JC-2, RJC-2	39	3.188	3.06
JC-3, RJC-3	57	—	3.5528
JC-4, RJC-4	66.2	4.25	3.71
JC-5, RJC-5	82.8	—	4.1798

Table 2. Parameter values of strengthened joints.

Category	Component	f_c (MPa)	Axial Compression Ratio	Stirrup Ratio	Angle Steel (mm)
Comparison component	J	22.572	0.15	0.092	—
	RJ				∠200 × 16
Concrete strength grade	JC-1	9.88	0.15	0.211	—
	RJC-1				∠200 × 16
	JC-2	29.64		0.070	—
	RJC-2				∠200 × 16
	JC-3	41.26		0.051	—
	RJC-3				∠200 × 16
	JC-4	50.312		0.041	—
	RJC-4				∠200 × 16
	JC-5	59.93		0.035	—
	RJC-5				∠200 × 16
Axial compression ratio	RJA-1	22.572	0	0.092	∠200 × 16
	RJA-2		0.3		
	RJA-3		0.45		
	RJA-4		0.6		
	RJA-5		0.75		
Stirrup characteristic value	RJT-1	22.572	0.15	0	∠200 × 16
	RJT-2			0.025	
	RJT-3			0.037	
	RJT-4			0.074	
	RJT-5			0.111	
Angle steel model	RJL-1	22.572	0.15	0.092	∠100 × 16
	RJL-2				∠120 × 16
	RJL-3				∠140 × 16
	RJL-4				∠160 × 16
	RJL-5				∠180 × 16
	RJL-6				∠220 × 16
	RJL-7				∠240 × 16
	RJL-8				∠300 × 16
	RJD-1				∠200 × 10
	RJD-2				∠200 × 12
	RJD-3				∠200 × 14
	RJD-4				∠200 × 18
	RJD-5				∠200 × 20
	RJD-6				∠200 × 22
	RJD-7				∠200 × 24
Spatial joint	JS	22.572	0.15	0.092	—
	RJS				∠200 × 16

5.2. Concrete Strength Grade

Under external loads, the node is subjected to various forces, and a portion of these forces is transmitted to the core region of the node through the longitudinal reinforcement of the beam and column in the form of bonding stress. The stirrup reinforcement within the node, together with the core concrete, bears these actions. Therefore, to study the changes in the performance of the node before and after strengthening, it is necessary to locate and analyze the stress–strain of the longitudinal reinforcement of the beam and column, the stirrup reinforcement within the node, and the strengthened angle steel throughout the entire loading process. The locations are shown in Figure 9.

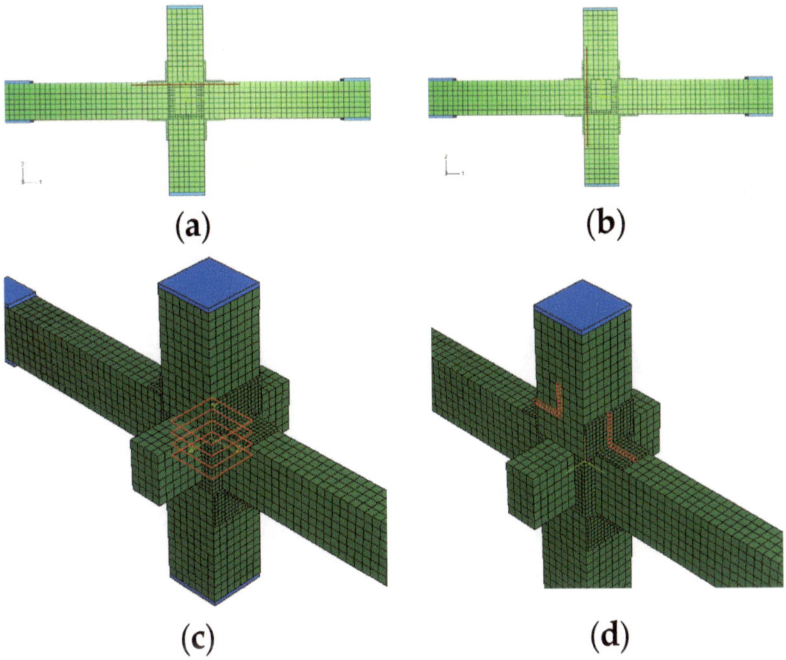

Figure 9. Location of analysis units. (**a**) Longitudinal reinforcement of the beam; (**b**) longitudinal reinforcement of the column; (**c**) stirrup reinforcement within the joint; (**d**) angle steel.

The skeleton curves of nodes with different concrete strength grades are shown in Figure 10. The node assemblies basically go through three stages: elastic, yield, and ultimate under low-cycle repeated loading. By comparing the curves, it can be observed that the yield load and ultimate load of the strengthened components have significantly increased. This indicates that the angle steel can simultaneously enhance the strength and stiffness of the strengthened section, leading to a change in the failure mode of the joint, forcing the failure to occur externally and protecting the core region of the node, thus improving the ductility of the assembly. However, for the strengthened components with lower concrete strength, their bearing capacity shows a significant decrease in the later stages of loading. This obviously cannot meet the requirements of the node for ductile seismic performance, which involves dissipating seismic input energy through stable plastic deformation of secondary components while ensuring the structural load-bearing capacity and stiffness without degradation. Therefore, for nodes with lower early strength, it is necessary to strengthen the surrounding beams and columns simultaneously. The ultimate loads of these six components are shown in Figure 10, and it can be observed that the bearing capacity of the strengthened components increases as the concrete strength grade increases.

Furthermore, it can be seen that the increase in the bearing capacity of the strengthened components is more pronounced with higher concrete strength due to the displacement of beam plastic hinges outward.

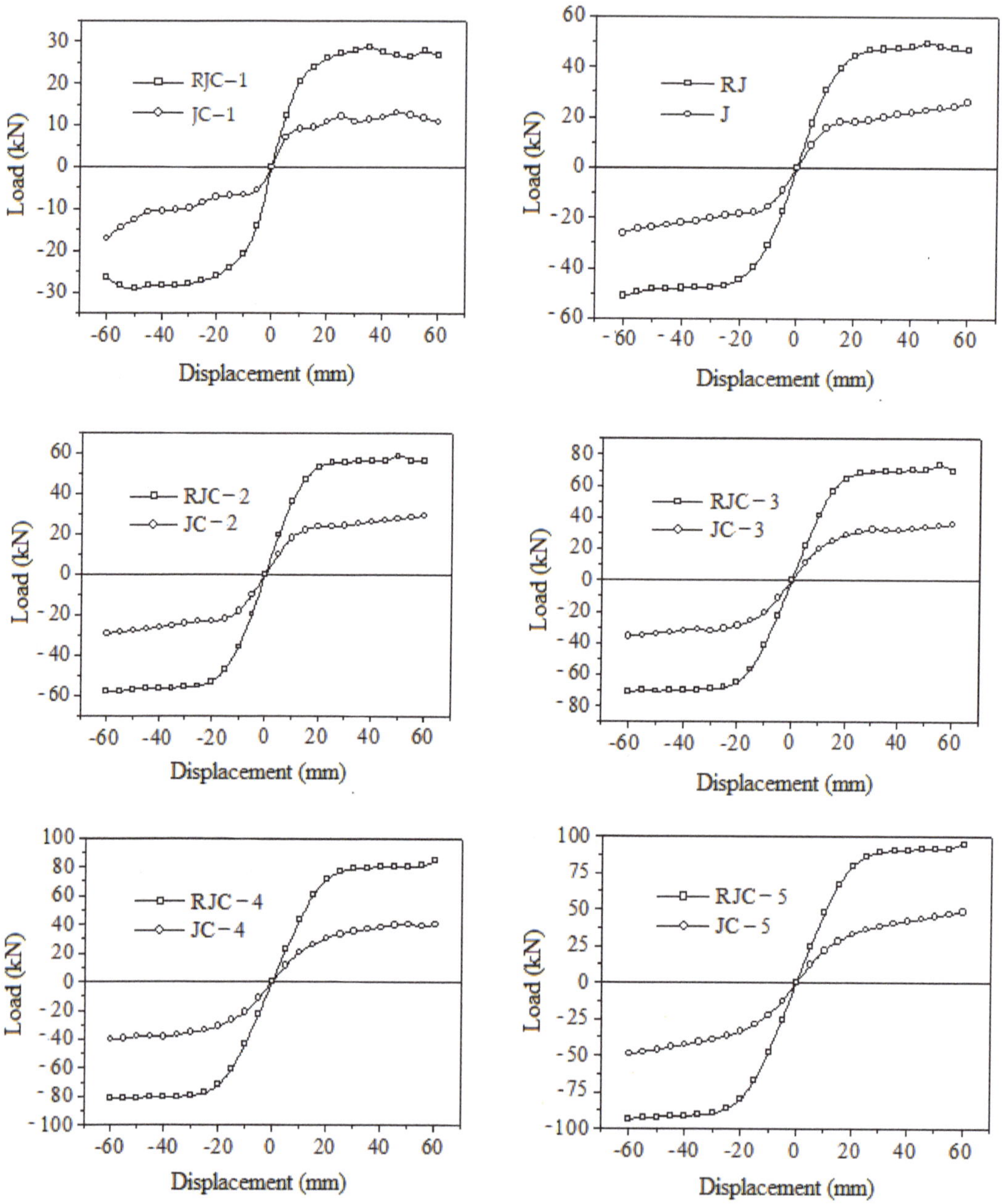

Figure 10. Skeleton curves of joints with different concrete strength grades.

5.3. The Axial Compression Ratio

The axial compression ratio is the factor that has the greatest disagreement among researchers regarding its impact on the structural performance of nodes. This is because the influence of the axial compression ratio varies depending on the different failure modes of nodes. Generally, an appropriate axial compression ratio can limit the development of cracks in the core region of the node and increase the bonding performance of the longitudinal reinforcement passing through the core region. However, a high axial compression ratio will increase the compressive stress in the inclined strut, leading to diagonal compression failure. Therefore, it is necessary to analyze the structural performance of strengthened components under different axial compression ratios.

The skeleton curves of nodes with different axial compression ratios are shown in Figure 11. It can be observed that the strengthened components do not exhibit a significant decrease in bearing capacity throughout the loading process, indicating that the strengthened components meet the requirements for seismic ductility. When the axial compression ratio is less than 0.6, the skeleton curves essentially overlap. However, when the axial compression ratio exceeds 0.6, the bearing capacity of the strengthened components significantly decreases. This phenomenon occurs because although an increased axial compression ratio causes an increase in compressive stress in the core region of the node, the strengthened angle steel also causes the compressive stress to shift towards the end of the angle steel, and the inclined strut enters the surrounding beams and columns, increasing in volume. Therefore, the strengthened angle steel can still make the plastic hinge move outward, so as to protect the joint. As the axial compression ratio continues to increase, the compressive stress in the core region gradually reaches the ultimate compressive stress of concrete. Eventually, concrete crushing occurs at the location of maximum compressive stress, resulting in the coexistence of beam plastic hinges and local node diagonal compression failure. The stiffness of the node deteriorates, and the bearing capacity decreases. Therefore, for components with high axial compression ratios, it is necessary to appropriately increase the reinforcement to withstand the higher compressive stress generated in the core region.

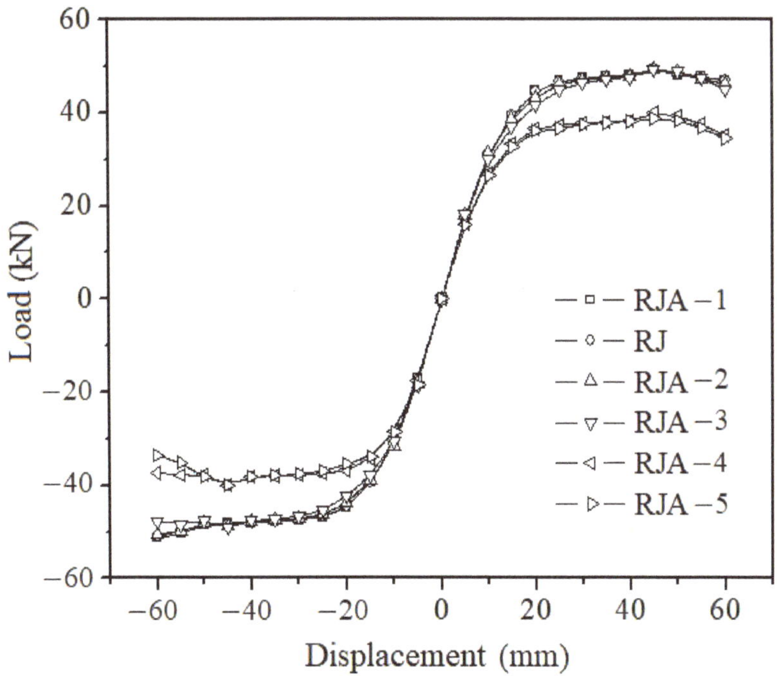

Figure 11. Skeleton curves of joints with different axial compression ratios.

5.4. Stirrup Ratio

The use of horizontal closed stirrups is widely recognized as the most common method to improve the shear strength of nodes. This is because the stirrups placed inside the node not only confine the core concrete of the node, increasing its ability to transfer axial loads, but also resist the horizontal shear in the truss mechanism, thereby enhancing the shear carrying capacity of the node. However, when the stirrup ratio is too high, the node may still experience shear failure and the stirrups may not fully play their role, resulting in a lower shear strength than the calculated theoretical value, which is unsafe. Therefore, in the process of reinforcing nodes, it is necessary to consider not only low-stirrup cases but also high-stirrup cases preventing premature diagonal compression failure in the core region of nodes.

The skeleton curves of nodes with different stirrup ratios are shown in Figure 12. It can be observed that the skeleton curves of the six different stirrup ratio cases are almost coincidental, indicating that the stirrup ratio has little influence on the strengthened nodes. This is because the angle steel plays a major role throughout the loading process, providing good protection to the strengthened concrete section, increasing the anchorage length of the longitudinal reinforcement passing through the core region, reducing the bond stress in the rebar, and weakening the truss mechanism. Moreover, the angle steel surrounding the node significantly improves its overall stiffness, minimizing the impact of increased stirrup quantities on the strengthened nodes. Therefore, it can be concluded that the strengthening method effectively solves the problem of configuring stirrups in the core region of joints.

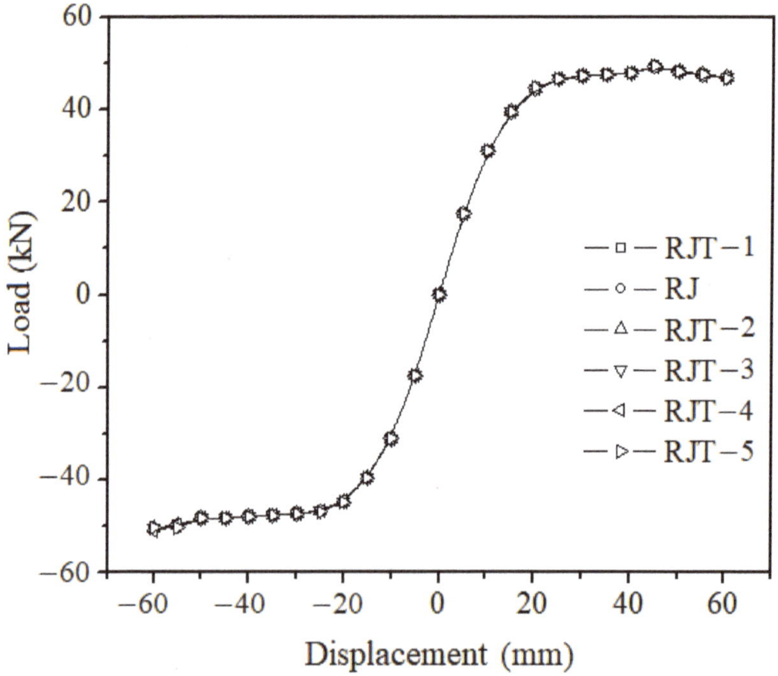

Figure 12. Skeleton curves of joints with different stirrup ratio.

5.5. Angle Steel Leg Length

Based on experimental measurements, the range of plastic hinge extension in the node is generally within 1.0 to 1.5 times the beam height from the column edge. Therefore, the length of the angle steel leg should be within this range. The skeleton curves of nodes with different leg lengths are shown in Figure 13. It can be observed that the strengthened components do not exhibit a significant decrease in load-carrying capacity throughout the loading process. This indicates that the slippage problem of the beam reinforcement has been effectively resolved, allowing the beam to continue to provide restraint to the core region of the node, thereby improving the shear strength of the node and enhancing its energy dissipation capacity.

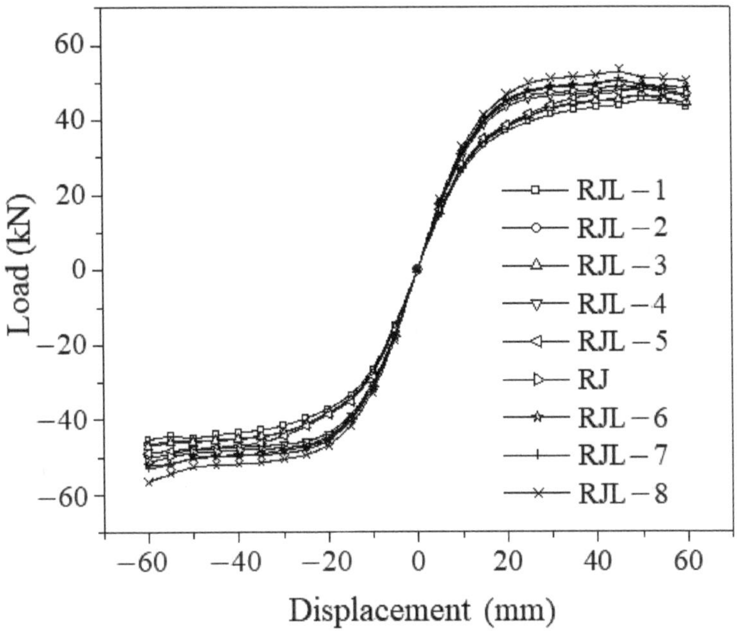

Figure 13. Skeleton curves of joints with different angle steel leg lengths.

As the leg length increases, the range covered by the skeleton curve gradually increases. When the leg length exceeds 180 mm (approximately 0.5 times the beam height), the skeleton curve shows significant growth. Afterward, the growth tends to be slower. This is because, initially, when the leg length is shorter, although the primary failure of the strengthened component occurs on the outside of the angle steel, cracks still extend to the column face, and even into the core region of the node, as demonstrated by the test results of specimen RJ-5. Furthermore, with a shorter reinforcement segment, the deformation of the plastic hinge region still concentrates near the column face, causing deformation in the node. These factors weaken the seismic performance of the strengthened component. Therefore, when selecting the size of the angle steel, the leg length should be ensured to be within the range of 0.5 hb to hb (where hb is the beam height).

5.6. Angle Steel Leg Thickness

In order to achieve plastic hinge displacement, the bending strength of the beam section at the column face must be at least 1.25 times higher than the expected bending strength at the beam hinge. Therefore, the selected cross-section of the angle steel should meet the strength ratio of the beam section. To ensure the bond performance of the beam reinforcement, the width of the angle steel section should not be smaller than the beam width, making the thickness of the angle steel leg to be a controlling factor. The skeleton curves of nodes with different angle steel leg thicknesses are shown in Figure 14. It can be observed that the selected angle steel leg thicknesses all meet the requirements and achieve plastic hinge displacement. However, increasing the leg thickness beyond the strength ratio for plastic hinge displacement only slightly improves the yield load and has little effect on other aspects. Therefore, the angle steel thickness can be appropriately increased on the basis of meeting the strength ratio to delay the strengthened component from entering the yielding stage.

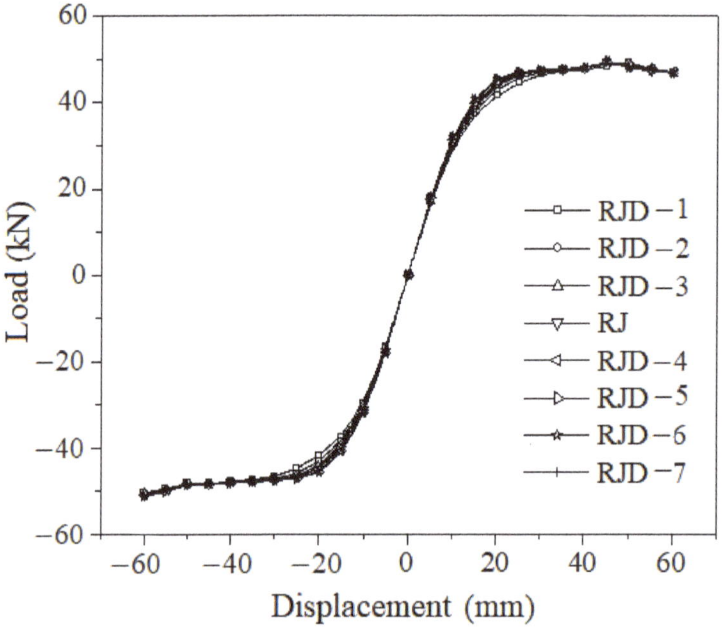

Figure 14. Skeleton curves of joints with different angle steel leg thicknesses.

5.7. Spatial Joints

In reality, the frames that bear loads are typically spatial frameworks (most nodes have two beams perpendicular to each other) and are equipped with floor slabs. Due to the randomness of seismic actions, most of the time there is an angle between the action and the building's axis. When subjected to forces in two directions simultaneously, the beams of the frame may yield in both directions, resulting in plastic hinges. After the reinforcement bars yield and enter the node, the level of constraint on the node significantly decreases, leading to increased cracking and deformation, and degradation of the seismic performance. Therefore, it is necessary to analyze the stress conditions of the spatial frame nodes before and after strengthening.

Due to the significant workload and difficulty involved in conducting experiments on spatial nodes, this study only uses numerical simulations. The beam–column dimensions and reinforcement are consistent with the experimental specimens, with only the perpendicular beams extended to match the length of the main beam. The main beam is designated as the north–south beam (NS), and the perpendicular beam is the east–west beam (WE). Loading is applied simultaneously on two beams in both directions, for example, loading is applied downward on the north and west beams simultaneously, and upward on the south and east beams simultaneously, and then reversed, resulting in a load direction at a 45° angle to the column axis. The loading procedure is the same as before.

The skeleton curves of the spatial components before and after strengthening are shown in Figure 15. It can be observed that the load-bearing capacity of both beams in the strengthened components has significantly increased without noticeable degradation. This indicates a change in the failure mode of the node, with no bond failure occurring in the reinforcement bars, resulting in improved shear strength and stiffness of the node, and an enhancement in seismic performance.

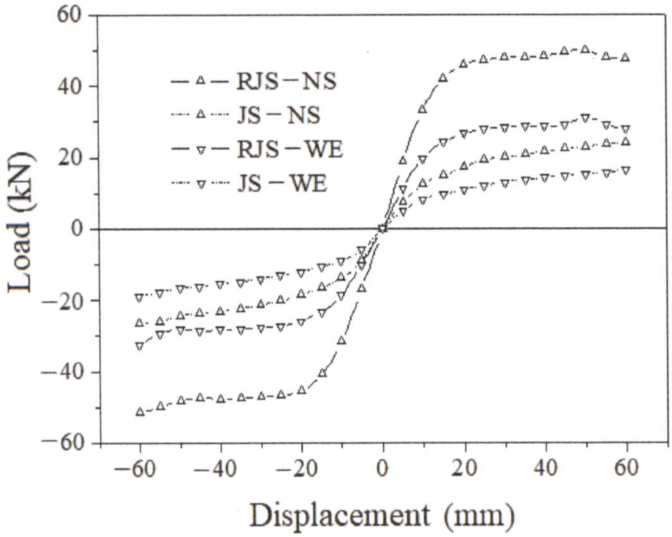

Figure 15. Skeleton curves of spatial joints.

The equivalent plastic strain between the node concrete and angle steel is shown in Figures 16–18. It can be seen that the node core area under bi-directional loading experiences high shear forces, and bond degradation along the reinforcement bars occurs early in the loading history, resulting in degradation of the beam plastic hinge region and failure concentrated in the node core area due to shear failure. In contrast, the failure of the strengthened components occurs in the relocated plastic hinge region, providing effective protection to the node core area and preventing its failure. The angle steel undergoes plastic deformation at the corner under bi-directional loading.

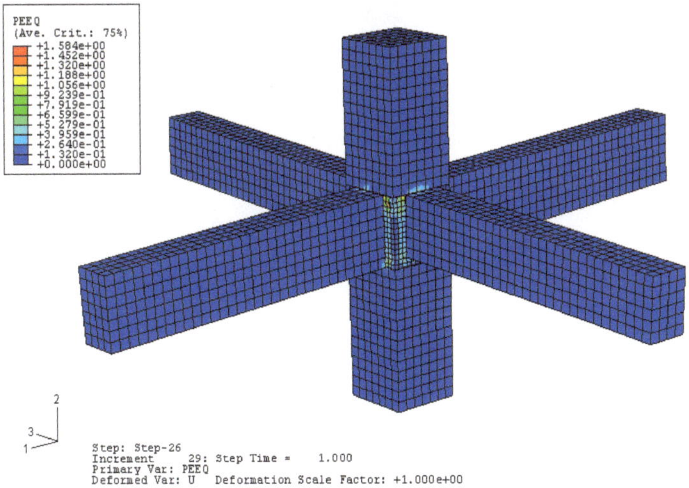

Figure 16. Equivalent plastic strain of JS concrete.

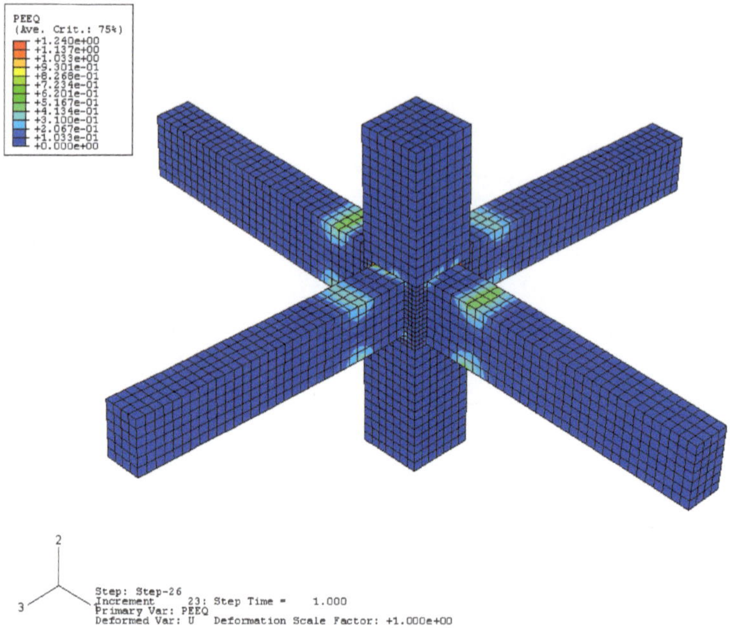

Figure 17. Equivalent plastic strain of RJS concrete.

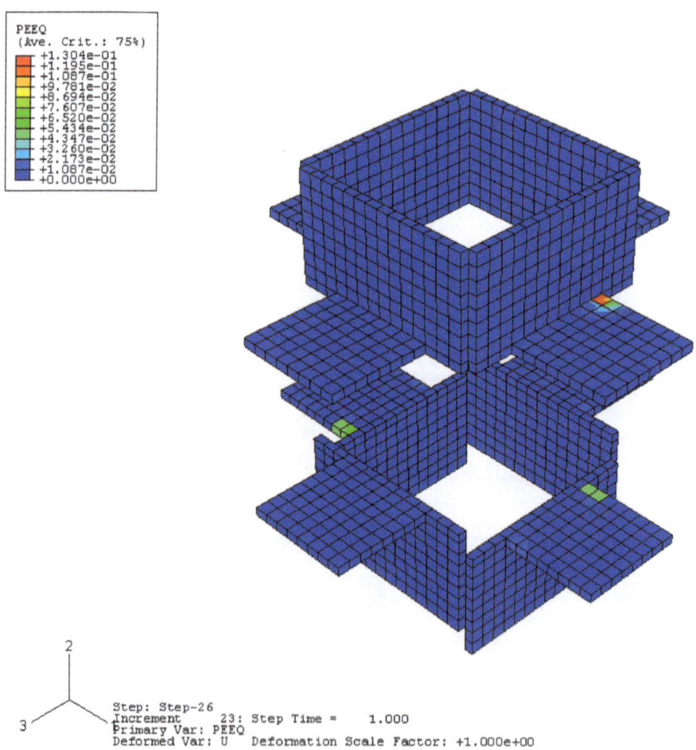

Figure 18. Equivalent plastic strain of RJS angle steel.

6. Conclusions

Through the simulation and analysis of 39 different conditions of joints, the effects of design parameters such as concrete strength grade, axial compression ratio, transverse reinforcement characteristic values, and angle steel type on the strengthening effectiveness were compared and analyzed. The changes before and after the strengthening of bi-directional loaded spatial nodes were simulated and compared. The following conclusions were drawn:

- The utilization of angle steel in concrete beam–column joints can effectively enhance both the strength and stiffness of the strengthened nodes. As the concrete strength grade rises, the load-bearing capacity of these strengthened nodes shows a corresponding increase. Nevertheless, once the concrete strength grade surpasses C60, the augmenting effect of the angle steel strengthening becomes notably less pronounced.
- When the axial compression ratio is below 0.6, the employment of angle steel to strengthen concrete beam–column joints leads to no remarkable change in the load-bearing capacity of the strengthened nodes. In contrast, once the axial compression ratio exceeds 0.6, the load-bearing capacity of the nodes strengthened with angle steel drops significantly. This indicates that within a certain range of the axial compression ratio, the strengthening effect of angle steel remains relatively stable, while beyond this critical value, the effectiveness of angle steel strengthening is severely affected, highlighting the importance of considering the axial compression ratio when using angle steel to enhance the performance of concrete beam–column joints.
- When using angle steel to strengthen the performance of concrete beam–column joints, the stirrup ratio is found to have a negligible impact on the shear resistance capacity of the strengthened nodes. This phenomenon indicates that, within the scope of enhancing the mechanical properties of beam–column joints with angle steel, simply increasing or decreasing the stirrup ratio does not lead to any substantial improvement or change in the shear resistance of these joints. As a result, it implies that during the design and construction process aiming at strengthening the joints with angle steel, the focus should perhaps be shifted away from the stirrup ratio and directed towards other more effective parameters or measures that can more significantly enhance the shear resistance of the strengthened nodes.
- Research shows that both the length and thickness of the angle steel legs prove to have a definite influence on the strengthening effectiveness. Specifically, these two geometric parameters of the angle steel, namely the leg length and thickness, interact with the concrete structure in a way that impacts how well the angle steel strengthen the joints. This understanding is crucial for optimizing the design of angle steel strengthened beam–column joints, ensuring that the angle steel is utilized most effectively to improve the mechanical properties of the joints, such as strength, stiffness, and ductility.
- By using angle steel to strengthen concrete beam–column joints, the plastic hinges at the joints can be relocated outward. This relocating method of plastic hinges can effectively enhance the seismic performance of bi-directional loaded spatial joints. Specifically, the outward relocation of plastic hinges caused by the strengthening of angle steel means that the stress concentration originally concentrated in the core area of the beam–column joints is dispersed. As a result, the core area of the beam–column joints is protected from severe damage during seismic actions. This protection mechanism not only strengthens the beam–column joints but also improves the overall seismic-resistant capacity of the structure, enabling it to better withstand seismic forces and reduce the risk of structural failure.

Author Contributions: Methodology, X.Y. and J.S.; validation, Y.L. and D.J.; investigation, S.D. and G.Z.; writing—original draft preparation, X.Y.; writing—review and editing, X.Y.; funding acquisition, J.S. All authors have read and agreed to the published version of the manuscript.

Funding: This research was funded by National Key Research and Development Program of China grant number 2022YFF0609200 and Shandong Provincial Housing and Urban-Rural Construction Science and Technology Program of China grant number 2022-K7-10 and Shandong Province Undergraduate Teaching Reform Research Project grant number M2022095.

Data Availability Statement: Data is contained within the article.

Conflicts of Interest: Author Shixin Duan was employed by the company Shandong Quality Inspection and Testing Center of Construction Engineering Co., Ltd. And author Guodong Zhao was employed by the company Shandong Academy of Building Research Co., Ltd. The remaining authors declare that the research was conducted in the absence of any commercial or financial relationships that could be construed as a potential conflict of interest.

References

1. Lu, L.; Zhang, J.; Zhang, G.; Peng, H.; Liu, B.; Hao, H. The Influence of Box-Strengthened Panel Zone on Steel Frame Seismic Performance. *Buildings* **2023**, *13*, 3042. [CrossRef]
2. Adil, W.; Rahman, F.U.; Ali, Q.; Papakonstantinou, C.G. An Improved Anchorage System for L-Shaped FRP Composites to Enhance the Seismic Response of Beam-Column Joints in a Low-Strength Substandard Reinforced Concrete (RC) Frame. *Buildings* **2024**, *14*, 721. [CrossRef]
3. Adane, M.; Chun, S.; Kim, J. Seismic retrofit of framed structures using a steel frame assembly. *Steel Compos. Struct.* **2023**, *46*, 857–865. [CrossRef]
4. Cao, Q.; Ju, H.; Jia, J.Q.; Zhang, Q.; Zhang, L.H. Seismic performance of reinforced concrete beam-column joint made of post-filling coarse aggregate concrete. *Struct. Concr.* **2023**, *24*, 1118–1133. [CrossRef]
5. Jiang, W.G. Experimental Study on Carbon Fiber Reinforcement of Reinforced Concrete Beam Column Node Composite Members. Ph.D. Thesis, Southeast University, Nanjing, China, 2004.
6. Ling, Y.H.; Xu, J.H.; Guo, Z.H.; Wen, X.G. Study on seismic behavior of reinforced concrete column-steel beam side joints. *J. Asian Archit. Build. Eng.* **2021**, *20*, 44–60. [CrossRef]
7. Tang, J.R. *Seismic Resistance of Reinforced Concrete Frame Joints*; Southeast University Press: Nanjing, China, 1989.
8. Jiang, P.F.; Zheng, H.; Xiong, J.G.; Wen, P.H. Nonlinear elastic-plastic analysis of reinforced concrete column-steel beam connection by RBF-FD method. *Eng. Anal. Bound. Elem.* **2021**, *128*, 188–194. [CrossRef]
9. Jiang, S.; Chen, Q.; Li, C.; Song, H.; Lin, E.; Fu, C. Assessment of Soft-First-Floor Structures Reinforced by Rocking Frame Based on Seismic Resilience. *Buildings* **2024**, *14*, 197. [CrossRef]
10. Shoukry, M.E.; Tarabia, A.M.; Abdelrahman, M.Z. Seismic retrofit of deficient exterior RC beam-column joints using steel plates and angles. *Alex. Eng. J.* **2022**, *61*, 3147–3164. [CrossRef]
11. Saghafi, M.H.; Ali, G. Seismic retrofit of deficient 3D RC beam–column joints using FRP and steel PT rods. *Mater. Struct.* **2022**, *55*, 210. [CrossRef]
12. Akhaveissy, A.H.; Permanoon, A.; Mirzaei, M. Seismic Retrofit of Defective RC Beam-Column Joints. *Period. Polytech. Civ. Eng.* **2018**, *62*, 596–611. [CrossRef]
13. Cai, Z.W.; Liu, X.; Li, L.Z.; Lu, Z.D.; Chen, Y. Seismic performance of RC beam-column-slab joints strengthened with steel haunch system. *J. Build. Eng.* **2021**, *44*, 103250. [CrossRef]
14. Masi, A.; Santarsiero, G.; Nigro, D. Cyclic Tests on External RC Beam-Column Joints: Role of Seismic Design Level and Axial Load Value on the Ultimate Capacity. *J. Earthq. Eng.* **2013**, *17*, 110–136. [CrossRef]
15. Santarsiero, G.; Manfredi, V.; Masi, A. Numerical Evaluation of the Steel Plate Energy Absorption Device (SPEAD) for Seismic Strengthening of RC Frame Structures. *Int. J. Civ. Eng.* **2020**, *18*, 835–850. [CrossRef]
16. Wu, P.; Li, D.; Yu, F.; Fang, Y.; Xiang, G.; Li, Z. Experimental study on bearing capacity of PFCC column-RC beam joint reinforced with CST. *Steel Compos. Struct.* **2023**, *47*, 19–36. [CrossRef]
17. Liu, W.; Zhang, J.; Liu, H.; Wang, F.; Liu, J.; Han, M. Seismic Fragility Analysis of Existing RC Frame Structures Strengthened with the External Self-Centering Substructure. *Buildings* **2023**, *13*, 2117. [CrossRef]
18. Zhao, G.D. Theoretical Analysis and Experimental Research on the Bending Stiffness of Prestressed Steel Reinforced Concrete Beams. Master's Thesis, Shandong Institute of Architecture and Engineering, Jinan, China, 2005.
19. Wu, C.L.; Liu, J.M.; Tan, W.Y.; Wang, P.F. Seismic behavior of composite interior joints of prefabricated H-shaped steel reinforced concrete column—Steel beam. *Structures* **2020**, *23*, 558–572. [CrossRef]

20. Zhang, Y.; Ma, W.; Li, X.; Li, K. Experimental Research on Seismic Behavior of Haunched Concrete Beam-Column Joint Based on the Bolt Connection. *Sustainability* **2022**, *14*, 15644. [CrossRef]
21. Sui, L.; Wu, H.; Tao, M.; Jia, Z.; Zhou, T. Seismic performance evaluation of steel moment frames with self-centering energy-dissipating coupled wall panels. *Steel Compos. Struct.* **2023**, *47*, 663–677. [CrossRef]
22. Santarsiero, G.; Masi, A. Seismic Upgrading of RC Wide Beam-Column Joints Using Steel Jackets. *Buildings* **2020**, *10*, 203. [CrossRef]
23. Jiang, H.; Li, J.; Cheng, Q.; Xiao, J.; Chen, Z. Axial behavior of RC column strengthened with SM-CFST. *Steel Compos. Struct.* **2022**, *43*, 773–784. [CrossRef]
24. Santarsiero, G. FE Modelling of the Seismic Behavior of Wide Beam-Column Joints Strengthened with CFRP Systems. *Buildings* **2018**, *8*, 31. [CrossRef]
25. Mou, B.; Yan, X.C.; Qiao, Q.Y.; Zhou, W.Q. Experimental and numerical investigation on exposed RCFST column-base Joint. *Steel Compos. Struct.* **2022**, *45*, 749–766. [CrossRef]
26. Chen, L.B. Analysis of Square Concrete Filled Steel Tube's Mechanical Behavior in Finite Element Method. Ph.D. Thesis, Wuhan University, Wuhan, China, 2004.
27. Luat, N.V.; Lee, H.; Shin, J.; Park, J.H.; Ahn, T.S.; Lee, K. Experimental and numerical investigation of RC frames strengthened with a hybrid seismic retrofit system. *Steel Compos. Struct.* **2022**, *45*, 563–577. [CrossRef]
28. Xiao, L.L. Nonlinear Finite Element Analysis of High-Strength Concrete Frame Joints. Ph.D. Thesis, Gansu University of Technology, Lanzhou, China, 2001.
29. Zhuang, Z. *ABAQUS Nonlinear Finite Element Analysis and Examples*; Science Press: Beijing, China, 2005.
30. Teng, Z.M.; Zou, L.X. Nonlinear finite element analysis of reinforced concrete components under cyclic loading. *China Civ. Eng. J.* **1996**, *29*, 19–27.
31. Lubliner, J.; Oliver, J.; Oller, S.; Onate, E. A plastic-damage model for concrete. *Int. J. Solids Struct.* **1989**, *25*, 229–326.
32. Lee, J.; Fenves, G.L. Plastic-damage model for cyclic loading of concrete structures. *J. Eng. Mech.* **1998**, *124*, 892–900.

buildings

MDPI

Article

The Roof of the Glòries Tower: The Design Process for a Steel Dome at a Height of 120 m

Ignacio Costales Calvo *, Oriol Muntane Raich *, Xavier Gimferrer Vilaplana and Pablo Garrido Torres

Department of Architectural Technology, ETSAV-UPC, 08005 Barcelona, Spain; xavier.gimferrer@upc.edu (X.G.V.); pablo.garrido@upc.edu (P.G.T.)
* Correspondence: nacho.costales@upc.edu (I.C.C.); oriol.muntane@upc.edu (O.M.R.)

Abstract: Twenty years after its construction, the Agbar Tower remains one of the most recognizable landmarks in Barcelona's skyline. This article analyzes the dome that serves as its roof. The design of this element required a year of development, during which more than ten iterations were analyzed, resulting in the final design. There was no similar precedent published, as it is a 30 m high dome with significant wind exposure. This study offers an analytical review of the design process and the dome's enclosure, providing a reference for similar architectural construction projects. This research focuses on describing the elementary loading hypotheses and obtaining numerical models for structural calculations, comparing the results of the five most representative models of the process. Various factors are considered, such as shape, enclosure, structural frequency, deformations obtained, fire protection, and cost. The conclusion focuses on explaining how a form that remains unchanged throughout the process slightly varies the structural solution while respecting the project, to adjust to all the final regulations, construction requirements, cost requirements, and project requirements.

Keywords: dome; steel structure; high-rise buildings; façade; pre-dimensioning; wind pressure calculation

1. Introduction

The *Glòries Tower* [1], formerly known as the *Agbar Tower*, is one of the tallest and most recognizable landmarks in Barcelona [2]. Located in Cerdà's Eixample district [3], it was among the first initiatives aimed at shifting the city's center of gravity [4] towards Plaça de les Glòries [5–7]. Soon after, this area saw the construction of the *Nou Mercat dels Encants* [8], the *Museu del Disseny* [9], *Glòries Park* [10], and several other less well-known urban development projects [11,12]. The tower, completed in 2005, was the result of an extensive design process. It was designed by the architectural firms *Ateliers Jean Nouvel* and *B720 Fermín Vázquez Arquitectos*. With a height of approximately 142 m, it is the third tallest service-sector building in the Barcelona metropolitan area, surpassed only by the twin towers of Port Olímpic, which rise to 154 m. The building was constructed to accommodate the offices of the renowned water management company that initially gave it its name. The company later moved to the *"L" building* in the Zona Franca, a project by *Arata Isozaki* and *CMT*. The *Glòries Tower* is currently owned by the *Merlín Properties*, which has converted the building into a multi-tenant property primarily designated for office use [13]. The structural design and calculations were undertaken by the firm *Brufau, Obiol i Moya*, known as *BOMA* [14], (now called Socotec) under the direction of architects *Dr. Agustí Obiol i Sánchez* and *Dr. Robert Brufau i Niubó* [15], who were, at the time, professors in the structural architecture department at *UPC* (*ETSAB* and *ETSAV*, respectively). The project was proposed between 2000 and 2001, with construction finalized between 2001 and 2005. The total cost was estimated to be approximately EUR 130 M.

The building (B720 Fermín Vázquez Arquitectos. Available online: https://b720.com/es/b720-projects/torre-agbar/ (accessed on 15 March 2025)), as can be seen in Figure 1, comprises four basement levels, extending to a depth of approximately 17 m, about seven meters below the water table [16]. Additionally, it includes an auditorium located below ground level, beyond the tower's footprint.

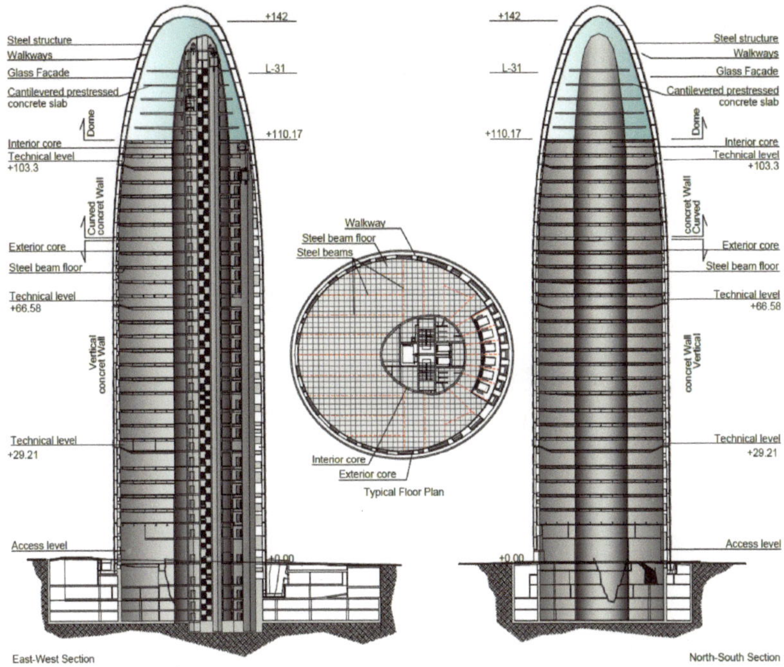

Figure 1. A typical office floor plan, where the absence of pillars is noticeable, creating a completely open space. Two sections of the tower, where the blue dome can be seen at the top, rest solely on the exterior wall (*Image created by the authors from CAD file downloaded from www.dwglab.com*).

The inner concrete core [17–20] houses the service elevators, utility risers, and restrooms, as well as elevators that ascend to the dome. The exterior curved wall, easily recognized by its predominantly white, gray, yellow, blue, and reddish colors, is illuminated at night by approximately 4500 LEDs. It encloses the building and supports the exterior maintenance galleries, which are covered by 52,744 glass slats. The colors are said to have been inspired by structural calculation maps of concrete stresses, although the final design was ultimately designed by the architects.

The basic project brief states the following: "It is a fluid mass that has pierced the ground like a geyser of permanent, measured pressure. The building's surface evokes water... The material is read in its depth, colored and uncertain, luminous and nuanced. This architecture comes from the earth, but lacks the weight of stone; although it could be a distant echo of ancient Catalan formal obsessions brought by the mysteries of the wind that blows from Montserrat...".

The space between the two walls, the exterior and the interior, is used for offices and access to the elevators, which are located in the narrowest area. Three technical floors [21,22] are provided, one for approximately nine office floors, to reduce thermal losses due to height.

The upper part was originally reserved for the executive offices of the Agbar company. With the change in ownership, a viewing platform is located in this space at level 125, on the 30th floor, where there is also a tensile construction called "Cloud cities", an interactive work by *Tomás Saraceno* (Studio Tomás Saraceno. Available online: https://studiotomassaraceno.org/ (accessed on 5 April 2025)), which has been open to the public since 2022. To obtain this new space, it was necessary to carry out a new project, in accordance with its new use and its new requirements [23], which is not the subject of this article.

Regarding the dome, as illustrated in Figure 2, we lacked clear calculation references that would allow us to approximate an approximate cross-section for its constituent elements. Its shape, based on an oval base and with a rather pointed cross-section, makes it rather unique. Its dimensions are similar to those of the Marble Church in Copenhagen, St. Paul's Cathedral in London, or the Capitol. Regarding similar steel typologies, the Bourse de Commerce dome and the Reichstag dome, although larger, are similar models to the one studied. Regarding pre-dimensioning formulas, there are formulas such as those proposed by Garde and Gehlot [24] for steel domes with meridians and parallels, or those for geodesic domes, as in various papers on the subject [25–28]. The formulas had to be carefully considered, because the shape, the actions, and the type of metal elements could decisively change the result.

Figure 2. Current view of the Glòries Tower dome. The top of the tower is a dome that has a height equivalent to a 10-story residential building. Photo © *Paulina Flores*, Barcelona.

2. Description of the Tower's Structure

The tower's structure primarily consists of two in situ concrete walls arranged in a ring shape for support:

- The curved outer wall, rising 110 m from ground level, has an irregular ovoid shape in plan, with main axis dimensions of 39.40 m and 35.40 m. Its thickness varies between

50 and 30 cm, defining the perimeter of the building. This wall incorporates the distinctive irregular windows that are visible from the exterior. At a height of 76 m, the wall begins to incline inward, gradually transitioning to a shape that aligns with and supports the base of the dome.

- The curved interior wall, which in plan takes the form of an ovoid with approximate dimensions of 16.3 by 15.90 m and a height of 132 m above ground level, is offset from the façade. This space houses the elevators and stairs, with wall thicknesses ranging from 50 to 30 cm. Its roof is a small reinforced-concrete dome located beneath the large steel dome.

These two walls, constructed in a continuous process even before the floor slabs, support metal beams that link and stabilize them, ensuring rigid diaphragm action in the horizontal plane. These beams also support the ceilings, which are made of ribbed reinforced-concrete slabs resting on a steel sheet that functions as permanent formwork. Above the outer wall, at an approximate height of 115 m, sits the dome, which serves as the building's roof. Though it might appear to be a secondary structure, in plan, it takes the form of an asymmetric ovoid measuring 28 by 32 m, with a height of 28 m. The dome is supported by twenty-six nearly equidistant metal meridians aligned with the façade surface. Originally, the company's executive offices were located at these levels.

Within the space enclosed by the dome, the original design includes three post-tensioned reinforced-concrete slabs, each with a thickness of up to 500 mm, supported exclusively by the inner core. These slabs extend 10 m outward toward the façade without making contact. Initially, the slabs were intended to rest on the meridians of the dome as well; however, this idea was soon discarded in favor of maximizing spatial flexibility.

Thus, the dome [29] follows a form closely resembling a catenary arch in section. As its designers explained during the development of the executive project, the building takes inspiration from Catalan architectural traditions [30]. Before defining the structure [31], the loads and forces it needed to support were first analyzed. Several reference models can be cited, including *Antoni Gaudí's* analytical studies using sandbag models for the Chapel of Colònia Güell, as well as the cable models employed in designing the diaphragm arches of Casa Milà. In three-dimensional structures, a notable example is the *dome of Vitoria* [32], designed by *Joan Margarit* and *Carles Buixadé*. Although this 76 m diameter spherical structure was built decades ago, it was calculated using a method developed by its authors in 1969, long before modern computational tools provided the reliability available today. Other similar structures were envisioned in the early 20th century by *Richard Buckminster Fuller* [33].

3. Dome Façade Description

The dome's enclosure is composed of a double-skin system, with an intermediate passageway accessible solely for maintenance and conservation, as illustrated in Figure 3. The outer skin is a continuation of the building's general façade, covered with 13 mm thick laminated glass louvers that allow for interior cleaning without requiring a gondola system. On the dome, these louvers have a greater inclination due to their function as part of the roof, yet they maintain the same configuration. The inner layer consists of a modular curtain wall with double 10 mm glass ((15) 5 + 5 mm; the thickness of the outer layer is 10 mm, the air chamber is 15mm, and the thickness of the inner layer is 5 + 5 mm), directly attached to the dome's metal framework. The glazed surface of the dome covers approximately 2350 m^2. The façade was built by *Permasteelisa* (Permasteelisa group. Available online: https://www.permasteelisagroup.com/project-detail?project=1984, accessed on 15 march 2025). The calculation was performed using Therm software). Their thermal transmittance analysis calculated a U-value (K coefficient) of 2.24 W/m^2K, considering an interior temperature of

20 °C and an exterior temperature of 0 °C. This calculation guarantees the prevention of surface condensation, assuming an interior humidity level of 50%.

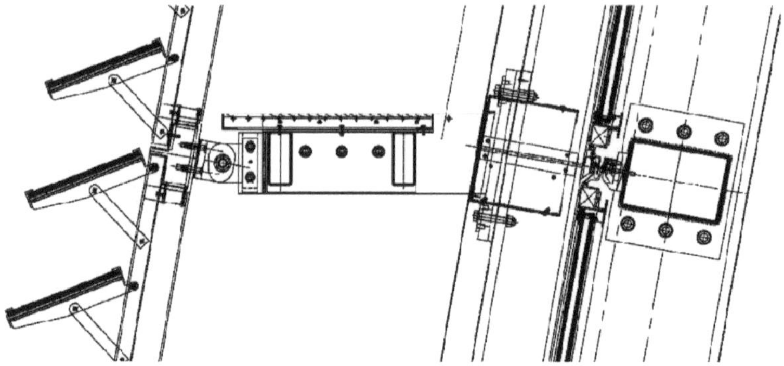

Figure 3. Detail of the façade corresponding to the construction project. On the left are the slats which are easy to clean from the inside. In the middle is the maintenance walkway with metal rebar serving as a pavement. On the right is the planned connection between parallels and meridians, which was ultimately welded rather than bolted.

4. Action Assessment

During the project phase, the weight of the glass panels was conservatively estimated at 40 kg/m^2, while the walkways were considered to weigh 15 kg/m^2, and the exterior louvers 10 kg/m^2. Snow load was not a significant concern, as surfaces with an inclination greater than 60° are generally exempt from consideration. Consequently, only the central section of the dome required consideration, with a projected load intensity of 40 kg/m^2. Even with climate change in mind, this snow load intensity is relatively high for a city like Barcelona, especially in comparison to other provinces across the Iberian Peninsula. Among Spain's coastal areas, only the Catalan coast, which includes cities such as Tarragona, must account for a 40 kg/m^2 snow load. By contrast, other coastal regions, such as the Cantabrian coast, are assigned lower values of 30 or 20 kg/m^2, depending on their specific location, as indicated in Figure 2 of the CTE-SE-AE Additionally, the louvers are set at an inclination of approximately 30° and overlap one another. Given these conditions, a generous estimate of 20 kg/m^2 was used to account for the projected snow load.

Thermal effects have minimal impact on this part of the building, as the metal columns or meridians can be considered to have low rigidity [34]. Additionally, with plan dimensions of around 35 m, there is no need for special calculation measures, given that these elements are located internally, behind the enclosure. It is also important to note that the dome rests on another structure—the exterior concrete wall—which will expand or contract in a similar manner.

This article does not delve into the topic of seismic activity, as its analysis would be too extensive.

5. Wind Analysis

Wind action poses the most significant challenge for several reasons. The first issue was the regulatory framework in place at the time, specifically NB-AE-88 [35]. The current technical code for structural actions [36] did not become mandatory until 29 March 2006, and had not yet been drafted during the project's development. Nevertheless, Eurocode 1 (EN 1991-1-4) [37] which was applicable on a voluntary basis, already outlined wind action considerations that differed from those mandated in Catalonia's regulations. Specifically, NBE-EA-88 (see Figure 4), states that for buildings located away from the coastline and exceeding 100 m in height, a dynamic wind pressure of 125 kg/m^2 can be applied. This dynamic pressure was then adjusted using the corresponding wind coefficient, as specified in Table 1 of the same regulation, for circular shapes and for rough or smooth curved surfaces.

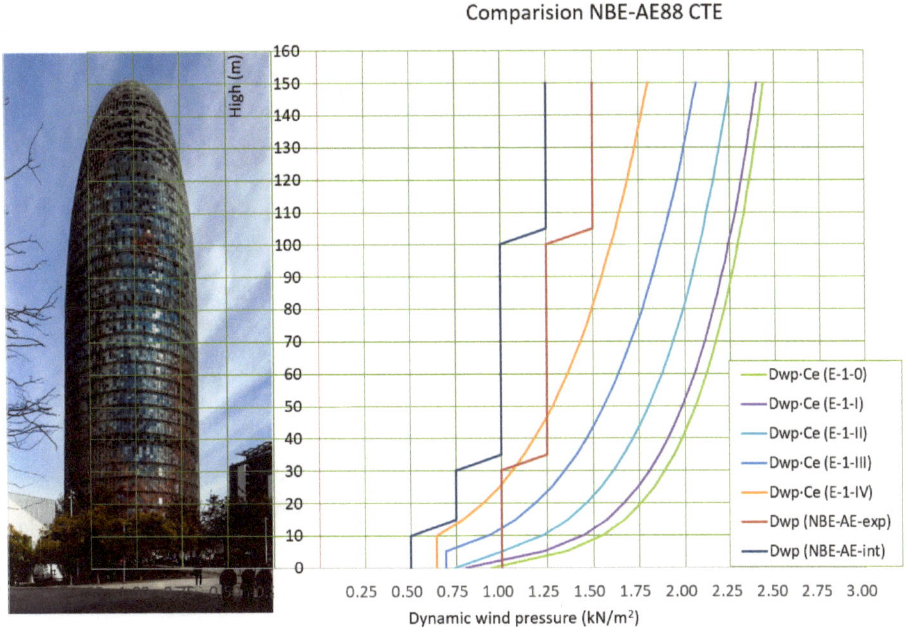

Figure 4. (**Left**) Current view of the Glòries Tower. Photo © Paulina Flores, Barcelona. (**right**) Comparison of wind pressure according to Europe EC Eurocode and Spanish CTE "Codigo Técnico de la Edificación" (curves) and Spanish code "Norma Básica de la Edificación" NB-AE-88 (broken lines). "Dwp" = dynamic wind pressure; "exp" = exposed situation; "int" = interior city situation; "Ce" = Spanish wind exposure coefficient.

The pressure exerted by the wind decreases slightly when the surface has roughness or when elements are introduced at height to disrupt or create asymmetries in the airflow. In the case of chimneys, incorporating a lightweight spiral staircase or propellers can positively contribute to structural stability. Therefore, the outer skin, composed of single glass panels arranged as louvers, proved to be more beneficial for the structure's stability [38] and for determining wind intensity, which is always lower than that of a perfectly smooth curved surface.

The Eurocode is both more precise and more demanding. It proposes five scenarios, the selection of which may be subject to controversy, but a thorough analysis clearly defines them. Curves are numbered from 0 to IV, with 0 representing the most unfavorable case (the current CTE adopts the same five curves but designates Curve I as the most unfavorable and Curve V as the least affected by wind action). When comparing these curves with those outlined in the former NBE-AE-88 regulation, the previous tables [39] can be derived.

Table 1. Comparison of Cp calculation according to NB-AE-88 (left) and CTE-AE (right). "Cp" = pressure wind coefficient, which depends on various factors, specifically the type of shape of the surface considered, as well as the finish.

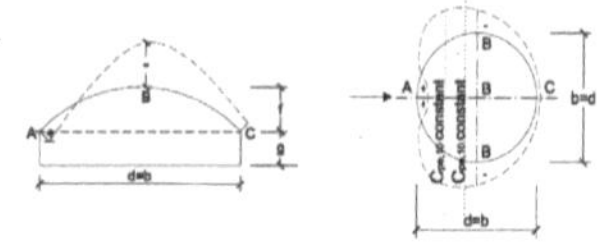

	Wind Load Coefficient in a Closed Structure			
Flow Wind	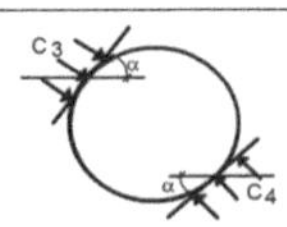			Wake
	Wind Coefficient on:			
Situation. Wind Incidence Angle α	**Rough Curved Surfaces**		**Very Smooth Curved Surfaces**	
	Windward c3	**Leeward c4**	**Windward c3**	**Leeward c4**
In Wake	*Intermediate values can be linearly interpolated*			
90-0°	+0.8	−0.40	+0.8	−0.40
In Flow	*Intermediate values can be linearly interpolated*			
90°	+0.8	−0.40	+0.8	−0.40
80°	+0.8	−0.40	+0.8	−0.40
70°	+0.8	−0.40	+0.4	−0.40
60°	+0.4	−0.40		−0.40
50°	−0.40	−0.40	−0.40	−0.40
40°	−0.80	−0.40	−0.80	−0.40
30°	−0.80	−0.40	−1.20	−0.40
20°	−0.80	−0.40	−1.60	−0.40
10°	−0.80	−0.40	−2.00	−0.40
0°	−0.40	−0.40	−2.00	−0.40

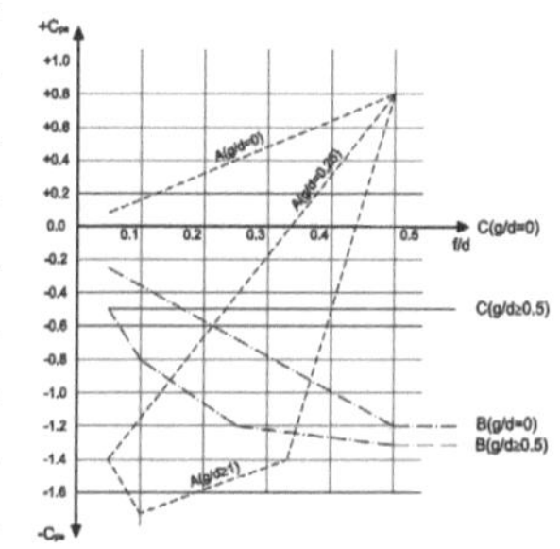

In Figure 4, the broken blue line represents the regulations in effect at the time for buildings in a "Normal" situation, while the red line applies to buildings classified as being in an "Exposed" situation. An exposed situation encompasses locations such as coastal areas, topographic ridges, narrow valleys, and plateau edges. Additionally, it was specified that in particularly exposed topographic conditions—for instance, high mountains, gorges, or cliffs—higher values may be necessary and should be determined through a specialized study. Consequently, the appropriate exposure classification to be applied in this case would be "Normal". The other curves represent the different options offered by the Eurocode, covering a wider range of positions. It can be observed that, under current regulations, almost any scenario results in higher wind pressure than what was in effect at the time. Although the NBE-AE-88 value was regulatory, a pressure of 180 kg/m^2 was adopted during the project phase for all parallels. This decision matched Eurocode 1, Curve IV, for the roughness grade associated with urban areas. The building is located a little over 1.5 km from the coast. This wind pressure, corresponding to the tower's maximum height, was applied uniformly from the base of the dome to the keystone. The center of gravity of a catenary's interior form is closer to the base than to its average height, it can be understood that the final wind pressure considered was even midway between curves IV and III. Annex A of EN 1991-1-4:2005 (Eurocode 1.4) provides images that support this decision. A wind tunnel was not commissioned, but basic checks were performed using a calculation program called GID, which confirmed these pressures. The knowledge that the structural design consultancy had acquired in the 1990s on another similar high building located in Barcelona (Agustí Obiol was the Catalan partner responsible for overseeing the calculations and subsequent construction management of the Hotel Arts in Barcelona, a project by SOM architects), which was previously compared with more advanced American standards, concluded that the wind loads, greater than 45% of the then-current Catalan standards, were adequate.

Finally, calculating the Cp coefficient for each point of the dome was necessary. Both regulations establish calculation values based on the angle between the considered wind direction and the tangent plane at the analyzed point. While the values are largely similar, Table 1 reveals slight variations. Specifically, the AE-88 reaches suction coefficients of 2.0 within a range of up to 30°, whereas the CTE, in certain specific cases, reaches a value of −1.7. In the building's nearly 25-year lifespan, there have been no problems related to wind exposure on this element. The wind speed required by current national regulations for multi-story buildings is excessively high; the zones on the national map should be adjusted.

Another issue in transferring the actions to the calculation model was the geometry of the construction itself. At the time of construction, matrix calculation programs for bar elements were not as advanced as they are today, with wind loads now able to be calculated automatically by the software. To streamline the analysis, a spreadsheet was used to generate the loads, enabling quick transfer into the calculation program. The pressure and suction coefficients were estimated based on the polar angle, as the surface is curved not only in plan but also in section. It is also worth noting that the wind should be considered with an angle variation of plus or minus ten degrees and that the exterior façade louvers could introduce inclinations that were either more or less unfavorable for the calculation.

The Microsoft Excel formula used to calculate the polar angle is shown below:

$$\alpha_{polar} = 90 - (\mathrm{ACOS}(\mathrm{COS}(\alpha_h \cdot \pi/180) \cdot (\mathrm{COS}(\alpha_v \cdot \pi/180)))) \cdot (180/\pi) \quad (1)$$

where

α_h is the plan angle of the surface;

α_v is the section angle of the surface.

Using this polar angle, the wind coefficient for a closed structure's surface was automatically obtained in Table 1 of the NBE-AE-88, considering the surface to be rough. To input the actions into the calculation program, over 400 action lines were generated for each scenario, making it easy to modify the values at any time if adjustments to the intensity, incidence angle, or orientation were needed.

Once the loads and shape were determined, the next step was to define the supporting structure that would enclose the building.

6. Searching for the Structural Form

Between 2001 and 2002, multiple meetings were held between the architectural and structural teams to determine the most effective way to ensure the roof's stability [40]. From the very beginning, it was clear that the oculus had to remain open. The best way to resolve the junction of so many bars is, without a doubt, to eliminate it altogether and replace it, with a compression ring positioned far enough away to prevent the meridians from converging at a single point. A common solution in the early 20th century for addressing this issue was to use a cast iron piece, onto which the meridians were bolted along its perimeter-bearing surface. This approach prevented excessive heating at a central point caused by the concentration of welds. Welding on cast iron can be performed effectively if an appropriate welding system is used [41].

Numerous proposals were analyzed, with the emphasis placed more on the structure than on the shape. This document presents only four models, as they are the most interesting and diverse for understanding the design and calculation process. Each version was thoroughly reviewed by all stakeholders, gradually refining the structure toward a final proposal.

The calculation software used was Robot V6, running under a licensed version in Microsoft's MS-DOS environment. The images included in this article were obtained using an educational license of Autodesk Robot [42], used exclusively to recover information from twenty years ago for the preparation of this article.

6.1. First Version, Trussed Dome: Basic Project

The first proposed design featured 20 spatial triangular meridians and 20 flat triangulated parallels arranged in a lattice structure. Additionally, three secondary meridians were inserted between each of the 20 main meridians, and one simple parallel was placed between each of the 20 main parallels. A simple element refers to a component consisting of a single profile with low inertia. By the end of the process, the final structure comprised a total of 11,980 bars. The use of such a large number of bars allowed for very thin profiles, specifically RHS-200 × 50 × 4 bars and solid diameters of 16 mm. The proposed structure had a total steel weight of 91,943 kg but posed challenges in ensuring adequate fire protection for the metal profiles [43], as the ϕ16 bars had a particularly high shape factor. (The "shape factor" is the ratio of the perimeter of a member exposed to fire to its cross-sectional area. The cross-sectional area will always be a constant value intrinsic to the section, while the perimeter can vary depending on the faces exposed to fire. The concept of "massiveness" assumes that the entire contour is exposed to the action of fire.) At the beginning of the project, the required fire resistance was estimated to be R-30. However, as we will see later, this requirement was increased following several discussions with the fire department. The final shape of this first proposal, shown in Figure 5, resembles the bow of a zeppelin but made of steel and more reinforced.

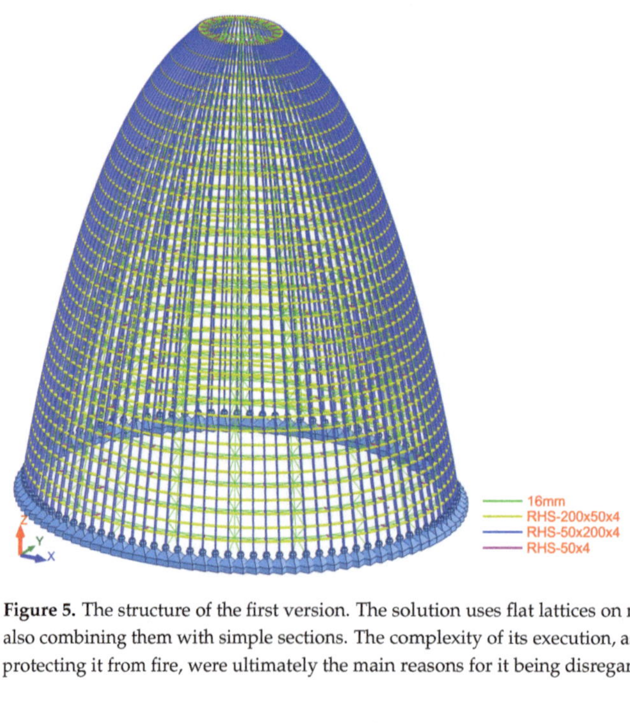

Figure 5. The structure of the first version. The solution uses flat lattices on meridians and parallels, also combining them with simple sections. The complexity of its execution, as well as its difficulty in protecting it from fire, were ultimately the main reasons for it being disregarded.

Furthermore, both the manufacturing cost and the installation process became overly complex. The structure needed to be assembled at a minimum height of 120 m, posing the usual challenges associated with working at height. Another issue with this initial solution was that the structure clashed with the prestressed cantilever slabs, as it reached a depth that was incompatible with the project requirements—specifically, 600 mm between axes and 708 mm in actual size—making it necessary to reduce them. This issue, along with fire safety considerations, was decisive in proposing a second, radically different structural solution, which is discussed later.

The proposed joint detail, though not meticulously developed since the solution was ultimately discarded, is shown below in Figure 6. A 16 mm diameter bar has a shape coefficient [44] (commonly known as massiveness) of

$$M\varphi 16 = 2/r = 250\ m^{-1} \quad (2)$$

where

r is the radius of the solid bar;

M is the shape factor (massiveness).

Figure 6. A diagram of the proposed joints for the first version. The compositional form of the structure can be understood as a simple dome reinforced locally with a second chord, and with vertical and diagonal elements.

For tubes, considering them exposed on all four sides,

$$M_{t=4mm} = 1/t_{RHS} = 1/0.004\ m = 250\ m^{-1} \quad (3)$$

where t_{RHS} is the tube's wall thickness in meters.

In general, achieving 60 min of protection with intumescent coatings was difficult for such high shape factors [45]. Moreover, considering that these are closed sections, which require a thicker protective layer than open sections, the challenge becomes even greater. Since the structure had to be visible, using gypsum panels or vermiculite was not an option. Applying such a thick layer of paint would not only increase material costs but also result in a poor esthetic finish for the dome. Intumescent paint tends to develop a textured finish known as "orange peel" when applied in thick layers and is also susceptible to impact damage [46].

Therefore, if this option was chosen, it would be necessary either to increase the thickness of the metal elements to reduce their shape factor—further increasing the cost of the structure—to conduct a performance-based CFD analysis [47], or to install sprinklers. One potential advantage is that, in the event of a hypothetical collapse, the failure of a joint could be managed more effectively and with greater control compared to the other solutions discussed below [48].

For this version, we propose a construction price of approximately EUR 500,000, updated to the current date, with an ideal construction period of four months.

6.2. Second Version, Fine-Shell Dome: Basic Design A

The second version examined in this article follows a shell-type solution. The parallels act as rings that independently prevent the buckling of the meridians, which transfer gravitational loads to the top of the concrete façade. The version presented—one of several—uses a single profile type, 200 × 50 × 5 mm, for both the parallels and the meridians. The 200 mm dimension was consistently oriented perpendicular to the surface. This solution addressed several issues identified in initial tests, including accommodating the passage of prestressed concrete slabs and simplifying the construction process, as the profile connections were easy to assemble. Additionally, it was significantly lighter than later versions. Nevertheless, some challenges remained, such as the external walkways, which generated significant moments at their connection points, necessitating the reinforcement of certain joints. Fire protection was also a concern, as a 5 mm tube wall had a shape factor of 200 m^{-1}.

The calculation model consisted of 14,836 bars and had a total weight of 158,424 kg. Once again, the increase in weight was an issue, but as will be seen later, an even greater weight was required. The bars created gaps of only 1 m × 1 m at best, which neither favored the view of Barcelona nor facilitated the installation of the glass enclosure. While this might have seemed like a regression in the project's evolution, it was actually a necessary transition toward the final version: a single "shell" with no additional thickness beyond that of the profile itself. As a result, it was key to use deeper profiles while increasing the spacing between them. The final shape of the dome in this version is shown in Figure 7.

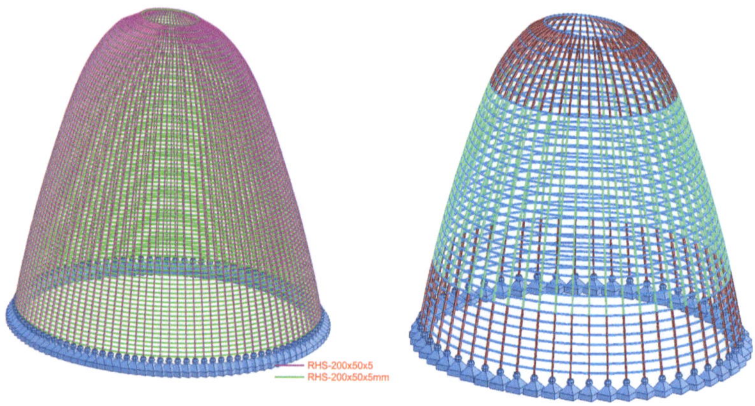

Figure 7. (**Left**) The 2A version described in this article, compared to the 2B version (**right**). The structural solution concept is the same in both proposals. In the second, half of the mid-radius and parallels are eliminated. To maintain rigidity, the cross-sections, as well as the wall and vertical dimensions of some elements, are increased.

The most significant difference between the first and second options is the behavior of the meridians. Initially, they functioned in compression and tension—common in triangulated truss structures. However, in the second option, they transitioned to working in combined compression, a characteristic feature of simple profiles.

In this case, a construction material price of approximately EUR 830,000 is proposed, updated to the current date, with an ideal construction period of 5.2 months.

6.3. Second Version, Medium-Shell Dome: Basic Design B

This second version A was excessively difficult to build due to the number of joints, so before analyzing the impact of fire protection and the admissible deformation of the glass, an intermediate solution was analyzed, which went from 106 meridians to 52, and from 69 parallels to 35. The comparison between the two solutions that can be seen in Figure 7 expresses the superiority of this second option. Material execution price: EUR 665,000; construction period: 3 months.

6.4. Third Version, Steel Frame: Executive Project

The third version described in this document is the one that finalized the executive project. At the time of construction, there was a more comprehensive understanding of the space, enclosures, walkways, and various constraints that the dome's structure needed to accommodate. The earlier versions can be seen as incomplete proposals that progressively evolved to the final solution.

A less dense and more open design was chosen to enhance the feeling of a private viewing platform, with glass panels reaching maximum dimensions of 3.70 × 1.90 m. This stage also marked the integration of the walkways, which had previously been considered only as loads on the joints but were now designed as cantilevers embedded into the shell. Another key factor was cost reduction, not necessarily in terms of the total steel weight but by minimizing the number of connections. Finally, the required 90 min fire resistance was addressed by considering the shape factor of the sections and increasing the tube wall thickness, ensuring better compatibility with the application of intumescent paint.

The final version, prior to construction, had a weight of 277,941 kg, with the model consisting of 910 bars, excluding the cantilevers. While the reduction from 14,836 to 910 bars does not represent exact data, it does illustrate the significant simplification achieved in terms of connections, as each bar in the calculation linked two structural nodes.

The main difference in this option lies in the use of two different profile types:

- The first type was a box section formed of two UPN-300 profiles welded together and attached to an IPE-330. This combined profile was positioned on the exterior of the shell, in the cold zone. It was also used for the meridians supporting the walkways, applied to every other one.
- The second type was a box section made of two UPN-300 profiles, similar to the first but without the IPE. This profile was placed on the interior side of the dome.

The decision to use a tube formed of two UPN profiles instead of a single RHS was driven by the slight advantage of bending a UPN profile before welding, as well as its sharp edges, which contrast with the rounded edges of an RHS. This approach significantly simplified the connections, where each profile joined another with the same characteristics, eliminating the need for stiffeners or additional plates. Since the profiles were curved, all connections remained tangential, preventing issues with section rotation—a frequent challenge in non-orthogonal spatial structures composed of planes. Bolted connections were generally avoided, except for potential joints between the meridians and the concrete façade. Architecturally, this solution also resulted in a highly refined connection. The final shape of this version is illustrated in Figure 8, while Figure 9 presents the typical joint detail.

The lower connections to the concrete façade were designed to be flexible with respect to bending moments. While full embedding could be achieved with proper concrete reinforcement, this approach provided advantages in terms of the final deformations and stress distribution in the profiles. Each meridian experienced a reaction force between 218 and 438 kN under ultimate limit state (ULS) conditions, always in compression. Bending moments were relatively minor, reaching a maximum of 55 kNm, also in ULS.

Material execution price: EUR 1,300,000; construction period: 1.7 months.

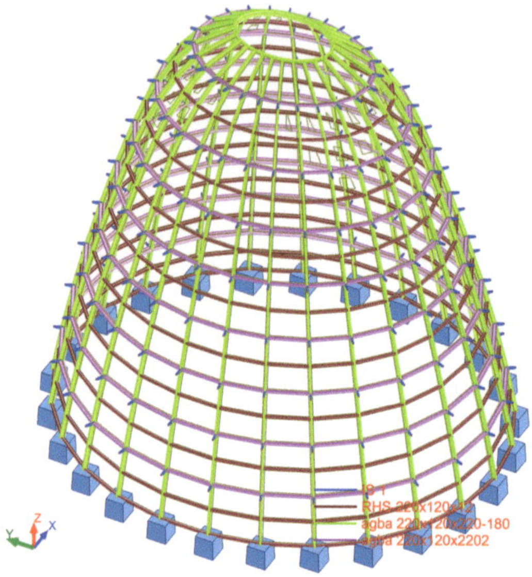

Figure 8. The final structural design of the executive project, featuring supports embedded in the ground. This structure turned out to be the first to meet all the design requirements. Despite the previous revision, it would undergo a final revision, changing the way each element worked, shifting responsibility for the meridians to the parallels.

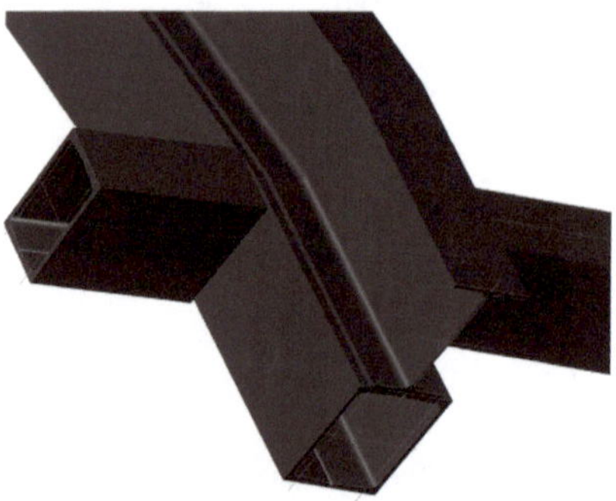

Figure 9. Executive project detail of a typical meridian and a simple parallel. The double UPNs were arranged on a single plane to facilitate the welded connection between them. The IPE, which reinforces the meridians, becomes essential to withstand the new deformability requirements and fire protection, which must be applied on site.

6.5. Fourth and Final Version, Braced Pillars: Built Project

The fourth and final version described corresponds to the as-built solution. In essence, it is a carefully refined adjustment of the sections used in the previous design, where closed profiles were replaced with open ones. This adjustment reduced construction costs, simplified assembly, and, most notably, facilitated the transportation of the structure from the workshop to the construction site. The reality is that, in the case of the third proposal, the need to join UPN profiles, sometimes requiring full-penetration butt welding, along with the bending of both sections forming each meridian, significantly increased the final cost. Closed profiles, on the other hand, required a greater paint thickness than open profiles. Moreover, a large portion of the steel section was positioned along the neutral axis, where it did not contribute to bending moment capacity. The issue was not element compression, but rather, insufficient bending inertia.

Considering all of the above, a simple open section was proposed consisting of a single double-T profile with a greater height, and with the flanges oriented parallel to the glass. For the meridians, a profile was used with exterior dimensions fitting within a 640 × 220 mm rectangle, while for the parallels, a hollow rectangular profile measuring 180 × 260 × 8 mm was utilized.

It can be noted that the response of this dome was noticeably different from that of the previous one. The meridians were reduced to allow the window frames to function properly. As a result, the parallels lost inertia, and the meridians had to be increased. Specifically, the total inertia of the meridians increased from 1,146,973 cm^4 to 2,503,437 cm^4 in version 4, representing an increase of 218%, while the inertia of the parallels decreased from 1,153,229 cm^4 in version 3 to 110,670 cm^4 in the final version, which accounted for just 9.5% of the inertia of the previous solution.

The curvature radius of the profiles was not particularly sharp, meaning the bending process would not present any problems. However, all sections of the meridians were ultimately built with a polygonal shape. This decision was made for three main reasons: the change would not be noticeable, construction costs would be reduced, and since the glass frames were positioned within the profiles, the frame would continuously rest on the profiles, simplifying the detail. A minor issue arose with the profile change: the structure would lose some torsional capacity, as a double-T profile has lower secondary inertia than a rectangular tube, and this loss might negatively impact the connection with the glass. This issue will be addressed in a later section. The final structure is shown in Figure 10.

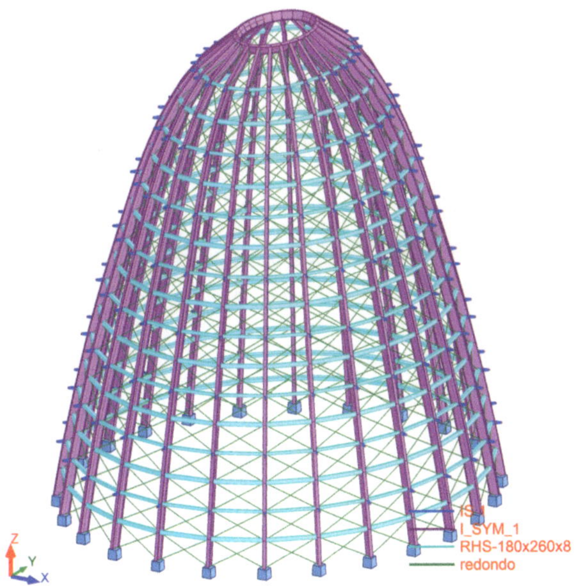

Figure 10. The structure of the final built version. The diagonal bars are fictitious elements used to analyze the angular distortion of the trapezoids that form the dome. The elongation of any diagonal should be less than 2 mm.

In this version, the profiles had double the inertia of the previous one, while the meridians significantly reduced their inertia, resulting in more refined detail for the façade.

Moreover, due to the open-profile design, the meridians required less intumescent paint for fire protection than would be needed for a closed profile with the same shape factor.

The current understanding of progressive collapse in welded metal connections on site is far greater than that considered at the beginning of the century. The collapse of the Twin Towers in New York, for example, occurred when construction had just begun, in 2001, and sparked a debate very similar to that of any high-rise building then under construction. Research on this problem began around the 1970s, but became more specific and gained greater relevance in the second decade of the 21st century, after the tower's construction [49,50]. The highest-quality steel grade for working at high heights with low temperatures and in unfavorable welding conditions is S-355-J0. Despite the above, the ductility of the connection was demonstrated, depending on the consumables used for the type of steel employed. Likewise, the welds performed on site were thoroughly reviewed. The self-weight of the steel structure, in the event of failure, can be supported by the lower floors in the event of an accident. The structure is independent of the concrete.

The support of the meridians (see Figure 11) on the concrete wall was resolved using an inverted U-shaped profile with a base matching the thickness of the upper part of the wall, made of steel plates and bolts. This profile, embedded in the wall during the concreting process, ensured a clean joint. The meridians were welded onto these plates, thus eliminating the need for stiffeners. The horizontal forces were transferred through the wall and the top slab supported by the façade.

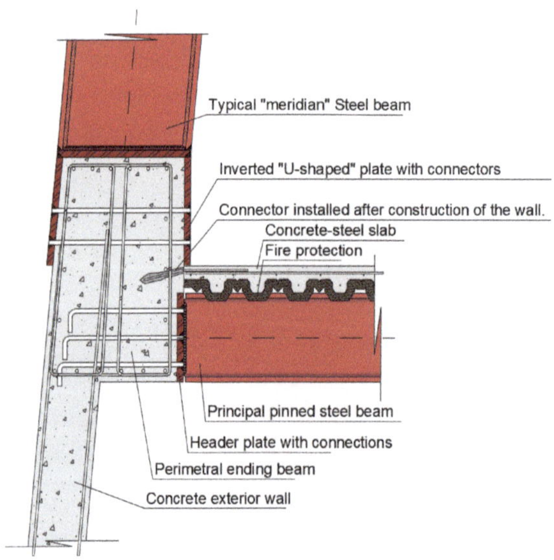

Figure 11. Typical detail of support for the meridians considered in the executive project. The perimeter beam acted not only as a support element for the beams of the last composite slab, but also as the lower tension ring of the dome.

To optimize the construction process, it was suggested to conduct an assembly test of the entire structure on an empty plot, away from the building, following the construction of a temporary foundation, which would be demolished after the test was completed. This approach aimed to mitigate potential issues at over 110 m in height while allowing workers to gain practice in the assembly process. It is worth noting that the dome stands at nearly 30 m high, equivalent to a 10-story building without intermediate columns. Given that the joints were ultimately welded—due to the challenges of using a bolted joint with a front plate—this test was not performed, partly due to the cost of the operation. The expertise of the metal workshop *ELTE* (*ELTEC*, also referred to as *ELTE* or later *ELFE*, was a renowned workshop that completed prestigious projects, including the *Fira de L'Hospitalet*, the roof of the *Liceu*, one of the grandstands at the *Montmeló F1 circuit*, the dome of the *Agbar Tower*, and the structure of the *Fórum 2004* building in Barcelona, among others), which was responsible for many high-profile metal works in Catalonia during that period, combined with the precision that BIM was beginning to enforce in the industry, helped minimize the challenges encountered during the construction phase.

The assembly system for the dome was proposed by the construction company *Dragados*. (*Dragados* was the construction company awarded the project. On 18 April 2002, while the project was already in progress, *Dragados* was acquired by *Florentino Pérez* and integrated into the *ACS group*. The construction and technical teams were part of *Dragados* before the merger.) Pairs of meridians were raised, connected by their parallel bars, forming structures resembling ladders. These ladders were then linked together by welding the intermediate horizontal elements or tubes. Each ladder had a height roughly equivalent to two-thirds [51] of the dome's total height. In the final stage, a small structure containing a cross-shaped oculus was assembled on the ground, then hoisted into place by a crane and carefully fitted at both ends, completing the building's enclosure.

Material execution price: EUR 1,000,000; construction period: 1.5 months.

7. Modal Analysis

Throughout the structure's calculation, various checks were performed on each model, taking into account the fundamental buckling modes. Up to 100 buckling modes were generated for each case, although this number was not strictly required due to the structure's highly regular nature. Figure 12 presents the first three modes in detail, which were used in a final comparison before finalizing the design.

The first version, which, as already mentioned, was not fully calculated, failed at the local level of a single bar, resulting in "progressive crushing." The value of 1.67 indicates that the dome was not sufficiently stiffened to achieve a higher safety factor, as is the case with the other four versions.

The solution that worked best was undoubtedly 2A, because the bars were very short. The failure occurred due to excessive deformation of several parallel bars in its plane.

Despite this, it was a localized failure known as "wrinkling" of the surface. Something similar happened to Solution 2B, but since the bars were longer, the buckling safety factor was reduced to half that of the previous version, despite remaining a completely acceptable factor. Finally, versions 3 and 4 also present a defect, not a general one, which could be understood as one that presents the same deformation in each parallel, but rather, a "wrinkled" one, but in a broader way.

Although the earthquake case is not addressed in this document, with regard to the modal calculation, the frequencies detected in the first three versions are shown in Figure 13.

The first two primary modes involved displacements along the two main axes, with the first mode, as expected, corresponding to displacement in a direction parallel to the shorter axis. The third mode was torsional, representing the rotation of the meridians in plan. In version 2, the frequencies were the highest among the models examined, with torsional deformation occurring in the fifth mode. This behavior is believed to be caused by the excess number of bars involved, even though it had the thinnest profile in plan. Mode four was characterized by the crushing of the oculus, along with the deformation of the meridians in their central section.

Modeling the supports at the base of each meridian as either a hinge or a fixed support influenced the deformation results and primary modes, with the fixed support providing a slightly more favorable response. The continuous lower wall, which supported the meridians, occasionally had windows beneath this level, making it necessary to consider the wall's torsional behavior to resist the fixed moments.

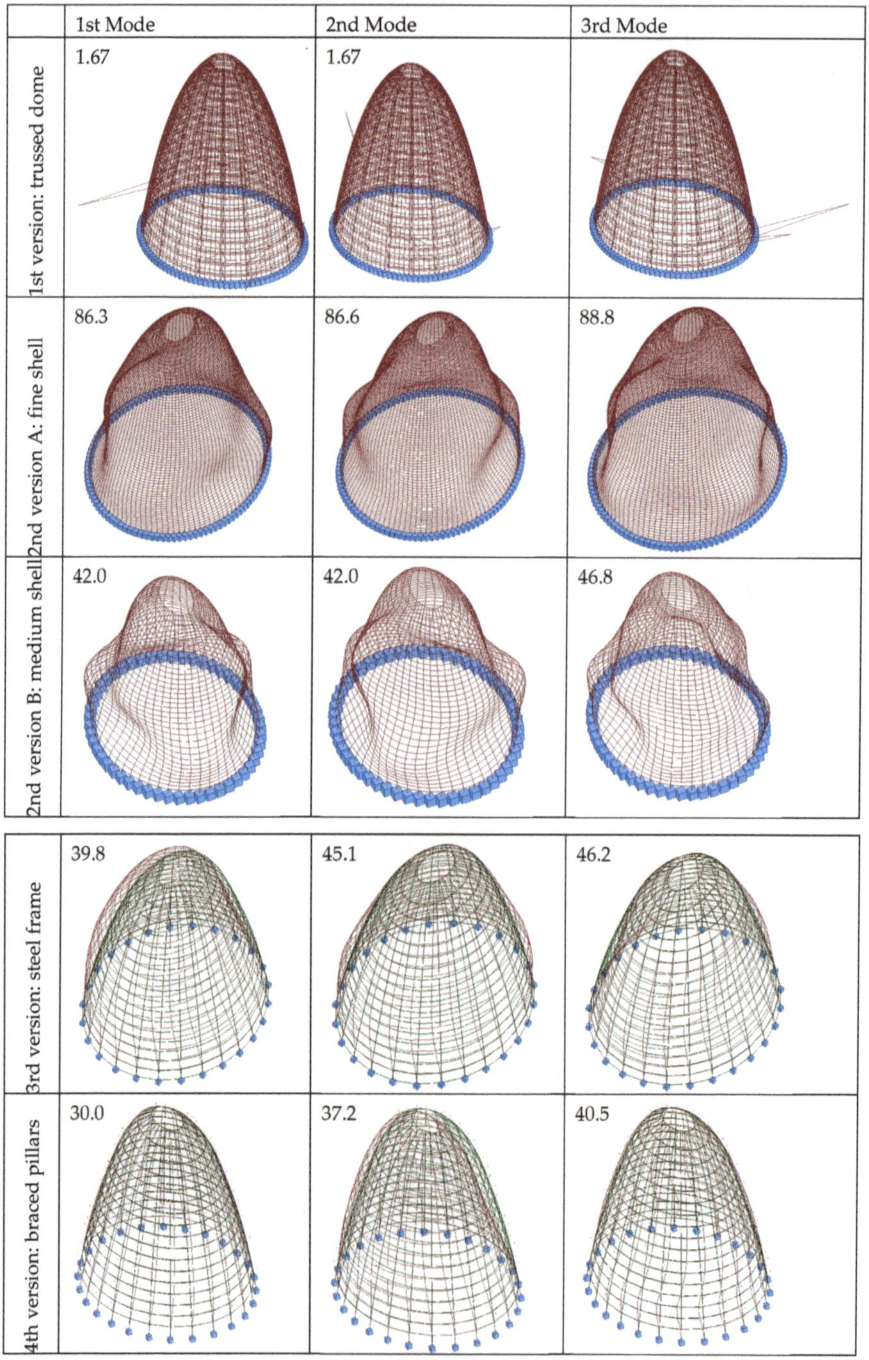

Figure 12. The first three buckling modes corresponding to the five proposed versions: The coefficient shown in each image corresponds to the shape utilization based on the buckling mode.

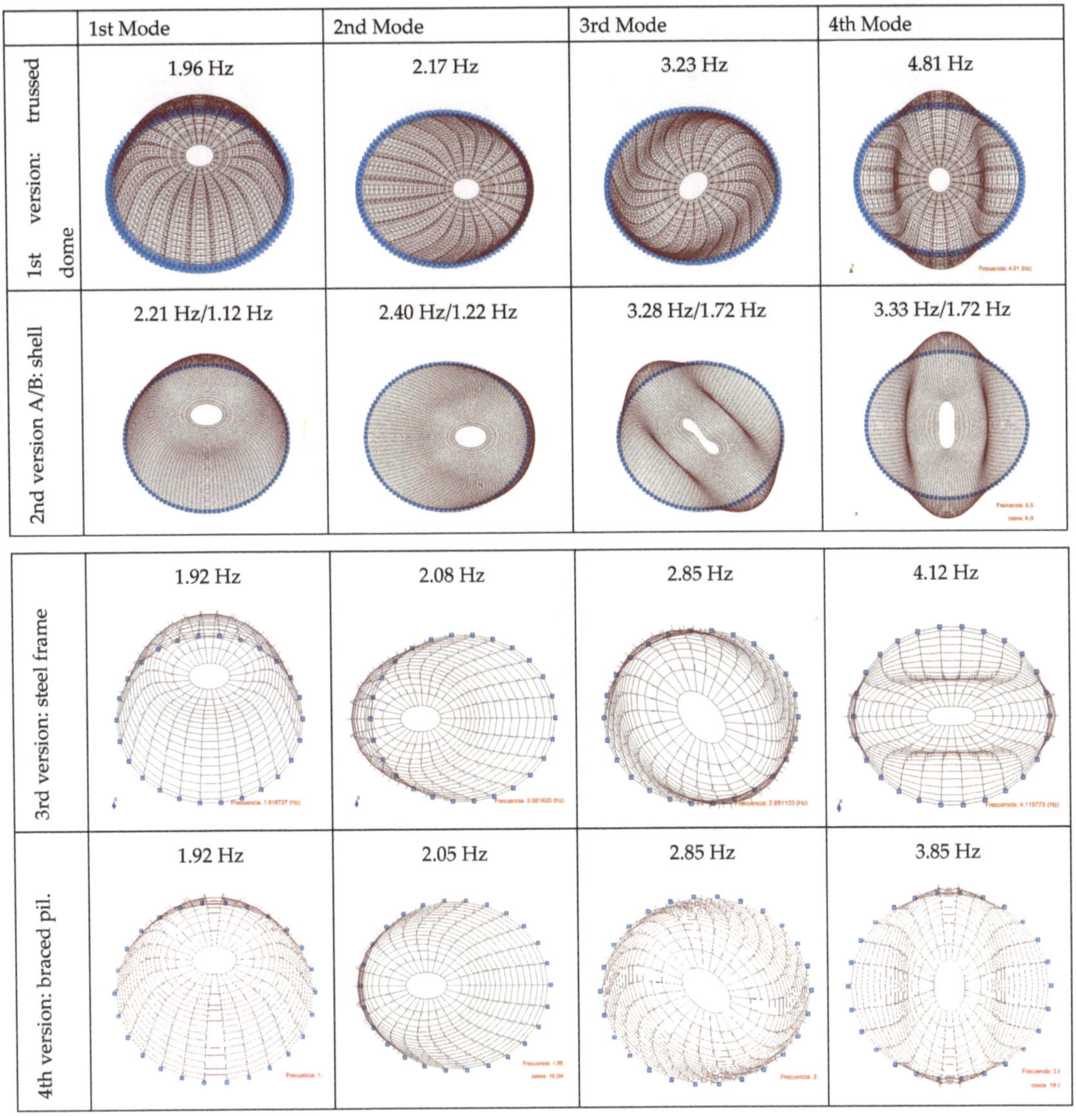

Figure 13. A Comparison of the first four modal modes across all four versions. The behaviors are generally similar, but slight differences are detected in some main modes. The frequency coefficient is shown in each case.

8. Verification of Angular Distortion on the Glass and Deformation Analysis

The dome's deformations were kept within limits to ensure the structure did not collide with the post-tensioned slabs. However, this was not the primary factor, as the slabs were sufficiently spaced to prevent such an effect. Instead, it was the stresses within the elements that ultimately determined the final section dimensions.

Once the dome's calculations were completed, a concern arose regarding the suitability of the glass for the designed structure. Although deformation had been considered at all stages of the project and the glass was not of a large format, there was still a need to adapt to the manufacturer responsible for installing the cladding. One of the key concerns was whether the angular distortion of the trapezoidal glass panels could potentially cause them to break. Typically, glass can accommodate deformations effectively if one edge moves continuously away from the opposite edge in a direction perpendicular to its plane, as the edges function like a hinge. With proper clearance in the joints, there should be no risk of breakage. The glass can also endure elastic deformations at its central point, similar to how a beam behaves, up to a limit of L/250, as per the consultant's guidelines, with the deformations considered in a direction perpendicular to the glass plane. However, the greatest risk to glass, aside from accidental impacts or corner damage, arises when one edge slides parallel to the plane of the glass in the opposite direction, creating tension along one diagonal and compression along the other. Since the glass is supported by two parallels and two meridians, this phenomenon could occur.

The need for this verification emerged during the construction phase and was developed by BOMA, with subsequent certification by civil engineer Francisco Quintero from Dragados, at a time when model 4 had not yet been finalized. Therefore, it was crucial to obtain results for each variant, automatically assessing every proposed change. Between two meridians, there are 17 trapezoids, with 26 spans and two diagonals per element, resulting in 882 diagonals, each with different calculation hypotheses. The only method provided by the calculation software for verification involved manually consolidating the data in a spreadsheet, comparing the deformations at the joints, subtracting them, measuring the diagonal, and then calculating the elongation in absolute terms. However, the program did not arrange the joints geometrically, making the process extremely slow. To expedite the process, fictitious bars were manually inserted into each trapezoid, corresponding to the positions of the two diagonals, within the program. Given the symmetry of the structure, it was sufficient to define only 25% of the diagonals and then apply two symmetry operations in the calculation program. The bars were defined using a material with a low elastic limit, specifically ceramic, and a small cross-section of φ4 mm. This configuration resulted in a model with stresses and deformations nearly identical to those of the previous one. This approach allowed us to determine the longitudinal deformation of each bar, effectively capturing the angular distortion. A clearer representation of these diagonals can be seen in Figure 10.

The clearance in the joints ultimately imposed the final restriction: the maximum allowable angular distortion was limited to 2 mm, representing the maximum possible elongation or shortening of the diagonals. Despite this strict limit, the dome performed correctly under each scenario. Only two diagonals exceeded the maximum deformation, and after applying symmetries, this resulted in a total of two glass panels across the entire façade. This situation raised some questions; as all the meridian profiles were identical, it seemed that all profiles would need to be increased, thereby unnecessarily increasing the final total. The section was upgraded to the next profile in the catalog range, and while the angular distortion result remained nearly the same, the weight increase amounted to roughly 15% more.

The maximum deformation obtained in this case was 2.02 mm. Achieving this result had been a particularly complex process, and it appeared that the expected cost savings would not materialize. If the steel weight turned out to be nearly identical to that of the executive project, the final calculation would have been in vain. To address this issue, we revisited the deformations in the spreadsheet. Francisco Quintero, upon reviewing the result, found the solution: he pressed the button to remove decimals, and the value displayed in the cell was 2.0 mm. While it might seem like an obvious or even unconscious decision, the situation made it clear that a 0.02 mm excess, equating to just a 1% increase, was not enough to risk breaking a glass, especially when wind pressures possibly greater than those that might occur in the future had been considered along with regulatory safety coefficients and multiple hypotheses. Thus, disregarding a deformation imperceptible to the human eye could never be deemed an imprudent action. We had reached a level of such precision and attention to detail that it seemed every minor aspect needed to be resolved and numerically justified. When the machines' responses end, human judgment must take over.

Another anecdote from the project is that, at one stage before entering the construction phase, there was consideration of using St. Andrew's crosses as structural elements. Undoubtedly, these elements, with a certain degree of rigidity, could have reduced the final weight of the dome, but not radically, as the shape of the profiles had to be carefully evaluated alongside the necessary intumescent coating. It is worth mentioning that diagonals have a strong engineering component and reduce the visibility from the viewing platform, while their removal is more appealing from an architectural standpoint. In the end, both alternatives would have been equally acceptable after thorough design analysis.

9. Future Research Directions

The article focuses on solving a real case from an initial form. To this end, a significant amount of time was invested in researching enclosure and structural solutions. Therefore, we consider this a research document of great value to the reader. The work was conducted under the conditions of the early 21st century; therefore, it does not explore other more current lines of research, such as modifying the choice of material. The use of more current glue-laminated timber or precast concrete to improve performance and on-site assembly may also be a more current factor. Another aspect to be addressed may be the choice of enclosure, more in line with current laws, focused on reducing climate control.

In the future, a new design study could be drafted considering the possibility of extreme situations, for example, in areas prone to earthquakes or in adverse areas with high wind turbulence. In Barcelona, when the executive project was drafted, the current NCSE-02 regulation was not yet mandatory; it is even expected that the regulation will change soon in Barcelona to an even more restrictive analysis scenario than the current one. Despite the above, Barcelona currently works with a basic acceleration of 0.04 g, much lower than other architectures in the same country, and much lower than other regions, such as Japan or South America, where certain architectural solutions are inadvisable. We do not believe this factor is sufficient to discourage a shape like the one analyzed, but it may be a determining factor in redefining joints, sections, and, in general, the ductility of the structure.

On the other hand, wind may be a determining factor in the future, depending on the climate change the world is experiencing. Wind damage is becoming more frequent and severe in Catalonia. The shape of a dome is better suited to reducing the effects of wind than other shapes formed of planes and edges, or smoother shapes. As mentioned at the beginning, a thorough analysis of the joints was not performed from the perspective of the progressive collapse of the dome, a design concern that was taken more seriously once the building's structure was completed.

Ultimately, the combination of extreme natural or human-made situations could be the subject of a future line of research.

10. Conclusions

This article describes the process of exploring the materiality of a building where both the structure and the enclosure play a fundamental role: a metal dome nearly 30 m high. The analysis was carried out based on a predefined volumetric form, the shape of which remained unchanged in the ten versions examined. Table 2 summarizes the main results obtained in this process and is undoubtedly the most notable contribution of this article.

The process described begins with an analysis of the building's requirements, including the design of the enclosure and perimeter maintenance platforms. It then progresses to defining the structural forces based on applicable current regulations. The final stage involves proposing multiple solutions, all maintaining the same external volumetric form while evolving and adapting at each iteration based on the knowledge gained from previous versions. This iterative design analysis is the most valuable contribution of this article, demonstrating how the final outcome emerges as a direct result of a structured form-finding process.

The first contribution drawn from this study consists of the proposal of multiple working forms of a dome—up to four—all maintaining the same external volumetric form. These forms evolve and adapt in each iteration based on the knowledge acquired in previous versions, recognizing problems, and providing solutions to the form.

Table 2. A Summary table of the structural characteristics obtained in this study.

	1st Version *	2nd Version A/B *	3rd Version	4th Version
Model	11,260 (elements) 4320 (unions)	14,836/3536 (elem.) 7526/1820 (unions)	910 (elements) 468 (unions)	910 (elements) 468 (unions)
Depth	600 mm (m) 600 mm (p) Grid max: 81 × 87 cm	200 (m) 200 mm (p) Grid max: 88 × 44.5 cmA/187 × 94 cmB	630 mm (m) 300 mm (p) Grid max: 362 × 188 cm	640 mm (m) 260 mm (p) Grid max: 362 × 188 cm
Inertia and Slenderness	1,048,783 cm^4 (m) 1,199,782 cm^4 (p)	110,982/116,116 cm^4 73,290/78,155 cm^4	1,146,973 cm^4 (m) 1,153,229 cm^4 (p)	2,503,437 cm^4 (m) 110,670 cm^4 (p)
	H/h = 51.03 RHS-200 × 50 × 4 (m and p) ϕ16 mm(p)	H/h = 153.10/153.1 RHS-200 × 50 × 5 (m and p)-A RHS-200 × 50 × 5(p)	H/h = 48.60 2CPN300 + IPE330 (mp) 2UPN-300 (p)	H/h = 47.84 I-640 × 220 (m) RHS-260 × 180 × 8 (p)
Maximum Deformation	Oculus X: 37.2 mm; Y: 10.1 mm Midpoint X: 93.7 mm; Y: 94.3 mm	Oculus X: 34.2 mm; Y:18.4 mm Midpoint X: 75.30 mm; Y: 96.0 mm	Oculus X: 3.26 mm; Y: 1.98 mm Midpoint X: 6.12 mm; Y: 7.82 mm	Oculus X: 19.9 mm; Y: 7.9 mm Midpoint X: 15.2 mm; Y: 35.1 mm
Weight	91,943 Kg (not considered)	158,424/137,447 kg 100%	277,941 Kg 175%	218,205 Kg 138%
Quantity	478 Kg/m^2 (plan) 39 Kg/m^2 (façade)	823/714 Kg/m^2 (plan) 67/58 Kg/m^2 (façade)	1444 Kg/m^2 (plan) 118 Kg/m^2 (façade)	1134 Kg/m^2 (plan) 93 Kg/m^2 (façade)
Fire Protection and Steel Cost	P = 3023 m^2 Fire Protec: EUR 139.572 ** Steel cost: EUR 367.772 **	P = 4202/2530 m^2 F.P: EUR 194.006/116.810 **	P = 4259 m^2 EUR 196.638 **	P = 2950 m^2 EUR 136.202 **
		S.c.: EUR 633.696/549.788 **	St. cost: EUR 1.111.764 **	Steel cost: EUR 872.820 **

Table 2. *Cont.*

Structure Detail	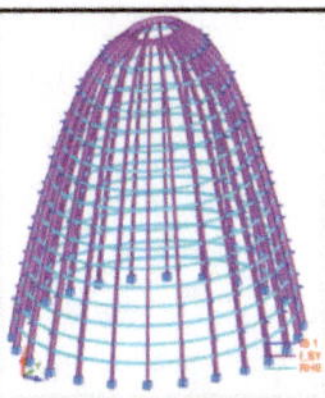	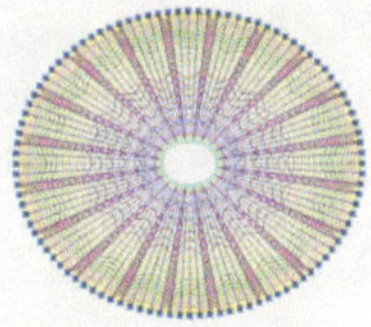	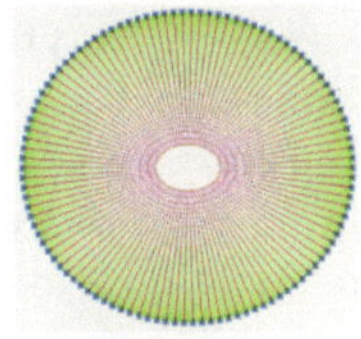	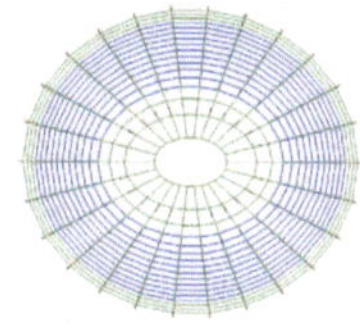
3D Geometry (1, 2A, 3, and 4)	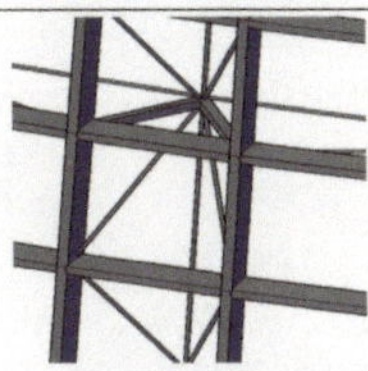			
Plan geometry (1, 2A, 3 and 4)	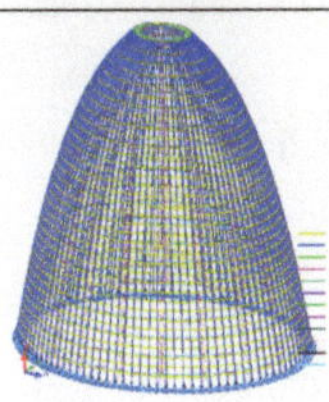	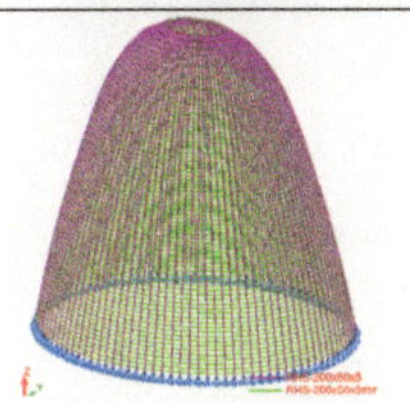		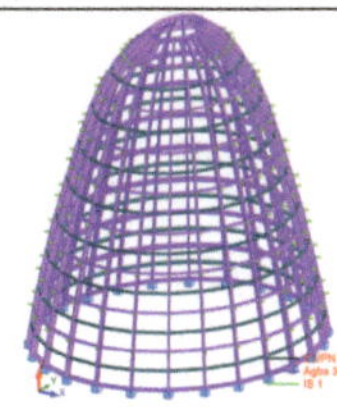
Construction time	4.1 months	5.2/2.7 months	1.7 months	1.5 months

(*) Versions 1, 2A, and 2B were not dimensioned in fire study case, nor were the glass deformation requirements, as they were preliminary options analyzed during the conceptual design phase.
(**) Material cost study updated to 2025.
("m" = meridians; "p" = parallels; "N.I." = no information.)

The second contribution focuses on the pre-dimensioning of a pointed dome. In cases like the one described in this article, the consultant must develop their own approach to offer a precise and cost-effective solution for the client. Table 2 illustrates that the meridians had a ratio of approximately H/50. To achieve higher ratios, i.e., to reduce profile thickness, it was necessary to place the profiles closer together, approximately 1 m apart.

The third contribution aimed to investigate the material quantities of the various models proposed in this paper. An approximate value of 100 kg/m^2 of S-355-J0 steel was obtained, using the square meters of the façade as a reference. A key aspect was the choice of structural type. However, the most decisive factor was undoubtedly the fire protection provided by the intumescent paint. Once the fire resistance requirement set by the fire department—although not excessively high at R-90—was established, its consideration in the calculations led to a 175% increase in the amount of steel required. However, by optimizing the node technology and adjusting the profiles, this increase was ultimately reduced to 138%. If the required fire resistance had been higher, a more precise calculation using CFD analysis would have been necessary. In 2002, numerical methods had not yet reached the level of accuracy, cost-effectiveness, or speed they offer today [52]. On the contrary, if a façade-type element supported by slabs had been considered and the fire department had accepted a lower stability requirement, close to REI-30, the amount of material would have been reduced to just under half of that used in the final solution.

The fourth and final aspect of our investigation focuses on the analysis of buckling modes and modal calculation. In all cases, except the first, local creasing failure was observed. The exception was the first version, which was not fully calculated. Regarding the modal calculation, in all versions, the dome showed a mass participation close to 90%, indicating no signs of local failure. The frequency results for the first four modes were largely consistent across all versions, with the exception of the second version, which exhibited noticeably higher values. The divergence was due to the use of a shell-type solution with an excessive number of elements, which increased the complexity of the construction. The first two modes were always translational: the first parallel to the minor axis and the second along the major axis. The third mode corresponded to torsion, while the fourth involved crushing of both the oculus and the dome. Notably, the second version was the only one that improved torsional behavior, making torsion the fifth buckling mode in this case.

Author Contributions: Software, I.C.C. and X.G.V.; Validation, I.C.C. and O.M.R.; Formal Analysis, O.M.R. and I.C.C.; Resources, I.C.C. and P.G.T.; Writing—Original Draft, I.C.C. and O.M.R.; Writing—Review and Editing, I.C.C. and X.G.V.; Supervision, P.G.T. All authors have read and agreed to the published version of the manuscript.

Funding: The Polytechnic University of Catalunya UPC funded this publication. R-02385_Ajut ALECTORS.

Data Availability Statement: The data are contained within the article.

Conflicts of Interest: The authors declare no conflicts of interest.

References

1. Vázquez, F. B720 Fermín Vázquez Arquitectos Arquitectura 1998–2011. *TC Cuad. Trib. Constr. Ser. Dédalo* **2011**, *99*, 182–197.
2. Gómez Hermoso, J. (Ed.) *Edificios Altos Españoles 2005–2012*; E.T.S. Ingenieros de Caminos, Canales y Puertos: Madrid, Spain, 2015; ISBN 978-84-7493-290-4.
3. Salvador Zurita, J. Eixample: Atles Operatiu 2024. Available online: https://upcommons.upc.edu/handle/2117/401441 (accessed on 3 February 2025).
4. El Origen de la Plaza de las Glòries. Available online: https://www.lavanguardia.com/participacion/cartas/20220525/8289518/origen-plaza-glories.html (accessed on 3 February 2025).
5. Vista de la Història Que Quedarà Soterrada: Del Nus de la Trinitat a la Plaça de les Glòries Catalanes. Available online: https://raco.cat/index.php/Finestrelles/article/view/214422/294114 (accessed on 3 February 2025).
6. Rui-wamba Sebastián Guerrero, J.E. La Plaça de les Glòries Catalanes la Plaza de las Glorias Catalanas Una Breve Historia Visual. Fundación Esteyco Eds. 2009. Available online: https://esteyco.com/wp-content/uploads/2017/08/L09_Glorias_Catalanas_r.pdf (accessed on 23 March 2025).
7. Aguiló, C.; Buil, P. *Lopez Other Glories: Proposal for the Renaturalization of Barcelona*; dpr-barcelona: Barcelona, Spain, 2016; ISBN 978-84-944873-0-9.
8. Obiol Sánchez, A.; Costales, N. Encants Barcelona: Nueva Plaza Comercial y Aparcamiento en Barcelona. *Quad. D'estructures (Dijous A l'ACE)* **2014**, *51*, 5–11.
9. Edifici Disseny Hub Barcelona | Disseny Hub Barcelona. Available online: https://www.dissenyhub.barcelona/ca/edifici-disseny-hub-barcelona (accessed on 3 February 2025).
10. Agence Ter Landscape Architects; Meta Engineering; Acpa Ana Coello Paisatge i Arquitectura. "Canopia Urbana", Proyecto de Espacio Libre Urbano de La Plaça de Les Glòries Catalanes: Barcelona. Spain. *Diseño* **2024**, *527*, 9.
11. Empieza El Proceso de Adjudicación de Viviendas Protegidas en la Illa Glòries | L'Eixample. Available online: https://ajuntament.barcelona.cat/eixample/es/noticia/empieza-el-proceso-de-adjudicacion-de-viviendas-protegidas-en-la-isla-glories_1470797 (accessed on 3 February 2025).
12. Congostrina, A.L. El Nuevo Tramo de Tranvía Entre Glòries y Verdaguer Entrará en Funcionamiento El 9 de Noviembre | Noticias de Cataluña | EL PAÍS. Barcelona, 21 Octubre 2024. Available online: https://elpais.com/espana/catalunya/2024-10-21/el-nuevo-tramo-de-tranvia-entre-glories-y-verdaguer-entrara-en-funcionamiento-el-9-de-noviembre.html (accessed on 3 February 2025).
13. Montañés, J.Á. *La Cúpula de la Torre Glòries Será Visitable en 2022*; El País: Madrid, Spain, 2021.
14. Brufau i Niubó, R.; Moya i Ferrer, L.; Obiol, A.; BOMA (Eds.) [s.n.]: S.l, 2010. Available online: https://discovery.upc.edu/discovery/fulldisplay?docid=alma991003710659706711&context=L&vid=34CSUC_UPC:VU1&lang=ca&search_scope=MyInst_and_CI&adaptor=Local%20Search%20Engine&tab=TOT3&query=any,contains,boma (accessed on 3 February 2025).

15. Sole, J.R.; Brufau i Niubó, R.; Obiol Sánchez, A.; Moya Ferrer, L. Proyecto y Ejecución de La Estructura de La Torre Agbar de Barcelona. In Proceedings of the X Congreso Internacional de Ingeniería de Proyectos, Valencia, Spain, 13–15 September 2006; pp. 3–14.
16. Terzaghi, K.; Peck, R.B.; Mesri, G. *Soil Mechanics in Engineering Practice*, 3rd ed.; Wiley: New York, NY, USA, 1996; ISBN 978-0-471-08658-1.
17. Özşahin, B. An Assessment of the Relation between Architectural and Structural Systems in the Design of Tall Buildings in Turkey. *Buildings* **2022**, *12*, 1649. [CrossRef]
18. Ilgın, H.E. Space Efficiency in Tapered Super-Tall Towers. *Buildings* **2023**, *13*, 2819. [CrossRef]
19. Al-Kodmany, K.; Du, P.; Ali, M.M. (Eds.) *Sustainable High-Rise Buildings: Design, Technology, and Innovation*; IET Built Environment Series; The Institution of Engineering and Technology: London, UK, 2022; ISBN 978-1-83953-281-8.
20. Goncikowski, M. Research by Design—Determining the Form of the Initial Architectural Concept of a Skyscraper Building in Warsaw. *Archit. Eng. Des. Manag.* **2024**, *20*, 1–31. [CrossRef]
21. Yaman, M. Different Façade Types and Building Integration in Energy Efficient Building Design Strategies. *Int. J. BES* **2021**, *8*, 49–61. [CrossRef]
22. Sha, H.; Qi, D. A Review of High-Rise Ventilation for Energy Efficiency and Safety. *Sustain. Cities Soc.* **2020**, *54*, 101971. [CrossRef]
23. Muñoz Blanc, C.; Fortea Navarro, I. El diseño basado en prestaciones frente al incendio de las estructuras de edificación. *Inf. La Construcción* **2017**, *69*, 1–10. [CrossRef]
24. Garde, R.J.; Gehlot, V. *Design of Steel Structures 1: Based on IS:800-2007 and Also on Newly Revised Code IS:883-1994*, 13th ed.; Scientific: Jodhpur, India, 2011; ISBN 978-81-7233-654-7.
25. Huybers, P. Dome-Type Space Structures of Ellipsoidal Form. *Int. J. Space Struct.* **1990**, *5*, 297–308. [CrossRef]
26. Huybers, P. Computer-Aided Design of Polyhedral Building Structures. *Des. Stud.* **1993**, *14*, 19–39. [CrossRef]
27. Opatowicz, D.; Radoń, U.; Zabojszcza, P. Assessment of the Effect of Wind Load on the Load Capacity of a Single-Layer Bar Dome. *Buildings* **2020**, *10*, 179. [CrossRef]
28. Camp, C.; Pezeshk, S.; Cao, G. Optimized Design of Two-Dimensional Structures Using a Genetic Algorithm. *J. Struct. Eng.* **1998**, *124*, 551–559. [CrossRef]
29. Bysiec, D.; Jaszczyński, S.; Maleska, T. Analysis of Lightweight Structure Mesh Topology of Geodesic Domes. *Appl. Sci.* **2023**, *14*, 132. [CrossRef]
30. Montaner, J.M. *La Torre Agbar, Emblema de Barcelona*; El País: Madrid, Spain, 2005.
31. Shen, X.; Zhang, Q.; Lee, D.S.H.; Cai, J.; Feng, J. Static Behavior of a Retractable Suspen-Dome Structure. *Symmetry* **2021**, *13*, 1105. [CrossRef]
32. Buxade, C.; Margarit, J. Cúpula Metálica en Vitoria—España. *Inf. Constr.* **1976**, *29*, 49–67. [CrossRef]
33. Lichtenstein, C.; Krausse, J. *Your Private Sky: R. Buckminster Fuller: Art Design Science*; Lichtenstein, C., Krausse, J., Eds.; Lars Müller: Baden, Switzerland, 1999; ISBN 3-907044-88-6.
34. *UNE-EN 1991-1-5*; Eurocódigo 1: Acciones en Estructuras: Parte 1–5. Acciones Generales. Acciones Térmicas. AENOR: Madrid, Spain, 2004.
35. Ministerio de Fomento; Espanya, M. *De AE-88: Acciones en la Edificación*; Ministerio de Fomento: Madrid, Spain, 2000; ISBN 84-498-0503-1.
36. *Código Técnico de la Edificación*; Tecnos: Madrid, Spain, 2014; ISBN 978-84-309-6139-9.
37. *UNE-EN 1991-1-4*; Eurocódigo 1: Acciones en Estructuras: Parte 1–4. Acciones Generales. Acciones de Viento/Asociación Española de Normalización y Certificación. AENOR: Madrid, Spain, 2007.
38. MeseguerRuiz, J. *Aerodinámica Civil: Cargas de Viento en las Edificaciones*; Meseguer Ruiz, J., Ed.; McGraw-Hill Profesional; McGraw-Hill/Interamericana de España: Madrid, Spain, 2001; ISBN 84-481-3332-3.
39. Muñoz Blanc, C.; Fortea Navarro, I. Evaluation of Horizontal Loads in the Concept Design of Tall Building Concrete Structures. *Hormig. Acero* **2023**, *75*, 51–64. [CrossRef]
40. Otto, F.; Schanz, S. *Frei Otto, Bodo Rasch: Finding Form: Towards an Architecture of the Minimal: The Werkbund Shows Frei Otto, Frei Otto Shows Bodo Rasch: Exhibition in the Villa Stuck, Munich, on the Occasion of the Award of the 1992 Deutscher Werkbund Bayern Prize to F*; Axel Menges: Stuttgart, Germany, 1995; ISBN 3930698668.
41. Molera Solá, P. *Soldadura Industrial Clases y Aplicaciones*; Marcombo: Barcelona, Spain, 1992; ISBN 1-4135-8372-5.
42. Gilemkhanov, R.; Bagautdinov, R.; Kankhva, V. Autodesk Revit and Robot Structural Analysis in Design of Framed Buildings. In *Proceedings of the International Scientific Conference Energy Management of Municipal Transportation Facilities and Transport EMMFT, Khabarovsk, Russia, 10–13 April 2017*; Murgul, V., Popovic, Z., Eds.; Springer International Publishing: Cham, Switzerland, 2018; pp. 1036–1045.
43. Sun, G.; Zhang, X.; Wu, J.; Xiao, S.; Xue, S. Theoretical and Numerical Simulation Research on Fire of Large-Span Spatial Structures. *Appl. Sci.* **2025**, *15*, 1154. [CrossRef]

44. Arnedo, A. (Ed.) *Guia per a la Comprovació de La Resistència al Foc D'estructures*; ASCEM, Asociación para la Construcción de Estructuras Metálicas: Sant Cugat del Vallès, Spain, 2004.
45. Rabajczyk, A.; Zielecka, M.; Popielarczyk, T.; Sowa, T. Nanotechnology in Fire Protection—Application and Requirements. *Materials* **2021**, *14*, 7849. [CrossRef]
46. Jiang, J.; Yazici, C. Mechanical Properties of S235 Steel Protected with Intumescent Coatings Under High Temperatures: An Experimental Study. *Buildings* **2024**, *14*, 1597. [CrossRef]
47. McGrattan, K.B.; Forney, G.P. *Fire Dynamics Simulator (Version 4)*; National Institute of Standards and Technology: Gaithersburg, MD, USA, 2004.
48. Chen, L.-M.; Huang, K.-Y.; Liu, Y.-J.; Zeng, Y.-H.; Li, Z.-B.; Zhou, Y.-Y.; Chen, L.-M.; Huang, K.-Y.; Liu, Y.-J.; Zeng, Y.-H.; et al. Optimisation of Cable Dome Structure Design for Progressive Collapse Resistance. *Appl. Sci.* **2023**, *13*, 2086. [CrossRef]
49. Meng, B.; Li, H.; Liew, J.-Y.R.; Li, S.; Kong, D.-Y. Enhancing the Collapse Resistance of a Composite Subassembly with Fully Welded Joints Using Sliding Inner Cores. *J. Struct. Eng.* **2024**, *150*, 04024085. [CrossRef]
50. Meng, B.; You, K.; Yang, B.; Kong, D.; Wen, J. Anti-Progressive Collapse Performance and Design Method of Novel T-Stub Connections. *Eng. Struct.* **2024**, *319*, 118880. [CrossRef]
51. Ortiz Herrera, J. *Manual de Uniones Atornilladas Frontales Pretensadas*; Hernando García, J.I., Cervera, J., Eds.; Práctica en el Proyecto de Estructuras de Acero en Edificación; Publicaciones APTA: Madrid, Spain, 2007; ISBN 978-84-690-6123-7.
52. Song, L.; Zhu, J.; Liu, S.; Qu, Z. Recent Fire Safety Design of High-Rise Buildings. *J. Urban Dev. Manag.* **2022**, *1*, 50–57. [CrossRef]

buildings

MDPI

Article

Seismic Performance Analysis for an Eccentrically Braced Frame (EBF) with an Innovative Self-Centering Shear Link

Xinyu Xu [1], Lifen Huang [1], Shangwen Liu [2], Bo Zhang [2,3] and Shujun Hu [2,*]

1 College of Intelligent Construction and Engineering, Minnan University of Science and Technology, Quanzhou 362700, China; xu_xin_yu4865@sina.com (X.X.)
2 School of Infrastructure Engineering, Nanchang University, Nanchang 330031, China
3 China Nerin Engineering Co., Ltd., Nanchang 330031, China
* Correspondence: hushujun@ncu.edu.cn; Tel.: +86-15579183858

Abstract: By integrating a very short shear link–shear slotted bolted connection (VSSL-SSBC) and two self-centering SMA braces (SCBs), a novel self-centering shear link (SC-SL) was developed for installation between a steel brace and steel beam in an eccentrically braced frame (EBF). The SC-SL can enhance the seismic performance and seismic resilience capacity of the EBF by achieving a high bearing capacity and low residual deformation. The mechanical properties of the VSSL-SSBC and SC-SL were designed and analyzed using both experimental and numerical methods. Subsequently, the seismic performances of EBFs equipped with VSSL-SSBC and SC-SL were analyzed under different earthquakes. Validated numerical methods were employed to investigate the deformation modes, stress nephograms, and hysteresis curves of the EBFs. The deformation mode and hysteresis curve of the VSSL-SSBC exhibit an initial frictional slip of the SSBC, followed by the load-bearing response of the VSSL. The skeleton curve of the VSSL-SSBC consists of elastic, slip, elastoplastic, and plastic stages, and the deformation and damage are significantly reduced at the same displacement. In the SC-SL, the SCB undergoes substantial deformation when the SMA is in tension, effectively minimizing residual deformation. Under frequent earthquakes, the stress and displacement of all components in both the EBF-VSSL-SSBC and EBF-SC-SL are essentially equivalent, and the VSSL-SSBC remains elastic, without significant yielding deformation. Under rare earthquakes, incorporating SCB in EBF-SC-SL significantly enhances the ultimate load capacity by 19.66% and reduces the residual deformation by 27.90%. This improvement greatly contributes to the seismic resilience of the EBF.

Keywords: eccentrically braced frame (EBF); self-centering shear link (SC-SL); very short shear link (VSSL); shape memory alloy (SMA); self-centering SMA brace (SCB); seismic performance

1. Introduction

Eccentrically braced frames (EBFs) are composed of steel columns, steel frames, steel braces, and links, representing a seismic design approach that effectively combines the high energy dissipation, ductility, and damage control capabilities [1–3]. The link, commonly known as the fuse, plays a crucial role in dissipating seismic energy and protecting the primary load-carrying elements of EBFs [4,5]. Links are classified based on their length ratios into three categories: shear link, flexural-shear link, and flexural link [6]. Very short shear links (VSSLs), characterized by a length ratio less than 1.0, exhibit obviously higher overstrength factors and inelastic rotation capacities, with values of 1.90 and 0.15 rad, respectively, both exceeding the standard code prescribed in building codes of 1.50 and

0.08 rad [7–9]. The ultimate shear force, deformation capacity, and lateral stiffness of VSSLs are superior to those of the other three types, making a substantial contribution to the overall seismic performance of EBFs [10]. In recent years, there has been growing interest in seismic-resilient structures within the field of civil engineering [11]. On 6 February 2023, at 04:17 local time (01:17 UTC), a magnitude (Mw) 7.7 earthquake struck the Pazarcık segment of the East Anatolian Fault (EAF) zone. This earthquake caused damage to residential houses, bridges, hospitals, and historical buildings, posing a challenge to the seismic performance of current structures [12–14]. However, the links in EBFs are susceptible to severe damage and residual deformation (Figure 1), leading to increased repair costs and the need for advanced rehabilitation techniques [15,16]. Therefore, enhancing the energy-dissipating and self-centering capabilities of VSSLs in EBFs is crucial to achieving higher seismic design standards.

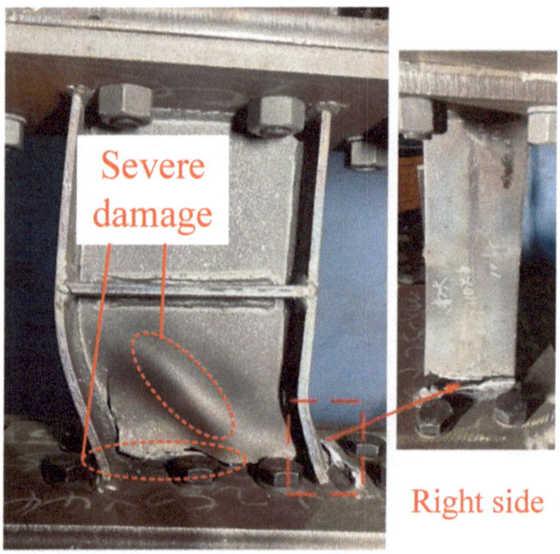

Figure 1. Damage and residual deformation of shear link in EBF.

Slotted bolted connections (SBCs) utilize the friction-slip mechanism generated by high-strength bolts slipping within the slot holes to achieve simultaneous energy dissipation and ductility improvement. This connection technology has found successful applications in steel frame components including beam–column joints, frame beams, and energy dissipation braces, substantially boosting the seismic performance of structures [17,18]. Research indicates that SBCs demonstrate three key advantages: stable hysteresis response, enhanced ductility, and cost-effective post-earthquake repair [19–21]. Liu et al. [19] proposed a prefabricated steel structural system with truss beams and bolted joints, and enlarged bolt holes were adopted in the joints for easier assembly and better energy dissipation. Zhang et al. [20] applied SBCs to replaceable steel beams in the proposed hybrid frame structure with energy dissipation frames (HF-EDFs) to improve seismic performance and recoverability. Zhai et al. [21] proposed a new type of bolted connection with a dual-slot hole and two kinds of connection-strengthening methods, including diagonal brace strengthening and flange strengthening. By integrating a VSSL with a shear slotted bolted connection (SSBC), a very short shear link with shear slotted bolted connections (VSSL-SSBCs) can be developed to further enhance the ductility and energy dissipation while minimizing damage [22]. However, VSSL-SSBCs may still experience high residual deformation, which can reduce the overall seismic capacity of the EBFs.

Shape memory alloy (SMA) is a category of intelligent material distinguished by its inherent material memory characteristics. These alloys possess distinctive shape recovery characteristics following repeated loading–unloading cycles, providing both energy-dissipating and self-centering capabilities. Experimental investigations have verified strain restoration levels reaching up to 8% under controlled deformation conditions [23]. Feng et al. [24] proposed an asymmetric self-centering steel connection equipped with SMA damper devices, which can experience a typical flag-shaped curve with high energy dissipation and self-centering capabilities. Qiu et al. [25] developed a NiTi-SMA BRB in knee braced frames, which can successfully eliminate residual drift ratios and has higher seismic resilience for the protected frames. Sun et al. [26] proposed an innovative sliding self-centering brace to reduce or eliminate residual deformation after strong earthquakes. Li et al. [27] conducted a study on a new self-centering coupling beam (SCCB) incorporating self-centering friction dampers (SCFDs) to improve the seismic resilience of the coupled wall structure. Additionally, Jia et al. [28] developed an innovative SCB with superior self-centering capacity and energy-dissipating capacity, and the hysteresis curve of the SCB was idealized as a flag shape with low residual deformation.

Existing research has indicated that enhancing the self-centering capacity of steel frames can effectively reduce residual deformation, which facilitates post-earthquake repair. Rahgozar et al. [29] studied the sensitivity analysis of controlled-rocking steel cores (CRSCs) under seismic loadings, and the results showed that CRSCs effectively prevent earthquake-induced residual damage but may suffer large lateral displacements. Chen et al. [30] investigated a novel K-type superelastic SMA self-centering eccentrically braced frame, and the simulation models of the SMA-based SC-EBF and a corresponding equal-stiffness traditional EBF counterpart were established. The results showed a much better resilience and represent a promising attractive alternative for future applications. Xiong et al. [31] proposed a novel self-centering frame with Y-eccentrically braced structures (SC-YEBFs) to solve the problem that the storey drift of the structure may exceed the specified limit in standards when self-centering beam–column joints are applied in high-rise and larger span buildings. The test and finite element results indicated that shear link has increased the lateral stiffness and energy dissipation capacity of the structure and maintained a good self-centering performance and replaceability.

In this paper, a novel SC-SL is proposed to achieve high bearing stiffness and ductility while minimizing residual deformation, as shown the Figure 2. The SC-SL includes a VSSL, two SCBs, and two vertical braces. Compared with the self-centering devices in existing studies, SC-SL exhibits both high energy dissipation and self-centering capacities, meeting the high requirements of seismic design [29–31]. The SC-SL can not only be applied to new buildings to improve the bearing capacity and energy dissipation capacity of the structure but can also be applied to existing structures through simple bolt connections to enhance the seismic performance and self-centering capacity of the structure. Cyclic loading tests were conducted on a VSSL-SSBC and an SC-SL specimen. The failure modes, hysteresis curves and skeleton curves were analyzed and their seismic performances compared. Furthermore, to access the structural response under frequent and rare earthquakes, two EBFs incorporating VSSL-SSBC and SC-SL were analyzed. Validated numerical methods were employed to investigate the deformation modes, stress nephogram, and hysteresis performance. The findings provide valuable insights for seismic design, offering reliable references to enhance the seismic resilience capacity of EBFs.

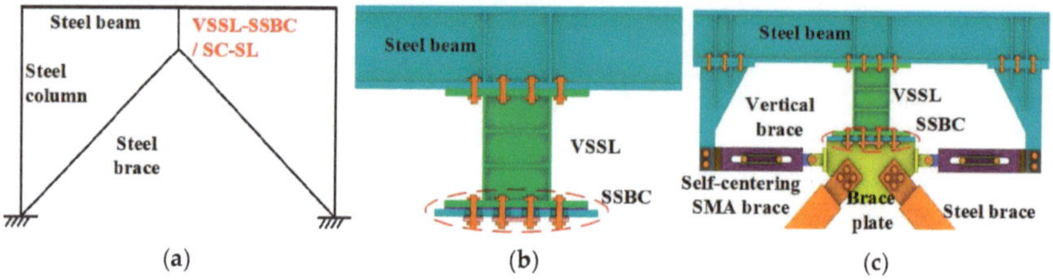

Figure 2. A novel eccentrically braced frame applied with SC-SL: (**a**) EBF; (**b**) VSSL-SSBC; (**c**) SC-SL.

2. Basic Performance

2.1. Basic Component

As shown in Figure 3, the SC-SL primarily consists of the VSSL-SSBC and two SCBs. The VSSL-SSBC is composed of a very short shear link (VSSL) and shear slotted bolted connection (SSBC), with the VSSL connected at both ends to the SSBC and the frame beam. The SSBC mainly consists of the VSSL, endplates I and II, base plates, and a metallic and rubber shim. The rubber shim is placed between endplate II (at the brace end) and the base plate, while the metallic shim is positioned between endplate I and endplate II. The self-centering SMA brace includes two slip plates, two slip bolts, a fixed plate, fixed bolt I, two fixed bolts II, and SMA wires. The two slip plates and the fixed plate each have two identically sized slot holes at corresponding positions on both sides. Initially, the two slip bolts are positioned at the ends of the left and right slot holes, near the opposite slot hole. Additionally, the vertical connection brace is secured to the beam by fixed bolt I, while the other end is attached to the one end plate of the SCB with two fixed bolts II. Meanwhile, the other end plate of the SCB is secured to the brace plate, thus forming the SC-SL.

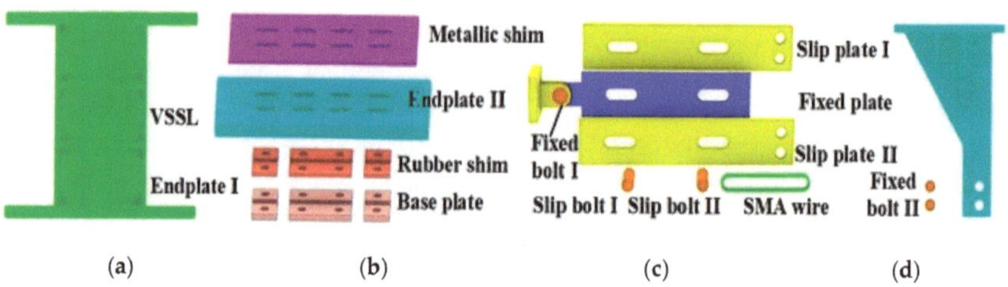

Figure 3. Basic components of an SC-SL: (**a**) VSSL; (**b**) SSBC; (**c**) SCB; (**d**) Vertical brace.

2.2. Working Mechanism

Figure 4 is the working mechanism of the SC-SL during cyclic loading. The SSBC and VSSL provide the initial and subsequent energy dissipation capacity of the structure, respectively, while the SCB provides the self-centering capacity during the entire process. Under cyclic loading, the frame beam in the SC-SL drives the VSSL-SSBC and the SCB to work in unison. Within the VSSL-SSBC, the SSBC first experiences frictional slip, while the VSSL remains elastic without undergoing shear yielding. Once the high-strength bolt reaches to the end of the slot hole, the VSSL begins to yield in shear, absorbing the load and deforming until failure. Throughout the entire loading process, one sliding bolt in the self-centering SMA brace remains fixed while the other moves, ensuring that the SMA wires remain continuously in tension. As a result, SCB I and SCB II effectively provide both energy-dissipating and self-centering capabilities during positive and negative moving. This synergistic mechanism enhances the energy-dissipating efficiency and minimizes residual deformation in the SC-SL.

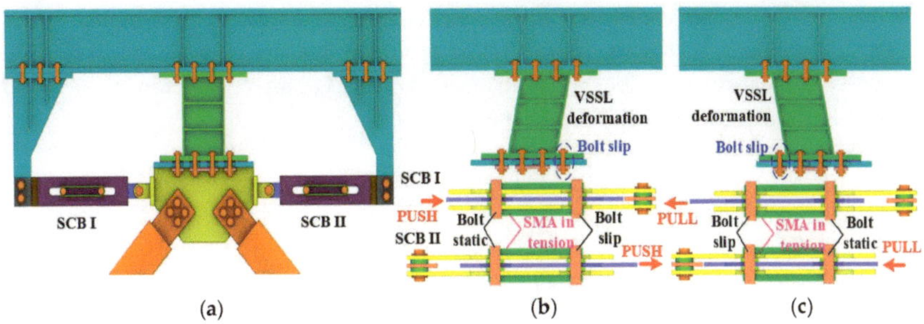

Figure 4. Working principle for EBF with SC-SL during cyclic loading: (**a**) SC-SL; (**b**) Positive moving; (**c**) Negative moving.

2.3. Design Method

At the ultimate limit state, both the VSSL-SSBC and the SCB within the SC-SL may experience failure, manifested as follows: (1) When the deformation of the SCB exceeds the deformation capacity of the SMA, while the VSSL-SSBC has not yet reached its limit state, failure of the SC-SL primarily occurs in the SCB, leading to the fracture of the SMA. (2) The SMA in the SCB is expected to remain in a superelastic state and continue to provide both energy-dissipating and self-centering capabilities; however, failure may occur in the VSSL once the SMA reaches its ultimate displacement, marking the ultimate limit state of the SC-SL. To ensure that the SC-SL achieves the expected energy-dissipating and self-centering performance, the design must adhere to the following requirements.

(1) VSSL-SSBC

The initial shear slip load, F_{slip}, of the SSBC must be lower than the yield shear force, V_y, of the VSSL. At the ultimate limit state, the load-bearing capacity, F_u, of the SSBC must be higher than the ultimate shear force, V_u, of the VSSL. These relationships can be expressed as follows [6,10]:

$$F_{slip} = n_s n_b P\mu \leq V_y = f_{yw} h_w t_w \tag{1}$$

$$F_u = n_b n_s \frac{\pi d^2}{4} f_v^b \geq V_u = 0.58\, f_{uw} h_w t_w + (3\, f_{uw} b_f t_f^2 + 1.8\, f_{uw} b_f t_s^2)/e \tag{2}$$

where n_s, n_b, P, μ, d, and f_{vb} are the number of sliding contact planes, number of sliding bolts, pretension force of the bolt, coefficient of friction, diameter of the slip bolt, and the yield stress of the slip bolt in the SSBC, respectively. f_{yw}, f_{uw}, h_w, t_w, b_f, t_f, t_s, and e are the web yield shear stress, web ultimate shear stress, web clear height, web thickness, flange width, flange thickness, stiffener thickness, and length of the VSSL-SSBC, respectively.

(2) SCB

The load-bearing capacity, F_{SCB}, of the SCB primarily consists of the load-bearing capacity of the SMA, F_{SMA}, and the force provided by the slotted bolt connection, F_f, which can be expressed as follows:

$$F_{SCB} = F_{SMA} + F_f = \sigma_{SMA} A_s + n_s n_b P_S \mu \tag{3}$$

where A_s and σ_{SMA} are the area and stress of the SMA wires, which can be calculated by the Graesser model [23], and n_s and n_b are the number of sliding contact planes and sliding bolts, which are equal to 6 and 2, respectively. P_S is the pretension force of the bolt, and μ is the rubber shim friction coefficient (0.075) [22].

(3) Steel brace

The seismic design code specifies the detailed design methods for eccentrically braced structural members. After the plastic failure occurs in the shear link, the steel brace is expected to remain in an elastic state at all times. Its axial bearing capacity can be calculated and designed according to the following equations:

$$\Omega = V_u / V_y \tag{4}$$

$$N_{br} = \Omega * N_{br,com} \leq 0.8\varphi f_{br} A_{br} \tag{5}$$

where Ω is the overstrength factor of the shear links. N_{br} and $N_{br,com}$ are the calculated and designed values of the steel brace axial force, and φ and A_{br} are the stability coefficient and cross-sectional area of the steel brace. f_{br} is the design value of the tensile strength of the steel brace.

(4) Frame beam and column

The seismic design code also provides detailed design methods for frame beams. The design values of internal forces can be expressed as follows:

$$M_b = \Omega * M_{b,com} \tag{6}$$

$$\frac{N_b}{A_{nb}} \pm \frac{M_b}{\gamma_x W_{nbx}} \leq f_b \tag{7}$$

where M_b and $M_{b,com}$ are the calculated and designed values of the bending moment of the frame beam. N_b, A_{nb}, and W_{nbx} are the axial force, net sectional moment of resistance, and net sectional area of the frame beam, respectively. γ_x and f_b are the plastic section development coefficient and yield strength of the frame beam, respectively.

(5) Column

The seismic design code has provided detailed design methods for columns in EBFs. The design values of the internal forces can be expressed as follows:

$$M_c = \Omega * M_{c,com} \tag{8}$$

$$N_c = \Omega * N_{c,com} \tag{9}$$

$$\frac{N_c}{A_{nc}} \pm \frac{M_c}{W_{nc}} \leq f_c \tag{10}$$

$$\frac{N_c}{\varphi A_c} \leq f_c \tag{11}$$

where M_c and $M_{c,com}$ are the calculated and designed values of the bending moment of the column, and N_c and $N_{c,com}$ are the calculated and designed values of the axial force of the column. A_{nc}, f_c, and W_{nc} are the axial force, net sectional moment of resistance, and net sectional area of the column, respectively. f_c and φ are the design value of the tensile strength and the stability coefficient of the column.

3. Experimental Study of VSSL-SSBC and SC-SL

3.1. Test Plan

3.1.1. Specimen Design

To explore the seismic performance and seismic resilience of the eccentrically braced frame (EBF) with self-centering shear link (SC-SL), a specimen of a very short shear link with shear slotted bolted connection (VSSL-SSBC-1) and a self-centering shear link specimen (SC-SL-1) are designed according to Equations (1)–(3). The mechanical properties and self-centering capacity of SC-SL can be effectively explored by comparing the hysteretic curves and failure modes of the VSSL-SSBC-1 and SC-SL-1 specimens. The specific dimensions of each specimen are presented in Figure 5, and the main parameters of each specimen are shown in Table 1. For the VSSL-SSBC-1 specimen, the cross-sectional dimensions of the VSSL are H210 × 110 × 6 × 8 mm, with a length of 300 mm. Stiffeners are symmetrically arranged on both sides, spaced at intervals of 100 mm, with a thickness of 10 mm. Endplate I has dimensions of 440 × 250 × 20 mm, while endplate II features eight slot holes measuring 34 mm in length and is connected using M22 high-strength bolts. A copper shim is placed between endplate I and endplate II to enhance the connection quality. For the SCB in the SC-SL-1 specimen, the fixed plate measures 590 × 90 × 10 mm, and both slip plates have dimensions of 548 × 90 × 10 mm. The SMA wires used in the SC-SL-1 specimen are arranged in 30 circles with an area of 94.20 mm^2. The bolt circular hole has a radius of 14 mm, and the slot hole measures 40 mm in length. The vertical brace in the SC-SL-1 specimen has a height, width, and thickness of 644 mm, 260 mm, and 10 mm, respectively. The lower end of the vertical brace contains two circular holes with a radius of 14 mm, and the upper end is reinforced with two stiffeners, symmetrically arranged on both sides. Note that all the units associated with the multiplication signs in this paper are in millimeters.

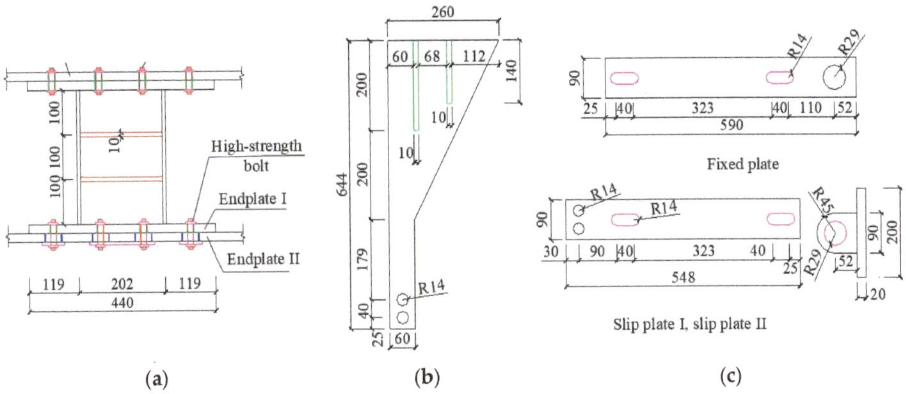

Figure 5. Dimensions of the VSSL-SSBC and SC-SL specimens: (**a**) VSSL-SSBC; (**b**) Vertical brace; (**c**) Self-centering SMA brace.

Table 1. Primary parameters for experimental specimens.

Specimen	P /kN	μ	VSSL /mm	SMA Area /mm^2	P_s /kN	Loading Displacement/mm	Slot Hole Length/mm
VSSL-SSBC-1	50	0.30	210 × 110 × 6 × 8	-	-	40	34
SC-SL-1	50	0.30	210 × 110 × 6 × 8	94.20	10	40	34

3.1.2. Material Properties

The structural elements in the VSSL-SSBC-1 and SC-SL-1 specimens were fabricated using Q355B grade steel plates with varying thicknesses ranging from 6 to 15 mm. Tensile tests were conducted on standardized test coupons to determine the material properties of the Q355B steel across different plate thicknesses [32], with the detailed results presented in Table 2. As for the 10.9 grade high-strength bolts, the yield and ultimate strength are 940 MPa and 1040 MPa, respectively.

Table 2. Mechanical parameters of steel plates.

Thickness t/mm	Yield Stress f_y/MPa	Ultimate Stress f_u/MPa	Elastic Modulus E/MPa	Elongation δ/%
6	356	535	203	21.7
8	361	547	206	22.2
10	368	552	208	23.3
15	372	556	209	23.2

All SMA wires employed in the experimental program maintained a consistent cross-sectional dimension of 1.0 mm in diameter. Tensile tests were conducted on a 375 mm long SMA wire with the same diameter to determine and validate the mechanical characteristics [24]. The key mechanical parameters are as follows, as illustrated in Figure 5: the starting stresses (σ_s^{AM}) and ending stresses (σ_f^{AM}) for the austenite-to-martensite phase transformation are 550 MPa and 625 MPa, respectively, while the starting stresses (σ_s^{MA}) and ending stresses (σ_f^{MA}) for the martensite-to-austenite phase transformation are 265 MPa and 175 MPa, respectively. The maximum phase transformation strain and the initial elastic modulus (EA) are 0.06 and 59,531 MPa, respectively. The SMA numerical model can be utilized to simulate the hysteresis curve of the SMA wire through the UMAT program in ABAQUS software 2021. Figure 6 presents the simplified relationship curve of the SMA.

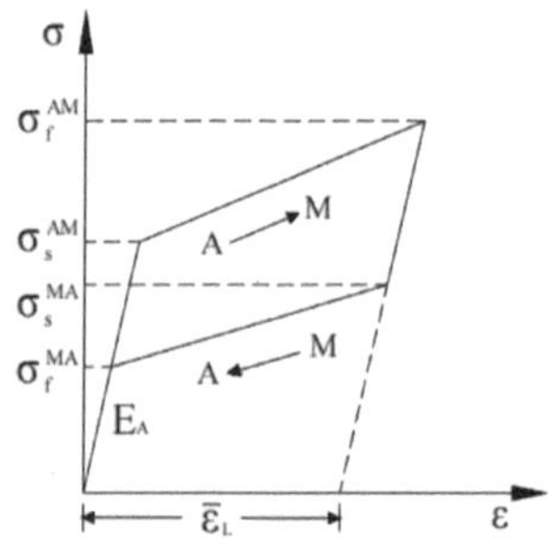

Figure 6. Constitutive model of SMA.

3.1.3. Test Setup and Loading Protocol

(1) Test setup

The experimental study of the VSSL-SSBC-1 and SC-SL-1 specimens was conducted in the structural engineering lab of Nanchang University (Figure 7). The test device primarily consists of VSSL-SSBC-1/SC-SL-1, a loading beam, upper and lower cushion beams, two vertical columns, and a ground beam. For the VSSL-SSBC-1 specimen, the upper end is secured to the upper cushion beam via eight high-strength bolts, which are also secured to the loading beam. The lower end is fastened to the lower cushion beam using eight high-strength bolts, with the lower cushion beam containing eight slot holes. In the SC-SL-1 specimen, the test setup of VSSL-SSBC is identical to that in VSSL-SSBC-1. The SCB is connected to the vertical brace using two high-strength bolts, while the other end is linked to the lower cushion beam via a pin connection. Additionally, the upper end of the vertical brace is secured to the upper cushion beam using six high-strength bolts, forming the corresponding loading setup.

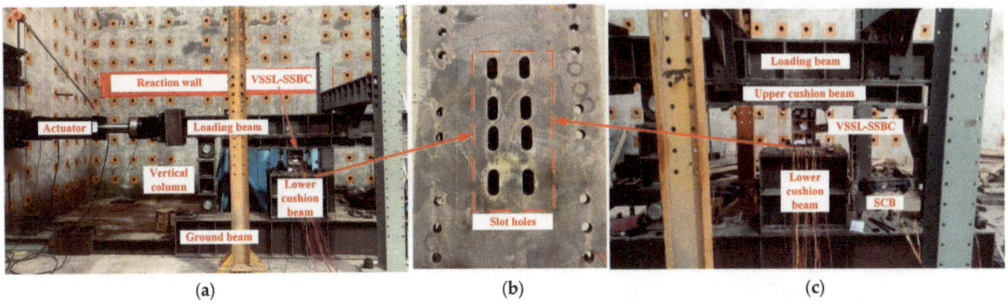

Figure 7. Test device of VSSL-SSBC-1 and SC-SL-1 specimens: (**a**) VSSL-SSBC-1; (**b**) Slot holes; (**c**) SC-SL-1.

(2) Loading protocol

The VSSL-SSBC-1 and SC-SL-1 specimens were subjected to displacement-controlled loading. At the beginning, displacements of ±6 mm, ±10 mm, and ±16 mm were applied. As the high-strength bolts slid to the ends of the slot holes, the displacement increments for each loading step were set as follows: $\pm 0.5S_1 \pm 0.01e$, $\pm 0.5S_1 \pm 0.03e$, $\pm 0.5S_1 \pm 0.05e$, $\pm 0.5S_1 \pm 0.07e$, and so on [22], until the specimens failed and loading was stopped. In this context, S_1 represents the length of the slot hole on the top plate of the lower cushion beam in the VSSL-SSBC, while e denotes the length of the VSSL. In the displacement loading scheme, the symbols "+" and "−" indicate the maximum positive and negative displacement values within a single cyclic loading, respectively.

3.1.4. Measurement Scheme

Throughout the loading process, the load values of the VSSL-SSBC-1 and SC-SL-1 specimens were directly recorded by the actuator, while the displacement and strain measurements were taken as shown in Figure 8.

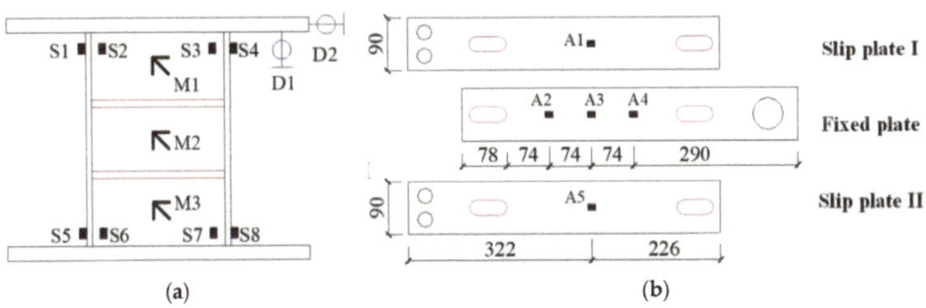

Figure 8. Location of displacement and strain gauges: (**a**) VSSL-SSBC; (**b**) SCB.

(1) Displacement measurement. To ensure synchronized displacement measurements for both the VSSL-SSBC and SCB, a horizontal displacement gauge (D1) and a vertical displacement gauge (D2) were installed on the right side of the upper endplate of the VSSL, as shown in Figure 8a.

(2) Strain Measurement. To monitor strain at critical sections of the VSSL-SSBC and the SCB, strain gauges were installed as follows: strain measurement points S1 to S8 and M1 to M3 were placed on the VSSL (Figure 8a), while A1 to A5 were arranged on the SCB (Figure 8b).

3.2. Failure Mode

3.2.1. VSSL-SSBC-1 Specimen

Figure 9 illustrates the failure mode of the VSSL-SSBC-1 specimen throughout the entire cyclic loading sequence. Red boxes in Figure 9 are center line of sliding plates. At the initial stage, all high-strength bolts were aligned with the centers of the slot holes (Figure 9a). During the first three loading steps, since the applied displacement was less than $0.5S_1$, the high-strength bolts only slid within the slot holes without reaching the ends. At this stage, only slippage occurred in the SSBC to dissipate the seismic energy, while the VSSL did not yield. After the fourth loading step, when the applied displacement exceeded $0.5S_1$, the high-strength bolts in the SSBC slid to the ends of the slot holes, and the VSSL began to yield in the positive direction, as shown in Figure 9b. At this stage, the bolts in the SSBC no longer slid, and the VSSL began to dissipate seismic energy through shear deformation. Subsequently, during the negative displacement cycle, the high-strength bolts in the SSBC slid to the negative end of the slot holes, causing the VSSL to yield in the negative direction, as shown in Figure 9d. Additionally, when the applied displacement exceeded $0.5S_1$, the VSSL experienced both web and flange buckling in the positive and negative loading directions, as illustrated in Figure 9c–e. Finally, when the applied displacement reached $-0.5S_1-0.09e$ in positive tension, a severe buckling and fracture developed between the flange and the endplate of the VSSL with a force of −404.45 kN, leading to specimen failure and termination of the loading process, as shown in Figure 9f. The SSBC remained undamaged throughout the entire test process.

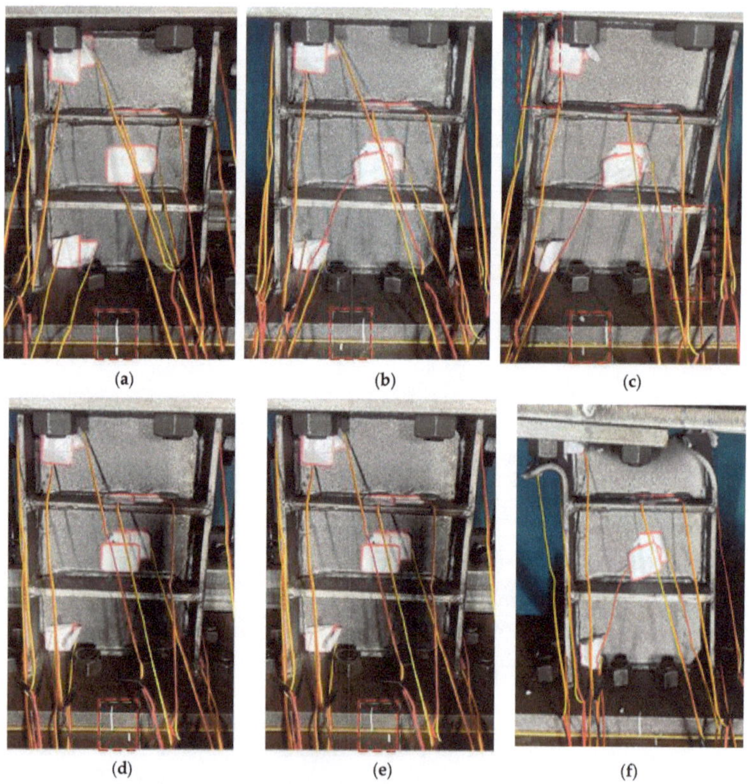

Figure 9. Failure model of VSSL-SSBC-1 specimen (+,− are the positive and negative loading): (**a**) Initial stage; (**b**) Web yielding (+); (**c**) Flange buckling (+); (**d**) Web yielding (−); (**e**) Flange buckling (−); (**f**) Failure mode.

3.2.2. SC-SL-1 Specimen

Figure 10 illustrates the deformation and failure modes of the SC-SL-1 specimen throughout the entire loading sequence. Red squares and ovals in Figure 10 are center line of sliding plates and concentrated damage, respectively. The primary failure modes include VSSL slip, yielding, and flange and web buckling. At the initial stage, the slip plates in the SCBs underwent notable deformation, with the SMA in tension. At this stage, the displacement of the slip plates matches that of the VSSL, as shown in Figure 10a. When the VSSL slipped to the end of the slot hole, the lateral deformation of the VSSL is zero (Figure 10d). At this stage, the SSBC and SCB exerted their energy dissipation capabilities, while the VSSL was not in an operating state. As the loading displacement was increased to 25.25 mm, the deformation of the slip plates within the SCB intensified, while the load-bearing and energy-dissipating capacities of the SMA wires were gradually enhanced, as shown in Figure 10b. Meanwhile, the VSSL transitioned to an elastic-plastic state, exhibiting pronounced inelastic deformation, as illustrated in Figure 10e. At this stage, the bolts in the SSBC slid to the ends, while the VSSL began to dissipate seismic energy through shear deformation and the SMA dissipated seismic energy through tensile deformation. At a loading displacement of 40.45 mm, the deformation of the slip plates in the SCBs became approximately equal to the loading displacement, indicating that the SMA wires contribute significantly to the load-bearing and restoring forces of the SC-SL system. Additionally, the vertical brace provides sufficient support to the SCB, as illustrated in Figure 10c. At this stage, both the web and flange of the VSSL entered a buckling state, resembling the failure mode of the VSSL observed in VSSL-SSBC-1, as shown in Figure 10f. At this point, the deformation capacity of the SMA wires reached its limit, leading to wire fracture and termination of the loading process. The vertical braces remained undamaged throughout the entire test process.

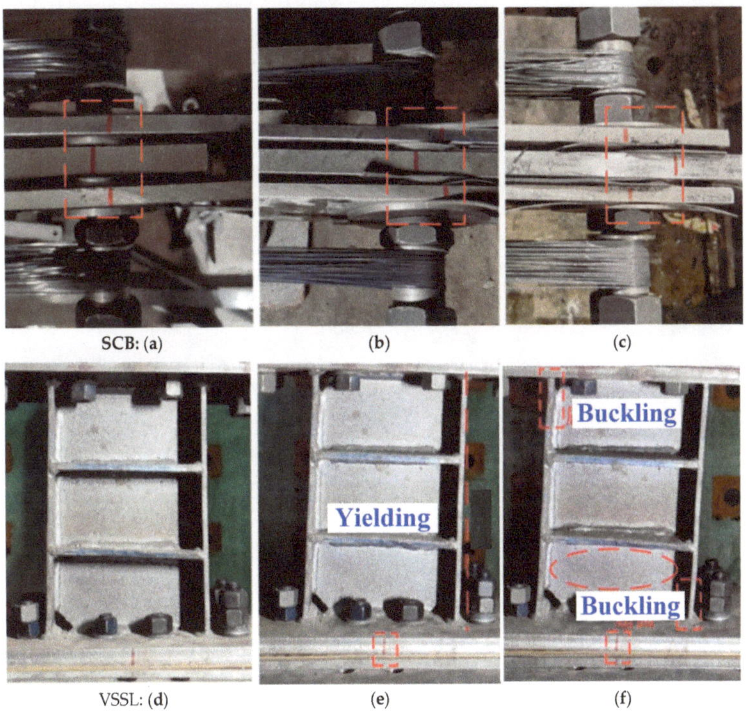

Figure 10. Failure model of SC-SL-1 specimen: (**a**) d = 16 mm; (**b**) d = 25.25 mm; (**c**) d = 40.25 mm; (**d**) Initial slip; (**e**) Web yielding; (**f**) Web and flange buckling.

3.3. Hysteresis Performance

Figure 11 presents the hysteresis curves of the VSSL-SSBC-1 and SC-SL-1 specimens, where F and D represent the shear force and loading-displacement of the specimen, respectively. For the VSSL-SSBC-1 specimen (Figure 11a), when the displacement is less than $0.5S_1$, the hysteresis curve exhibits an ideal rectangular shape. As the displacement exceeds $0.5S_1$, the hysteresis curve becomes increasingly full, with the load gradually increasing as the displacement grows. Throughout the entire loading process, the specimen maintains its load-bearing capacity without any noticeable decline before failure. However, as the displacement increases, the slip force of the specimen gradually decreases, primarily due to the reduction in the friction coefficient of the shims.

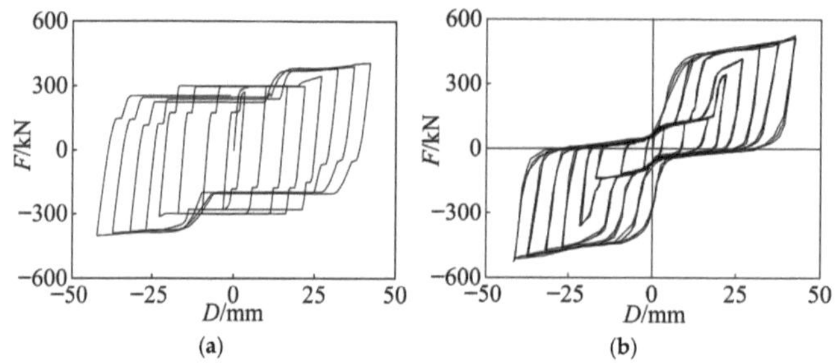

Figure 11. Hysteresis curves of VSSL-SSBC-1 and SC-SL-1 specimens: (**a**) VSSL-SSBC-1; (**b**) SC-SL-1.

For the SC-SL-1 specimen (Figure 11b), at an ultimate displacement of 40.45 mm, the high-strength bolts within the SSBC reach the end of the slot holes, marking the transition of the VSSL into the energy-dissipating and load-bearing stage. The hysteresis curve is formed by the contributions from the VSSL-SSBC and the SCB. At this stage, the load-bearing capacity of SC-SL-1 shows a positive correlation with the displacement, reaching a peak load of 525.25 kN, with a residual deformation of 26.58 mm. Compared to the corresponding VSSL-SSBC-1 specimen, the SC-SL-1 specimen exhibits a remarkable improved load-bearing capacity and reduced residual deformation, indicating that the self-centering shear link possesses superior load-bearing and self-centering capabilities.

3.4. Skeleton Curves

Figure 12 shows the skeleton curves of the VSSL-SSBC-1 and SC-SL-1 specimens. The VSSL-SSBC-1 specimen undergoes four stages, as shown in Figure 12a: elasticity, slip, elastic-plasticity, and plasticity. The slip displacement is 17 mm, corresponding to a slip load of 175.00 kN, which is mainly caused by the SSBC. The load capacity of the VSSL-SSBC-1 specimen is composed of the VSSL and SSBC, which is 442.30 kN, closely matching the calculated ultimate shear force for the VSSL in Equation (2).

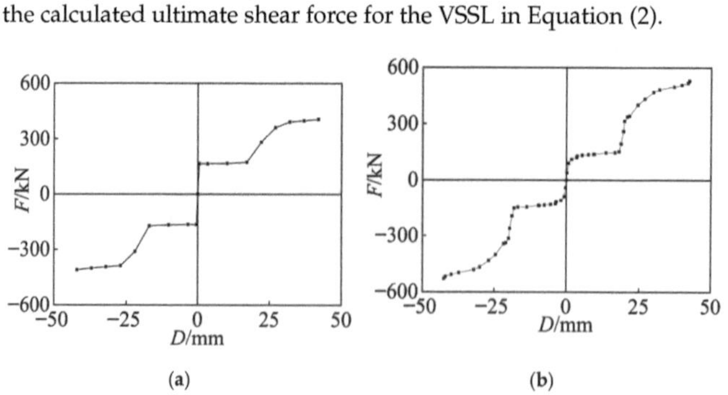

Figure 12. Skeleton curves of VSSL-SSBC-1 and SC-SL-1 specimens: (**a**) VSSL-SSBC-1; (**b**) SC-SL-1.

The skeleton curve of the SC-SL-1 specimen, as shown in Figure 12b, includes three stages: elasticity, elastic-plasticity, and strengthening. After the high-strength bolt slides to the end of the slot hole, the load capacity increases significantly. The load capacity of the SC-SL-1 specimen is composed of the VSSL, SSBC, and SCBs, which is 529.18 kN and is notably higher than the load capacity of the VSSL-SSBC-1 specimen. This increase is mainly due to the enhanced load capacity provided by the self-centering SMA braces.

4. Numerical Study of VSSL-SSBC and SC-SL

4.1. Numerical Analysis Method

In the finite element model, it is assumed that the SMA remains in a superelastic state before fracture, and the friction coefficient of the shims in the SSBC remains constant under cyclic loading. The material properties, boundary conditions, etc., are consistent with those in the experiment.

4.1.1. Verification of the SMA

The UMAT program was used to simulate the constitutive curve of the SMA material. An SMA bar with a diameter of 6 mm (No. 10/6-450-30) tested by Li [33] was used to validate the SMA numerical model, as shown in Figure 13a. One end of the SMA finite element model is fixed, and a cyclic displacement load is applied to the other end. The hysteresis curves of the experiment and the finite element analysis are shown in the figure. The hysteresis curves from both the experiment and the finite element analysis are flag-shaped, and their mechanical behaviors are consistent, indicating that the SMA constitutive model is reliable.

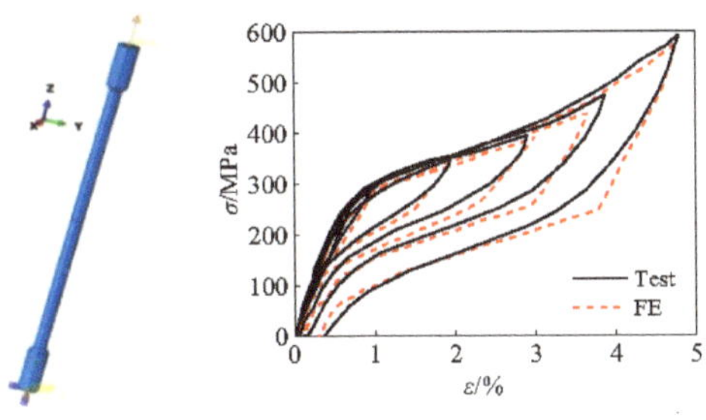

Figure 13. Verification of the SMA material: (**a**) Finite element model; (**b**) Hysteresis curves.

4.1.2. Numerical Method of VSSL-SSBC-1 Specimen

(1) Material properties. In the VSSL-SSBC model, the material properties of the steel plates with different thicknesses are listed in Table 2. The constitutive curves of all steel models are shown in Figure 14a, which is bilinear constitutive model.

(2) Mesh quality. The numerical analysis of the VSSL-SSBC was performed using ABAQUS software. The C3D8R element, an eight-node three-dimensional solid element, is capable of modeling large deformation and strains, enabling the accurate simulation of various structural components. The mesh quality of the VSSL-SSBC-1 specimen is shown in Figure 14b.

(3) Boundary conditions. For the VSSL-SSBC-1 specimen, the degrees of freedom at both end sections of endplate II are constrained. The friction coefficient between endplate I in VSSL and endplate II is set to 0.30, while the coefficient between endplate II and the base plate is 0.075. Additionally, the displacement in the X direction on the upper surface of the top plate is released, while rotation in the Z direction is constrained. Figure 14c shows the boundary conditions of the VSSL-SSBC-1 specimen.

(4) Loading protocol. In the numerical analysis, applying a horizontal displacement to the top plate of the VSSL ensures that the bending moment and shear force align with the loading conditions in an eccentrically braced frame. The specific loading protocol is detailed in Section 3.1.3. Based on this approach, the VSSL-SSBC-1 specimen can be effectively analyzed using numerical analysis methods.

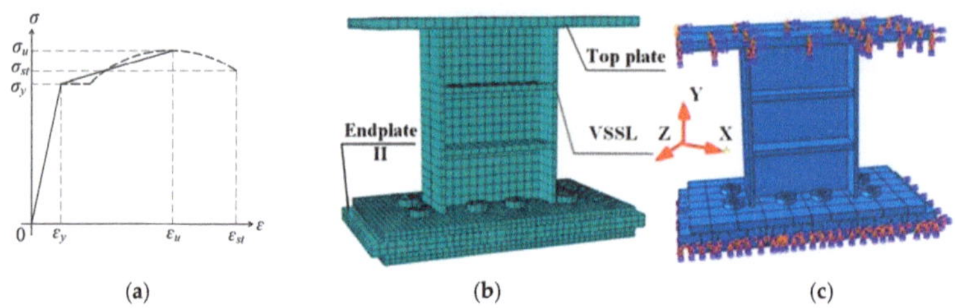

Figure 14. Numerical model of VSSL-SSBC-1 specimen: (**a**) Constitutive curve; (**b**) Analytical model; (**c**) Boundary condition.

4.1.3. Numerical Method of SC-SL-1 Specimen

The self-centering shear link primarily consists of two SCBs, a VSSL-SSBC, a loading beam, and two vertical braces, as shown in Figure 14. In the numerical model, all components are simulated using C3D8R elements in the ABAQUS software, consistent with the VSSL-SSBC-1 specimen.

(1) Material properties. The material properties of each component are detailed in Table 2, and the constitutive curves of all steel models are shown in Figure 14a.

(2) Mesh quality. The mesh quality of the SC-SL-1 specimen is shown in Figure 15.

(3) Boundary conditions. In the SC-SL-1 specimen, the displacements in all directions of all nodes at surfaces 1 and 2 of the two SCBs must be constrained. Likewise, the displacements in all directions of all nodes at surfaces 3 and 4 on both sides of endplate II in the VSSL-SSBC must also be constrained. Additionally, all components must have their out-of-plane displacement restricted in the XY plane.

(4) Loading protocol. A cyclic displacement in the X direction is applied to the left loading surface of the loading beam, with displacement increments following those described in Section 3.1.3. With the appropriately defined elements, material properties, boundary conditions, and loading protocols, a comprehensive numerical simulation analysis can be conducted.

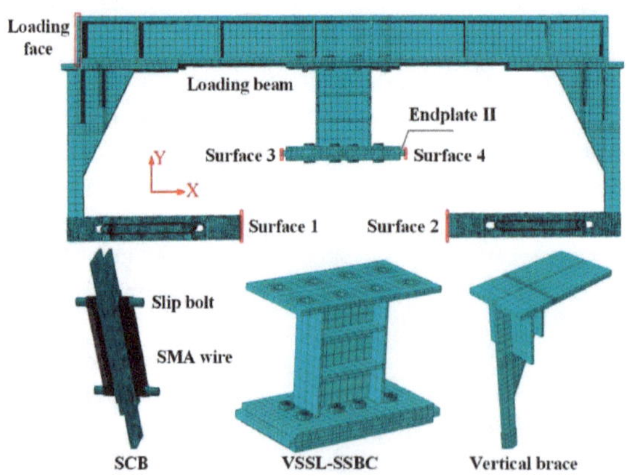

Figure 15. Numerical model of SC-SL-1 specimen.

4.2. Comparison of VSSL-SSBC-1 Specimen

4.2.1. Hysteresis Curves

Figure 16a presents the hysteresis curves obtained from the test and numerical analysis for the VSSL-SSBC-1 specimen. In the numerical results, the hysteresis curve also follows four stages: elastic, slippage, elastoplastic, and plastic. A discrepancy in the bearing capacity is observed between the test and numerical analysis results during the slippage stage, primarily due to the cyclic friction in the shim under cyclic loading, which gradually reduces the friction coefficient. However, during the load-bearing stage of the VSSL, both methods yield nearly identical bearing capacities and displacement. At ultimate displacement, the maximum loads recorded from the test and numerical analysis are 404.45 kN and 409.97 kN, respectively, with a marginal error of 1.36%. This demonstrates that the numerical results closely align with the test results in terms of load-bearing capacity, stiffness, and overall structural response across different loading stages.

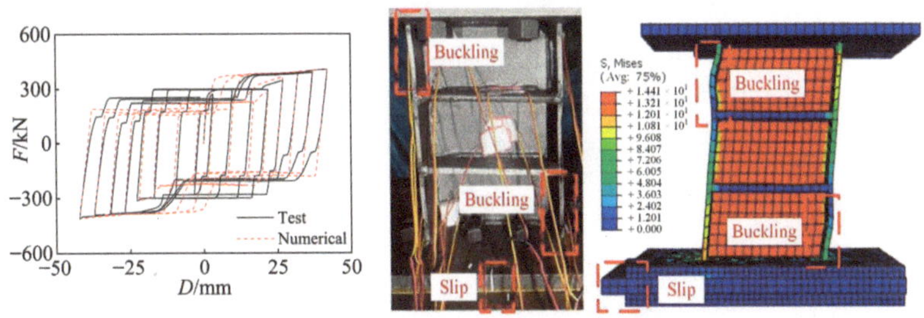

(a) (b)

Figure 16. Comparison of hysteresis curves and failure modes for VSSL-SSBC-1 specimen: (**a**) Hysteresis curves; (**b**) Failure modes.

4.2.2. Failure Modes

Figure 16b illustrates the failure mode of the VSSL-SSBC-1 specimen under cyclic loading, as determined by numerical analysis. The results indicate that the initial deformation is attributed to the frictional slippage of the high-strength bolts. As loading progresses, the VSSL begins to yield and sustain the load, followed by buckling of the flange and web, which is consistent with expectations. Additionally, the highest stress concentrations in VSSL-SSBC-1 occur at the connection between the flange and web, where failure is most likely to initiate. This confirms that the failure modes in the test and numerical analysis are highly consistent.

4.3. Comparison of SC-SL-1 Specimen

4.3.1. Hysteresis Curves

A numerical model of the SC-SL-1 specimen was developed and analyzed, and the resulting hysteresis curve was compared with the test results, as shown in Figure 17a. The loading displacement, stiffness, and bearing capacity at each loading step exhibit close agreement. When SC-SL-1 reaches the maximum loading displacement, the ultimate load values obtained from the numerical analysis and test are 535.23 kN and 519.28 kN, respectively, with a minimal discrepancy of only 3.07%. This confirms that the numerical analysis method for the self-centering shear link can accurately predict its bearing capacity.

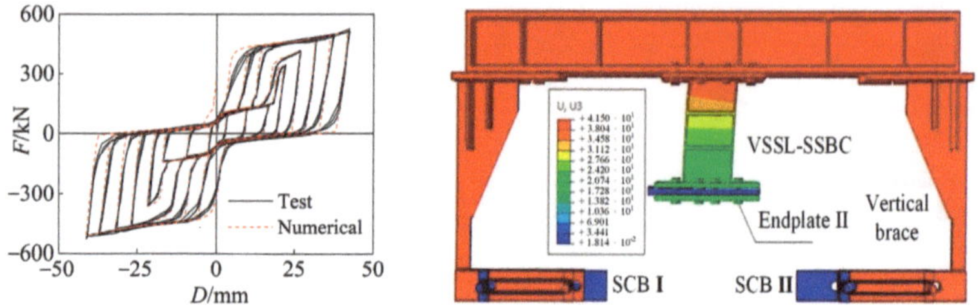

(a) (b)

Figure 17. Comparison of hysteresis curves and failure modes for SC-SL-1 specimen: (**a**) Hysteresis curves; (**b**) Failure modes.

4.3.2. Failure Modes

As shown in Figure 17b, endplate II of the VSSL-SSBC is fixed, and the VSSL undergoes inelastic deformation, bearing, and energy dissipation under cyclic loading. The displacement of the VSSL-SSBC matches that of the SC-SL, aligning with Figure 10d–f. Driven by the vertical brace, the SCB induces movement and superelastic deformation in both the slip plate and SMA. As the displacement increases, the SMA's bearing capacity and restoring force progressively increase, contributing to both energy-dissipating and self-centering functionality in the SC-SL, consistent with the deformation mode depicted in Figure 10a–c.

5. Seismic Performance of EBF with Self-Centering Shear Link

5.1. Analysis Model

5.1.1. Modeling

An eccentrically braced steel frame (EBF) with a VSSL-SSBC system (Model 1: EBF-VSSL-SSBC) and another EBF structure with an SC-SL system (Model 2: EBF-SC-SL) are designed. Both models are composed of steel columns, steel beams, steel braces, the VSSL-SSBC/SC-SL component, and brace plates, as illustrated in Figure 18a. The main distinction between the two models lies in the fact that in Model 2, two self-centering SMA braces (SCBs) and two vertical braces are installed on either side of the brace plate. As shown in Figure 18b, both models have a span of 4800 mm and a height of 3600 mm. The steel columns are H320 mm × 320 mm × 10 mm × 12 mm, the steel beams are H300 mm × 250 mm × 10 mm × 12 mm, and the steel braces are square steel tubes measuring 300 mm × 10 mm.

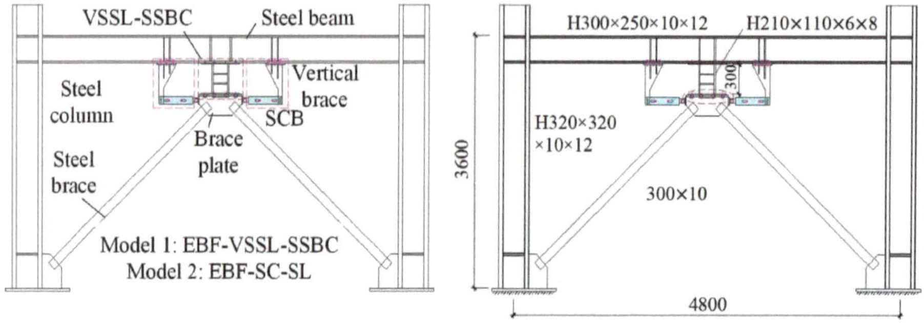

(a) (b)

Figure 18. Analysis models of EBF-VSSL-SSBC and EBF-SC-SL: (**a**) Basic components; (**b**) Details.

In the VSSL-SSBC, the VSSL component has dimensions of H210 mm × 110 mm × 6 mm × 8 mm with a length of 210 mm. It is connected at both the upper and lower ends using eight M22 high-strength bolts. A pretension force of 155 kN is applied at the upper end and 30 kN at the lower end. The length of the slot hole is 50 mm. The friction coefficient between the upper end of the brace plate and the lower end of the VSSL is 0.20, while the coefficient between the upper end of the brace plate and the base plate is 0.075. Additionally, in Model 2, the SCB has a length of 700 mm, with an SMA area of 219.80 mm^2. Other parameters for the vertical brace and the SCB are the same as those outlined in Section 3.1.1.

5.1.2. Numerical Analysis Method

The EBF-VSSL-SSBC and EBF-SC-SL specimens were comprehensively investigated using the ABAQUS software. The steel columns, steel beams, steel braces, VSSL-SSBC/SC-SL, brace plates, high-strength bolts, and self-centering SMA braces were all modeled using C3D8R elements. Additionally, the friction behavior between the plates in the VSSL-SSBC and the SCB was simulated by defining the appropriate friction coefficient, as shown in Section 4.1. The material properties of all components are detailed in Section 3.1.2.

As shown in Figure 19, the boundary conditions of the two EBF models are as follows: the column bases are constrained, and out-of-plane displacement of the EBFs is restricted throughout the entire loading process. A displacement-controlled load protocol is applied to the top ends of the two steel columns, following the same loading method as that of the VSSL-SSBC described in Section 3.1.3.

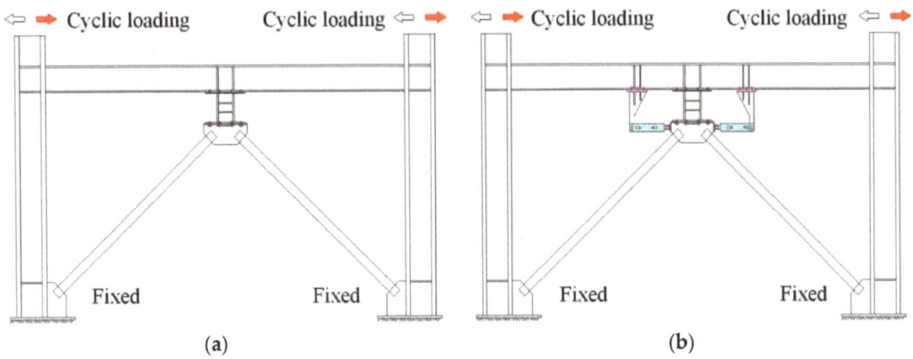

Figure 19. Numerical analysis method of EBF-VSSL-SSBC and EBF-SC-SL: (**a**) EBF-VSSL-SSBC; (**b**) EBF-SC-SL.

5.2. Frequent Earthquakes

The loading displacements of the designed EBFs during frequent and rare earthquakes were determined at the inter-storey drifts of 0.4% and 2.0% [9], respectively.

5.2.1. Deformation Modes

Under frequent earthquakes, the expected displacement values of both EBF-VSSL-SSBC and EBF-SC-SL are 14.40 mm. Figure 20 illustrates the deformation modes of the two steel frames and braces, revealing that the maximum displacement occurs at the top end of the steel frame and braces. Due to the low friction coefficient and small pretension of the SSBC, the bearing capacity during the slippage of VSSL is relatively low. Consequently, when the frames reach the expected displacement value, the deformation of the braces remains minimal, with the maximum displacement of the braces in both EBF-VSSL-SSBC and EBF-SC-SL being only 0.22 mm. This suggests that the deformation modes of the frames and braces in both EBFs are essentially the same.

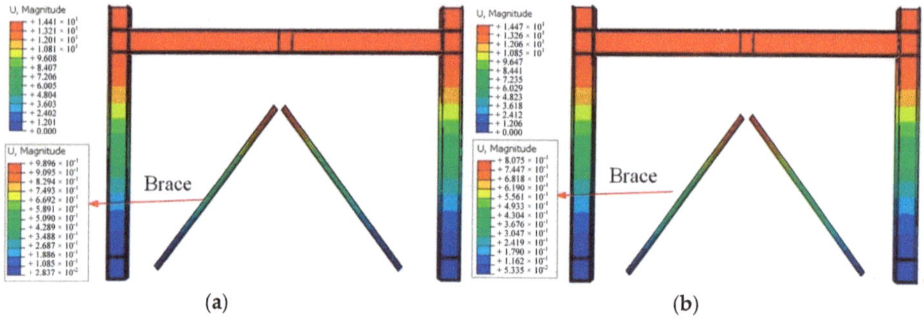

Figure 20. Deformation modes of the steel frame and brace during frequent earthquakes: (**a**) EBF-VSSL-SSBC; (**b**) EBF-SC-SL.

Figure 21 presents the deformation modes of the VSSL and the SCBs. Due to the slippage of the high-strength bolts in the VSSL-SSBC, the overall slippage values of the VSSL in both EBFs are 14.40 mm, meaning that there is no significant relative slippage between the two ends of the VSSL, which is in line with expectations. Additionally, when the EBF-SC-SL reaches the expected maximum displacement, the vertical brace drives the slip plates and slip bolts to move, leading to the tension of the SMA wires with an elongation of 14.40 mm. Therefore, the deformation modes of the VSSL in both EBFs are completely identical, and the deformation mode of the SCBs also meets the expectations.

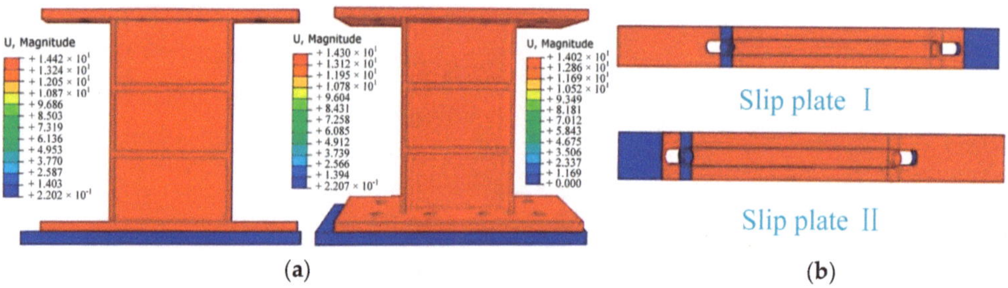

Figure 21. Deformation modes of the VSSL-SSBC and SCB during frequent earthquakes: (**a**) VSSL-SSBC; (**b**) SCB.

5.2.2. Stress Nephograms

Figure 22 illustrates the stress nephograms of the steel frame and braces in both EBFs under the expected maximum displacement. For the EBF-VSSL-SSBC and EBF-SC-SL, the maximum stresses in the steel frame are 275.20 MPa and 275.60 MPa, respectively, and the maximum stresses in the braces are 16.13 MPa and 17.79 MPa, respectively. These values are remarkably close to each other. All of these stress values are below the yield stress of the steel and braces, indicating that these components stay within an elastic range. Additionally, owing to the low slip resistance of the SSBC, the initial slip load and stiffness of the VSSL-SSBC are relatively low, leading to reduced stress levels in the structural components.

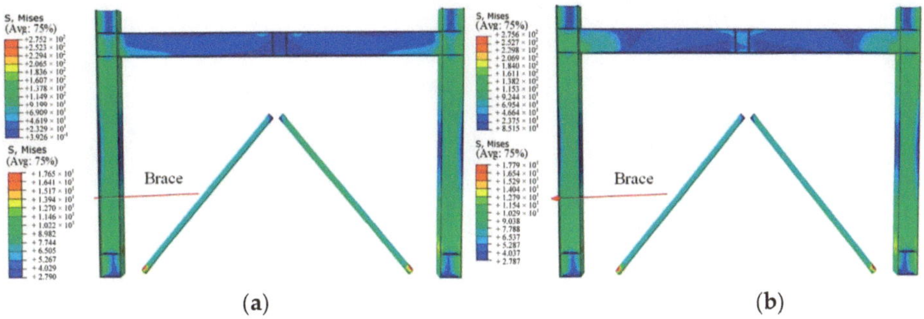

Figure 22. Stress nephograms of the steel frame and brace during frequent earthquakes: (**a**) EBF-VSSL-SSBC; (**b**) EBF-SC-SL.

Figure 23 presents the maximum shear stress of the VSSL and the stress nephograms of the SCB under frequent earthquakes. Since the SSBC has a low initial frictional slip load and the VSSL remains in the slip stage, the maximum shear stress of the VSSL-SSBC in both EBFs is 73.10 kN, which is significantly lower than the yield shear stress of the VSSL. In the EBF-SC-SL, the maximum stress of the SMA reaches 308.00 MPa, indicating that the SMA has entered its working state, effectively contributing to the energy dissipation and self-centering capabilities.

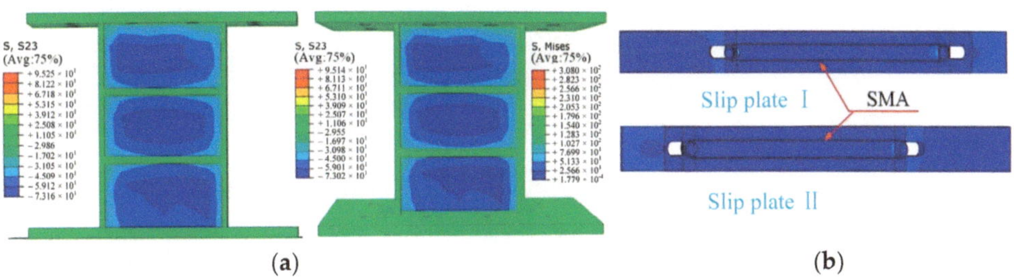

Figure 23. Stress nephograms of the VSSL-SSBC and SCB during frequent earthquakes: (**a**) VSSL-SSBC; (**b**) SCB.

5.2.3. Hysteresis Curves

Figure 24 presents the hysteresis curves of the EBFs under frequent earthquakes. When the displacement reaches 14.40 mm, the maximum bearing capacities of EBF-VSSL-SSBC and EBF-SC-SL are 420 kN and 504 kN, respectively. This indicates that the EBF with SC-SL has a higher bearing capacity than the one with VSSL-SSBC, along with a significantly improved energy dissipation capacity. Additionally, the residual deformation values of EBF-VSSL-SSBC and EBF-SC-SL are 2.99 mm and 1.32 mm, respectively, demonstrating that the EBF equipped with SCBs exhibits a superior self-centering capability.

Based on the above analysis, under frequent earthquakes, the stress and displacement values of the steel frame, braces, and VSSL in both EBF-VSSL-SSBC and EBF-SC-SL are essentially the same. However, the incorporation of SCBs effectively enhances the load-bearing capacity and reduces residual deformation in the EBF.

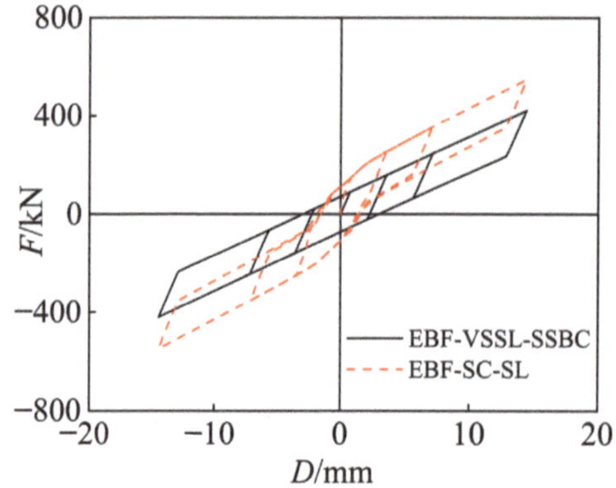

Figure 24. Hysteresis curves of the EBFs during frequent earthquakes.

5.3. Rare Earthquakes

5.3.1. Deformation Modes

Under a rare earthquake, when both EBF-VSSL-SSBC and EBF-SC-SL reach the expected displacement of 72 mm, the deformation of the corresponding main frame and braces is shown in Figure 25. The deformation modes of the main frames in both EBFs are nearly identical. The maximum displacement at the upper end of the braces in the EBF-VSSL-SSBC and EBF-SC-SL are 1.26 mm and 1.29 mm, respectively, with similar deformation modes. However, the maximum displacement of the braces is significantly smaller than that of the frame, primarily because the braces are much stiffer than the VSSL-SSBC, causing the deformation to concentrate mainly in the VSSL.

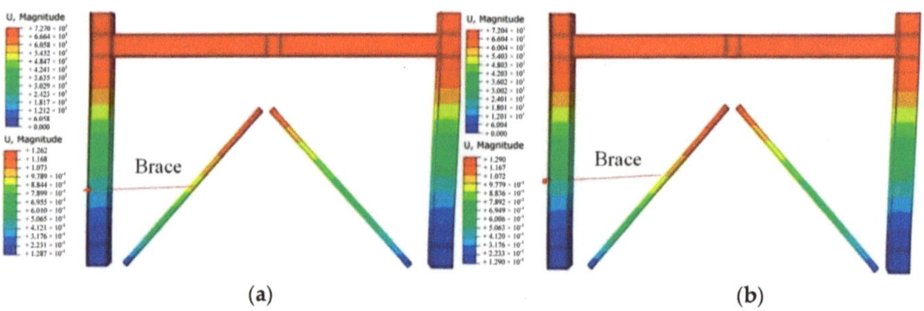

Figure 25. Deformation modes of the steel frame and brace during a rare earthquake: (**a**) EBF-VSSL-SSBC; (**b**) EBF-SC-SL.

The deformation modes of the VSSL-SSBC and SCB in EBF-VSSL-SSBC and EBF-SC-SL under a rare earthquake are shown in Figure 26. Since the braces in both EBFs provide sufficient stiffness and bearing capacity for the corresponding VSSL, the displacement of the brace end plates connected to the lower end plates of the VSSL remains relatively small. The lower end plates experience a displacement of 25 mm (bolts sliding to the end of the slot hole), while the upper end plate displacement is approximately 72 mm, effectively reducing the relative deformation of the VSSL. Additionally, in the EBF-SC-SL, the side plates and sliding bolts on the SCB move approximately 72 mm, matching the elongation of the SMA wires.

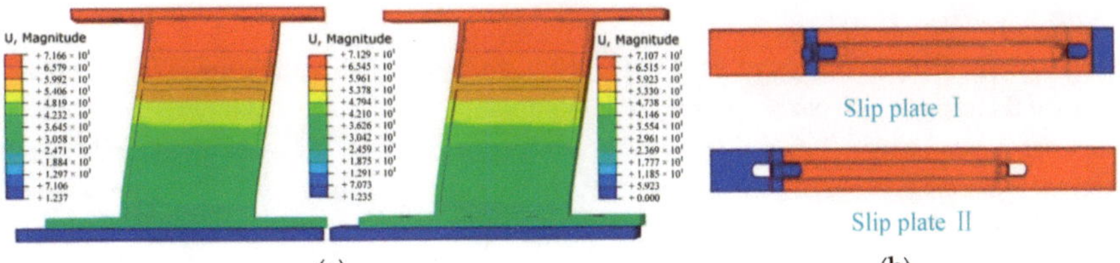

Figure 26. Deformation modes of the VSSL-SSBC and SCB during a rare earthquake: (**a**) VSSL-SSBC; (**b**) SCB.

5.3.2. Stress Nephograms

Under a rare earthquake, the stress nephograms of the main frame and braces in EBF-VSSL-SSBC and EBF-SC-SL are shown in Figure 27. The maximum stress values in the steel frames of the two EBFs are 521.10 MPa and 520.80 MPa, respectively, primarily concentrated at the beam–column joints and the bases of the columns, reaching the ultimate stress limit of the steel. The maximum stresses in the braces are 70.85 MPa and 85.28 MPa, respectively, which remain below the yield stress. Additionally, the brace stress in EBF-SC-SL is relatively higher due to the increased load-bearing capacity of the SCB, which amplifies the stress of the brace plates and the steel brace itself.

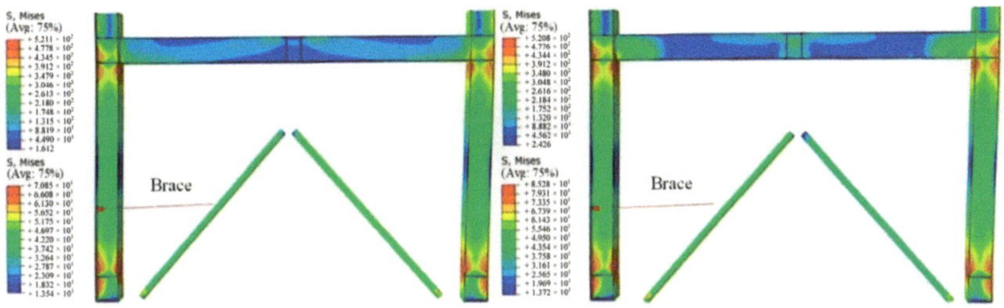

Figure 27. Stress nephograms of the steel frame and brace during a rare earthquake: (**a**) EBF-VSSL-SSBC; (**b**) EBF-SC-SL.

The stress nephograms of the VSSL and SCBs in EBF-VSSL-SSBC and EBF-SC-SL under a rare earthquake are shown in Figure 28. The shear stress distribution in the web of the VSSL is nearly identical in both EBFs, with a maximum stress of 311.70 MPa, reaching the ultimate shear stress limit and experiencing shear failure, which aligns with expectations. Furthermore, in the EBF-SC-SL, all components of the SCB remain in the elastic state, and the stress of the SMA reaches 597.20 MPa, staying within the superelastic range. This enables the structure to provide both energy-dissipating and self-centering capabilities.

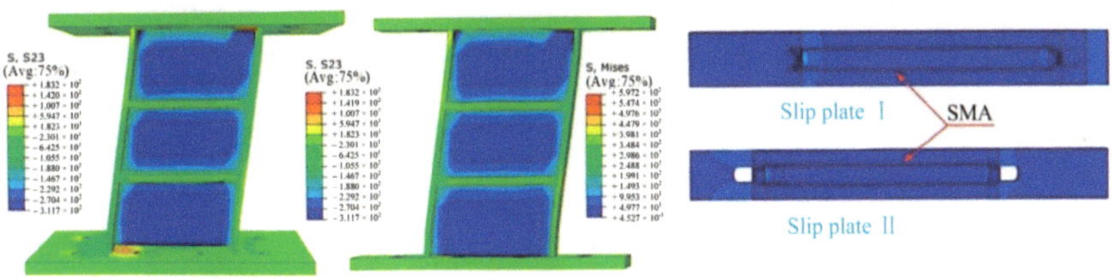

Figure 28. Stress nephograms of the VSSL-SSBC and SCB during a rare earthquake: (**a**) VSSL-SSBC; (**b**) SCB.

5.3.3. Hysteresis Curves

The hysteresis curves of EBF-VSSL-SSBC and EBF-SC-SL are shown in Figure 29, where their maximum bearing capacities are 1408.93 kN and 1685.94 kN, respectively, with an increase of 19.66%. This indicates that the incorporation of SCBs has a significant impact on the ultimate bearing capacity of the EBF. The hysteresis loop area at each loading step in EBF-VSSL-SSBC is larger than those in EBF-SC-SL, and the energy dissipation values of the last hysteresis loops of the two specimens are 186.7 kJ and 193.6 kJ, respectively, with an increase of 3.70%. Additionally, the maximum residual deformations of the EBF-VSSL-SSBC and EBF-SC-SL are 35.28 mm and 25.44 mm, respectively, with a reduction of 27.90% suggesting that the EBF-SC-SL exhibits better load-bearing and self-centering capabilities.

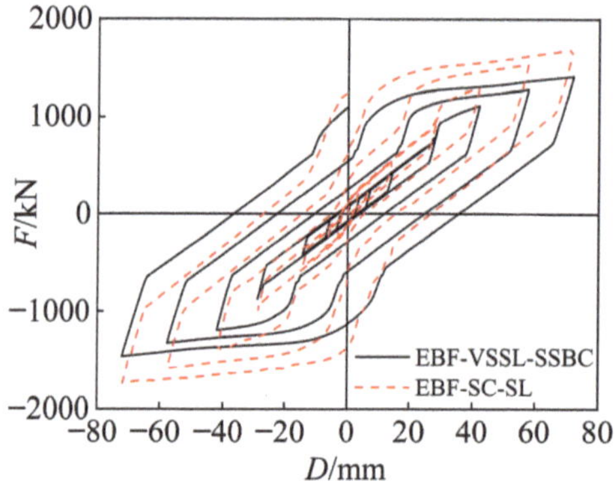

Figure 29. Hysteresis curves for the EBFs during a rare earthquake.

Under a rare earthquake, the steel frames in both EBF-VSSL-SSBC and EBF-SC-SL yield, while the braces remain in the elastic state. The VSSL in both EBFs enters the plastic stage. However, in the EBF-SC-SL, the VSSL experienced smaller residual deformation after the earthquake, effectively enhancing the seismic resilience capacity of the EBF. In conclusion, a comparative analysis reveals that the EBF with a novel SC-SL demonstrates superior stiffness, load-bearing capacity, and self-centering performance.

5.4. Seismic Results Analysis

Table 3 presents the key analysis results of the EBFs during different earthquakes. Here, η represents the residual deformation, σ_S denotes the stress of the SMA wire, D_F is the maximum displacement of the steel frame, D_V refers to the relative deformation of VSSL, σ_F indicates the maximum stress of the steel frame, τ_V is the maximum shear stress of the VSSL, and D_B and σ_B correspond to the maximum displacement of the brace and stress, respectively. Under various earthquake conditions, the frame stress and displacement, brace stress and displacement, and the VSSL stresses and displacements in EBF-VSSL-SSBC and EBF-SC-SL remain identical. However, under the same conditions, the residual deformation of EBF-SC-SL is significantly smaller than that of EBF-VSSL-SSBC, which is beneficial for the seismic resilience of the EBF.

Table 3. Comparison with the results for the EBFs during different earthquakes.

	EBF-VSSL-SSBC							EBF-SC-SL							
	D_F /mm	D_V /mm	D_B /mm	σ_F /MPa	τ_V /MPa	σ_B /MPa	η	D_F /mm	D_V /mm	D_B /mm	σ_F /MPa	τ_V /MPa	σ_B /MPa	η	σ_S /MPa
Frequent earthquake	14.56	0	0.22	275	73	16.13	2.99	14.38	0	0.22	275	73	17.79	1.32	308
Rare earthquake	72.70	46.66	1.26	521	312	70.85	35.28	72.04	46.29	1.29	521	312	85.28	25.44	597

6. Conclusions and Future Prospects

6.1. Conclusions

In this study, an eccentrically braced frame (EBF) with a novel self-centering shear link (SC-SL) was developed and investigated through test and numerical methods to enhance seismic resilience. The mechanical properties of VSSL-SSBC and SC-SL, along with the corresponding seismic performance of the EBFs were analyzed. The key findings are summarized as follows:

(1) Under seismic loading, the VSSL-SSBC mainly experienced elastic, slip, elastoplastic, and plastic stages, while the SC-SL experienced VSSL slip, yielding, and flange and web buckling, with the SMA wires in tension. The ultimate bearing capacities of the two devices are 442.30 kN and 529.18 kN, respectively.

(2) The experimental and finite element results demonstrate that the SC-SL has a significant improvement in both the ultimate bearing capacity and energy dissipation capacity compared with the VSSL-SSBC. Moreover, due to the function of the SCB, it exhibits an excellent self-centering capacity.

(3) Under frequent earthquakes, the frame, brace, and VSSL stresses and displacements in EBF-VSSL-SSBC and EBF-SC-SL are basically equivalent. Both frames dissipate energy through the slippage of high-strength bolts, and VSSL-SSBC remains elastic without significant yielding deformation.

(4) Under rare earthquakes, the bearing force and stress of the brace of the EBF-SC-SL are slightly increased compared with EBF-VSSL-SSBC. Its ultimate load capacity is enhanced by 19.66%, and the residual deformation is reduced by 27.90%. This improvement contributes to the seismic resilience capacity of the EBF.

6.2. Outlook

This paper effectively demonstrates that SC-SL has excellent self-centering ability and seismic performance. To conduct a further exploration, development, and application of this technology, the following contents can serve as future research topics:

(1) In this paper, a method combining experiments and finite element verification is adopted to study the mechanical properties of SC-SL, which provides a reference for the design of self-centering structures. However, it is necessary to carry out frame design and experimental research for the EBF with SC-SL in the future.

(2) Based on future experiments on the EBF with SC-SL, the influencing factors on its self-centering capacity, such as the circles and diameter of the SMA wires, can be investigated.

(3) This paper only investigated the application of SC-SL in Y-shaped eccentrically braced frames, whereas the application of SC-SL in other types of EBFs can be expanded in the future.

Author Contributions: Conceptualization, S.H.; methodology, B.Z. and S.H.; validation, L.H.; formal analysis, X.X. and S.L.; writing—original draft preparation, X.X., L.H. and S.L.; writing—review and editing, L.H.; supervision, S.H.; project administration, B.Z.; funding acquisition, S.H. and B.Z. All authors have read and agreed to the published version of the manuscript.

Funding: This research was funded by the National Nature Science Foundation of China (No. 52468025), National Nature Science Foundation of Jiangxi Province (No. 20224BAB204062), and Science and Technology Research Project of Jiangxi Bureau of Geology (2022JXDZKJKY10). The funders of the above funds are all corresponding author Shujun Hu.

Data Availability Statement: No new data were created or analyzed in this study. Data sharing is not applicable to this article.

Conflicts of Interest: Author Bo Zhang was employed by the company China Nerin Engineering Co., Ltd. The remaining authors declare that the research was conducted in the absence of any commercial or financial relationships that could be construed as a potential conflict of interest.

References

1. Okazaki, T.; Arce, G.; Ryu, H.C.; Engelhardt, M.D. Experimental study of local buckling, overstrength, and fracture of links in eccentrically braced frames. *J. Struct. Eng.* **2005**, *131*, 1526–1535. [CrossRef]
2. Bosco, M.; Rossi, P.P. Seismic behaviour of eccentrically braced frames. *Eng. Struct.* **2009**, *31*, 664–674. [CrossRef]
3. Azad, S.K.; Topkaya, C. A review of research on steel eccentrically braced frames. *J. Constr. Steel Res.* **2017**, *128*, 53–73. [CrossRef]
4. Okazaki, T.; Engelhardt, M.D. Cyclic loading behavior of EBF links constructed of ASTM A992 steel. *J. Constr. Steel Res.* **2007**, *63*, 751–765. [CrossRef]
5. Hu, S.J.; Zeng, S.Z.; Zhou, Q.; Zhi, Q.; He, K. Mechanism and seismic performance of a novel precast concrete beam-to-very short shear link joint: Experimental and numerical simulation. *Structures* **2023**, *56*, 104938. [CrossRef]
6. Hu, S.J.; Liu, S.W.; Zeng, S.Z.; Shao, T.F. Mechanical properties and influence factors of ordinary shear links. *Buildings* **2024**, *14*, 160. [CrossRef]
7. *AISC 341*; Seismic Provisions for Structural Steel Buildings. American Institute of Steel Construction: Chicago, IL, USA, 2010.
8. Ji, X.D.; Wang, Y.D.; Ma, Q.F.; Okazaki, T. Cyclic behavior of very short steel shear links. *J. Struct. Eng.* **2016**, *142*, 04015114. [CrossRef]
9. *GB 50011-2010*; Code for Seismic Design of Buildings. Chinese Building Industry Press: Beijing, China, 2016.
10. Hu, S.J.; Xiong, J.G.; Zhou, Q.; Lin, Z.B. Analytical and numerical investigation of overstrength factors for very short shear links in EBFs. *KSCE J. Civ. Eng.* **2018**, *22*, 4473–4482. [CrossRef]
11. Chen, Z.P.; Zhu, S.; Yu, H.; Wang, B. Development of novel SMA-based D-type self-centering eccentrically braced frames. *Eng. Struct.* **2022**, *260*, 114228. [CrossRef]
12. Kazaz, I.; Bilge, I.H.; Gurbuz, M. Near-fault ground motion characteristics and its effects on a collapsed reinforced concrete structure in Hatay during the February 6, 2023 Mw7.8 Kahramanmaras, earthquake. *Eng. Struct.* **2024**, *298*, 117067. [CrossRef]
13. Dedeoglu, I.O.; Yetkin, M.; Tunc, G.; Ozbulut, O.E. Evaluating earthquake-induced damage in Dogansehir, Malatya after 2023 Kahramanmaras Earthquake sequence: Geotechnical and structural perspectives. *J. Build. Eng.* **2025**, *104*, 112266. [CrossRef]
14. Damci, E.; Temur, R.; Kanbir, Z.; Sekerci, C.; Koroglu, E.O. Comprehensive investigation of damage due to 2023 Kahramanmaras, Earthquakes in Türkiye: Causes, consequences, and mitigation. *J. Build. Eng.* **2025**, *99*, 111420. [CrossRef]
15. Hu, S.J.; Liu, S.W.; Zeng, S.Z.; Zhang, B.; Xu, Z.H. Investigating seismic performance of a novel self-centering shear link in EBF utilizing experimental and numerical simulation. *J. Constr. Steel Res.* **2025**, *224*, 109129. [CrossRef]
16. Liu, S.W.; Chen, S.Y.; Zeng, S.Z.; Zhang, B.; Hu, S.J. Seismic performance analysis of K-shaped EBF with an innovative crack-resistant composition beam. *J. Build. Eng.* **2025**, *101*, 111818. [CrossRef]
17. Loo, W.Y.; Quenneville, P.; Chouw, N. A new type of symmetric slip-friction connector. *J. Constr. Steel Res.* **2014**, *94*, 11–22. [CrossRef]
18. Kolbadi, S.M.S.; Piri, H.; Keyhani, A.; Kolbadi, S.M.S.; Mirtaheri, M. Seismic performance evaluation of slotted-web and bolt-flange plate moment connection. *J. Build. Eng.* **2021**, *20*, 655–667. [CrossRef]
19. Liu, X.C.; Chen, M.L.; Chen, X.S.; Li, Y.M.; Wang, Y.; Xu, L. Seismic behavior of bolted truss-to-column joint with oversized or slotted bolt hole. *Eng. Struct.* **2021**, *247*, 113110. [CrossRef]
20. Zhang, Q.; Men, J.; Wang, J.; He, P.J.; Huang, C.H.; Li, J.F. Seismic performance evaluation of a replaceable steel beam with slotted bolted connections. *J. Constr. Steel Res.* **2022**, *198*, 107553. [CrossRef]
21. Zhai, S.Y.; Lyu, Y.F.; Cao, K.; Li, G.Q.; Wang, W.Y.; Chen, C. Seismic behavior of an innovative bolted connection with dual-slot hole for modular steel buildings. *Eng. Struct.* **2023**, *279*, 115619. [CrossRef]
22. Hu, S.J.; Liu, S.W.; Xu, H.W.; Zeng, S.Z.; Zhang, B.; Yu, Y.J. Experimental investigation of an innovative very short shear link with shear slotted bolted connection in eccentrically braced frames. *Structures* **2024**, *66*, 106890. [CrossRef]
23. Zhang, B.; Zeng, S.Z.; Tang, F.H.; Hu, S.J.; Zhou, Q.; Jia, Y.G. Experimental and numerical analysis of the mechanical properties of a pretreated shape memory alloy wire in a self-centering steel brace. *Processes* **2021**, *9*, 80. [CrossRef]
24. Feng, W.K.; Fang, C.; Wang, W. Behavior and design of top flange-rotated self-centering steel connections equipped with SMA ring spring dampers. *J. Constr. Steel Res.* **2019**, *259*, 315–329. [CrossRef]
25. Qiu, C.X.; Jiang, T.Y.; Liu, J.W.; Du, X.L. Seismic performance of knee-braced frames equipped with NiTi BRBs. *J. Constr. Steel Res.* **2022**, *197*, 107480. [CrossRef]
26. Sun, G.H.; Liu, H.; Liu, W.Y.; Yang, W.X. Development, simulation, and validation of sliding self-centering steel brace with NiTi SMA wires. *Eng. Struct.* **2022**, *256*, 114069. [CrossRef]
27. Li, Y.; Zhang, H.X.; Yu, H.F.; An, J.H. Experimental study on seismic performance of a new self-centering coupling beam. *J. Constr. Steel Res.* **2024**, *216*, 108572. [CrossRef]

28. Jia, Y.G.; Zhang, B.; Zeng, S.Z.; Tang, F.H.; Hu, S.J.; Chen, W.P. Effect of loading rate and initial strain on seismic performance of an innovative self-centering SMA brace. *Materials* **2022**, *15*, 1234. [CrossRef]
29. Rahgozar, N.; Pouraminian, M.; Rahgozar, N. Reliability-based seismic assessment of controlled rocking steel cores. *J. Build. Eng.* **2021**, *44*, 102623. [CrossRef]
30. Chen, Z.P.; Zhu, S.Y. Seismic responses and resilience of novel SMA-based self-centring eccentrically braced frames under near-fault ground motions. *Front. Struct. Civ. Eng.* **2022**, *16*, 962–975. [CrossRef]
31. Xiong, J.G.; Xiong, M.Q.; Liu, Y.Q.; Li, X.; Hu, S.J. Lateral Load Behavior of Self-centering Frame with Y-Eccentrically Braced Substructure: An Experiment and Numerical Analysis. *KSCE J. Civ. Eng.* **2024**, *28*, 715–731. [CrossRef]
32. *GB/T 228.1-2021*; Metallic Materials—Tensile Testing. China Standards Press: Beijing, China, 2021.
33. Li, P.C. *Study on Seismic Performance of Friction Composite Damper Based on SMA Rod*; Hebei University of Science of Technology: Shijiazhuang, China, 2022.

MDPI AG
Grosspeteranlage 5
4052 Basel
Switzerland
Tel.: +41 61 683 77 34

Buildings Editorial Office
E-mail: buildings@mdpi.com
www.mdpi.com/journal/buildings

www.ingramcontent.com/pod-product-compliance
Lightning Source LLC
Chambersburg PA
CBHW042014260926
51850CB00020B/3382
* 9 7 8 3 7 2 5 8 4 3 0 5 3 *